Discrete Oscillator Design

Linear, Nonlinear, Transient, and Noise Domains

For a listing of recent titles in the
Artech House Microwave Library,
turn to the back of this book.

Discrete Oscillator Design

Linear, Nonlinear, Transient, and Noise Domains

Randall W. Rhea

ARTECH HOUSE
BOSTON | LONDON
artechhouse.com

Library of Congress Cataloging-in-Publication Data
A catalog record for this book is available from the U.S. Library of Congress.

British Library Cataloguing in Publication Data
A catalogue record for this book is available from the British Library.

Cover design by Igor Valdman

ISBN 13: 978-1-60807-047-3

© 2010 ARTECH HOUSE
685 Canton Street
Norwood, MA 02062

All rights reserved. Printed and bound in the United States of America. No part of this book may be reproduced or utilized in any form or by any means, electronic or mechanical, including photocopying, recording, or by any information storage and retrieval system, without permission in writing from the publisher.
    All terms mentioned in this book that are known to be trademarks or service marks have been appropriately capitalized. Artech House cannot attest to the accuracy of this information. Use of a term in this book should not be regarded as affecting the validity of any trademark or service mark.

To my mother:

always caring, always loving, encouraging independence,
explaining integrity with each of my missteps,
and many lessons that wisdom is part patience,
thank you.

Contents

Preface   xv

1 Linear Techniques   1
  1.1 Open-Loop Method   1
  1.2 Starting Conditions   2
    1.2.1 Match Requirements   3
    1.2.2 Aligning the Maximum Phase Slope   10
    1.2.3 Stable Amplifier   11
    1.2.4 Gain Peak at Phase Zero Intersection   11
    1.2.5 Moderate Gain   11
  1.3 Random Resonator and Amplifier Combination   12
  1.4 Naming Conventions   13
  1.5 Amplifiers for Sustaining Stages   15
    1.5.1 Bipolar Amplifier Configurations   15
    1.5.2 Stabilizing Bipolar Amplifiers   19
    1.5.3 Stabilized FET Amplifier Configurations   21
    1.5.4 Basic Common Emitter Amplifier   23
    1.5.5 Statistical Analysis of the Amplifier   25
    1.5.6 Amplifier with Resistive Feedback   26
    1.5.7 General-Purpose Resistive-Feedback Amplifier   29
    1.5.8 Transformer-Feedback Amplifiers   33
    1.5.9 Monolythic Microwave Integrated Circuit Amplifiers   34
    1.5.10 Differential Amplifiers   35
    1.5.11 Phase-Lead Compensation   37
    1.5.12 Amplifier Summary   39
  1.6 Resonators   40
    1.6.1 R-C Phase Shift Network   40
    1.6.2 Delay-Line Phase-Shift Network   41
    1.6.3 L-C Parallel and Series Resonators   43
    1.6.4 Loaded Q   45
    1.6.5 Unloaded Q   46
    1.6.6 Resonator Loss   47
    1.6.7 Colpitts Resonator   49
    1.6.8 Resonator Coupling   51
    1.6.9 Matching with the Resonator   52
    1.6.10 Measuring the Unloaded Q   55
    1.6.11 Coupled Resonator Oscillator Example   56

1.6.12 Resonator Summary	56
1.7 One-Port Method	57
1.7.1 Negative-Resistance Oscillators	58
1.7.2 Negative-Conductance Oscillators	69
1.8 Analyzing Existing Oscillators	74
1.9 Optimizing the Design	76
1.10 Statistical Analysis	78
1.11 Summary	80
References	81
2 Nonlinear Techniques	**83**
2.1 Introduction	83
2.2 Harmonic Balance Overview	83
2.3 Nonlinear Amplifiers	84
2.3.1 Quiescent Current and Compression	86
2.3.2 Impedance Shift	87
2.3.3 Phase Shift	88
2.3.4 Output Spectrum	89
2.3.5 Time Domain Waveform	91
2.3.6 Conversion Efficiency	92
2.3.7 Operating Class	93
2.3.8 Power Amplifier Case Study	99
2.4 Nonlinear Open-Loop Cascade	109
2.4.1 Nonlinear Open-Loop Cascade Example 1	109
2.4.2 Nonlinear Open-Loop Cascade Example 2	111
2.5 Nonlinear HB Colpitts Example	112
2.5.1 Closing the Loop and Excitation	112
2.5.2 Harmonic Balance Colpitts Output Spectrum	114
2.5.3 Excitation Current Versus Oscillator Parameters	115
2.6 Nonlinear Negative-Resistance Oscillator	116
2.7 Output Coupling	120
2.7.1 Coupling Node	120
2.7.2 Load Pulling	121
2.7.3 Loaded Q and Load Pulling	122
2.7.4 Degree of Coupling	123
2.7.5 Loaded Q and Coupling	124
2.7.6 Coupling Reactance and Load Pulling	125
2.7.7 Coupling Reactance and Harmonics	125
2.7.8 Output Coupling Example 2	126
2.7.9 Coupling Summary	128
2.8 Passive Level Control	128
2.9 Supply Pushing	132
2.10 Spurious Modes	134
2.10.1 Unstable Amplifiers	134
2.10.2 Multiple Phase Zero Crossings	135

2.10.3 Bias Relaxation Modes ... 135
2.10.4 Parametric Modes ... 136
2.10.5 Multiple Resonance Modes ... 137
2.10.6 Spurious Mode Summary ... 138
2.11 Ultimate Test ... 139
References ... 140

3 Transient Techniques ... 143
3.1 Introduction ... 143
3.2 Starting Modes ... 144
   3.2.1 Noise Mode of Starting ... 146
   3.2.2 Transient Mode of Starting ... 147
   3.2.3 Time Constant of the Supply Step ... 148
3.3 Starting Basic Example ... 149
3.4 Simulation Techniques ... 151
   3.4.1 SPICE ... 152
   3.4.2 Cayenne ... 152
3.5 Second Starting Example ... 155
3.6 Starting Case Study ... 157
3.7 Triggering ... 160
3.8 Simulation Techniques for High Loaded Q ... 162
3.9 Steady-State Oscillator Waveforms ... 164
   3.9.1 Clapp Oscillator Waveforms ... 164
   3.9.2 The Resonator Voltage ... 166
   3.9.3 Varactor Coupling ... 168
3.10 Waveform Derived Output Spectrum ... 169
References ... 171

4 Noise ... 173
4.1 Definitions ... 173
   4.1.1 Vector Representation of the Oscillator Output ... 174
   4.1.2 Jitter ... 174
   4.1.3 The Output in the Frequency Domain ... 174
   4.1.4 SSB Phase Noise ... 176
   4.1.5 Residual FM and Residual PM ... 177
   4.1.6 Two-Port Noise ... 178
   4.1.7 Acoustic Disturbances ... 179
4.2 Predicting Phase Noise ... 179
   4.2.1 Linear Time Invariant Theory ... 180
   4.2.2 Extensions to LTI-Based Theory ... 181
   4.2.3 Linear Time Variant Theory ... 184
4.3 Measuring Phase Noise ... 186
   4.3.1 Direct Method with a Spectrum Analyzer ... 186
   4.3.2 Selective Receiver Method ... 188
   4.3.3 Heterodyne/Counter Method ... 189
   4.3.4 Reference Oscillator Method ... 190

4.3.5 Frequency Discriminator Method	192
4.3.6 Example Phase-Noise Measurement System	194
4.4 Designing for Low Phase Noise	196
4.4.1 Estimating the Predominant Noise Source	197
4.4.2 Reducing Leeson Noise	197
4.4.3 Reducing Pushing Induced Noise	201
4.4.4 Reducing Buffer Noise	202
4.4.5 Reducing Varactor Modulation Noise	203
4.4.6 Reducing Oscillator Noise Summary	204
4.5 Nonlinear Noise Simulation	205
4.5.1 Negative Resistance Oscillator Noise Example	206
4.5.2 Linear Oscillator Phase Noise Example	210
4.6 PLL Noise	213
References	216
5 General-Purpose Oscillators	**219**
5.1 Comments on the Examples	219
5.2 Oscillators Without Resonators	220
5.2.1 R-C Oscillators	220
5.2.2 Wien Bridge	226
5.2.3 Multivibrators	227
5.2.4 Ring Oscillators	230
5.2.5 Twin-T Oscillators	232
5.3 L-C Oscillators	235
5.3.1 Colpitts	236
5.3.2 Clapp	242
5.3.3 Seiler	243
5.3.4 Hartley	243
5.3.5 Pierce	249
5.3.6 Coupled Series Resonator	251
5.3.7 Rhea	253
5.3.8 Coupled Parallel Resonator	254
5.3.9 Gumm	256
5.3.10 Simplified Gumm	257
5.4 Oscillator Topology Selection	258
References	261
6 Distributed Oscillators	**263**
6.1 Resonator Technologies	263
6.2 Lumped and Distributed Equivalents	264
6.3 Quarter-Wavelength Resonators	268
6.3.1 The Quarter-Wavelength Resonator	268
6.3.2 Ceramic-Loaded Coaxial Resonators	268
6.3.3 Capacitor-Loaded Quarter-Wavelength Resonator	271
6.4 Distributed Oscillator Examples	273
6.4.1 Negative-Resistance Hybrid Oscillator	273

Contents

    6.4.2 Negative-Resistance High-Power 1 GHz Oscillator   274
    6.4.3 Quarter-Wavelength Hybrid Oscillator   276
    6.4.4 Simple Hybrid Coaxial Resonator MMIC   277
    6.4.5 Probe-Coupled Coaxial Resonator Bipolar   279
    6.4.6 End-Coupled Hybrid Half-Wavelength Bipolar   281
    6.4.7 Helical Transmission Line Resonator Bipolar   282
  6.5 DRO Oscillators   284
    6.5.1 Dielectric Resonator Basic Properties   284
    6.5.2 Dielectric Resonator Resonant Frequency   286
    6.5.3 Dielectric Resonator Unloaded Q 286
    6.5.4 Dielectric Resonator Coupling   287
    6.5.5 DRO Examples   289
    6.5.6 Coupling Test by Modulation   293
  References   294

7 Tuned Oscillators   **295**
  7.1 Resonator Tuning Bandwidth   295
  7.2 Resonator Voltage   297
  7.3 Permeability Tuning   298
  7.4 Tunable Oscillator Examples   299
    7.4.1 Permeability Tuned Colpitts JFET   299
    7.4.2 Vackar JFET VCO   300
    7.4.3 Hybrid Negative Resistance VCO   301
    7.4.4 Capacitor-Transformed Negative-Resistance VCO   304
    7.4.5 Negative-Resistance VCO with Transformer   304
    7.4.6 Negative-Conductance VCO   305
    7.4.7 Hybrid Coaxial Resonator MMIC   308
    7.4.8 Loaded Quarter-Wavelength MMIC   312
    7.4.9 Seiler Coaxial-Resonator CC VCO   314
  7.5 YIG Oscillators   316
  References   319

8 Piezoelectric Oscillators   **321**
  8.1 Bulk Quartz Resonators   321
    8.1.1 Quartz Blank Cuts   322
    8.1.2 Crystal Resonator Model   324
    8.1.3 Calculating Crystal Resonator Parameters   325
    8.1.4 Crystal Resonator Frequency Pulling   326
    8.1.5 Inverted-Mesa Crystal Resonators   328
    8.1.6 Crystal Oscillator Operating Mode   329
    8.1.7 Crystal Oscillator Frequency Accuracy   332
    8.1.8 Temperature Effects on Crystal Oscillators   334
    8.1.9 Crystal Resonator Drive Level   334
    8.1.10 Crystal Resonator Spurious Modes   337
    8.1.11 Crystal Resonator Aging   338
    8.1.12 Crystal Resonator $1/f$ Noise   338

8.1.13 Crystal Resonator Acceleration Effects 339
8.1.14 Crystal Resonator Standard Holders 341
8.2 Fundamental Mode Crystal Oscillators 341
  8.2.1 Miller JFET Crystal 343
  8.2.2 Colpitts Bipolar Crystal 344
  8.2.3 Colpitts JFET Crystal 346
  8.2.4 Pierce Bipolar Crystal 347
  8.2.5 Pierce MMIC Crystal 350
  8.2.6 Pierce Inverter TTL Crystal 350
  8.2.7 Pierce Inverter CMOS Crystal 351
  8.2.8 Butler Dual Bipolar Crystal 355
  8.2.9 Driscoll Bipolar Crystal 357
  8.2.10 Inverted-Mesa Pierce Bipolar Crystal 359
8.3 Overtone Mode Crystal Oscillators 361
  8.3.1 Colpitts Overtone Bipolar Crystal 361
  8.3.2 CB Butler Overtone Bipolar Crystal 364
  8.3.3 CC Butler Overtone Bipolar Crystal 366
8.4 Crystal Oscillator Examples Summary 366
8.5 Oscillator with Frequency Multiplier 369
8.6 Crystal Oscillator Starting 371
8.7 Surface Acoustic Wave Resonators 371
  8.7.1 SAW Resonator Models 372
  8.7.2 SAW Resonator Frequency Stability 373
8.8 SAW Oscillators 374
  8.8.1 SAW 1-Port Colpitts Bipolar 375
  8.8.2 SAW 1-Port Butler Bipolar 377
  8.8.3 SAW 2-Port Pierce MMIC 378
8.9 Piezoelectric Ceramic Resonators 380
  8.9.1 Ceramic Resonator Models 380
  8.9.2 Ceramic Resonator Accuracy and Stability 380
  8.9.3 Ceramic Resonator Oscillators 381
8.10 MEMS and FBAR Resonators 386
References 387

Appendix A: Modeling **389**
A.1 Capacitors 389
  A.1.1 Capacitor: First-Level Model 389
  A.1.2 Capacitor: Second-Level Model 390
  A.1.3 Capacitor: Third-Level Model 391
A.2 Varactors 393
A.3 Inductors 394
  A.3.1 Single-Layer Wire Solenoid 395
  A.3.2 Toroid 400
  A.3.3 Ferrite Beads 401
  A.3.4 Mutually Coupled Inductors 401

A.4 Helical Transmission Lines	402
A.5 Signal Control Devices	404
A.5.1 Bifilar Transformer Operating Modes	404
A.5.2 Ruthroff Impedance Transformer	405
A.5.3 Wire-Wound Couplers	407
A.6 Characteristic Impedance of Transmission Lines	409
A.6.1 Coax	409
A.6.2 Coax with Square Ground	410
A.6.3 Rod over Ground	411
A.6.4 Rod over Flat Ground with Dielectric Layer	411
A.6.5 Rod Between Ground Planes	411
A.6.6 Stripline	412
A.6.7 Microstrip	413
A.6.8 Twisted-Pair Transmission Line	414
A.7 Helical Resonators	415
References	416
Appendix B: Device Biasing	**419**
B.1 Biasing Bipolar Transistors	419
B.1.1 Bipolar Model for Biasing	419
B.1.2 Common Emitter Bias Networks	421
B.1.3 Bias 7 Network with Base Diode	427
B.1.4 Bias 8 Network with Zener	427
B.1.5 Bias 9 Active Network	428
B.1.6 Bias 10 Dual Supply	430
B.1.7 Bias 11 Common Collector Network	431
B.1.8 Bipolar Bias Network Summary	432
B.1.9 Saturated Output Power and Biasing	433
B.2 FET Bias Networks	433
B.2.1 Bias 15 Simple FET Network	434
B.2.2 Bias 16 Gate Voltage	434
B.2.3 Bias 17 Source FB	435
B.2.4 Bias 18 Dual-Gate FET	436
B.3 Bias 19 MMIC Gain Block	437
References	438
Constants and Symbols	**439**
About the Author	**443**
Index	**445**

Preface

Discrete Oscillator Design: Linear, Nonlinear, Transient, and Noise Domains covers the practical design of high-frequency oscillators with lumped, distributed, dielectric, and piezoelectric resonators. Linear, nonlinear harmonic-balance, nonlinear transient, and nonlinear noise analysis techniques are described in detail, with many examples. All of these domains are integrated into an easy-to-understand unified theory of oscillator design. By grasping these concepts you will create unique designs that elegantly match specifications. This understanding also demystifies oscillators, a primary goal of all my books. The book assumes the reader has a basic understanding of HF terminology, measurement techniques, and *S*-parameters.

This is my third book on the subject. The first is *Oscillator Design and Computer Simulation* published by Prentice-Hall in 1990. The second edition was a complete rewrite and was copublished by McGraw-Hill and Noble Publishing in 1995. The current book is motivated by a decade of advancement in the state-of-the-art, my additional design experience, and more importantly, the questions and feedback of a thousand attendees to seminars that I taught at trade shows, companies worldwide, and the Georgia Institute of Technology.

A legitimate criticism of my first two books is their over-simplification of oscillator design. This book addresses oscillators in more depth, specifically in the areas of nonlinear and transient behavior. Nevertheless, this book is also not a rigorous, mathematical treatment of oscillator theory. For example, Barkhausen's rather than Nyquist's criterion is used for starting, and Leeson's method is used extensively for noise analysis. These methods are intuitive and useful, but not rigorous. An excellent rigorous book is *RF and Microwave Oscillator Design* edited by Michal Odyniec. The style of prose of this book is not formal, but rather is chosen for clarity and brevity. For example, rather than writing "noise quantified by Leeson's equation" I write "Leeson noise," and the term "line" may be substituted for "transmission line." While the fundamental concepts presented in this book are useful in integrated circuit (IC) oscillator design, differential active devices and integrated resonators that are central to IC oscillator design are covered only briefly.

I strongly encourage the reader to begin design with an understanding of oscillator behavior rather than by copying designs. However, many designers chose to begin by modifying existing designs. Therefore, this book includes a greater number of examples than my earlier books. Over forty example oscillators are described, many with measured data to confirm the theories presented.

The first four chapters introduce linear, nonlinear, transient, and noise theory, respectively. I recommend reading them in that order. The remaining chapters apply those techniques to examples of general purpose R-C and L-C oscillators, distributed oscillators, wide-tuning oscillators, and piezoelectric oscillators. These chapters may be read as the need arises. Appendix A offers models for lumped elements including a seldom used but elegant distributed inductor model. Appendix B covers biasing of active devices.

As with my previous books, *Discrete Oscillator Design* relies heavily on the use of computer simulation to analyze circuit behavior and present the results graphically. The specific programs used belong to the GENESYS software suite developed by the Eagleware group of Agilent Technologies. However, the purpose of using the software is to advance the understanding of fundamental concepts and to predict the performance characteristics of oscillators. GENESYS is not required to benefit from this book.

Some of the references at the end of each chapter are URLs to Internet sites that contain a wealth of additional information. Because the internal structure of sites often changes, the URL of only the home page is cited.

I would like to thank Mark Walsh and Lindsey Gendall of Artech House for their professional help and for expediting the publication of this book. I would also like to thank Dr. Stephen A. Maas for his review of the book and many helpful suggestions.

1 Linear Techniques

Linear analysis is the critical first step for oscillator design. While linear analysis does not predict all oscillator characteristics, it establishes initial starting conditions and forms the foundation of the design. Initial oscillator design using closed-loop, nonlinear techniques may indicate oscillation, but it does not reveal starting criteria margin or loaded Q.

This book covers two linear analysis methods: the 2-port open-loop response and the 1-port negative-resistance or negative-conductance response. Historically, the open-loop method was used for piezoelectric-resonator and low-frequency oscillator design while the 1-port method was used for microwave oscillator design. The open-loop method provides better insight and it is suitable for microwave design as well, so it is preferred in this book. The 1-port negative resistance method is important and is also covered.

1.1 Open-Loop Method

Consider the amplifier-resonator cascade depicted in Fig. 1.1. The oscillator is formed by connecting the output to the input. The open-loop analysis considers the characteristics of the cascade open-loop rather than as an oscillator. The amplifier provides energy for the system and is sometimes called the sustaining stage. In this example, power is extracted through a coupler. More often power is taken from a node of either the amplifier or resonator using a coupling capacitor. The resonator controls the operating frequency and this critical component determines many performance parameters of the oscillator.

The amplitude of the forward scattering parameter, S_{21}, of this cascade is plotted in Fig. 1.2 with circular symbols on the trace. The gain peaks at 1000 MHz at about 5 dB. The small-signal open-loop gain at the phase-zero crossing is called the gain margin. The angle of S_{21} is plotted with square symbols. The phase-zero crossing is called ϕ_0 in this book. The magnitude and phase of S_{21} together constitute the open-loop Bode response. The loaded Q, defined in a later section, is plotted with triangular symbols.

2 Discrete Oscillator Design

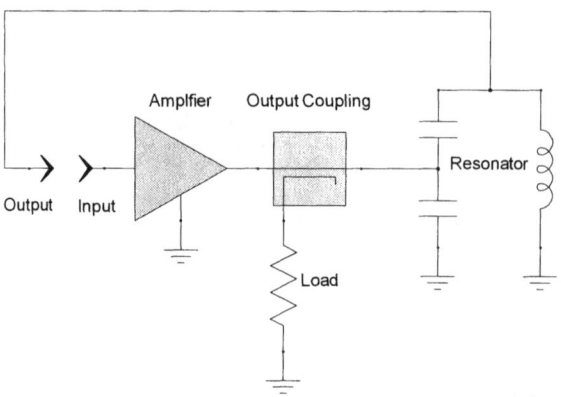

Figure 1.1 Open-loop amplifier-resonator cascade with output power coupling.

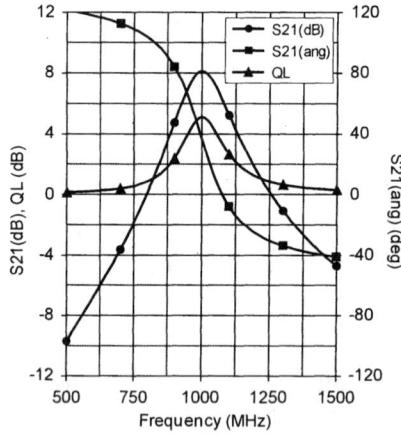

Figure 1.2 Forward S-parameter S_{21} amplitude for the amplifier-resonator cascade (round symbols), the angle of S_{21} (square symbols), and the loaded Q (triangular symbols).

1.2 Starting Conditions

When the loop is closed, a signal builds to a level where nonlinearity establishes a steady-state condition with 0 dB gain at ϕ_0. Nonlinear behavior reduces the gain, shifts the phase, and modifies the impedance characteristics of the amplifier. These effects raise questions as to the validity of small-signal oscillator analysis. However, the resonator establishes many oscillator performance characteristics and it remains linear. With moderate gain margin, the nonlinearity is not as onerous as it might seem. Chapter 2 deals with nonlinear behavior in detail so this discussion is delayed. When oscillation begins, signal levels are low and the system is linear. Although starting is time-variant, initially it is linear. Only if starting occurs and oscillation builds does the system become nonlinear.

The *necessary* starting conditions determined from the open-loop Bode response are:

1) the frequency of phase zero crossing, ϕ_0, is the oscillation frequency;
2) the initial gain must be greater than 0 dB at ϕ_0;
3) the phase slope at ϕ_0 must be negative and if there are multiple ϕ_0, the quantity with a negative phase slope must exceed the quantity with a positive slope.

These necessary conditions are called Barkhausen's criterion. They are intuitive and simple to implement. Condition 3 is added for generality. An alternative and more rigorous technique is called the Nyquist stability criterion. The Nyquist contour encompasses the right-half of the complex plane. This contour mapped through the open-loop transfer function produces a Nyquist plot. P is the quantity of poles encircled by the plot and Z is the number of encircled zeros. The Nyquist stability criterion states that the number of right-half plane zeros must be zero and the number of counterclockwise encirclements of $-1+j0$ must be equal to P. Bode responses and Barkhausen's criterion is used in the remainder of this book.

Oscillation does not occur at the gain peak. It occurs at ϕ_0. The initial gain at ϕ_0 is referred to as the gain margin. Additional *objectives* of the open-loop cascade are:

1) the maximum $\partial\phi/\partial\omega$ occurs at ϕ_0;
2) the amplifier is stable;
3) S_{11} and S_{22} are small;
4) the maximum gain margin occurs at ϕ_0;
5) the gain margin should be moderate, typically 3 to 8 dB.

These desirable conditions are considered next in additional detail.

1.2.1 Match Requirements

Ideally, the open-loop gain and phase responses for a potential oscillator are first simulated by computer and then confirmed using a vector network analyzer. Analyzers utilize a resistive reference impedance such as 50 or 75 ohms. S_{11} and S_{22} of the open-loop cascade are not necessarily matched to this reference impedance. Therefore, when the loop is closed, the self-terminated condition does not comply with the simulated or measured condition, thus resulting in a prediction error. This situation is considered in the following sections. First, match and port impedance relationships are reviewed.

1.2.1.1 Match and Impedance Relationships

The impedance and the magnitude of the VSWR at a port are related to the linear port S-parameters by

$$Z_n = Z_0 \frac{1+S_{nn}}{1-S_{nn}} \qquad (1)$$

$$VSWR_n = \frac{1+|S_{nn}|}{1-|S_{nn}|} \qquad (2)$$

where n is the port number and Z_0 is the reference impedance. The magnitude of the port S-parameters, the reflection coefficient, and the reflection loss are related by

$$|S_{nn}|(dB) = 20\log|\rho_n| \qquad (3)$$

$$\rho_n = \frac{VSWR-1}{VSWR+1} \qquad (4)$$

$$L_A(dB) = -10\log(1-|\rho_n^2|) \qquad (5)$$

where ρ_n is the reflection coeffient at a port and L_A is the transmission loss due to reflection and ignoring dissipation.

Table 1.1 includes representative values relating these parameters. These parameters are called radially scaled parameters because they are functions of the distance from the center of the Smith chart to an impedance point. The linear scale is ρ_n, which ranges from zero for a point at the center of the chart to 1 for a point at the circumference of a conventional Smith chart.

Table 1.1 Representative values of radial scaled parameters

ρ_n	S_{nn}(dB)	$VSWR_n$	L_A(dB)
0.010	-40.00	1.020	0.0004
0.032	-30.00	1.065	0.0043
0.056	-25.00	1.119	0.0138
0.100	-20.00	1.222	0.0436
0.158	-16.00	1.377	0.1105
0.200	-14.00	1.499	0.1764
0.251	-12.00	1.671	0.2830
0.316	-10.00	1.925	0.4576
0.333	-9.54	2.000	0.5118
0.398	-8.00	2.323	0.7494
0.500	-6.02	3.000	1.2496
0.600	-4.44	3.997	1.9365
0.707	-3.01	5.829	3.0106
0.794	-2.00	8.724	4.3292

Port S-parameters are frequenctly displayed on the Smith chart. To conserve space in this book, Smith chart plots are printed in reduced size. This is a minor compromise because as is demonstrated match plays only a secondary role in oscillator design. However, so that the reader may interpret the scales of smaller charts, a larger unity-radius, impedance-grid, normalized Smith chart is shown in Fig. 1.3.

1.2.1.2 Mismatch Error

If the open-loop cascade input and output impedances are not matched to the analysis reference impedance then the analysis does not represent the closed-loop match. This results in an error in the analyzed gain and the phase. The magnitude of this error increases with increasing open-loop cascade input and output mismatch.

Provided the reverse transmission, S_{12}, is small, the open-loop mismatch error with the loop output driving the loop input is given by

$$Mismatch\ Error = 20\log\frac{1}{1 - S_{11}S_{22}} \qquad (6)$$

S_{11} and S_{22} are complex so the mismatch error is potentially positive or negative. Table 1.2 lists the maximum possible error for representative values of S_{11} and S_{22}.

If either S_{11} or S_{22} is a good match then the mismatch error of the open-loop gain margin is rather small. Even if both S_{11} and S_{22} are only -6 dB the maximum mismatch error in under reporting the gain is 1.9 dB. Unless the gain is marginal, the cascade match is not critical. However, not only is the magnitude of the open-loop gain uncertain with mismatch but the frequency of ϕ_0, and the phase slope, $\partial\phi/\partial\omega$, are affected as well.

1.2.1.3 Sign of the Mismatch Error

The mismatch error listed in Table 1.2 is the maximum potential error associated with the magnitude of mismatch and with a small S_{12}. The gain margin predicted by open-loop simulation or measurement can be either optimistic or pessimistic.

Consider case A in Fig. 1.4 where the magnitudes of S_{11} and S_{22} are -6.02 dB with a 50 ohm reference impedance. The port S-parameter angles are 0° and 180° representing port impedances of 150 and 16.67 ohms. When the loop is closed, a 16.67 ohms source drives a 150 ohm load. This 9:1 VSWR is a port reflection S-parameter of -1.94 dB. This mismatch is significantly worse than the -6.02 dB of the measurement. The predicted gain margin for case A is optimistic by 1.94 dB.

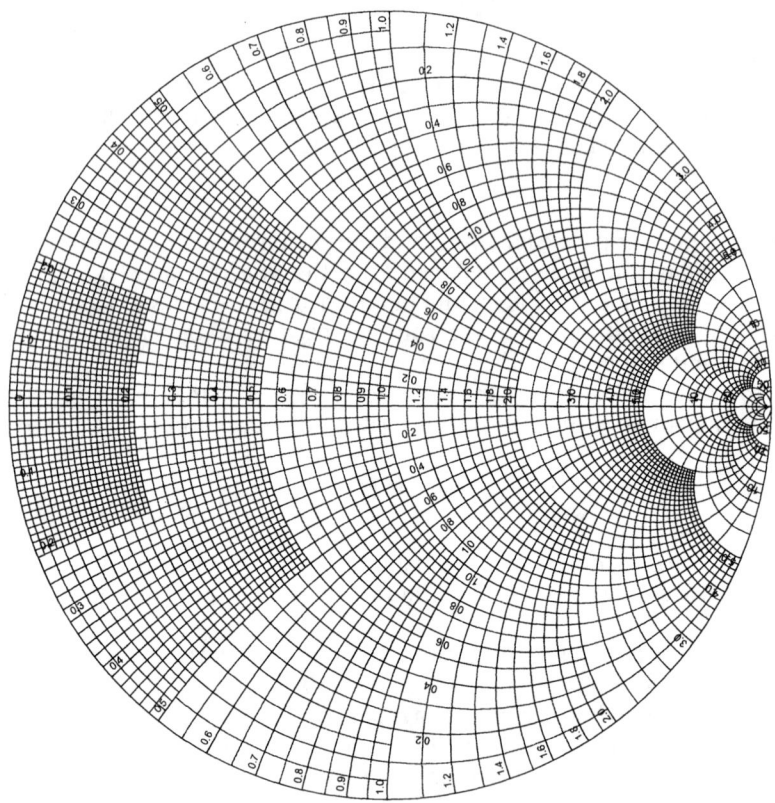

Figure 1.3 The normalized, unity-radius, impedance grid Smith chart.

Table 1.2 Representative values of the maximum mismatch error when the open-loop cascade is mismatched

Snn(dB)	Smm(dB)	Error(dB)
-20	-20	+0.087,-0.086
-20	-10	+0.279,-0.270
-20	-6	+0.447,-0.425
-20	-3	+0.638,-0.594
-10	-10	+0.915,-0.828
-10	-6	+1.499,-1.278
-10	-3	+2.201,-1.755
-6	-6	+2.513,-1.946
-6	-3	+3.806,-2.638
-3	-3	+6.041,-3.529

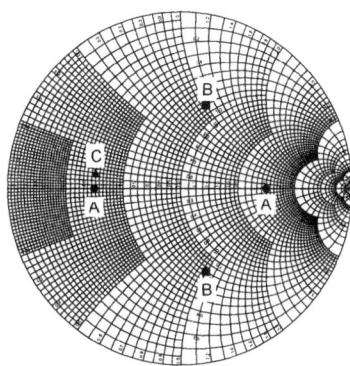

Figure 1.4 Three cases (A, B, and C) for the angles of S_{11} and S_{22} all with a magnitude of 0.5 or -6.02 dB.

Consider next case B, also with S-parameter magnitudes of -6.02 dB. The port S-parameter angles are 90° and -90°. When the loop is closed, the input and output port impedances are complex conjugates and the loop is matched. In this case, the predicted gain margin is pessimistic by 2.5 dB. Finally, consider case C with equal port S-parameters of -6.02 dB at 170°. As with case B, the open-loop measurement is mismatched to the 50-ohm reference impedance but the port resistances are equal with only a small reactive component. The predicted gain margin is pessimistic by 2.34 dB.

When the port S-parameters are similar and near the real axis of the Smith chart, the reference impedance can be adjusted to better center S_{11} and S_{22} on the chart, thus improving the accuracy of the gain margin prediction. This is easy to accomplish with simulation but is more difficult with a vector network analyzer.

In summary, the predicted gain margin is either low (pessimistic) or high (optimistic) depending on the phase of the port S-parameters. In addition, not only is the gain margin affected by the open-loop mismatch, but also are the frequency of ϕ_0 and the slope of the phase response.

1.2.1.4 Match Example

Consider the Colpitts oscillator schematic in Fig. 1.5. This example uses a common collector bipolar sustaining stage and a capacitive tap to transform the low impedance of the emitter to the high impedance of the base. For analysis, the loop is opened between the capacitive tap and the emitter. Oscillator output power is delivered to a 50-ohm load at port 3. For this analysis, the measurement reference impedance is 50 ohms.

The linear open-loop cascade responses are conveniently computed using S-parameter data for the active device. This technique has been successfully used for linear circuit simulation for decades. The S-parameter data is measured at specific device voltage and current, and for simulation the device is assumed biased at this quiescent point.

Here, the active device, a 2N3904, is characterized by a Spice-type model. Bias circuitry must then be included in the schematic description of the circuit. DC biasing is computed and the device is linearized by the GENESYS simulator [1] to compute the small signal characteristics of the circuit. This supports convenient simulation at different supply and bias configurations. Capacitor C_5 is not required in the final oscillator. It is required for simulation to avoid shorting the emitter bias by the 50-ohm load termination of the simulator.

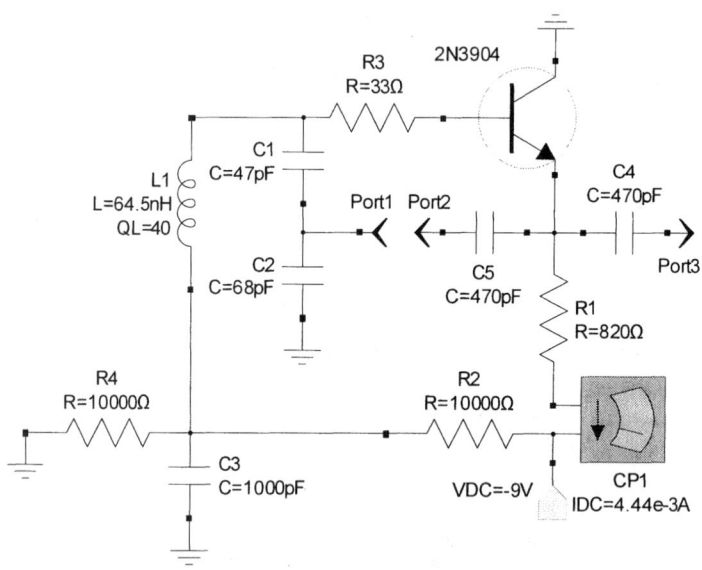

Figure 1.5 Colpitts oscillator with a bipolar sustaining stage. For analysis, the loop is opened between the capacitive tap and the emitter.

Fig. 1.6 shows the magnitude and angle of S_{21} on the left plotted with solid traces and the complex S_{11} and S_{22} plotted on a 50-ohm unity-radius Smith chart on the right for the open-loop cascade of Fig. 1.4. Inductor L_1 is tuned until phase zero-intersection occurs at 100 MHz. The predicted gain margin is 3.98 dB.

At 100 MHz the cascade S_{11} is -15.72 dB at -126.56° and S_{22} is -4.49 dB at 119.87°. The expected error in the gain margin estimated by Eq. 6 is 0.88-dB pessimistic; therefore, the self-terminated open-loop gain margin, assuming S_{12} is small, is 4.86 dB. Arcs of reflection coefficient rotate counter clockwise with increasing frequency. In Fig. 1.5, S_{11} crosses the real axis slightly above 100.0 MHz.

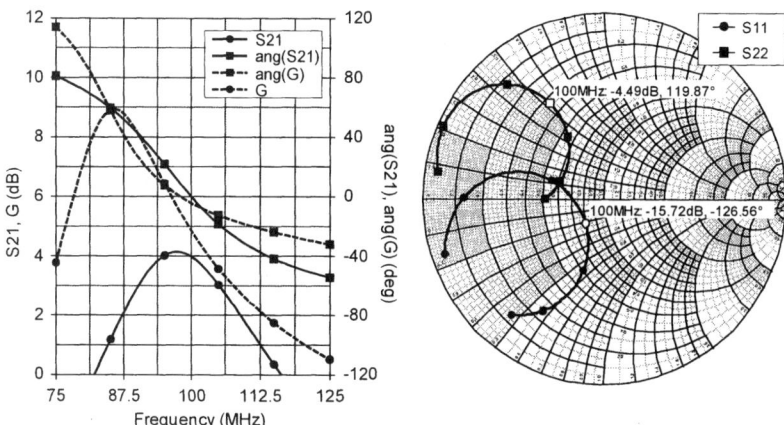

Figure 1.6 S_{21}(dB) and angle of S_{21}(deg) of the open-loop cascade(left) and S_{11}(dB) and S_{22}(dB)plotted on a Smith chart(right).

1.2.1.5 Randall/Hock Correction

The simplest method to reduce the prediction error is to design the open-loop cascade for a reasonable match. This is not accomplished by adding matching networks to the cascade. Reactive components other than those required for the resonator potentially introduce ambiguous ϕ_o. One function of the capacitive tap in the resonator of a Colpitts oscillator is to match the low emitter impedance of a common collector sustaining stage to the high impedance at the base. The Colpitts resonator is used as a matching device.

When it is difficult to match the open-loop cascade, or when a more accurate prediction of the gain margin is required (because of low gain margin), the following techniques are used.

Harada used a flow graph and Mason's rule [2] to derive a simple formula for G, the self-terminated open-loop transfer function of the amplifier-resonator cascade. His formula is

$$G = \frac{S_{21} - S_{12}}{1 - S_{11}S_{22}} \qquad (7)$$

Notice that G includes a term for the reverse transmission S-parameter S_{12}. If S_{12} is zero and S_{11} and S_{22} are small, then $G = S_{12}$. Despite the fact that Harada's formula includes a term for S_{12}, the expression is not exact unless S_{12} is 0.

In June of 2001, Mitch Randall and Terry Hock improved the derivation for G and thus published an elegant method for oscillator analysis [3]. Their equation for the true complex gain of a self-terminated cascade is

$$G = \frac{S_{21} - S_{12}}{1 - S_{11}S_{22} + S_{21}S_{12} - 2S_{12}} \qquad (8)$$

G is computed from the S-parameters of the open-loop cascade. The equation is exact. When the loop is opened at any point the same results are obtained. The loop may even be opened within the resonator. From Eq. 8 the missing terms in Harada's expression are evident. The Randall/Hock expression was truly an essential contribution to the art.

Given in Fig. 1.6 on the left, plotted with dashed traces, are the magnitude and angle of G computed using the Randall/Hock equation. The simulated S-parameters of the open-loop cascade are used in the GENESYS post processor to compute G. The rather poor match of S_{22} of -4.49 dB suggested the use of the Randall/Hock expression. ϕ_0 occurs at approximately 98 MHz, 2% lower in frequency than the uncorrected open-loop prediction. At this frequency, the gain margin is approximately 5.45 dB, 1.47 dB higher than the uncorrected open-loop prediction of 3.98 dB. Recall that Eq. 6 predicted a gain 0.88 dB higher than the uncorrected gain, the difference due to S_{12} not equaling zero. Notice also that the slope of the phase response at ϕ_0 is lower than the uncorrected prediction. The importance of the phase slope at ϕ_0 is considered later.

1.2.1.6 Match Summary

An objective of the initial oscillator design process is to use the resonator as a matching network to achieve port S-parameters as small as possible. With computer simulation, the user is free to select the reference impedance that best centers the port S-parameters on the Smith chart. Confirmation of the open-loop design using a vector network analyzer is desirable and designing for a reference impedance of 50 ohms is convenient. It is shown later, when the design of the sustaining stage is covered, that matching to 50 ohms is more feasible than might be assumed.

When port S-parameters of -10 dB or better are achieved, the error in the predicted cascade gain and phase are small. If the S-parameters at one port are small, a poorer match at the remaining port is tolerated. When the gain margin is small, which occurs with some sustaining stage devices, or when greater prediction accuracy is required, the Randall/Hock equation is used.

1.2.2 Aligning the Maximum Phase Slope

Achieving the maximum slope versus frequency of the transmission phase is critical for oscillator performance. Disturbances such as bias changes, temperature variation, noise and termination impedance changes may shift the transmission phase. If the phase of G in Fig. 1.6 shifts up 10°, then ϕ_0 occurs at approximately 103.2 MHz rather than 98 MHz. However, if the phase slope is steep, a phase disturbance has less effect on the oscillation frequency. Steep phase slope is a critical measure of oscillator performance.

Consider the phase of S_{21} plotted in Fig. 1.2. Notice that ϕ_0 occurs at 1075 MHz but that the maximum phase slope occurs at 1000 MHz. There is ample gain margin at the oscillation frequency of 1075 MHz but the phase slope is steeper at 1000 MHz. Ideally, the maximum phase slope occurs at ϕ_0.

1.2.3 Stable Amplifier

The oscillator is formed by closing the loop. If the open-loop initial conditions are correct, oscillation occurs. The open-loop cascade must have a stable sustaining-stage amplifier. If not, spurious oscillation caused by the unstable amplifier may occur at a frequency other than or in addition to the desired mode at ϕ_0. Techniques that are used to stabilize a conventional amplifier are used to stabilize the oscillator sustaining stage prior to beginning the design of the open-loop cascade.

The situation is different for negative resistance and negative conductance oscillators. In this case, an active device is selected and a topology is used that encourages instability. This instability is then controlled to occur at the desired frequency. These techniques are considered later in this chapter.

1.2.4 Gain Peak at Phase Zero Intersection

It is desirable that the gain peaks near ϕ_0. This utilizes all the available gain of the sustaining stage. This goal is less critical than the other design objectives. This goal tends to naturally occur if ϕ_0 occurs at the maximum phase slope.

1.2.5 Moderate Gain

A common mistake is designing the cascade for maximum gain margin, or in the case of negative resistance analysis, designing for maximum negative resistance. A linear oscillator can be constructed by using AGC to limit the signal level so the sustaining stage operates in a nearly linear mode. In a conventional oscillator, limiting caused by nonlinear sustaining-stage behavior is simply an economic method for establishing the steady-state gain at 0 dB.

High gain margin leads to excessive compression thus resulting in high harmonics, spurious oscillation modes, and degraded noise performance. A gain of 3 to 8 dB is a typical goal. If the sustaining-stage gain is well controlled, as with feedback-controlled amplifiers, if the match is good so the analysis error is small, and if noise is of prime consideration, then gain near the low end of the range is indicated. If high output power, high conversion efficiency, and fast starting are important, then higher gain is indicated. These topics are covered in detail in Chapter 2.

1.3 Random Resonator and Amplifier Combination

Consider the open-loop resonator-amplifier cascade shown in Fig. 1.7. The 100-MHz resonator is a parallel L-C and the sustaining stage is a UHF bipolar transistor. The analysis ports are 1 and 2, which are closed to form the oscillator. Output power to a 50-ohm load is taken at port 3. Resistor R_2 provides shunt feedback to reduce the input and output impedance of the amplifier to near 50 ohms. R_2 and R_3 form a voltage divider to bias the base near 0.8 volts with approximately 3.8 volts at the collector. The voltage dropped across collector resistor R_1 increases with increasing collector current, thus reducing the bias drive to counteract the increasing collector current.

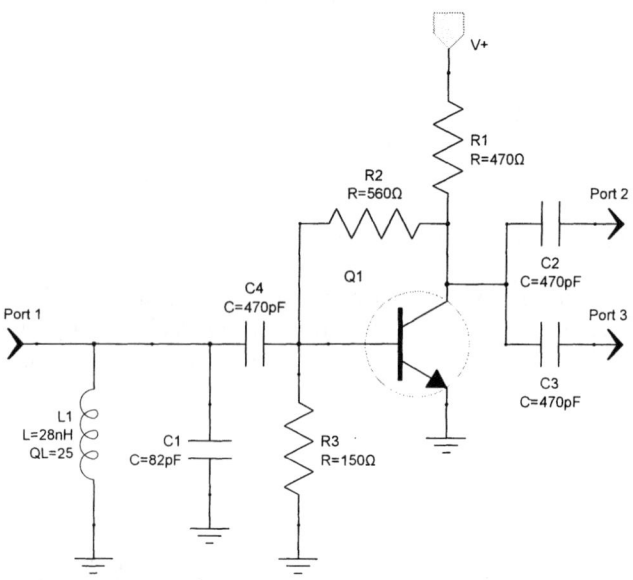

Figure 1.7 Unsuccessful open-loop cascade with a parallel resonator and a common-emitter bipolar sustaining stage.

The open-loop gain and phase responses are shown on the left in Fig. 1.8 and the port S-parameters are plotted on a Smith chart on the right. Notice that an excellent match to 50 ohms is achieved at both ports 1 and 2.

Also notice that the gain peaks at just over 8 dB near the design goal frequency of 100 MHz, thus satifying several design objectives. However, the open-loop transmission phase is 146° at the gain peak and does not intersect 0° at any frequency where the gain margin is greater than 0 dB. This is an unsuccessful open-loop design for an oscillator. The designer must manage not only the cascade gain margin and match, but also the transmission phase. A random selection of a resonator and a sustaining stage does not guarantee a successful design. The transmission phase shift of the active device at the desired operating frequency, the phase shift of the resonator at the transmission amplitude peak, and the impact of PWB parasitics and cascade electrical path length must all be considered when designing the oscillator.

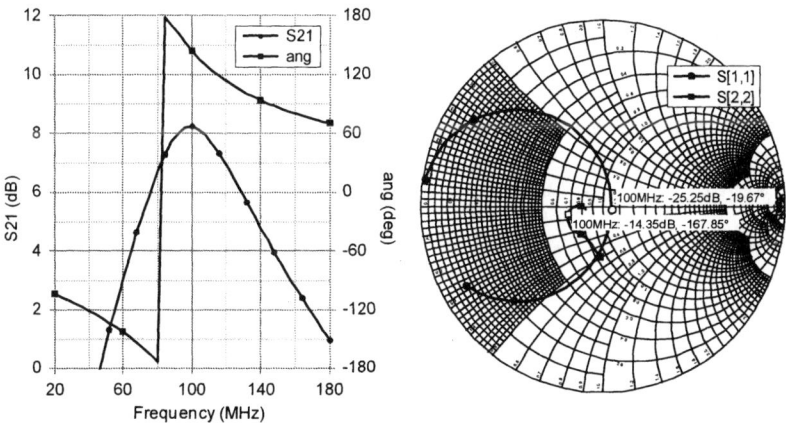

Figure 1.8 Open-loop cascade gain and phase response (left) for the unsuccessful schematic shown in Fig. 1.7 and the port S-parameters plotted on a Smith chart (right).

1.4 Naming Conventions

Because a random combination of a resonator and a sustaining stage does not necessarily satisfy the required initial conditions, historically, when a successful design was found, the topology became a standard design and it was named after the discoverer.

Because successful oscillator design requires satisfying multiple goals, the habit of designing oscillators by starting with a named topology persists to this day. This is a satisfactory method, provided the open-loop cascade is carefully analyzed. Suboptimum results occur when the design is not adjusted to best satisfy the initial conditions. A few decades ago, analysis was mathematically tedious and required overly simplified models for both active and passive devices. This encouraged an approach based on standard designs. Modern simulation tools offer convenient analysis of the open-loop cascade including excellent models or measured data for active and passive devices. This facilitates convenient exploration of design alternatives, the optimization of initial conditions, and even the discovery of new or modified topologies that achieve the required oscillator performance with fewer components.

The presentation of the open-loop cascade as an oscillator design methodology may be disconcerting to the designer accustomed to working with named standard designs. While names are a convenient way to describe and categorize topologies, this habit is often abused and misleading.

Compare an early Colpitts vacuum tube oscillator on the left in Fig. 1.9 with a modern Colpitts JFET oscillator on the right. In the early Colpitts the loop is closed by feeding signal from the plate back to the grid through the bypass capacitor C. The resonator is essentially a pi lowpass network with C_1 in shunt, L_o in series, and C_2 in shunt with the signal path. In the modern Colpitts the signal is fed from the source to the gate through a capacitive tapped parallel bandpass resonator with C_2 in shunt, C_1 in series, and L_o in shunt with the signal path. The early Colpitts topology realized with solid-state devices is often referred to as a Pierce! To make matters worse, careless authors may refer to any oscillator as a Colpitts!

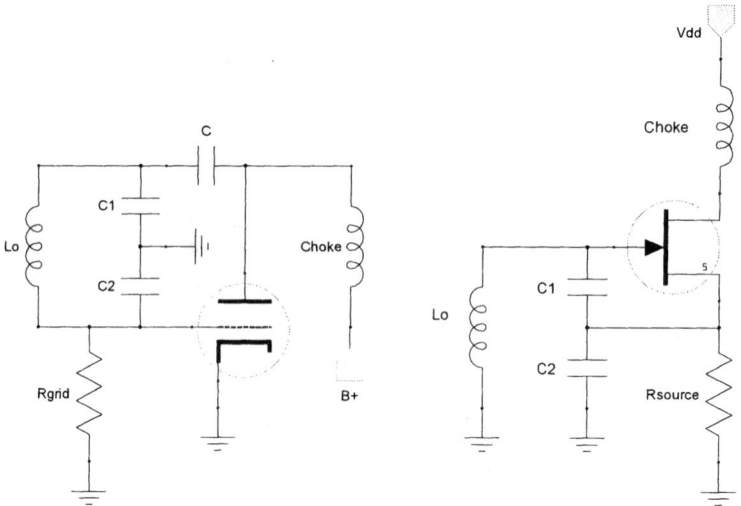

Figure 1.9 An early Colpitts vacuum tube oscillator (left) and a modern Colpitts JFET oscillator (right).

Consider the simplified schematic of a modern Colpitts oscillator at the upper left in Fig. 1.10. At the upper right, the grounds are connected and a new ground reference is chosen at the tap point of the resonator capacitors. Notice the resemblance to the early Colpitts oscillator. At the lower left, the schematic is simply redrawn with grounds. This schematic is often referred to as a Pierce oscillator. The Colpitts and Pierce are essentially the same circuit with a different ground reference. At the lower right, the loop is opened to facilitate the open-loop analysis method being described in this chapter. The shifting of the ground reference is used later in the chapter to develop a better understanding of the negative resistance oscillator.

1.5 Amplifiers for Sustaining Stages

The sustaining stage of an oscillator provides gain and RF power to sustain oscillation. The ideal amplifier used as a sustaining stage is stable and has moderate gain and low noise. Ideally, the amplifier characteristics do not change from device to device, with the supply voltage fluctuations or with temperature. Finally, the design process is simpler, particularly for wide-tuning oscillators, if the gain, transmission phase, and impedance characteristics are flat with frequency. These characteristics are achievable only up to a small fraction of the gain-bandwidth product of a given device. Oscillators are realizable above this frequency limit but design adjustments may be required. Such adjustments may include compensation of the transmission phase, reduction of cascade losses to increase available gain, or tuning control of additional parameters in the design. Basic device characteristics and amplifier techniques important in sustaining-stage amplifiers are considered in the following sections.

1.5.1 Bipolar Amplifier Configurations

There are three basic amplifier topologies available for three-terminal active devices such as transistors. Fig. 1.11 at the upper left shows the common-emitter (CE) bipolar transistor amplifier, at the upper left the common-collector (CC) amplifier, and at the lower left the common-base (CB) amplifier. The transistor is simulated using a Spice type model from Avago Technologies for an AT41486, a general purpose 8-GHz F_t transistor with 13 interdigital emitter fingers of 4-micron pitch [4]. The bias networks shown for the CE and CC topologies were manually adjusted to achieve 11-mA collector current. The CE and CC bias networks are not suitable for production use because the desired bias point is not maintained with device and temperature variation. The CB bias circuit requires two power supply polarities. It is stable with temperature and device beta. The forward beta parameter in the model is approximately 68. Ideal bypass capacitors and inductive chokes are used in these simulations.

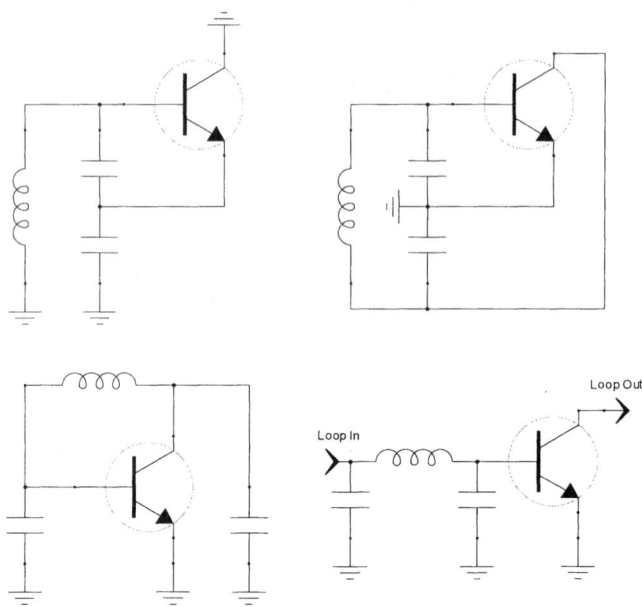

Figure 1.10 Simplified Colpitts schematic at the upper left, the Colpitts redrawn with the grounds connected and moved to the tap at the upper right, redrawn again with grounds thus appearing as a Pierce at the lower left and with the loop opened at the lower right.

Fig. 1.12 shows the forward transmission and reflection S-parameters for the three bipolar transistor amplifier topologies. The responses for the common-emitter topology are given at the top. At low frequency the phase of S_{21} is 180° and thus the amplifier is inverting. However, even with an F_t of 8 GHz, the phase lags 30° by 150 MHz and 60° by 500 MHz. At 5.7 GHz, S_{21} of the AT41486 is noninverting.

As shown on the Smith chart for the common-emitter topology at the upper right in Fig. 1.11, at low frequency, both S_{11} and S_{22} are right of center and therefore higher in impedance than 50 ohms. The resistive component of S_{11} passes through the 50-ohm circle at approximately 160 MHz while the resistive component of S_{22} passes through 50 ohms at 6 GHz. Both the input and output impedance are capacitive except the input is inductive above 1300 MHz.

Next consider the characteristics of the AT41486 common-collector amplifier in the middle row of Fig. 1.12. At low frequency the angle of S_{21} is 0° so the amplifier is non-inverting. The gain is just over 5 dB. The low gain is caused by a high input impedance and a low output impedance. The output impedance approaches 50 ohms at high frequency. The gain could be significantly increased by matching the device. However, notice that the input impedance crosses the circumference of the unity-radius Smith chart at the highest frequency. Therefore the real component of the input impedance is negative which indicates potential amplifier instability. This is also evident in the magnitude of S_{21} with a small peak at about 3000 MHz.

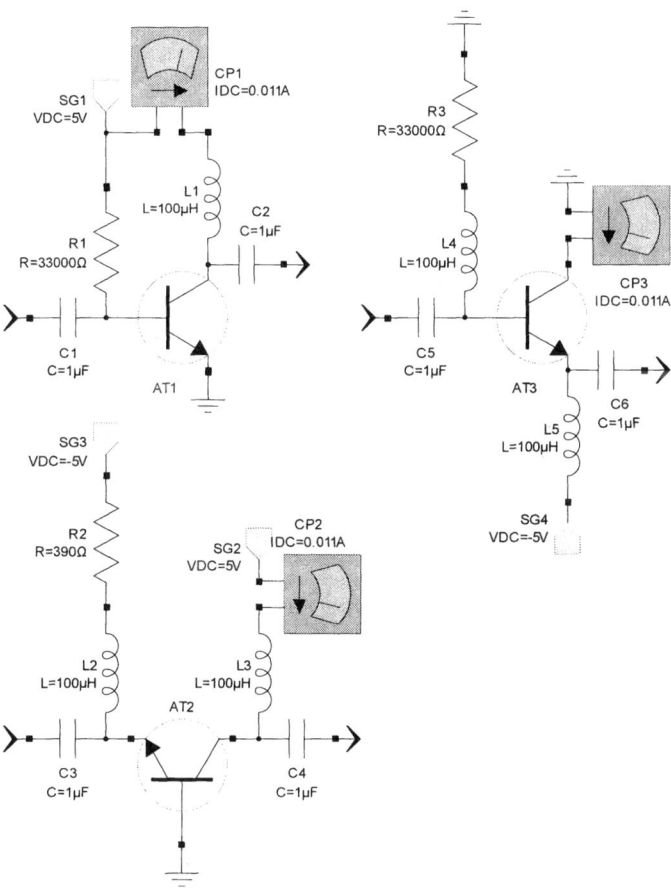

Figure 1.11 Bias and test circuits for common-emitter (upper left), common-collector (upper right) and common-base bipolar (lower right) transistor amplifiers.

The bottom plots are the responses for a common-base topology. Again at low frequency the gain is approximately 5 dB and the amplifier is noninverting. However, the instability that was noticeable in the common-collector topology is extreme with the common base topology. The peaking in the magnitude of S_{21} is severe and both the input and output impedances have a negative resistive component over much of the frequency range.

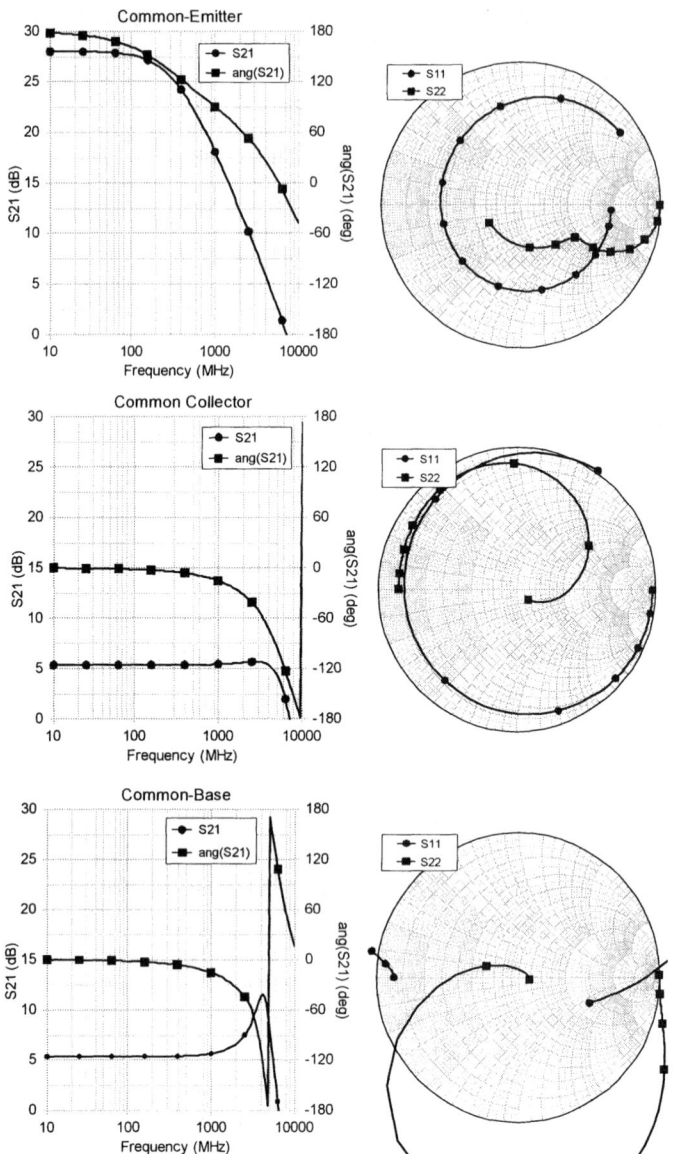

Figure 1.12 S_{21} magnitude and angle (left) and S_{11} and S_{22} (right) for common-emitter (top), common-collector (middle) and common-base (bottom) AT41486 bipolar topologies.

The characteristics of transistors considered for use as a sustaining stage should be evaluated before commencing the design of the cascade. The stability should be assessed over a wide frequency range and not just the range of desired operation.

1.5.2 Stabilizing Bipolar Amplifiers

Consider the effect of adding resistance in series with the base of the common-collector and common-base topologies as shown in Fig. 1.13. Since little DC current flows in the base circuit, the 47-ohm resistor has little effect on the bias.

The responses of the common-collector topology with 47-ohm base resistance added are given in the top row of Fig. 1.14. The input impedance of the common-collector topology is high, so the 47-ohm resistor has little effect on the gain or the input impedance below about 500 MHz. Above 500 MHz the 47-ohm resistor establishes the resistive component of the input impedance at approximately 50 ohms. Notice that both the input and output S-parameters are well within the circumference of the Smith chart and that the gain peaking in S_{21} is absent. The high-frequency bandwidth is reduced but the circuit is stabilized.

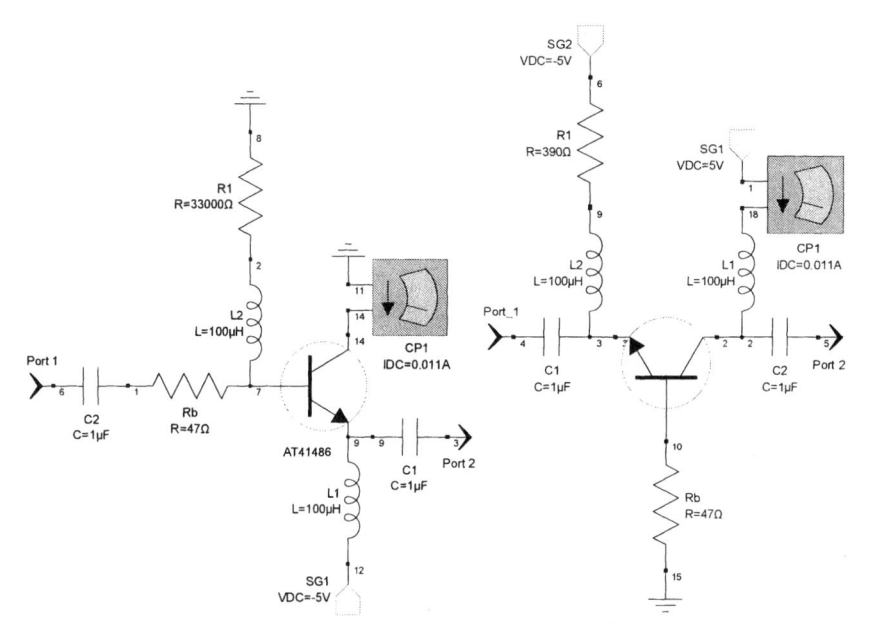

Figure 1.13 AT41486 bipolar common-collector (left) and common-base (right) topologies with 47 ohm resistors in series with the base.

The characteristics of the common-base topology with 47-ohms added base resistance are given in the bottom row of Fig. 1.14. The series base resistance stabilized this topology also. S_{11} and S_{22} are well within the circumference of the Smith chart and the severe peaking in S_{21} is completely eliminated.

The resistance in series with the base is implemented either as a conventional resistor or as a ferrite bead. Ferrite beads are placed either directly on the base of a leaded transistor, as a bead with through wire, or as a surface mount chip. Because of the high input impedance of a CC bipolar amplifier, the value of resistance is not critical. Any value that stabilizes the device up to approximately 200 ohms may be used. For the CB amplifier topology, either a conventional resistor or a ferrite bead with a large resistive component is used. The resistance rather than the inductance of the ferrite bead is the stabilizing factor.

It is important that before the resonator is added and the characteristics of the cascade are evaluated that the stability of the amplifier topology is evaluated and corrected. This typically also aids in achieving a good match in the resonator-amplifier cascade.

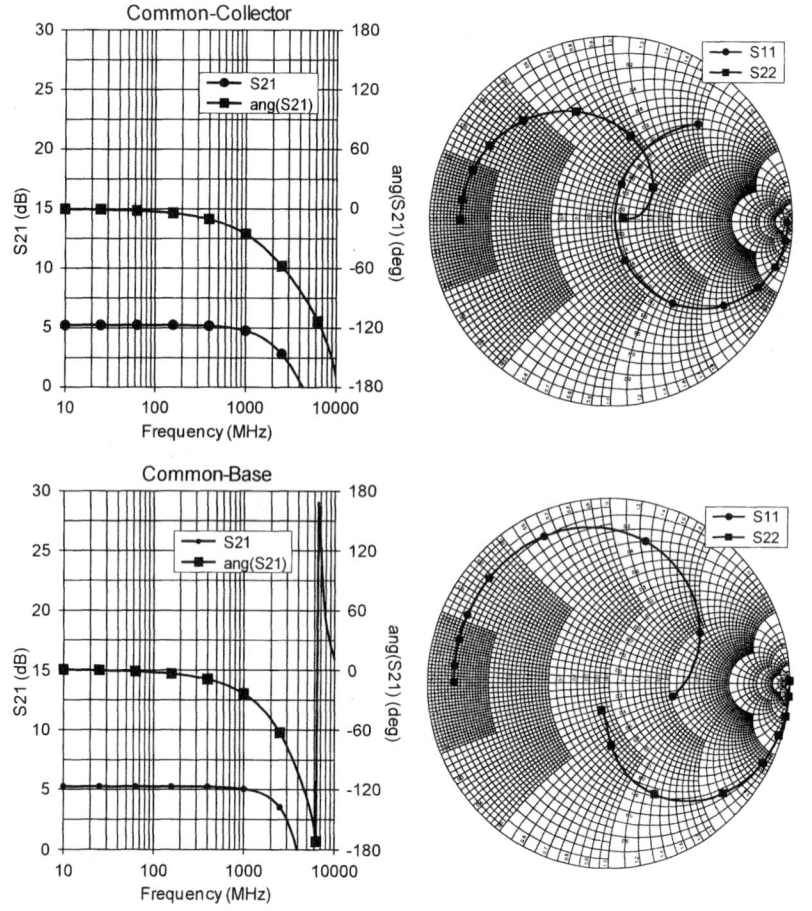

Figure 1.14 Response of the common-collector topology (top) and the common-base topology (bottom) with 47 ohms added resistance is series with the base.

1.5.3 Stabilized FET Amplifier Configurations

Shown in Fig. 1.15 at the upper left is the common-source (CS) FET transistor amplifier, at the upper left the common-drain (CD) amplifier, and at bottom the common-gate (CG) amplifier. The transistor is simulated using a Spice-type model from California Eastern Laboratories for a NE33200 heterojunction GaAsFET chip, a low-noise microwave small-signal transistor with a relatively high I_{dss} [5]. The bias networks include a 33-ohm series source resistor to reduce and stabilize the quiescent drain current to 11 mA. The I_{dss} in the model is 48 mA. Ideal bypass capacitors and inductive chokes are used in the simulations.

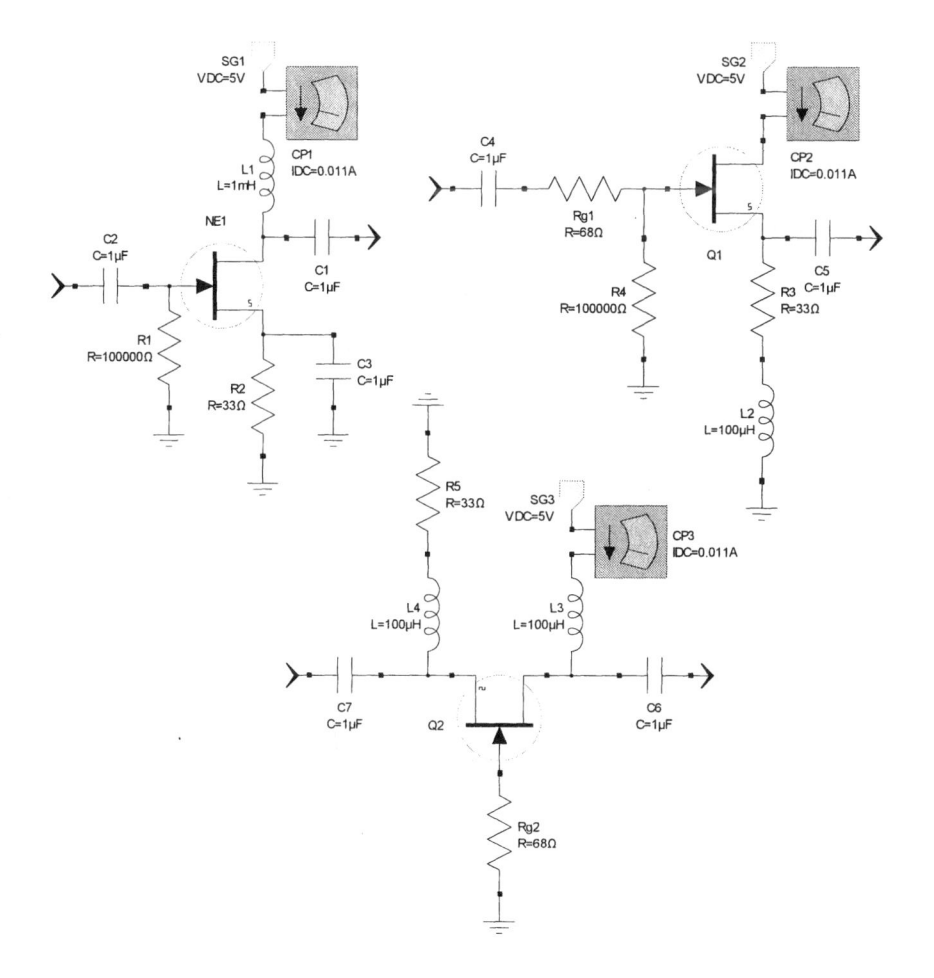

Figure 1.15 Bias and test circuits for common-source (upper left), common-drain (upper right), and common-gate GaAsFET (bottom) transistor amplifiers.

As is the case with bipolar transistor amplifiers, the common-source topology is the more stable of the three, and the common-drain and common-gate topologies are the least stable. The CS amplifier includes an ideal, high-impedance choke at the drain to apply supply power. The source resistor is bypassed to obtain the highest available gain. The top row in Fig. 1.16 is the magnitude and angle of S_{21} and the input and output S-parameters of the CS topology. Despite high input impedance, the gain of this device at low frequency in a 50-ohm system is 16.4 dB, partially due to a moderate output impedance. The CS amplifier is inverting at low frequency and this device remains within 15° of inverting up to 1000 MHz.

A stabilizing resistance of 68 ohms is placed in series with the gate of the CD and CG topologies to stabilize them. As with the bipolar devices, this resistance is implemented either as conventional resistors or as ferrite beads. The common-drain topology includes resistance at the gate to hold the gate at low DC voltage for biasing. A high value of resistance is used to avoid shunting signal at the gate. A high-impedance choke is added in series with the source resistor to avoid shunting the output signal. The characteristics of the common-drain topology are given in the middle row of Fig. 1.16. The input impedance is extremely high and the output impedance is low. The low-frequency gain at 50 ohms is 3.7 dB and the gain remains above 0 dB above 10 GHz. The gain in a 50-ohm system is rather low because of mismatch at primarily the input port. Provided the resonator is used to match the device input and output impedances, the CD NE33200 has adequate gain for use as a sustaining stage through 10 GHz. This CD topology is noninverting within 150° up to 2500 MHz.

The characteristics of the common-gate topology are given in the bottom row of Fig. 1.16. Again, a high-impedance choke is placed in series with the source resistor to avoid shunting the signal. The characteristics are similar to the common-drain topology except the bandwidth is slightly less and the input impedance is low while the output impedance is high.

As is the case with bipolar amplifiers, the CS topology is best suited for use with noninverting resonators with similar input and output impedance. The optimum cascade impedance level is generally higher for FET devices than with bipolars. The CG and CD topologies are best suited for inverting resonators that require dissimilar terminating impedance such as a Colpitts resonator. Because it is inherently more stable than the CG and CD topologies, the CS topology is poorly suited for use with negative resistance and negative conductance oscillators. The CD topology is well suited for negative-resistance oscillators and the CG topology is well suited for negative-conductance oscillators (see Section 1.7).

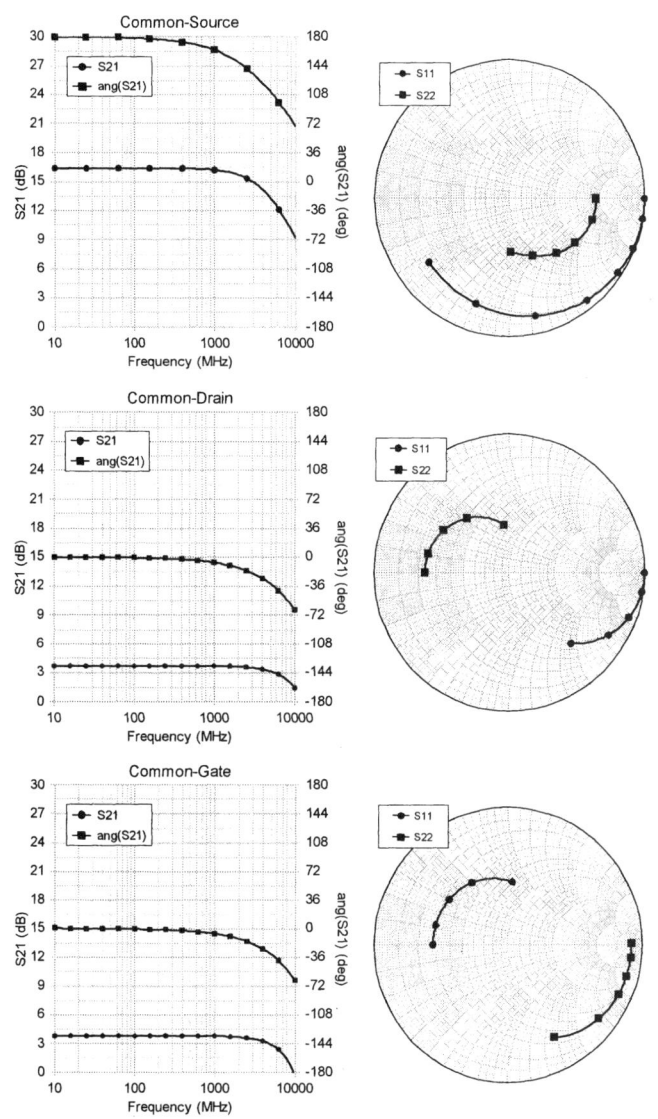

Figure 1.16 S_{21} magnitude and angle (left) and S_{11} and S_{22} (right) for CS (top), CD (middle), and CG (bottom) FET topologies of the NE33200 FET.

1.5.4 Basic Common Emitter Amplifier

The test circuits shown in Fig. 1.11 are not suitable for production use because the bias networks are not designed for temperature or device parameter variation. Consider the one-battery biased amplifier shown in

Fig. 1.17. This circuit has improved bias stability and is the basic building block for many discrete amplifiers.

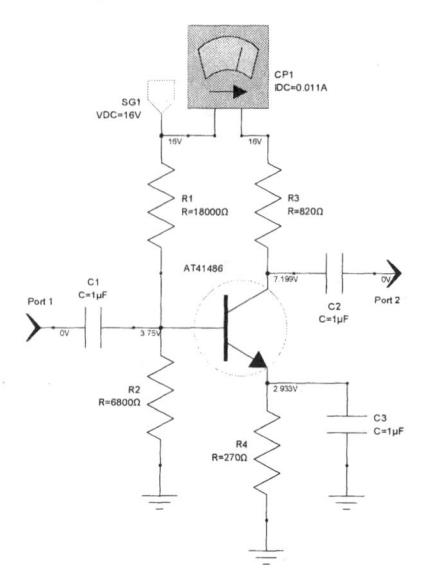

Figure 1.17 Basic bipolar amplifier stage with one-battery biasing.

The responses for this basic amplifier using the Avago Technologies AT41486 are given in Fig. 1.18. The responses are similar to the CE test circuit except, at the lowest frequencies, the collector resistor R_3 shunts the output impedance thus reducing the output impedance at low frequency.

The voltage divider formed by resistors R_1 and R_2 establish the base bias voltage. This voltage minus the base-emitter voltage (V_{be}) forms the emitter voltage that establishes the current in R_4 and therefore the emitter current. Because V_{be} is temperature-dependent, the emitter voltage should be set greater than V_{be}. Moderate emitter current then requires R_4 be large. This emitter degeneration reduces the gain so R_4 is bypassed with capacitor C_3. The collector resistor R_3 must be small enough to avoid excessive voltage drop and reduce the available collector-emitter voltage.

This amplifier will function as an oscillator sustaining stage, but it is not the best choice for a number of reasons:

1) the topology is not economic;
2) gain is excessive at low frequency;
3) the response bandwidth is a small fraction of the device F_t;
4) it has no features for improving stability;
5) the responses are device-dependant.

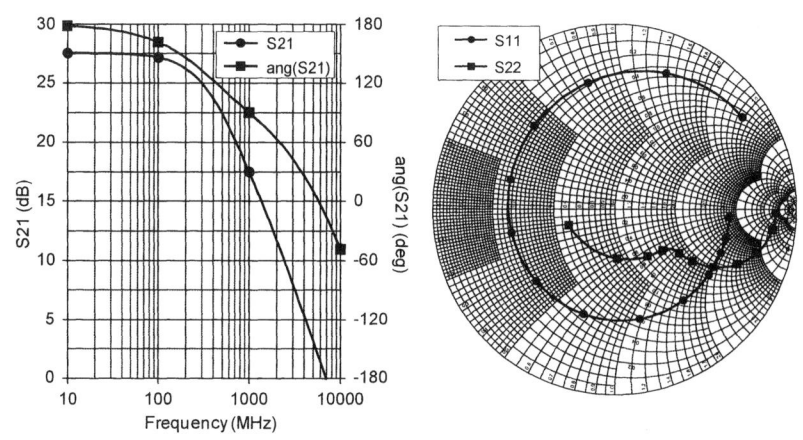

Figure 1.18 Responses of the basic bipolar amplifier stage with one-battery biasing.

1.5.5 Statistical Analysis of the Amplifier

Statistical analysis reveals the device-dependent nature of the responses of the basic common-emitter amplifier. Table 1.3 shows selected device and circuit parameters, their nominal values, and the tolerance applied during analysis. The first seven parameters are characteristics of the device, and the resistor parameters are values of the four bias resistors in the amplifier.

Table 1.3 Selected parameters, their nominal values and tolerance range used for statistical analysis of the basic CE amplifier

Param	Description	Nominal	% down	% up
Bf	Ideal forward beta	68	50	100
Cje	B-E zero-bias capacitance	3.5 pF	20	20
Cjc	B-C zero-bias capacitance	0.37 pF	20	20
Tf	Ideal forward transit time	2E-3 nS	20	20
Lb	Base inductance	1.5 nH	20	20
Lc	Collector inductance	0.6 nH	20	20
Le	Emitter inductance	0.1 nH	20	20
R1	Supply to base bias resistance	18,000 ohms	5	5
R2	Base to ground bias resistance	6800 ohms	5	5
R3	Collector load resistance	820 ohms	5	5
R4	Emitter degeneration resistance	270 ohms	5	5

Fig. 1.19 shows the multiple responses of a statistical run with parameters varied as specified in Table 1.3. At low frequency, the gain ranges from 25.4 to 31 db. At 500 MHz, the phase shift varies from 108° to 124°, a variation of 16°. The influence on device and external component tolerances on the transmission phase is significant at any frequency above a small fraction of F_t. Variation in the transmission phase of the sustaining

stage translates into shift in the oscillation frequency. This impact is quantified later in this chapter.

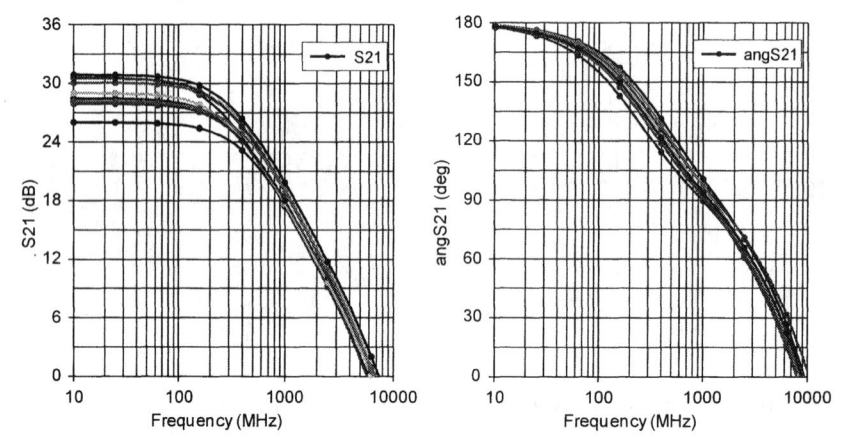

Figure 1.19 Multiple responses of an 11-sample statistical analysis of the basic CE topology with device and circuit parameters varied as given in Table 1.3.

1.5.6 Amplifier with Resistive Feedback

The behavior of the CE amplifier is improved for use as a sustaining stage by applying resistive feedback. In Fig. 1.20, resistor R_2 from the collector to the base provides negative feedback because, at low frequency, the signal at the collector is 180° out of phase with the base signal. This feedback is referred to as shunt feedback. It lowers the gain and reduces both the input and output impedance of the amplifier.

Resistor R_3 in the emitter also provides negative feedback and reduces the gain, but it increases both the input and output impedance. It is called series feedback.

When used in combination, these two resistors set the magnitude of S_{21} near a desired value G_f and set the input and output impedance near a desired impedance Z_o. With R_2 designated R_f and R_3 designated R_e

$$G_f(dB) = 20\log\left[\frac{(R_f/Z_0)^2 - 1}{R_f/Z_0 + 1}\right] \qquad (9)$$

or solving for R_f

$$R_f = 0.5 Z_0 \left(10^{G_f/20} + \sqrt{10^{G_f/10} + 4\left(1 + 10^{G_f/20}\right)}\right) \qquad (10)$$

and

$$R_e = \frac{Z_0^2}{R_f} \qquad (11)$$

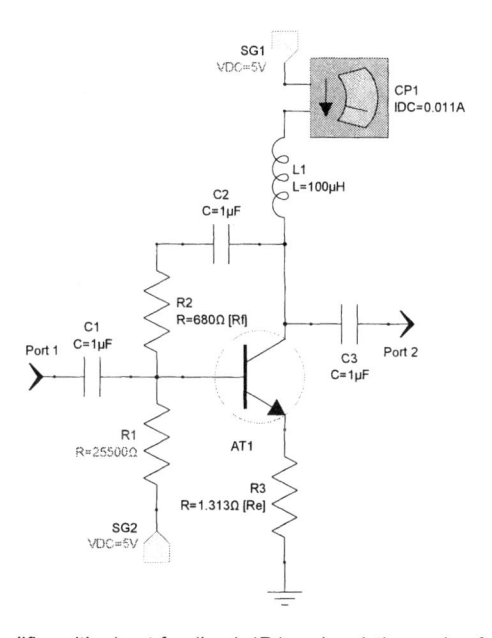

Figure 1.20 CE amplifier with shunt feedback (R_2) and resistive series feedback (R_3).

Also in series with the external series feedback resistor R_e is internal emitter resistance. This resistance is emitter current dependant and is equal to approximately 26 x $10^{-3}/I_e$ where I_e is the emitter current. Therefore, the corrected external series feedback emitter resistor is

$$R_e = \frac{Z_0^2}{R_f} - \frac{26 \times 10^{-3}}{I_e} \qquad (12)$$

Fig. 1.21 shows the responses of the CE amplifier with three sets of shunt and external series feedback resistor values. First, consider the gain, the middle three traces on the left. When biased at 5 volts and 11 mA, the open-loop gain of this AT41486 amplifier without feedback as seen in Fig. 1.21 is approximately 28 dB and the 3-dB bandwidth is approximately 300 MHz.

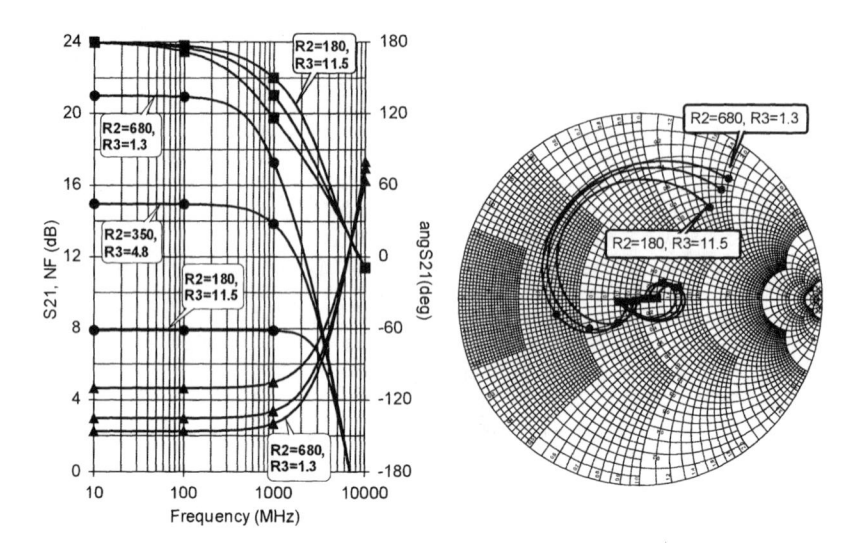

Figure 1.21 Responses of the CE amplifier with three sets of shunt and series feedback resistor values.

The shunt feedback resistor is adjusted and the series feedback resistor is computed using Eq. 12. With 680-ohm shunt feedback resistance, the series feedback resistor is 1.31 ohms, and the computed gain is approximately 21 db and the 3-dB bandwidth is 725 MHz. With shunt feedback of 350 ohms the low-frequency gain is approximately 15 dB, and with shunt feedback of 180 ohms the low frequency gain is 8 dB. With this heavy feedback, the 3-dB bandwidth is 4000 MHz. As seen in the Smith chart plots on the right, even the lightest feedback significantly improves the input and output match. At low frequency, the matches are excellent.

Next, consider the transmission phase shift with feedback. Without feedback, the phase shift at 1000 MHz is 90° rather than the ideal 180° at low frequency. With R_2 equal to 680 ohms, the phase is improved to 120° and with R_2 equal to 180 ohms the phase is improved to 150°. Feedback significantly improves the phase characteristics of the amplifier at all frequencies below the amplifier bandwidth.

One characteristic that is degraded by the application of resistive feedback is the noise figure. The bottom three traces on the left in Fig. 1.21 are the noise figure of the amplifier with the three levels of feedback. With no resistive feedback, the low-frequency noise figure is approximately 2.2 dB. With R_2 equal to 680 ohms, the low-frequency noise figure is approximately 2.3 dB, a minor degradation. However, with R_2 equal to 180 ohms the noise figure is degraded to 4.7 dB.

Linear Techniques

Another feature of resistive feedback is that it tends to improve the stability of the CE amplifier. To insure stability, the Rollet stability factor, K, must be greater than 1 and the factor B_1 must be positive [6].

$$K = \frac{\left(1 - |S_{11}|^2 - |S_{22}|^2 + |\Delta|^2\right)}{2|S_{12}||S_{21}|} \quad (13)$$

and

$$B1 = 1 + |S_{11}|^2 - |S_{22}|^2 - |\Delta|^2 \quad (14)$$

where

$$\Delta = S_{11}S_{22} - S_{12}S_{21} \quad (15)$$

Given in Fig. 1.22 are the stability factors K for the common-emitter amplifier with no resistive feedback plotted with circular trace symbols, a 1000-ohm shunt feedback resistor R_2 plotted with square symbols, and a 180-ohm shunt feedback resistor plotted with triangular symbols. R_e is set using Eq. 12 with R_e set to 0 ohms if R_e calculates to less than 0 ohms. The stability measure B_1 is positive for this device over the entire frequency range.

With no feedback, K is less than 1 up to 800 MHz, thus representing a significant threat to amplifier stability. With a shunt feedback resistance of 1000 ohms, K is greater than 1 through the entire frequency range. The resistive feedback significantly improved stability. With heavy feedback of 180 ohms, K is less than 1 from 3 to 6 GHz. Nevertheless, the stability situation is improved.

1.5.7 General-Purpose Resistive-Feedback Amplifier

Some of the limitations of the basic CE amplifier block given in Fig. 1.17 for use as a sustaining are improved by using resistive feedback. Consider the general-purpose resistive-feedback amplifiers in Fig. 1.23. The Type A amplifier on the left uses an inductive choke to bias the collector, and Type B uses a more economical resistor.

The following procedures may be used to design these amplifiers. The inductor value is chosen for a reactance much higher than the load impedance. The feedback factors for Type A and Type B amplifiers were found empirically and are approximately

$$FB_A \cong 0.5\left(10^{G_f/20} + \sqrt{10^{G_f/10} + 4\left(1 + 10^{G_f/20}\right)}\right) \quad (16)$$

$$FB_B \cong 0.5\left(10^{(G_f+2)/20} + \sqrt{10^{(G_f+2)/10} + 4\left(1 + 10^{(G_f+2)/20}\right)}\right) \quad (17)$$

where G_f is the desired gain in decibels. G_f must be less than the open-loop gain of the device at the desired operating frequency. If G_f is not less than the open-loop gain, then the final gain is less than G_f.

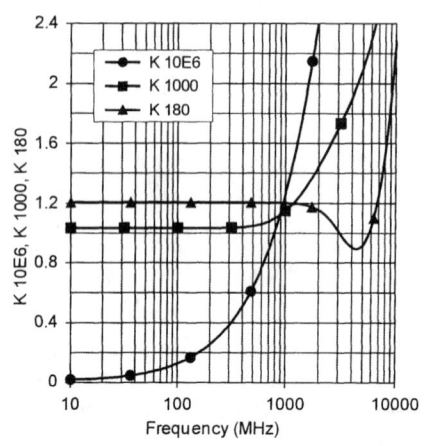

Figure 1.22 Stability factors K for the common-emitter amplifier with no resistive feedback (circular trace symbols), 1000 ohm shunt feedback resistor (square symbols), and 180-ohm shunt feedback resistor (triangular symbols). R_e is set using Eq. 12.

The resistors R_1 are set to 3 times Z_o, the desired amplifier input and output resistance. Then the shunt feedback resistors from the collector to the base are

$$R_{2A} = 0.5 R_{1A} FB_A \qquad (18)$$

$$R_{2B} = 0.5 R_{1B} FB_B \qquad (19)$$

Then the emitter series-feedback resistors are

$$R_{3A} = \frac{Z_0}{FB_A} - \frac{26 \times 10^{-3}}{I_e} \qquad (20)$$

$$R_{3B} = \frac{Z_0}{FB_B} - \frac{26 \times 10^{-3}}{I_e} \qquad (21)$$

where I_e is the desired emitter current. If R_3 is negative, then R_3 is not used (R_3 is 0). The voltage at the base of the device is approximately

$$V_b = V_{be} + R_3 I_e \cong 0.79 + 3.8 I_e + R_3 I_e \qquad (22)$$

The resistors R_1 and R_2 form a voltage divider that defines the required voltage at the collector and thus the power supply voltage.

Linear Techniques

$$V_{ccA} = \frac{V_{bA}(R_{1A} + R_{2A})}{R_{1A}} \quad (23)$$

$$V_{ccB} = \frac{V_{bB}(R_{1B} + R_{2B})}{R_{1A}}\left(1 + \frac{R_{4B}}{R_{1B} + R_{2B}}\right) + R_{4B}I_e \quad (24)$$

where R_{4B} is 3 times Z_0.

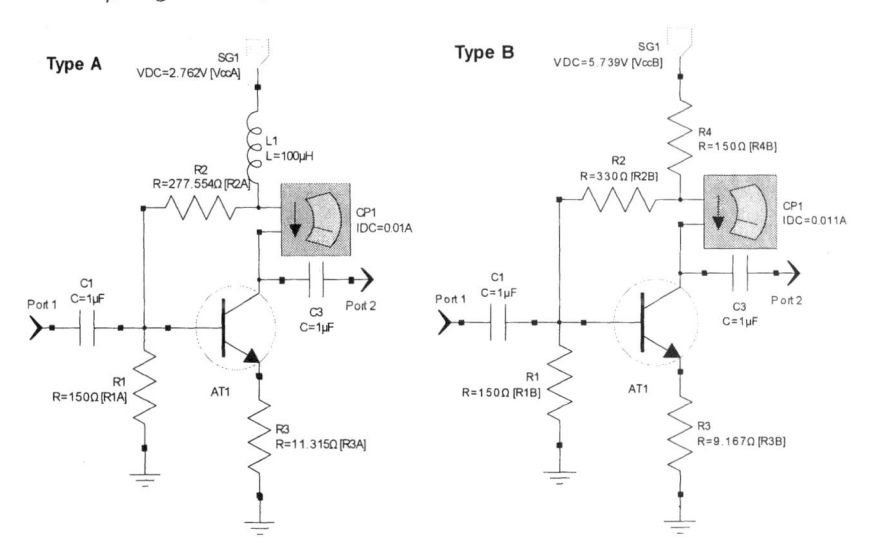

Figure 1.23 General-purpose resistive-feedback CE amplifiers of Type A (left) with an inductive collector load and Type B (right) with a resistive collector load.

The bias point in the Type A amplifier is stable only if R_3 is 10 ohms or more. This amplifier should be used only for low gain, which tends to require larger R_3, and when the operating temperature is relatively constant. The resistors R_1 and R_2 not only act as a voltage divider for the bias voltage but for the feedback as well. This causes the output impedance of the Type A amplifier to be higher than Z_0.

Since R_{4B} is in parallel with the output, the output impedance is lowered, thus compensating for the higher output impedance caused by voltage dividing the feedback. Type B is the preferred topology since it is more economical, provides a better match at the input and output, and has better bias stability. The responses of a Type B amplifier designed for 9 dB gain and 11-mA emitter current using an AT41486 transistor are given in Fig. 1.24. The 3-dB gain bandwidth is 2.6 GHz, the low-frequency noise figure is less than 5 dB, and the amplifier is within 30° of inverting to 800 MHz.

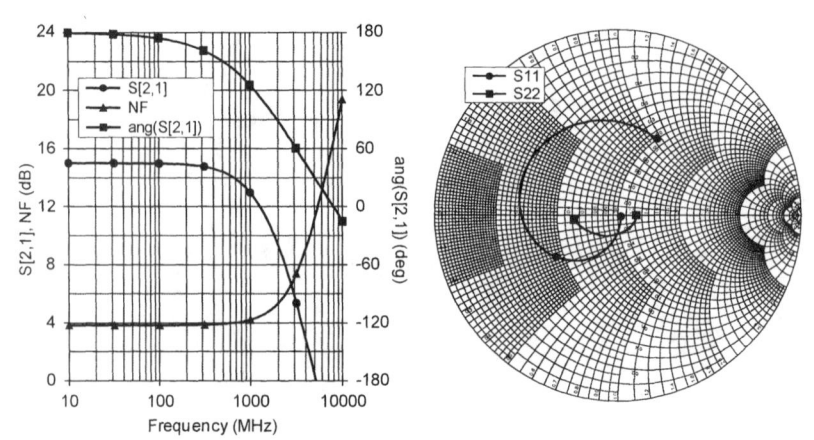

Figure 1.24 The noise figure and S_{21} magnitude and angle (left) and S_{11} and S_{22} (right) for a 9-dB gain Type B amplifier using an AT41486 transistor.

Certain values of gain and emitter current result in R_{3B} equal to zero, thus saving a resistor. This is because the entire resistance required to satisfy Eq.12 is provided by the internal emitter resistance. This is recommended for the Type B amplifier only. Table 1.4 lists the approximate gain and emitter current combinations that eliminate resistor R_{3B} for an AT41486 transistor used for a 50-ohm input and output impedance amplifier. Because the impedance is not critical for sustaining stages, emitter currents of 1/3 to 3 times the current listed in Table 1.4 are appropriate.

Table 1.4 Gain and emitter current combinations that eliminate resistor R_{3B} for a 50-ohm Type B amplifier using an AT41486 transistor

Gain (dB)	Current (mA)
18	6.7
15	4.6
12	3.2
9	2.3
6	1.6

With the elimination of R_{3B}, the amplifier consists of only the active device and three resistors. With the proper selection of a resonator, the coupling capacitors are not required since they can be part of the resonator. The lack of additional capacitors and chokes is not only economical but it eliminates potential spurious resonances and oscillation modes. The Type B amplifier is a simple but effective amplifier for use as an oscillator sustaining stage.

1.5.8 Transformer-Feedback Amplifiers

A disadvantage of resistive feedback is degradation of the amplifier noise figure. This disadvantage is eliminated by using transformer feedback rather than resistive feedback. An amplifier that uses transformer feedback is shown in Fig. 1.25. It is sometimes called a Norton amplifier. Norton amplifiers were introduced by Q-Bit and were later available from Spectrum Controls [7].

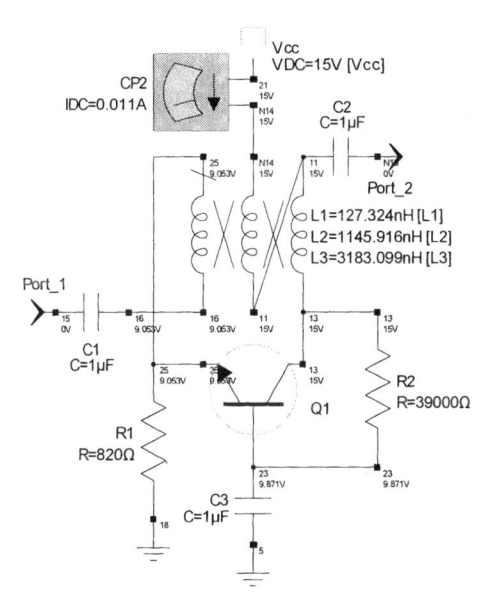

Figure 1.25 Common-base bipolar amplifier with transformer feedback.

The design of the triple-winding transformer is critical. Setting the turn ratio of the primary on the left to 1, then the number of turns of the middle winding is m

$$m = \sqrt{10^{G_f/10}} \qquad (25)$$

where G_f is the desired gain in decibels. The closest value of m that is a realizable multiple of the number of turns of the primary is then chosen as the new value of m. Then the number of turns on the right winding is then

$$n = m^2 - m - 1 \qquad (26)$$

The inductance of the primary, L_1, is chosen so its reactance is 10 to 20 times Z_0 at the lowest frequency of interest. Provided the reactance of the coupling capacitors is small, the transformer reactance determines the low-frequency range of the response. Then the inductance of the middle and right winding are

$$L_2 = m^2 L_1 \qquad (27)$$

$$L_3 = n^2 L_1 \qquad (28)$$

For example, if the desired G_f is 9 dB, then m is 2.828. Setting m equal to 3, then n is 5. If the primary is three turns on a core with an inductance of 127 nH, then the middle winding is nine turns with an inductance of 1146 nH and the right winding is 15 turns with an inductance of 3183 nH.

The leakage inductance of the windings typically determines the high-frequency response. High gain increases m and especially n, thus complicating the design of the transformer. This is generally not an issue for amplifiers used as sustaining stages because the required gain is low.

The responses of the amplifier in Fig. 1.25 using an AT41486 are given in Fig. 1.26. The coefficient of coupling between windings, K, used for the simulation is 0.98. At midband the amplifier is noninverting. At low frequency the phase leads and at high frequency the phase lags. The noise figure is less than 2 dB up to 500 MHz.

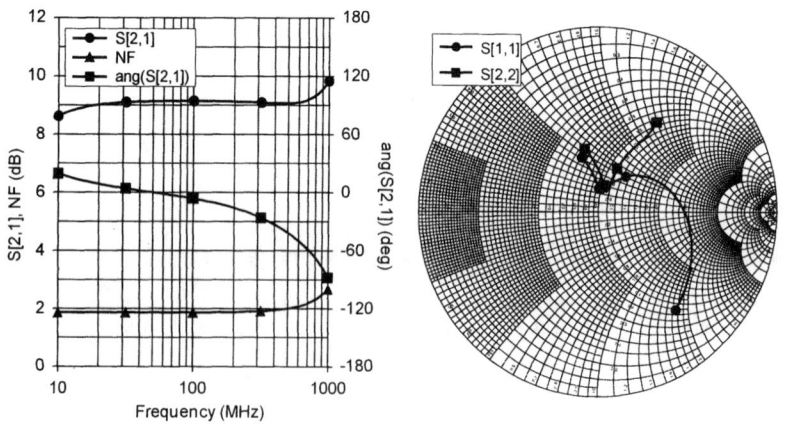

Figure 1.26 Responses of a AT41486 common-base amplifier employing transformer feedback.

1.5.9 Monolythic Microwave Integrated Circuit Amplifiers

Many popular gain blocks use a topology similar to Type B in Fig. 1.23 except the single, active device is replaced with a Darlington pair. The resulting higher open-loop gain provides improved performance. Table 1.5 is a list of monolithic microwave integrated circuits (MMIC) employing Darlington pairs and resistive feedback from Avago Technologies [4], Mini-Circuits [8] and Sirenza Microwave [9] and hybrid amplifiers with transformer feedback from Spectrum Microwave [7]. This list is only representative of a larger number of devices available from these and other manufacturers.

Table 1.5 Representative list of monolithic and hybrid gain blocks

Model	Freq (MHz)	Gain (dB)	NF (dB)	P (1dB)	Vcc	I (mA)
Avago Technologies monolithic InGaP HBT (also Mini-Circuits MAR series)						
MSA-0286	DC-2500	12.5	6.5	4.5	5	25
MSA-0386	DC-2400	12.5	6	10	5	35
MSA-0686	DC-800	20	3	1	5	16
MSA-0786	DC-2000	13.5	5	5.5	4	22
Mini-Circuits monolithic InGaP HBT						
GALI-S66	DC-3000	17.3	2.7	2.8	3.5	16
GALI-21+	DC-8000	13.1	4.0	10.5	3.5	40
GALI-4F+	DC-4000	13.4	4.0	13.8	4.4	50
GALI-6F+	DC-4000	11.6	4.5	15.8	4.8	50
Sirenza Microdevices monolithic SiGe						
SGA-0163(Z)	DC-4500	12.7	4.6	-1.8	2.1	8
SGA-1163(Z)	DC-6000	11.5	3.1	-3.3	4.6	12
SGA-2263(Z)	DC-5000	14.7	3.2	7.5	2.2	20
SGA-4263(Z)	DC-3500	14.0	3.4	14.2	3.2	45
SGA-6386(Z)	DC-5000	15.4	3.6	21.0	4.9	80
Spectrum Microwave hybrid Norton amplifiers TO-8 package						
TM5118	3-100	16.3	1.5	6.5	15	21
TM101	5-500	13	2.4	7	15	18
TM3085	100-500	11.5	1.5	19.5	15	90
TM3019	500-1300	13.5	2.2	22	15	105
TM9711	1000-2000	12	2.2	16	6	62

All values in the table are typical and not guaranteed. *Freq (MHz)* is the useable frequency range of the device. If the low-frequency limit is DC then the response is limited by external coupling capacitors. *Gain(dB)* is the gain at low frequency or midband. The gain may be somewhat less at the extremes of the specified frequency range. *P(1dB)* is the output power at 1 db compression. *I(mA)* is the typical current drawn when the device is operated at the specified *Vcc*.

Several trends may be noted from this list. In general, with resistive feedback amplifiers, higher gain offers better noise figure and higher current provides higher output power. Transformer feedback limits the bandwidth and improves the noise figure.

1.5.10 Differential Amplifiers

Fig. 1.27 shows a differential amplifier using discrete bipolar transistors. It is driven single-ended at the base of AT414b, and the output is taken single-ended at either the collector of AT414b (port 2) or the collector of AT414c (port 3). In IC design, the bases of AT414b and AT414c are typically driven differentially, increasing the voltage gain by a factor of 2. The outputs may also be taken differentially for another voltage gain of 2. In discrete design, differential inputs and outputs may be implemented using transformers. To avoid a DC offset, the devices must have similar properties or the base bias must be adjusted. In this implementation, the balance is adjusted via any of the four-base bias resistors. Differential devices offer power supply noise immunity. With IC implementation, the use of three devices is not an economic disadvantage and the devices on a given wafer have similar properties.

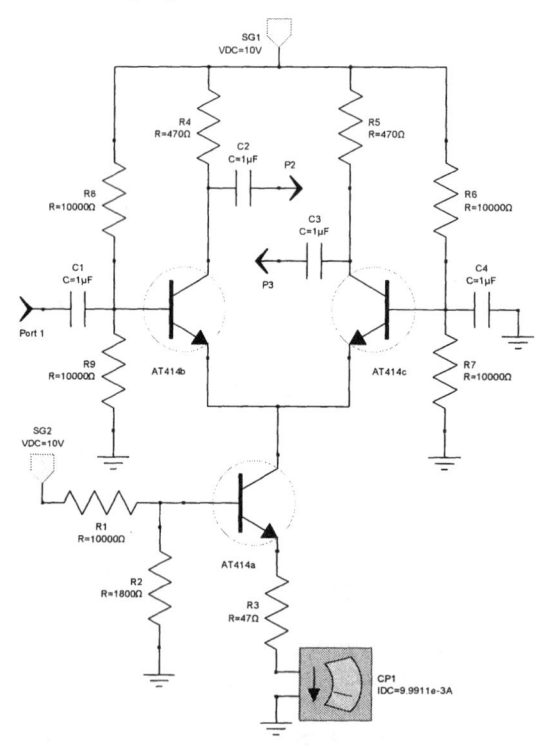

Figure 1.27 Differential amplifier using discrete devices.

Transistor AT414a is a source of current controlled by the voltage across resistor R_3. This current is shared by AT414b and AT414c. Positive-going signal at the base of AT414b increases the current in that device and decreases the current in AT414c. At low frequency, the signal output at the collector of AT414b is inverted and the output at AT414c is not inverted.

The input impedance of differential amplifers is relatively high. To achieve reasonable voltage gain, collector resistor values and the load termination impedance must be high. Therefore, the device junction and stray capacitance significantly limit the bandwidth. Given in Fig. 1.28 are the responses of the differential amplifier given in Fig. 1.27. The input and output reference impedances are 500 ohms. Notice the bandwidth is lower than the simpler, lower impedance single-ended AT41486 CE amplifier. Notice that the differential phase of ports 2 and 3 is 180° at all frequencies. Two choices are therefore available to adjust the cascade open-loop phase.

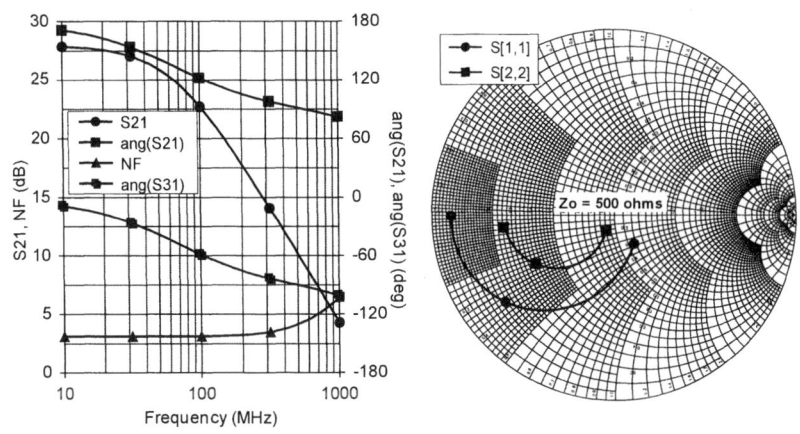

Figure 1.28 Responses of the differential amplifier shown in Fig. 1.26.

A differential amplifier that uses a resistor as a current source is shown in Fig. 1.29. This simpler circuit is better suited for discrete oscillator design. Sharing a voltage source for base bias, coupled via inductor L1, eliminates the need to balance the bias resistors.

The responses of this amplifier are given in Fig. 1.30. The excessive gain at low frequency of the high-gain AT414 transitors is moderated by using low values of collector resistance, in this case 51 ohms. The lower collector impedance also increases the bandwidth. The input impedance remains high, near 500 ohms at low frequency. The resistive component falls to 50 ohms at approximately 300 MHz. The noise figure is excellent.

While differential circuits are well suited for IC design, the additional complexity discourages their use in discrete oscillator design. However, if economy is not a factor, the noise immunity, benign limiting characteristics, and differential interface to ICs are attractive for certain applications. Devices with a high F_t are required and PWB parasitics may require intervention to stabilize the devices.

1.5.11 Phase-Lead Compensation

Consider the MSA0386 MMIC gain block amplifier in Fig. 1.31. Inductor L_l is commonly used to supply power to the gain block. The inductor is normally a high reactance choke that has little influence on the response of the amplifier. The responses of this amplifier circuit with a large inductor value are given in Fig. 1.32 as solid traces. At low frequency, the amplifier is inverting and the transmission phase shift is near 180°. However, by 1100 MHz, the phase has rolled off to approximately 130°.

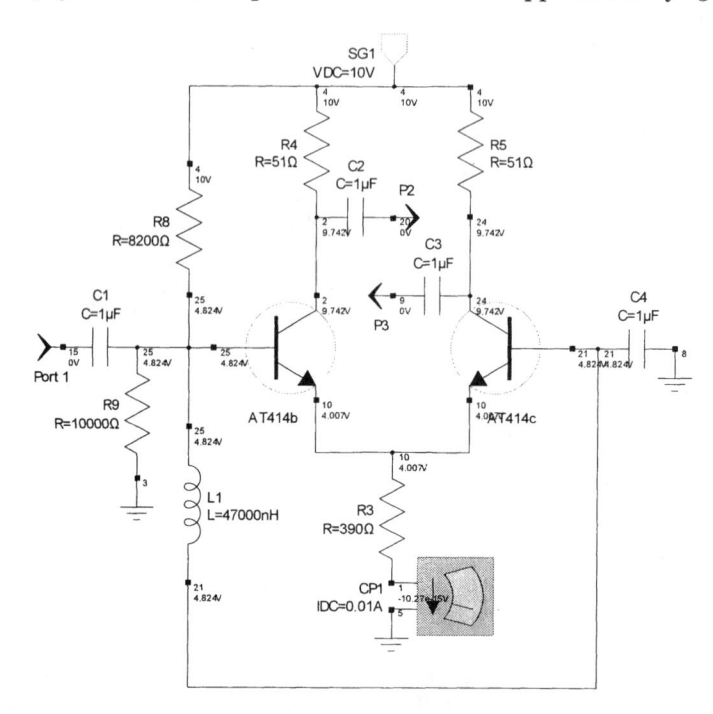

Figure 1.29 Differential amplifier using a resistor (R_3) as a current source.

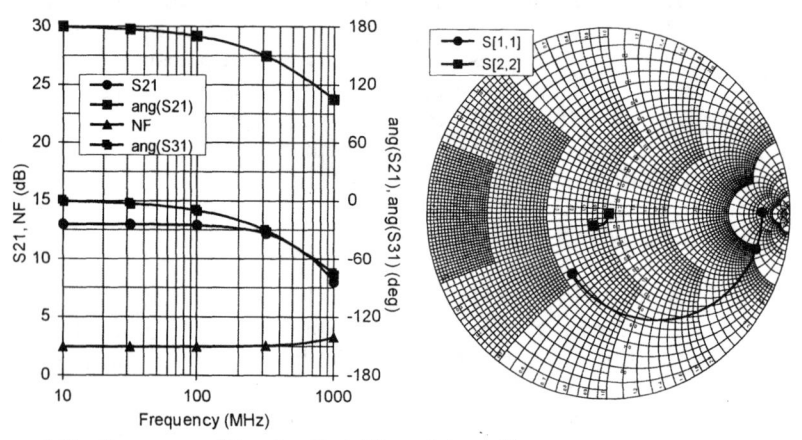

Figure 1.30 Responses of the simplified differential amplifier.

A shunt inductor has the property of leading the phase. If inductor L_1 is reduced in value, the phase rolloff is reduced. With an inductor value of 3.8 nH the transmission phase is returned to 180° at 1100 MHz as shown by the dashed traces in Fig. 1.32. This also shunts signal amplitude and disturbs the amplifier input and output impedance. However, if the phase lead is moderate this technique is a useful tool for adjusting the cascade phase shift to align ϕ_0 at the maximum phase slope. Notice that the amplifier phase slope is increased. This makes it more difficult to control the phase over a wide tuning range. The inductive lead network is suitable for narrow tuning oscillators only.

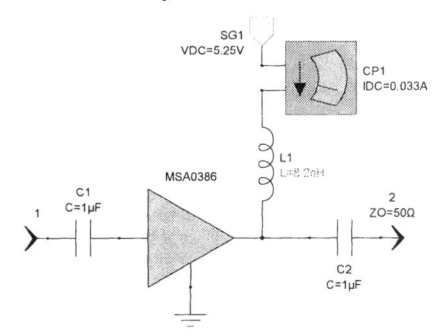

Figure 1.31 An MSA0386 MMIC gain block amplifier with an inductor to supply power.

1.5.12 Amplifier Summary

The sustaining stage provides the gain margin and supplies oscillator output power. The CE and CS topologies are inverting at low frequency, have similar input and output impedance, and tend to be the more stable. The remaining topologies are noninverting and have dissimilar input and output impedance. The phase shift and impedance characteristics of the amplifier dictate the resonator topologies required to satisfy the desired initial conditions of the cascade.

The amplifier should be stable and low-noise. Gain requirements are moderate and F_t should be 4 to 10 times the operating frequency. Excessive gain is undesirable and excessive F_t may cause instability. Ideally, the open-loop characteristics of the cascade are independent of the active device characteristics, temperature, and the power supply voltage. In other words, the ideal amplifier provides only gain and power and does not influence the transfer phase, which determines the oscillator frequency, noise, and stability. These objectives are better satisfied when feedback is utilized in the amplifier.

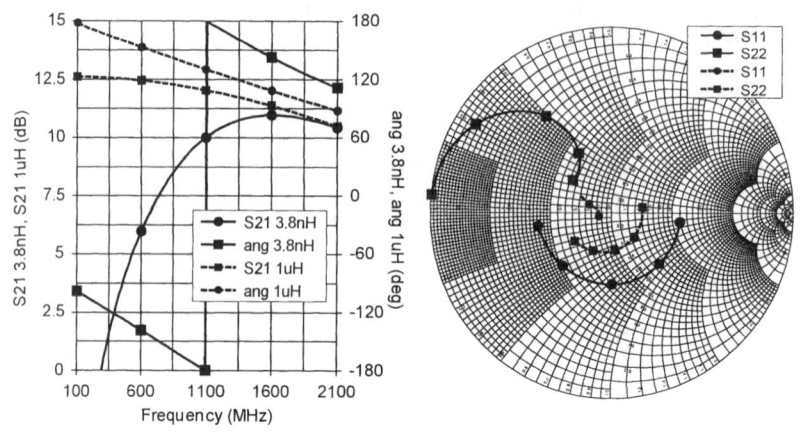

Figure 1.32 Responses of the MSA0386 MMIC gain block amplifier with phase-lead compensation (solid) and without (dashed).

The amplifier is the nonlinear component during oscillation and it is considered in more detail in Chapter 2. However, the cascade component that critically determines the oscillation frequency, the long-term stability, and the phase noise is the resonator. It is considered next.

1.6 Resonators

Fundamental resonator concepts are introduced using primarily resistor-capacitor (R-C) and inductor-capacitor (L-C) elements. Other resonators such as transmission-line (distributed), dielectric, and various piezoelectric forms are covered in detail in later chapters.

1.6.1 R-C Phase Shift Network

Fig. 1.33 shows an inverting CE amplifier cascaded with a resistor-capacitor (R-C) phase-shift network. R_1 is a bias resistor with voltage taken from the collector. The collector voltage drops with increasing collector current thereby reducing the base current. This simple bias scheme guarantees the device is biased in the active region. R_1 is too large to provide significant signal feedback so the amplifier gain is high at low frequency.

The phase-shift network consists of three R-C sections: the amplifier output resistance driving C_2, R_3 driving C_3, and R_4 driving C_4. Each section provides approximately 60° of phase shift at 1 MHz. The total cascade phase shift is then 360°, or 0° at 1 MHz. Each section also attenuates the signal at 1 MHz, so high gain is required in the amplifier to provide gain

Linear Techniques 41

margin at 1 MHz. Additional R-C sections result in less attenuation for a given total phase shift, but diminishing returns are realized with more than four sections. Output may be taken by sampling signal at the collector of the transistor. When the loop is closed to form the oscillator, capacitor C_5 is omitted.

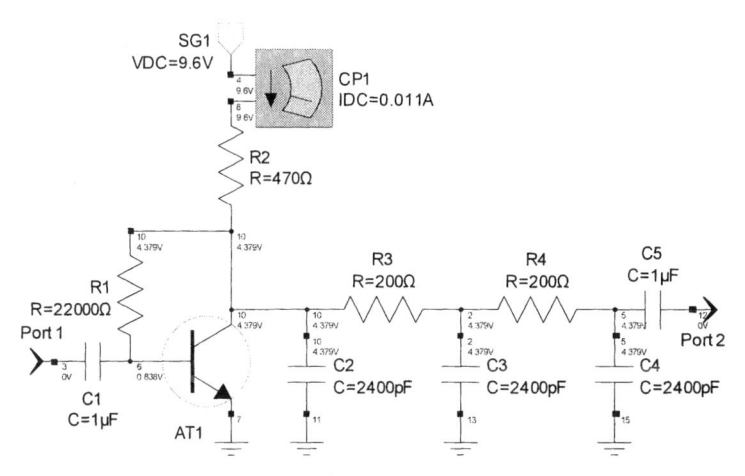

Figure 1.33 Open-loop cascade of amplifier and R-C phase-shift network.

The responses of the cascade are shown in Fig. 1.34. Because the gain margin is low and the matches are poor, the transmission responses are given as G, the true open-loop cascade gain computed by Randall and Hock, Eq. 8. The gain margin is 4.5 dB. With four R-C sections and the shunt capacitors reduced to 1240 pF, the gain margin increases to 6.9 dB.

R-C oscillators are often used at very low frequency where inductors are expensive and values are large. R-C resonators are also used in IC designs where inductors are difficult to implement on chip. R-C oscillators are tunable over a wide frequency range by adjusting the series resistance.

The difficulty with R-C oscillators is that steep phase slopes cannot be achieved, thus limiting the oscillator stability and phase-noise performance. As shown in Chapter 4, with other factors being equal, the phase-noise of an oscillator degrades with the square of frequency. At low frequency, phase-noise performance is naturally better. This places a lower demand of phase slope and is another reason R-C oscillators are more common and practical at low frequency.

1.6.2 Delay-Line Phase-Shift Network

A method that achieves phase shift without introducing insertion loss is the use of a transmission line. The transmission line may be considered a phase-shift network or a delay line. Fig. 1.35 shows the schematic of the cascade of a Mini-Circuits MAR 3 MMIC gain block and a 50-ohm transmission line.

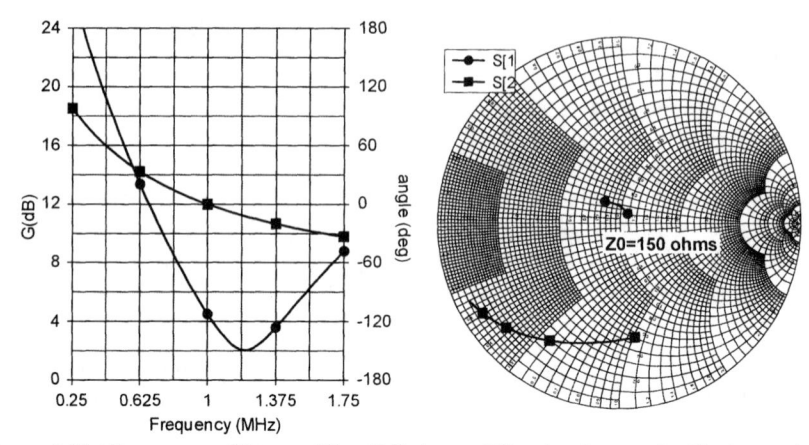

Figure 1.34 Responses of the amplifier, *R-C* phase-shift network cascade. The transmission magnitude and angle are *G*, the true open-loop gain.

The MAR 3 is the same device as an Avago Technologies MSA03. The MAR 3 is biased using a 330-ohm resistor to a 13.6-volt supply. Capacitor C_2 is eliminated when the loop is closed to form the oscillator. At low frequency the MAR 3 is inverting so a 180° line length would be used. At higher frequency the MAR 3 has additional phase shift and a 136.45° long line sets ϕ_0 at 1000 MHz.

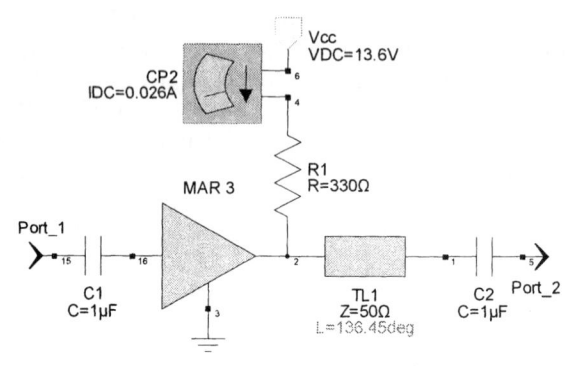

Figure 1.35 Open-loop cascade of an MAR 3 and a 136.45° length of 50-ohm line.

The transmission amplitude and phase of the MAR 3 and transmission line are shown in Fig. 1.36. The cascade exhibits approximately 11-db gain at 1000 MHz. The solid traces are with a transmission line length of 136° at 1000 MHz, the desired oscillation frequency. The phase slope is increased by using a longer length of line. The dashed trace is with a line length of 497°. Again ϕ_0 coexists at 1000 MHz and approximately 333 MHz. This cascade is likely to oscillate at 333 MHz when the loop is closed. Therefore, to realize a 1000-MHz oscillator, the shorter line length and lower phase slope is used. Ring oscillators are closely related to delay-line oscillators. Ring oscillators are covered in Chapter 5.

1.6.3 L-C Parallel and Series Resonators

It is now evident that the function of the resonator in the cascade is threefold. First, the resonator insures that ϕ_0 exists at one and only one frequency. This is accomplished by amplitude selectivity. However, in other respects, it is not the amplitude characteristics of the cascade that are important but rather the phase characteristics. Second, the resonator is used to create a steep slope in the cascade transmission phase. This improves the long- and short-term stability of the oscillator by making the frequency at ϕ_0 as immune as possible to perturbations in the phase shift of the amplifier caused by device variation, temperature, the power supply voltage, and device noise. Third, the resonator may be used as a matching device to correct for dissimilar impedances at the input and output of the cascade amplifier. This eliminates the need for additional matching networks that would only add to the cost and complexity of the oscillator and that would potentially introduce spurious resonant modes.

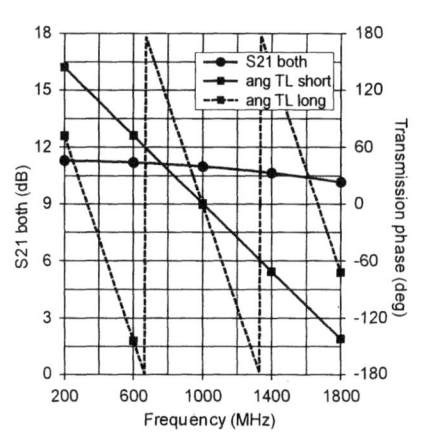

Figure 1.36 Responses of the MAR 3 and 50-ohm transmission line cascade with a 136.45° line at 1000 MHz (solid traces) and a 497° line (dashed trace).

The simplest L-C networks that satisfy the first and second functions are the basic series L-C and parallel L-C resonators shown in Fig. 1.37. At resonance, the reactance of the series reactors L_1 and C_1 are equal and cancel. If the reactors are lossless, at resonance, the insertion loss is zero.

Similarly, at resonance the reactance of L_2 and C_2 cancel. These networks have no ability to match dissimilar source and load resistance.

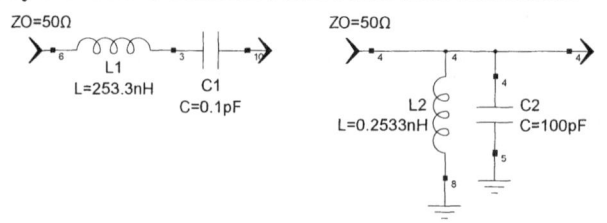

Figure 1.37 Basic L-C series resonator (left) and parallel resonator (right) inserted in transmission paths with 50-ohm characteristic impedance.

The transmission magnitude and angle of these series and parallel resonators are identical and are shown in Fig. 1.38. The amplitude peaks at 0-dB insertion loss and the transmission phase is 0° at the resonant frequency. Below resonance the circuits are inductive and the phase leads. Above resonance the circuits are capacitive and the phase lags.

With lossless reactors, at resonance the impedance at the input is equal to the load impedance, so when the output of the resonator is terminated in the reference impedance, the input S-parameter reflection coefficient magnitude is zero and the trace passes through the center of the chart as shown on the right in Fig. 1.38. The inverse is true for the output, and the plot of S_{22} is identical to S_{11}.

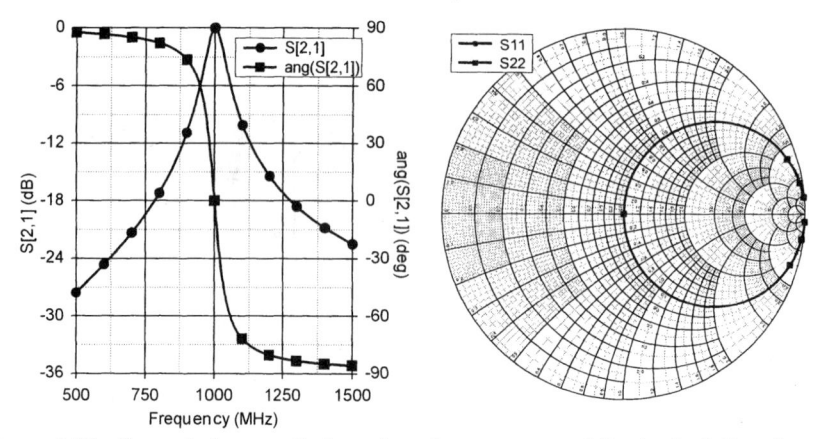

Figure 1.38 Transmission amplitude and angle responses of the basic L-C series and parallel resonators (left) and S_{11} and S_{22} (right).

While the responses are identical, the reactor values required to realize these identical responses are not equal. To realize a steep phase slope, the reactance of series elements must be greater than the system characteristic impedance and the reactance of parallel elements must be less than the system characteristic impedance. In this case, the series inductor is 253.3 nH, which is too large to realize easily at 1000 MHz, and the series capacitor is 0.1 pF, which is too small. The parallel inductor is 0.253 nH and the parallel capacitor is 100 pF, extreme values in the opposite direction. A solution to this difficulty is given later, but the solution description is described best using new terminology.

1.6.4 Loaded Q

The concept of Q is presented next. There are many forms of Q and the design of not only oscillators but also other high-frequency circuits critically requires an intuitive understanding of each form. Perhaps less confusion would result if the nomenclature for the various forms were labeled Q, R, S, and so forth. However, the forms are closely related and they share similar defining equations. The first form considered is loaded Q, Q_L. The definition of Q_L introduced in circuit fundamentals courses is

$$Q_L = \frac{f_0}{BW_{3dB}} \qquad (29)$$

where f_0 is the resonant frequency and BW_{3dB} is the 3-dB down bandwidth of the amplitude transmission response. The 3-dB bandwidth of the amplitude response in Fig. 1.38 is 62.8 MHz and Q_L is therefore 15.9. For a basic L-C resonator, the 3-dB bandwidth equals the $+-45°$ bandwidth.

For the basic, lossless series resonator, Q_L may be defined also by

$$Q_L = \frac{X}{R_{total}} = \frac{X}{2Z_0} \qquad (30)$$

where X is the reactance of either the inductor or capacitor at the resonant frequency and R_{total} is the total series resistance. For the basic, lossless parallel resonator, Q_L may be defined by

$$Q_L = \frac{R_{parallel}}{X} = \frac{Z_0}{2X} \qquad (31)$$

Narrow bandwidth results in steep phase slope. Thus, the phase slope is improved with increased Q_L. However, the transmission phase in the cascade, and not the amplitude, controls the oscillation frequency. The above definitions exactly represent the phase slope only if the resonator is a basic two element series or parallel type and the cascade gain peak occurs at ϕ_0. This is often not precisely achieved and a better definition of Q_L for oscillator design is based on the cascade phase characteristics. Therefore,

$$Q_L = -\frac{\omega_0}{2} \frac{\partial \phi}{\partial \omega} \qquad (32)$$

where the angle unit is the radian. Because the definition of group delay, t_d, is

$$t_d = -\frac{\partial \phi}{\partial \omega} \qquad (33)$$

then

$$Q_L = \frac{\omega_0 t_d}{2} \qquad (34)$$

The last equation is convenient because vector network analyzers often display t_d. To measure and display Q_L, the GENESYS software suite employs a discrete computation of the phase of S_{21} using Eq. 32. Because the computation is discrete, more accurate results are obtained by using a large number of points in the frequency sweep. A plot found by this method for the loaded Q of the resonators in Fig. 1.37 is given in Fig. 1.39.

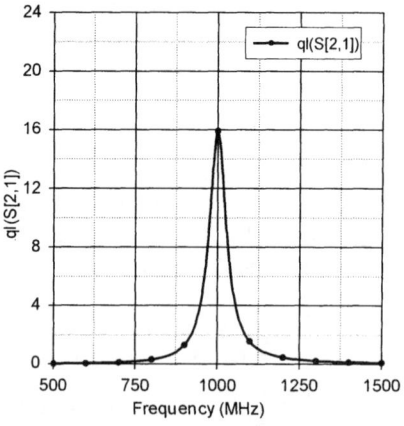

Figure 1.39 A plot of Q_L computed using Eq. 32 for the basic series and parallel resonators in Fig. 1.36.

1.6.5 Unloaded Q

The second form of Q to consider is the unloaded Q, Q_U. It is also referred to as component Q since it is a measure of the quality of a reactor. The definition of unloaded Q is 2π times the energy stored in a reactor divided by the energy dissipated per cycle. For loss mechanisms that are in series with a reactor, this reduces to

$$Q_U = \frac{X}{R_S} \qquad (35)$$

Linear Techniques

where R_S is the series loss resistance of the reactor. Examples of series loss mechanisms are metal loss in wire-wound inductors and metal loss in the leads and plates of capacitors. Notice the similarity of Eqs. 35 and 30. In Eq. 30 the loss is the series combination of the terminations and in Eq. 35 the resistance is loss resistance.

For loss mechanisms that are in parallel with the reactor the definition reduces to

$$Q_U = \frac{R_P}{X} \qquad (36)$$

where R_P is the parallel loss resistance of the reactor. Examples of parallel loss resistance are core losses in an inductor and dielectric loss in a capacitor. Notice the similarity of Eqs. 36 and 31.

For inductors wound on air, ceramic, or plastic cores, the predominant source of loss is winding metal loss and the series model is used. For inductors wound on ferrite and powdered-iron cores, the loss is often a combination of series and parallel loss and a more complex model using Q versus frequency data is required. The loss resistance is often specified by inductor manufacturers by providing the unloaded Q. Inductor models are considered in Appendix A, Models.

Despite the fact that the dielectric is an important material in capacitor construction, modern dielectrics are low loss and the predominant source of loss is resistance in the metal leads and plates, so the series model is also used for capacitors. The series resistance is specified by capacitor manufacturers as the effective series resistance (ESR). Capacitor models are also considered in Appendix A.

The resonator consists of both an inductor and a capacitor. The unloaded Q of a resonator, Q_R, is given by

$$Q_R = \frac{1}{1/Q_{ind} + 1/Q_{cap}} \qquad (37)$$

where Q_{ind} is the unloaded Q of the inductor and Q_{cap} is the unloaded Q of the capacitor.

1.6.6 Resonator Loss

The responses of the basic series L-C resonator on the left in Fig. 1.37 but constructed using a lossy inductor with an unloaded Q of 50 and a capacitor with an unloaded Q of 400 are given in Fig. 1.40. Components with these finite unloaded Q result in an insertion loss of 2.66 dB and an input S-parameter of -11.58 dB at resonance.

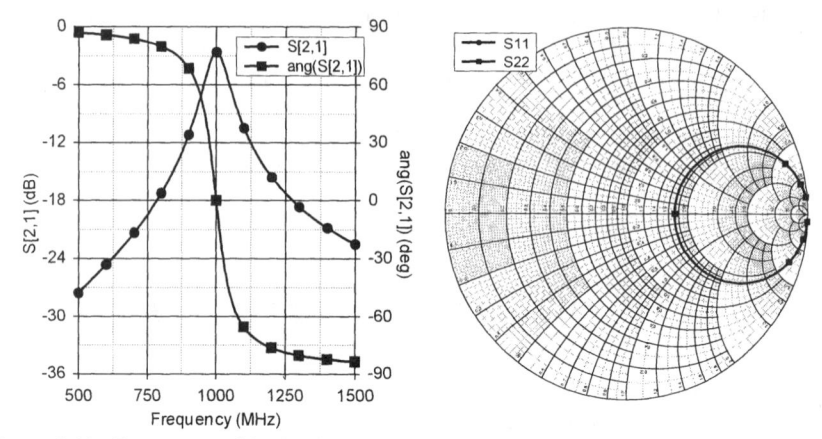

Figure 1.40 Responses of the basic series resonator with an inductor with an unloaded Q of 50 and a capacitor with an unloaded Q of 400.

The insertion loss of a single-pole resonator is

$$IL = 20\log\frac{Q_R}{Q_R - Q_L} \qquad (38)$$

This insightful equation reveals that the *IL* of a resonator is known based only on the loaded *Q* and unloaded *Q* of the resonator. Notice when the unloaded resonator *Q* is much greater than the loaded *Q* that the insertion loss is low. On the other hand, as the loaded *Q* approaches the unloaded *Q* the *IL* approaches infinity. With the loaded *Q* equal to ½ of the unloaded *Q*, the *IL* is 6.02 dB.

With an inductor *Q* of 50 and a capacitor *Q* of 400, the resonator unloaded *Q* is 44.4. With a loaded *Q* of 15.91, the *IL* from Eq. 38 is -3.85. This disagrees with the *IL* in Fig. 1.39 because the finite *Q* components result in a reduced loaded *Q* of the resonator. The resulting loaded *Q* with a lossy resonator is

$$Q_L = \frac{1}{1/Q_R + 1/Q_{L-lossless}} \qquad (39)$$

where $Q_{L-lossless}$ is the loaded *Q* with a lossless resonator. The resulting loaded *Q* for this example is 11.72 and the resulting insertion loss is 2.66 dB, in agreement with the response in Fig. 1.40.

The insertion loss of the resonator is important because the sustaining stage must overcome this loss to realize the desired cascade gain margin.

1.6.7 Colpitts Resonator

The Colpitts resonator was introduced in Section 1.2.1.4. The primary features of the Colpitts resonator are phase shift near 0° at resonance and dissimilar input and output termination impedances. These features are compatible with CC, CB, CD, and CG sustaining stages. This section provides formulas for determining inductor and capacitor values in the Colpitts resonator.

Fig. 1.41 shows a common-gate J309 JFET amplifier and Colpitts 250-MHz resonator cascade. Resistor R_1 helps stabilize the J309 and returns the gate to ground for biasing. Resistor R_2 stabilizes and reduces the drain current to 5.3 mA, well below I_{dss} for the device. Capacitor C_3 avoids DC termination of the source by the simulator.

The reactance of capacitor C_4 at 250 MHz is high so the 50-ohm termination at port 3 loads the cascade only lightly. When the common-gate J308 is terminated in 50 ohms at the input and output, the real part of the parallel input impedance at the source is 52.7 ohms and the real part of the parallel output impedance at the drain is 2277 ohms. The high drain impedance terminates the top of the Colpitts resonator and the 52.7-ohm source impedance terminates the capacitive tap at the junction of C_1 and C_2. The Colpitts resonator matches the source and drain. The parallel reactances at the device terminals are absorbed into the resonator and influence the frequency of the cascade ϕ_0.

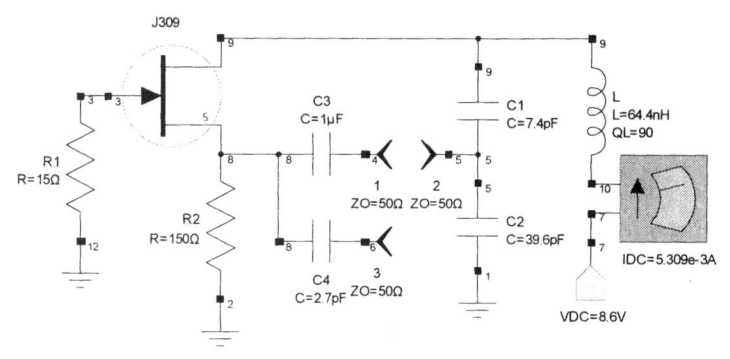

Figure 1.41 Common-gate J309 JFET amplifier and Colpitts resonator cascade for a 250-MHz oscillator.

Given a desired loaded Q, resonator unloaded Q and known termination resistances, the resonator reactor values are found using the following equations.

$$L = \frac{R_{top}}{2Q_L \omega_0} \quad (40)$$

where R_{top} is the real part of the parallel drain impedance, ω_0 is the desired resonant frequency in radians/second, and Q_L is the desired loaded Q of the resonator. Because Q_R reduces the resonator loaded Q, Q_L used in Eq. 40 must be set higher than the desired final Q_L. Using Eq. 39, with a desired loaded Q of 10 and Q_R equal to 90, Q_L in Eq. 40 must be 11.25. The factor of 2 in the denominator is because ½ of the terminating load of the resonator is due to the drain and ½ is due to the source.

With a drain parallel termination resistance of 2277 ohms, L equals 64.4 nH. Then, the capacitive tap must transform the real part of the parallel impedance at the source, R_{tap}, up to the resistance R_{top}. Therefore,

$$C_2 = \frac{1}{\omega_0} \sqrt{\frac{R_{top}^2 + X_L^2 - X_L^2 R_{top}/R_{tap}}{X_L^2 R_{top} R_{tap}}} \quad (41)$$

$$C_1 = \frac{1}{\omega_0} \left(\frac{R_{top}^2 X_L}{R_{top}^2 + X_L^2} + \frac{R_{tap}^2 X_{C2}}{R_{tap}^2 + X_{C2}^2} \right)^{-1} \quad (42)$$

Given in Fig. 1.42 are the responses of the common-gate JFET amplifier and Colpitts resonator cascade with resonator values determined by the above formula. The frequency of ϕ_0 is 226 MHz, significantly lower than the desired 250 MHz. This is caused by the reactive components of the impedance at the source and drain that are ignored in Eqs. 40 to 42. The frequency error is easily compensated by decreasing the value of either the inductor or both capacitors in the resonator. The resulting match is excellent and use of the Randall/Hock correction is unnecessary.

At resonance, the loaded Q is 14.5, which is higher than the design goal of 10. Because the device is not unilateral, and because the drain is terminated in 2277 ohms by the resonator rather than the 50 ohms used to measure the device impedances, the actual impedances terminating the resonator differ from the design values. Increasing the frequency by reducing the value of the inductor would increase the loaded Q. Both the frequency and Q are compensated by reducing the capacitor values. Adjusting C_1 to 4.17 pF and C_2 to 20 pF centers the frequency and lowers the loaded Q to 9.8.

The design of the Colpitts resonator is not exact for nonunilateral devices. However, the procedure provides a starting solution, which is easily compensated for by small tuning or optimization of resonator values using measured data or computer simulation.

Linear Techniques 51

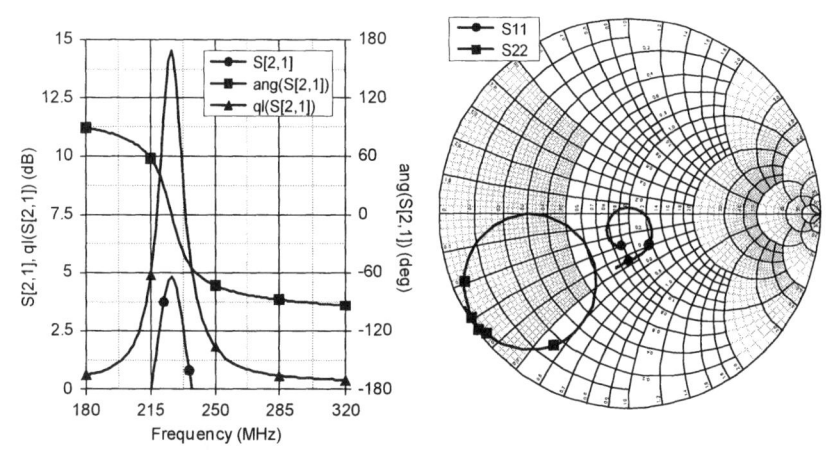

Figure 1.42 Responses of the common-gate JFET amplifier and Colpitts resonator cascade with resonator values determined by Eqs. 40-42.

1.6.8 Resonator Coupling

It is apparent from Eq. 31 that achieving a high loaded Q for a given inductive reactance in a parallel resonator requires a high load resistance. Fortunately, the inputs of the CS and CC topologies and the outputs of the CG and CB topologies are high impedance. The more moderate impedance of the other ungrounded terminal of these devices is transformed up by the capacitive tap. Therefore the Colpitts resonator is compatible with these topologies.

The input and output impedance of the CS and CE topologies is often too low to achieve good unloaded Q with parallel resonators. This is particularly true with popular gain blocks designed for 50-ohm input and output impedance. Furthermore, these topologies are inverting so the 0° phase shift of the Colpitts parallel resonator is not suitable.

These issues are resolved by the use of another important design concept referred to as coupling. This term is used frequently in filter literature, but although coupling is important for oscillator design, the term is used less frequently. Either coupling reactors in series with the source and load are used with a parallel resonator or coupling reactors in shunt with the source and load are used with a series resonator. The former is illustrated next.

Shown at the left in Fig. 1.43 is a coupling reactance X_s in series with a load termination, R_s. To determine the loading of this circuit on a parallel resonator, the parallel equivalent resistance, R_p, and reactance, X_p, are

$$R_p = \frac{R_s^2 + X_s^2}{R_s} \qquad (43)$$

$$X_p = \frac{R_s^2 + X_s^2}{X_s} \qquad (44)$$

The desired resonant frequency is used to compute the reactance since the equivalence holds only at a single frequency. Notice that for any coupling reactance greater than zero, the parallel resistance is greater than the series resistance. For example, if the coupling reactance is 3 times R_s, R_p is 10 times R_s. The coupling reactance transforms upward the resistance presented to the resonator, thus increasing the loaded Q. Good loaded Q is achieved with more reasonable resonator parallel inductor and capacitor values. This concept is similar to the capacitive tap used in the Colpitts resonator except that it may be applied to both resonator terminations as shown on the right in Fig. 1.43 for a 1000-MHz parallel resonator with 50-ohm terminations, a loaded Q of 9.42, and a resonator unloaded Q of infinity.

The coupling reactors may be either capacitors or inductors. They not only transform the resistance presented to the resonator, they also load the resonator with reactances given by Eq. 44, thus shifting the resonant frequency down. This is compensated by decreasing resonator component values.

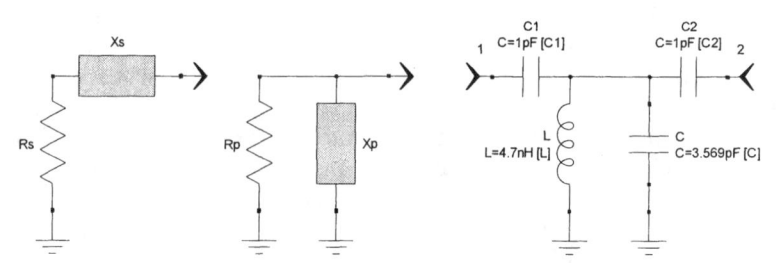

Figure 1.43 Coupling reactance X_s in series with the load (left), the equivalent parallel load resistance and reactance (center), and a parallel resonator with input and output coupling capacitors (right).

1.6.9 Matching with the Resonator

For maximum power transfer, if the resonator terminations are equal then the coupling reactor values are equal. The resonator may be used as a matching device by using dissimilar values of coupling reactance. Therefore, the reactor coupled parallel resonator may be used with sustaining stages with equal or unequal input and output impedance. With given termination resistances and a desired loaded Q there are three known parameters. The top-C coupled parallel resonator depicted on the right in Fig. 1.42 has four component values. In the following design formula, this degree of freedom is used to choose the resonator inductor value to achieve practical resonator values. This degree of freedom may also be used to realize the resonator with a tuning varactor of a specific value. Then

$$R_{total} = \frac{\omega_0 Q_R Q_L L}{(Q_R - Q_L)} \qquad (45)$$

where R_{total} is the total resistance loading the parallel resonator caused by both terminations and L is the chosen resonator inductor value. To match dissimilar terminations, each coupling reactance must transform its corresponding termination to twice this total parallel load resistance. Therefore

$$C_1 = \frac{1}{\omega_0 \sqrt{2 R_{total} R_1 - R_1^2}} \qquad (46)$$

$$C_2 = \frac{1}{\omega_0 \sqrt{2 R_{total} R_2 - R_2^2}} \qquad (47)$$

where R_1 and R_2 are the termination resistances and C_1 and C_2 are the corresponding coupling reactors. The values of capacitance loading the resonator are then

$$C_{1p} = \frac{X_{C1}}{\omega_0 (R_1^2 + X_{C1}^2)} \qquad (48)$$

$$C_{2p} = \frac{X_{C2}}{\omega_0 (R_2^2 + X_{C2}^2)} \qquad (49)$$

and the final corrected resonator capacitor value is

$$C = C_r - C_{1p} - C_{2p} \qquad (50)$$

where C_r is the capacitance that resonates with L at the desired frequency. Certain parameters of inductor value, loaded Q, resonator Q, and termination resistance may result in negative component values. The ratio of source to load resistance must be within a range that narrows with decreasing loaded Q. The inductor value may need to be adjusted to avoid negative component values. As is discussed in Section 3.9.2, smaller values of inductance result in lower RF voltage across a tuning varactor. This may be required to avoid varactor breakdown. This is weighed against the fact that achieving high inductor unloaded Q requires a minimum value of inductance.

Because of the resistance introduced in parallel with the resonator by the finite unloaded Q of the resonator, the port S-parameters are degraded. Nevertheless, the above equations provide the best match.

A parallel resonator with series coupling capacitors designed using these equations is shown in Fig. 1.44. The selected inductor value is 4.7 nH, the design loaded Q is 12.7, the resonator unloaded Q is 90, the source resistance, R_1, is 50 ohms, and the load resistance, R_2, is 186 ohms.

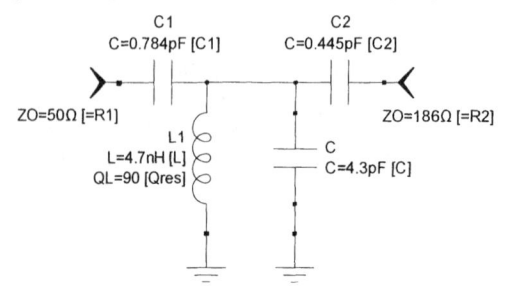

Figure 1.44 Example parallel resonator with series coupling capacitors matching a 50-ohm source to a 186-ohm load.

The responses for the resonator in Fig. 1.44 are given in Fig. 1.45. The insertion loss at the peak in the amplitude response caused by the finite resonator Q is -1.32 dB. The S-parameter magnitudes at the input and output are -17 dB. The loaded Q (not shown) is 12.7.

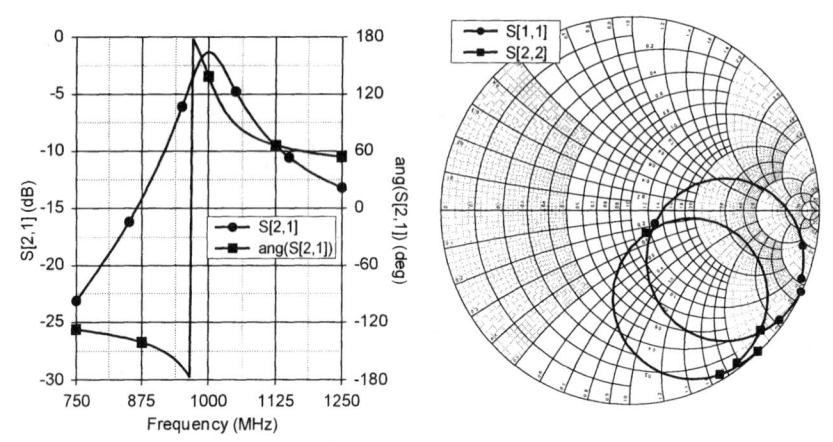

Figure 1.45 Responses of the parallel resonator with series coupling capacitors matching a 50-ohm source to a 186-ohm load.

The transmission phase shift of this example resonator is 138.6° at the gain peak. The phase shift of the capacitive coupled parallel resonator varies from 180° at infinite loaded Q to 0° at a loaded Q of zero. For a given

Linear Techniques

loaded Q, the phase shift approaches 0° with decreasing inductor values. The phase shift is given by

$$\phi = 180^0 - \tan^{-1}\left(\frac{-R_1}{X_{C1}}\right) - \tan^{-1}\left(\frac{-R_2}{X_{C2}}\right) \qquad (51)$$

1.6.10 Measuring the Unloaded Q

Knowledge of component and resonator Q is critical in the design of high-performance oscillators. Unloaded Q data from the manufacturer is a good starting point. Also, models for unloaded Q are provided in later chapters covering oscillators with specific resonator technologies. But the best practice is to verify Q by measurement of the actual devices being used.

The unloaded Q of a component may be computed from the one-port S-parameter of the component with the opposite terminal grounded.

$$Q_U = \frac{2 \times im(S_{11})}{1 - re(S_{11})^2 - im(S_{11})^2} \qquad (52)$$

where $re(S_{11})$ and $im(S_{11})$ are the real and imaginary components of S_{11}. However, S-parameters are mathematically ill-conditioned for determining the unloaded Q of highly reactive components. S-parameters were devised for characterizing transmission systems and components such as active devices that generally have resistive input and output impedances.

For example, the 1-port S_{11} for an 18 nH inductor with an unloaded Q of 120 is -0.054 dB at 47.7°. A measurement error of only 0.02-dB magnitude with no error in the angle yields S_{11} equal to -0.034 dB and a resulting inductor Q of 190, an error of 58%. Instruments specifically designed for component measurement are better suited for determining component Q than are vector network analyzers.

Unloaded Q can be more accurately measured with a scalar or vector network analyzer by using the resonator in Fig. 1.44 as a test circuit. The series coupling capacitors have little impact on the resonator Q because they are coupling rather than resonator components. Another way of looking at this is that the terminations are a significant series resistance and so the small loss resistance of these capacitors is insignificant. Therefore, the unloaded Q of the test circuit is approximately the unloaded Q of the resonator. If a measurement of only the inductor Q is desired, then the resonating capacitor unloaded Q must be much higher than the inductor unloaded Q. If this is not the case, and the ESR of the capacitor is known or estimated, then Eq. 37 may be used to estimate the inductor Q.

To determine the resonator unloaded Q, the amplitude transmission response of the test circuit is measured with a network analyzer. The loaded Q is measured using Eq. 29 or 34. The 3-dB bandwidth is defined with respect to the peak. The coupling capacitors are adjusted for an IL of 3 to 10 dB and a loaded Q of greater than 20. The resonator unloaded Q is then

$$Q_R = \frac{Q_L}{1-10^{-IL/20}} \qquad (53)$$

1.6.11 Coupled Resonator Oscillator Example

Fig. 1.46 shows a coupled-resonator 800-MHz oscillator example using an MSA0386 MMIC sustaining stage and a top-C coupled parallel resonator. Capacitor C_3 is not required when the loop is closed. Capacitor C_4 couples output power to a 50-ohm load at port 3. Inductor L_1 is not a high reactance. It leads the phase shift of the MSA0386 at 800 MHz as described in Section 1.5.11 so that the open-loop cascade ϕ_0 occurs at maximum slope.

The responses of the example coupled resonator oscillator are given in Fig. 1.47. The gain margin is 6.7 dB and the loaded Q is 14. The matches are excellent and the Randall/Hock equation is not required.

1.6.12 Resonator Summary

The resonator is the critical component in the design of high-performance oscillators. It strongly affects the phase-noise performance and the immunity of the oscillator from sustaining-stage characteristics and external disturbances. By far the most critical parameter in resonator design is the loaded Q. Unloaded Q plays a secondary role except it must exceed the loaded Q and it causes resonator insertion loss.

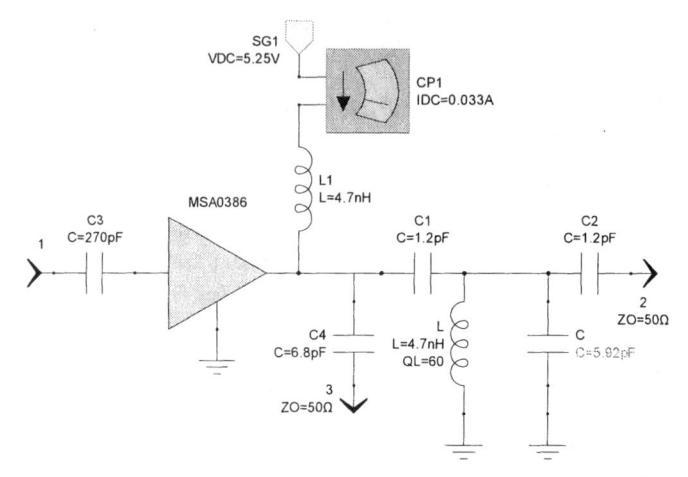

Figure 1.46 Example of a coupled-resonator 800-MHz oscillator using an MSA0386 sustaining stage.

Linear Techniques   57

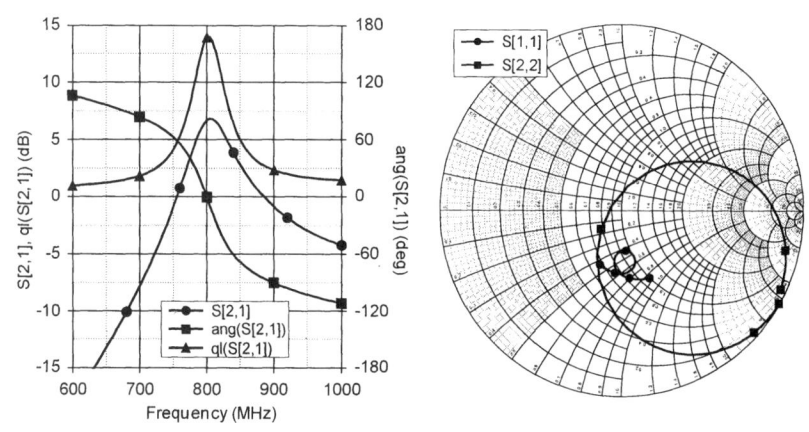

Figure 1.47 Responses of the coupled resonator 800-MHz oscillator open-loop cascade.

A resonator must be chosen to complement the sustaining stage phase shift so that the total cascade transmission phase passes through 0° near the gain peak and maximum phase slope. The $L\text{-}C$ Colpitts with a phase shift at resonance near 0° and the capacitor-coupled parallel resonator with a phase shift at resonance between 180° and 0° are considered in this chapter. Many other resonators are covered in later chapters that cover specific resonator technologies.

The resonator may also serve as a matching device for interfacing the sustaining stage input and output. Additional matching devices are neither required or desirable in oscillator design.

1.7 One-Port Method

The one-port method for the design of microwave oscillators is popular and well published. Unfortunately, a misconception exists that results in nonoptimum designs. Specifically, there are two basic forms of one-port oscillators. Both forms are often referred to as negative-resistance oscillators. However, one form is a true negative-resistance oscillator while the other is a negative-conductance oscillator. The distinction is more than semantic because the negative-resistance form requires a series resonator while the negative-conductance type requires a parallel resonator.

In this book, both forms of oscillator designed by the one-port method are referred to as one-port oscillators. This is a matter of convenience but it is not rigorous since some oscillators may be analyzed using either the one-port or the open-loop cascade method. As with the two-port cascade oscillator method, an additional port is used to extract power. The two forms of one-port oscillators have some design aspects in common. Nevertheless, the two forms are covered in different sections that follow.

1.7.1 Negative-Resistance Oscillators

The sustaining stage for an oscillator designed using the open-loop cascade method requires a device and topology that is stable. However, the active device and topology for a negative-resistance oscillator must be unstable. This instability is then controlled by adding a resonator to establish oscillator properties.

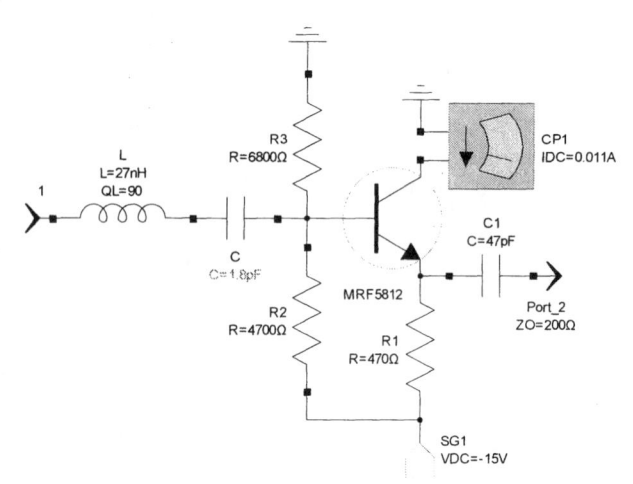

1.7.1.1 Negative-Resistance Oscillator Basic Circuit

Fig. 1.48 shows a basic negative-resistance oscillator. The device is an MRF5812 bipolar NPN transistor with an F_t of approximately 5 GHZ. With the common-collector topology, bipolar NPN transistors are often unstable at frequencies below F_t. The device is biased at 11 mA from a -15-V supply. Output power is taken at the emitter into a 200-ohm load impedance (port 2) through a coupling capacitor.

The series L-C network forces oscillation at the desired frequency. Port 1 is used for analysis of the circuit. This port is grounded to form the oscillator.

Figure 1.48 Negative-resistance oscillator basic circuit.

Given in Fig. 1.49 on a rectangular graph at the left are the reflection coefficient (S_{11}) magnitude and angle at port 1 of the negative-resistance oscillator circuit shown in Fig. 1.48. The reflection coefficient peaks at 1.46 at 1000 MHz and the phase angle passes through 180° at that frequency. S_{11} greater than 1 indicates there is return gain rather than return loss.

Linear Techniques 59

At the right, the reflection coefficient is plotted on a normal unity radius Smith chart. S_{11} greater than 1 plots outside the circumference of the chart. Any impedance with a positive real component plots inside the circumference. If S_{11} plots outside the Smith chart, the port impedance has a negative-resistance component. The net resistance of the resonator loss and the active device is negative.

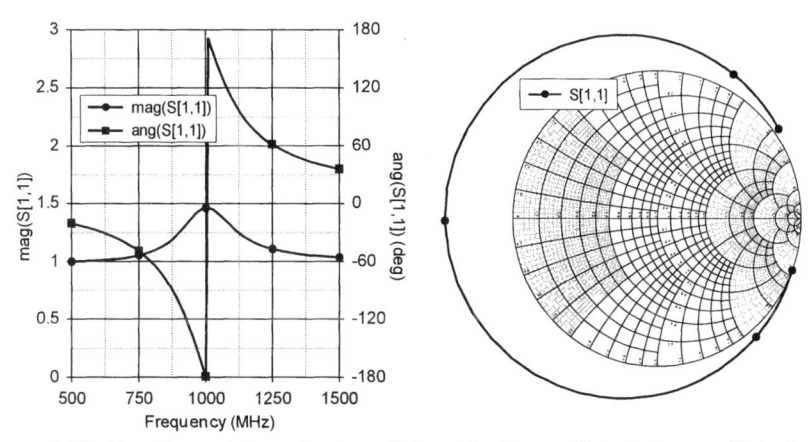

Figure 1.49 Negative-resistance basic oscillator reflection coefficient S_{11} magnitude (round symbols) and angle (square symbols) plotted on a rectangular graph and S_{11} plotted on a Smith chart (right).

1.7.1.2 Starting Criteria for the Negative-Resistance Oscillator

When the series resonator is grounded on the left, the net negative resistance in the resonator-device path causes a signal to build until device nonlinearity reduces the net path resistance to 0 ohms. The sufficient starting conditions at the desired frequency are:

1) the angle of S_{11} must be either zero or 180°;

2) the small-signal magnitude of S_{11} must be > 0 dB at $\phi=0°$ or $\phi=180°$.

Just as excessive gain margin is undesirable in open-loop cascades, excessive negative resistance is undesirable because its absorption requires a high, nonlinear signal level, which degrades noise performance, increases harmonics, and may cause spurious modes. Appropriate values of negative resistance range from -50 to -5 ohms. The higher magnitude of negative resistance is used for fast starting and high output power. The lower magnitude of negative resistance provides better stability and noise performance. The reason the angle of S_{11} is either 0° or 180° is discussed next.

1.7.1.3 Analysis Measurement Alternatives

Plotted with a dashed trace in Fig. 1.50 is S_{11} of the negative-resistance basic oscillator with a reference impedance of -50 ohms. Normalizing the Smith chart with a negative resistance causes all impedances with a negative real component to plot inside the circumference of a unity radius chart and positive resistances plot outside the circumference. Using a negative reference impedance is mathematically equivalent to plotting $1/S_{11}$ on a Smith chart. This measurement is available with vector network analyzers by exchanging the reference cables.

In Fig 1.50, the dashed trace of S_{11} crosses the real axis left of center near -10 ohms. This is an angle of S_{11} of 180°.

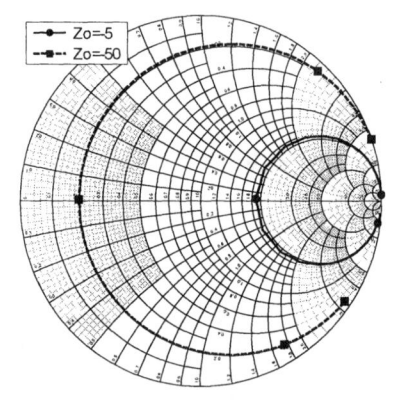

Figure 1.50 S_{11} of the negative-resistance basic oscillator with a reference impedance of -50 ohms (dashed trace) and with a reference impedance of -5 ohms (solid trace).

S_{11} is also plotted as a solid trace in Fig. 1.50 using a reference impedance of -5 ohms. With this reference impedance, S_{11} crosses the real axis right of center near the circle representing two times the reference impedance or again -10 ohms. In this case, S_{11} crosses the real axis at an angle of 0°. Whether the angle of S_{11} at resonance is 180° or 0° depends merely on the reference impedance. It has no other real significance.

Yet another alternative for analyzing a negative resistance oscillator is the impedance corresponding to S_{11}, or simply the input impedance at port 1. Plotted with solid traces in Fig. 1.51 are the real component of the input impedance (resistance) and the imaginary component (reactance) for the negative-resistance basic oscillator. Resonance and the oscillation frequency are indicated by the reactance passing through 0-ohm reactance near 1000 MHz. The input resistance is -10 at this frequency.

The oscillator is normally tuned by adjusting the series capacitor that shifts the reactance trace left or right thus decreasing or increasing the resonant frequency. The dashed trace on the left is with the series resonator capacitor increased to 3.6 pF. The dashed trace on the right is with the

Linear Techniques 61

series resonator capacitor decreased to 1.16 pF. When using an impedance plot to analyze a negative resistance oscillator, assessing the magnitude of the negative resistance over the desired tuning range is straightforward.

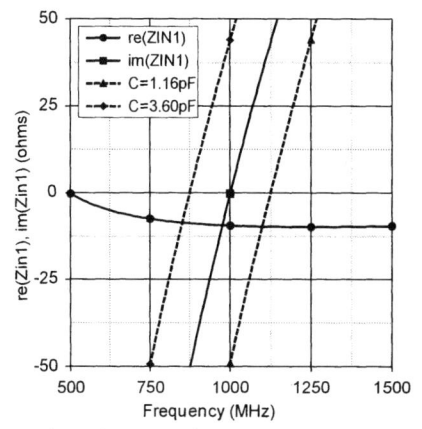

Figure 1.51 Port 1 input impedance real component (resistance) plotted with circular symbols and imaginary component (reactance) plotted with square symbols.

From Fig. 1.51 it is evident that oscillation is not possible below 500 MHz because the net resistance is positive. Above 625 MHz, the negative resistance is -5 ohms or greater. This circuit may be tuned from 625 MHz to higher frequency by adjusting the resonator capacitor.

Typically, literature covering negative-resistance oscillators uses the reflection coefficient plotted on the left in Fig. 1.49 as the analysis measurement. Because the reflection coefficient magnitude is a function of the selected reference impedance, because the reflection coefficient magnitude has little intuitive or physical significance, and because of the 0°/180° ambiguity, the input impedance is the preferred analysis measurement for negative-resistance oscillators in the remainder of this book. Odyniec discusses the starting conditions in terms of the Nyquist criterion [10].

1.7.1.4 Device Selection for Negative-Resistance Oscillator

The two-port open-loop design method utilizes the active device to merely supply gain and energy to the oscillator. Effective open-loop oscillator design isolates device characteristics from oscillator performance. On the other hand, device instability is intrinsically a function of the device. Therefore, device selection is critical in one-port oscillator design and effective oscillator design does not begin with an oscillator circuit, but rather with the device.

Columns 2 and 3 of Table 1.6 are the real and imaginary components of the base input impedance (resistance and reactance) versus frequency of the common-collector MRF5812 NPN bipolar transistor biased at I_c=11 mA and V_{ce}=9.8 volts. Column 4 is the effective input series capacitance that produces the measured reactance at each frequency. Notice that although the measurement is taken over a frequency range of 3 to 1 that the effective capacitance is more constant. This suggests that an approximate model for the input impedance of this device at this bias is negative resistance in series with approximately 2 pF of capacitance.

Table 1.6 Real and imaginary components of the input impedance (resistance and reactance) of the CC MRF5812 NPN bipolar transistor

Freq (MHz)	re[Zin]	im[Zin]	Cap (pF)
500	-1.2	-192.5	1.65
600	-5.5	-157.8	1.68
700	-8.0	-132.5	1.72
800	-9.6	-113.1	1.76
900	-10.6	-97.6	1.81
1000	-11.3	-84.9	1.88
1100	-11.7	-74.2	1.95
1200	-12.0	-65.0	2.04
1300	-12.2	-56.9	2.15
1400	-12.3	-49.8	2.28
1500	-12.4	-43.3	2.45

This capacitance is significant in negative-resistance oscillators. Consider the oscillator schematic in Fig. 1.48. The external series resonator capacitor is 1.8 pF. The net series capacitance of the external and internal series capacitors is 0.92 pF. This is confirmed by the fact that 0.92 pF and 27 nH resonate at 1000 MHz. Unfortunately, it also indicates the oscillation frequency is as dependant on the device as the resonator. This is clearly undesirable.

The ideal device for a negative resistance oscillator has a moderate negative resistance of -50 to -5 ohms that is flat with frequency and no reactance, or at least low reactance, as is the case for a large value of capacitance.

Given in Fig. 1.52 are the input resistance on the left and the effective series capacitance on the right of a common-collector 2N3866 biased at V_{ce}=15 volts and I_c=80 mA (round symbols), an MRF901 biased at 8 volts and 10 mA (square symbols), and an AT41486 biased at 10 volts and 15 mA (triangular symbols). No output load is placed at the emitter simulating very loose power output coupling. These plots are generated from S-parameter data provided by the manufacturers.

The 2N3866 develops suitable negative resistance from below 100 MHz to approximately 800 MHz, the MRF901 from approximately 200 MHz to 2 GHz, and the AT41486 from approximately 250 MHz to 7 GHz.

The effective series capacitance, plotted on the right in Fig. 1.52, ranges from 2 to 7 pF for the 2N3866 transistor. A very small series resonator capacitor used with this transistor results in a fair degree of isolation of the device from the oscillation frequency. At higher frequency, higher F_t devices such as the MRF901 and AT41486 are required to develop negative

Linear Techniques 63

resistance. But higher F_t devices have extremely small values of effective series capacitance. At 1000 MHz, the capacitance of the AT41486 transistor is 0.25 pF. This makes it difficult, if not impossible, to build negative-resistance oscillators at higher frequency using the basic circuit given in Fig. 1.48 because the resonator capacitor must be extremely small and the inductor extremely large in value. To understand how to improve the basic circuit, consider the following technique.

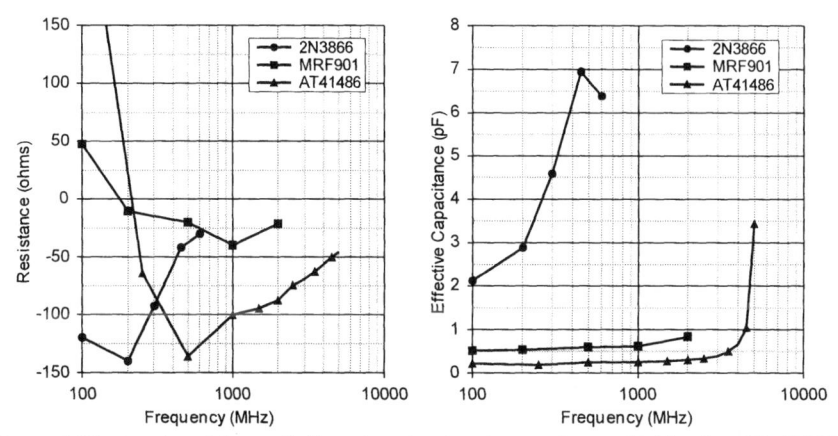

Figure 1.52 Input resistance (left) and series input capacitance (right) of a CC 2N3866 (round symbols), MRF901 (square symbols), and an AT41486 (triangular symbols).

1.7.1.5 Alechno's Technique

Kurokawa was an early contributor to negative-resistance oscillator theory [11-13]. Hamilton reviewed this work using more accessible terminology [14]. A model of a negative-resistance oscillator is given in Fig. 1.53. The device is modeled as negative resistance in series with an effective capacitance. The resonator consists of external reactors L_{reso} and C_{reso} and loss resistance R_{loss}.

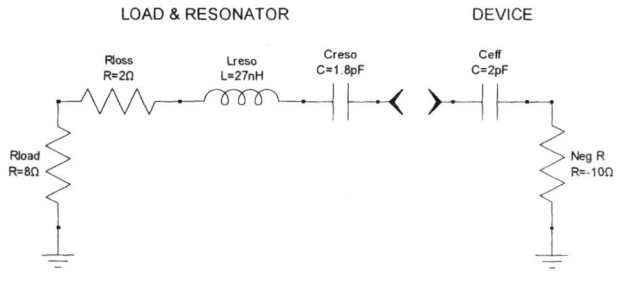

Figure 1.53 Negative resistance oscillator model with load and resonator on the left and the active device on the right.

In the cited references, the load was assumed in series with the resonator, as was often the case with tunnel, Gunn, and IMPATT diodes popular at the time. A reactive network was sometimes employed between a 50 ohm load and the resonator so as to present a lower resistance to the resonator. In this example, the load resistance presented to the resonator is 8 ohms.

The basic form of the model did not include C_{reso} and the inductor resonated directly with C_{eff}. From a performance standpoint, this is the worst possible scenario, with the minimum Q (Q_{ext}) and the frequency directly dependent on the device. The performance is enhanced by adding C_{reso} with as small a value as is practical. The resonating capacitance is then the series combination of C_{eff} and C_{reso} (C_{total})

Using this terminology, the loaded Q due to Kurokawa theory is

$$Q_L = \frac{1}{\omega_0 C_{total} |R|} \tag{54}$$

where $|R|$ is the absolute value of the negative resistance. There are difficulties in extending these concepts to the design of transistor oscillators using linear theory. Kurokawa states the model applies "for a steady-state free-running oscillation" [12]. The model assumes nonlinear action has occurred that absorbed excess negative resistance so $|R|=R_{load}+R_{loss}$. Eq. 54 does not apply to the negative resistance found by small-signal linear analysis of the circuit. Furthermore, modern negative-resistance oscillators often extract output power at the device emitter, source, collector, or drain rather than in series with the resonator. The load resistance is then represented in the device negative resistance. In this scenario, the circuit Q determining the noise performance is equal to the unloaded resonator Q. This is a delightful but dubious outcome.

The negative-resistance technique is a popular and well proven technique for establishing oscillation conditions, selecting devices, and determining the tuning characteristics of an oscillator, but it offers little intuitive insight into how to improve the circuit topology or how to estimate the loaded Q.

Recognizing these dilemma, in 1999 Stanislov Alechno published a clever trick for a more insightful analysis of the negative-resistance oscillator [15]. He manipulated the ground reference point to translate the negative-resistance topology into an open-loop cascade.

Shown in Fig. 1.54 at the upper left is a simplified schematic of a negative-resistance oscillator. The resistor that supplies power to the emitter is included because for lower values of resistance it affects the impedance seen at the base. The power supply is a voltage source at RF ground potential. At the upper right, the original grounds are removed, each original ground point is connected by wire, and a new ground reference is placed at the emitter. The circuit is redrawn at the lower left

Linear Techniques   65

with the output connected to the input via a wire. At the lower right, the loop is opened to create a resonator/amplifier cascade.

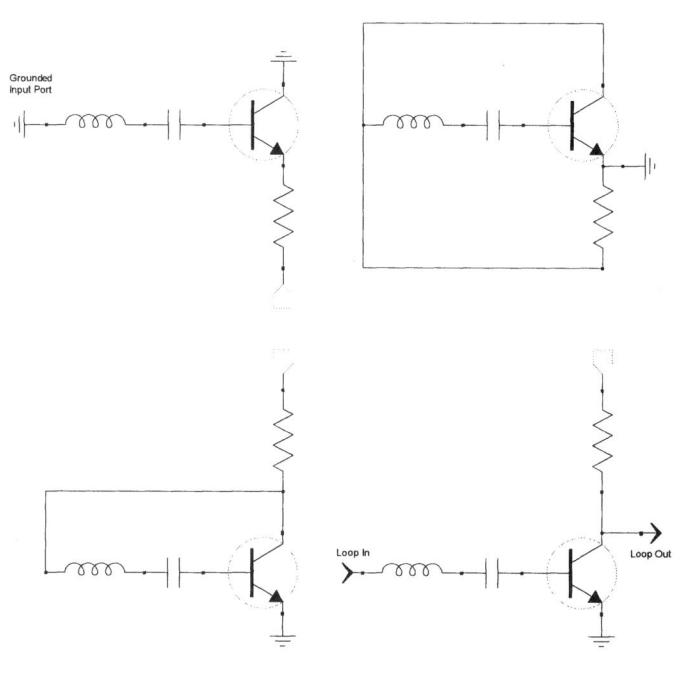

Figure 1.54 The Alechno transform of a negative-resistance oscillator to an open-loop cascade.

The negative-resistance oscillator is converted to an open-loop cascade by moving the ground reference and redrawing the circuit. The equivalence is not exact unless the active device is ideal. For example, the oscillation frequency of the negative resistance oscillator at the upper left is affected by the effective series input capacitance of the device. The oscillation frequency of the open-loop cascade is affected by the phase shift of the device. The effective series input capacitance and the device transmission phase shift are not analogous. Nevertheless, Alechno's transform is sufficiently accurate to assist in the design of the negative-resistance oscillator.

The advantage of Alechno's transform is that the measurement techniques and insight available with the open-loop cascade method are now available for analyzing the negative resistance oscillator. For example, consider the open-loop cascade at the lower right. The ideal CE amplifier is inverting or 180°. The phase shift of the series resonator at resonance is 0°. Therefore, the total phase is not 0° at resonance and the cascade is not an oscillator candidate with an ideal device. The operating frequency must be sufficiently high so the phase shift of the device lags until the device is nearly noninverting. This is why the input resistance plotted in Fig. 1.52 of the high F_t devices is not negative below a certain frequency and the circuit would not oscillate below this frequency limit.

An additional insight provided by the transform to an open-loop circuit is that because the real part of the series input and output impedance of CE bipolar devices is moderate, this high loading resistance requires a very large series resonator inductor reactance to achieve reasonable loaded Q. The loaded Q of the basic negative-resistance oscillator is typically poor.

1.7.1.6 Improving the Negative-Resistance Oscillator

The insight gained from the open-loop cascade provides potential solutions to the performance issues of the basic negative-resistance oscillator. Fig. 1.55 shows an open-loop cascade similar to the transformed negative-resistance oscillator at the lower right in Fig. 1.54. Biasing and output coupling are added.

The open-loop cascade S_{21} magnitude and angle, and the loaded Q, are given in Fig. 1.56 as solid traces. Notice that the cascade phase shift is inadequate to align the maximum phase slope at ϕ_0 but rather it occurs at 60°. At this frequency the device is too ideal and too close to inverting. At ϕ_0 the gain margin is small, approximately 0.7 dB. Because ϕ_0 does not occur at the maximum phase slope, the loaded Q is 1.7, significantly below the available loaded Q of 6.0. The nonoptimum characteristics of this design are not evident from the negative-resistance analysis.

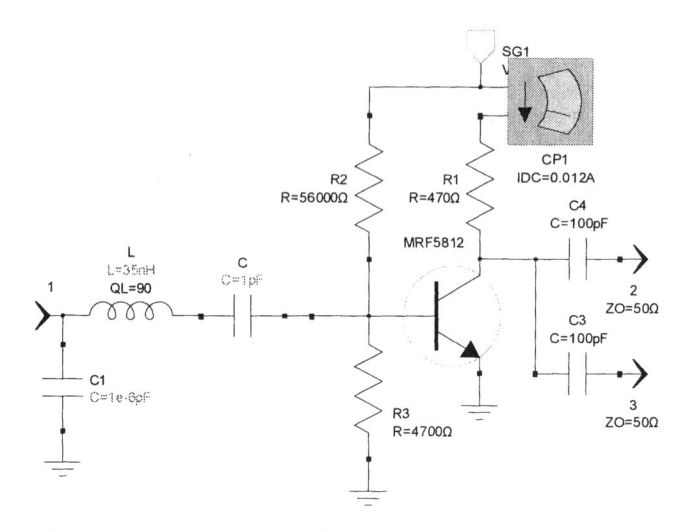

Figure 1.55 Open-loop cascade of a potential oscillator derived using Alechno's transform.

Linear Techniques 67

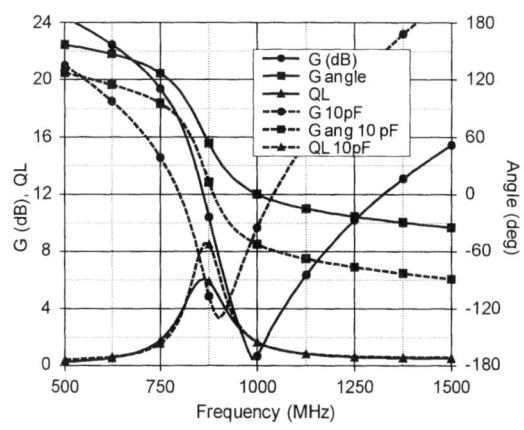

Figure 1.56 Responses of the open-loop cascade with the initial topology from Alechno's transform (solid traces) and with 10 pF shunt capacitance at the input (dashed traces).

To align ϕ_0 at the maximum phase slope, additional cascade phase shift is required. This is provided by a shunt capacitor at the input of the cascade to lag the phase by about 50°. To accomplish this, capacitor C_1 in Fig. 1.55 is increased from the initial NIL value to 10 pF. The resulting responses are given as dashed traces in Fig. 1.56. Notice that ϕ_0 is shifted down to approximately 890 MHz, but it occurs much closer to the maximum phase slope. The shifting down of the frequency is an indication that the effective input capacitance of the negative-resistance device has increased. This is a beneficial consequence of adding C_1. Also, the loaded Q is increased to 7.7 and the gain margin is increased to 3.8 dB.

Capacitor C_1 is critical for improving the performance of a negative-resistance oscillator. When the open-loop cascade is transformed back into the negative-resistance topology, capacitor C_1 connects from the emitter to ground.

The dual of the series capacitor-coupled parallel resonator in Fig. 1.42 is a shunt capacitor-coupled series resonator. In Fig. 1.57, an additional capacitor, C_2, to ground at the input of the transistors forms this dual resonator. When the open-loop cascade is transformed back into the negative-resistance topology, capacitor C_2 connects from the base of the transistor to ground. The improvement of the negative-resistance oscillator by the addition of these two capacitors is considered next.

1.7.1.7 Negative-Resistance Oscillator Coupling Capacitors

Consider the negative-resistance test circuit in Fig. 1.57 with added emitter coupling capacitor C_1 and base coupling capacitor C_2. The resonator is removed and the broadband characteristics of the circuit are evaluated versus the values of these coupling capacitors. A model rather than S-parameter data characterizes the active device. Capacitor C_3 isolates the DC bias at the base from the simulator 50-ohm source impedance. An output load is added later.

Fig. 1.58 shows the input negative resistance and the effective input series capacitance from 500 to 1500 MHz for the test circuit in Fig. 1.57. The solid traces are with no added coupling capacitance. The negative resistance is approximately 9 ohms at 500 MHz and it stabilizes near 20 ohms above 750 MHz. The effective series input capacitance slowly increases from 1.7 pF at 500 MHz to 2.2 pF at 1500 MHz. The dash-dot traces are with an emitter coupling capacitance of 2.7 pF without base-coupling capacitance. The negative resistance is increased. The negative resistance is not as flat with frequency and is somewhat excessive at low frequency. Using this circuit with a resonator to tune the oscillator would lead to higher output power and excessive harmonics at low frequency. However, the effective series input capacitance is significantly increased, particularly at the higher frequencies. C_1 is recognized by many authors as critical in the design of negative-resistance oscillators.

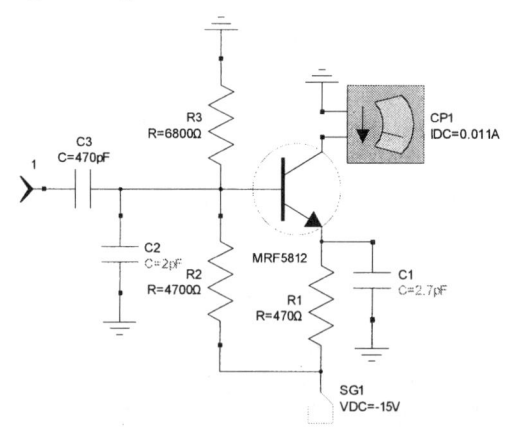

Figure 1.57 Biased device to be used with a negative-resistance oscillator with emitter coupling capacitor C_1 and base-coupling capacitor C_2.

Observed less frequently in negative-resistance oscillator circuits is the base coupling capacitor C_2 although it is equally helpful. The dashed traces in Fig. 1.58 are with C_1 equal to 2.7 pF and C_2 equal to 2 pF. The negative resistance is flat versus frequency at approximately 15 ohms. The effective series input capacitance is increased to a minimum value of 3.6 pF, more than double the capacitance without coupling capacitors. C_1 increases the absolute value of negative resistance, particularly at lower frequencies. C_2 decreases the absolute value of negative resistance and flattens the curve with frequency. When used together, they are particularly important when using devices with a high F_t relative to the operating frequency.

Linear Techniques

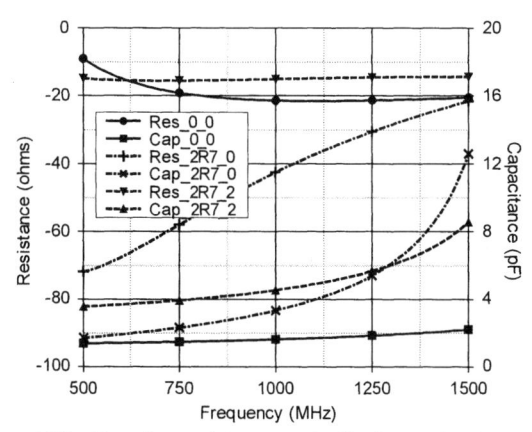

Figure 1.58 Negative resistance and effective series input capacitance with various values of C_1 and C_2 for the test circuit in Fig. 1.57.

1.7.1.8 Negative-Resistance Oscillator Summary

Negative-resistance oscillators designed with common-collector and common-drain devices are among the simplest of UHF and microwave voltage-controlled oscillators (VCOs). A tuning bandwidth of up to an octave is achievable. The loaded Q tends to be rather poor so stability and phase-noise performance is moderate. Successful design follows a few simple rules.

1) Begin the design without a resonator, evaluating devices for flat negative resistance and high effective series input capacitance. Devices with F_t from 1 to 3 times the highest operating frequency work well. Do not use devices with extremely high F_t. Do not design for maximum negative resistance.

2) Optimize the value of C_1 and C_2. C_2 may not be required for moderate F_t devices.

3) Use a resonator with as high a value for L and as low a value for C as is practical given component and PWB parasitics.

4) When computing the oscillator frequency, the effective series input capacitance is in series with the resonator capacitance. Simulators consider this automatically via the device model.

1.7.2 Negative-Conductance Oscillators

As with the negative-resistance oscillator, the design of negative-conductance oscillators begins with the selection of a device and topology that is unstable. This instability is then controlled by adding a resonator to control oscillation. For the negative-resistance oscillator the resonator must be series mode while for the negative conductance oscillator the resonator must be parallel mode.

1.7.2.1 Negative-Conductance Oscillator Basic Circuit

The common-emitter and common-source circuits are more stable than the other topologies. The common-collector and common-drain circuits are used for negative-resistance oscillators. The common-base and common-gate topologies are used for negative-conductance oscillators. Fig. 1.59 shows a basic circuit for a negative-conductance oscillator. The device is an MRF5812 biased at 6-volts V_{ce} and 11-mA I_e. In this circuit, the parallel resonator and test port are connected to the emitter. To form the oscillator, the test port is removed. Output power is taken at the collector.

The input conductance, G, and susceptance, B, are plotted in Fig. 1.60. The conductance is -0.024 mhos at 1000 MHz where the susceptance is 0 mhos. The reflection coefficient, or the input admittance plotted on a Smith chart normalized to a negative conductance, or the input conductance and susceptance plotted on a rectangular graph are used to analyze negative-conductance oscillators. As was the case with the negative-resistance oscillator, plotting G and B on a rectangular graph is straightforward and insightful and is used in the remainder of the book.

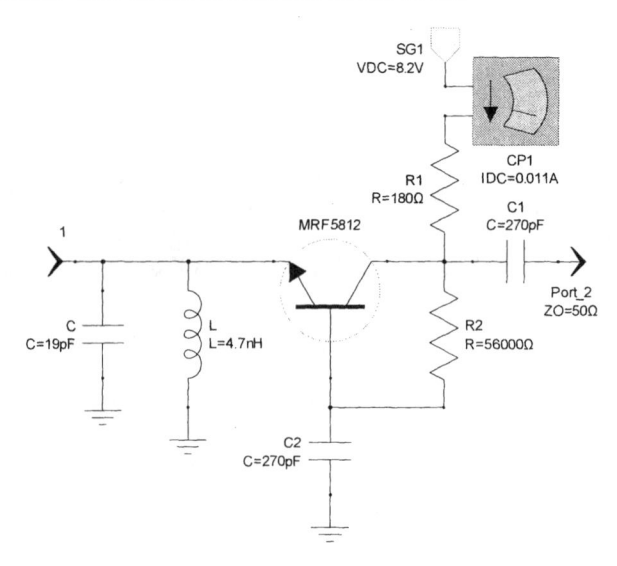

Figure 1.59 Negative-conductance basic oscillator circuit using an MRF5812.

1.7.2.2 Starting Criteria for the Negative-Conductance Oscillator

When the test port across the parallel resonator is removed, the net negative conductance causes a signal to build until device nonlinearity reduces the net path conductance to 0 mhos. The signal builds at the

Linear Techniques

frequency where the net susceptance is zero mhos. The sufficient starting conditions are:

1) oscillation occurs at the frequency where B is zero;
2) $G<0$.

Just as excessive negative resistance is undesirable in negative-resistance oscillators, excessive negative conductance is undesirable because it degrades noise performance, increases harmonics, and may cause spurious modes. The appropriate negative conductance values range from -0.02 to -0.2 mhos. Higher magnitudes of negative conductance are used for fast starting and high output power. Lower magnitudes of negative conductance provide better stability and noise performance.

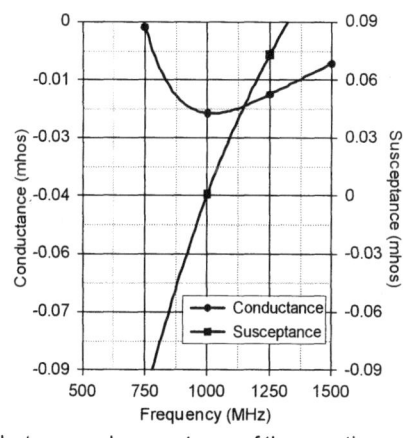

Figure 1.60 Input conductance and susceptance of the negative-conductance oscillator.

1.7.2.3 Negative-Conductance Device Selection and Improvement

The performance of the negative-conductance oscillator is heavily dependant on the device characteristics and device selection is critical. Design should begin with characterization in a test circuit such as Fig. 1.59 with the resonator removed. Here, the MRF5812 device is used with the collector current increased to 21 mA.

Alechno analysis of the negative-conductance basic oscillator reveals that an inductor in series with the base to ground, L_1, and inductance in series with the emitter, L_2, improve the phase characteristics and increase the loaded Q.

A model for the active device in a negative conductance oscillator is negative conductance in parallel with an effective shunt input inductor. Plotted in Fig. 1.61 are the negative conductance and effective shunt input inductance of the test circuit with three sets of values for L_1 and L_2.

With inductor L_1 and L_2 equal to zero (solid traces), the input conductance becomes negative above approximately 700 MHz. At

1000 MHz, the effective shunt input inductance is 1.36 nH. With L_1 equal to 1.5 nH and L_2 equal to zero (dot-dash traces), the magnitude of the negative conductance is significantly increased and the effective shunt inductance is increased to 1.6 nH at 1000 MHz. Finally, with L_2 increased to 1 nH (dashed traces), the conductance is flattened to near -0.02 to -0.028 mhos and the effective shunt inductance is 2.6 nH at 1000 MHz.

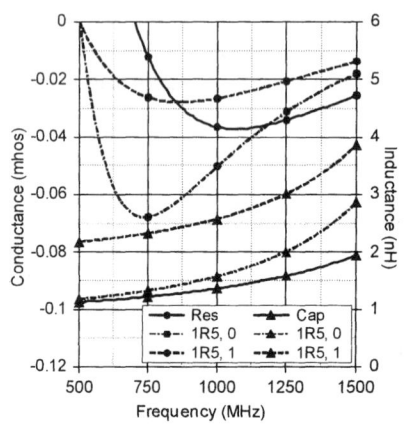

Figure 1.61 Negative conductance and effective shunt input inductance with various values of L_1 and L_2 for the test circuit in Fig. 1.58.

The addition of L_1 and L_2 are important because they moderate and flatten the negative conductance and because they increase the effective shunt input inductance. The former results in an appropriate level of nonlinear drive with flat output power versus frequency when tuned.

The latter is important for isolating the characteristics of the active device from the oscillating frequency. Because the emitter input of a typical negative conductance circuit appears inductive, a shunt capacitor would be all that is required to form a parallel resonator and a successful oscillator. However, in this case the oscillation frequency is directly dependant on the inductance generated by the active device. Device to device variation, bias variation and noise perturbations in the device then influence the oscillation frequency.

To mitigate this problem, external shunt inductance is typically added at the emitter. To be effective, this inductor must be smaller than the effective device inductance. The device inductance is normally extremely small, so the increase of this effective shunt inductance caused by L_1 and L_2 is beneficial.

1.7.2.4 Negative Resistance and Negative Conductance Distinction

As stated previously, the distinction between negative-resistance and negative-conductance oscillators is not merely semantic. A device

Linear Techniques 73

developing negative resistance must use a series resonator to form an oscillator and a device developing negative conductance must employ a parallel resonator. The common-collector and common-drain topologies are negative resistance and the common-base and common-gate topologies are negative conductance. How is this confirmed and how is it determined for other devices?

When the resistance or conductance is negative at a low drive level, the locus of points plotted on a unity radius Smith chart are outside the circumference of the chart. As happens when the oscillation signal builds, if the test signal drive level is increased, device nonlinearity causes the locus of points to approach the circumference of the chart. When the test signal drive level equals the oscillator final steady-state level, the reflection coefficient plots on the circumference. As the test-signal drive level is increased, the real part of the input impedance or admittance becomes positive and plots inside the circumference. Example sets of points versus input power are plotted in Fig. 1.62 on superimposed impedance and admittance Smith charts.

If the circuit develops negative resistance, the locus of points for increasing drive level approximately follow arcs of constant reactance as shown with circular symbols for a negative resistance topology. If the topology develops negative conductance, the locus of points for increasing drive level approximately follow arcs of constant susceptance as shown with square symbols for a negative conductance topology. These tests may be conducted using either computer simulation or a vector network analyzer.

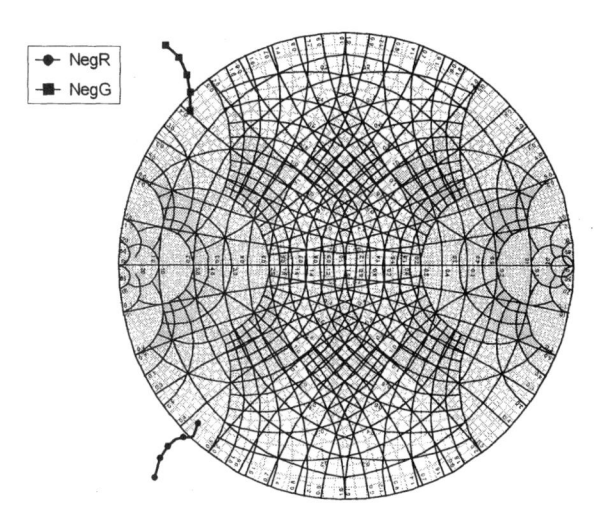

Figure 1.62 The input reflection coefficient versus drive level for a negative resistance device (round symbols) and a negative-conductance device (square symbols).

1.8 Analyzing Existing Oscillators

Thus far, oscillator design has been considered from the viewpoint of synthesis. However, when the problem is correcting an existing design that does not start or does not meet specifications, the following steps may be helpful. The first step is determining an appropriate analysis technique. The flow diagram in Fig. 1.63 is a basic list of troubleshooting and performance optimization procedures.

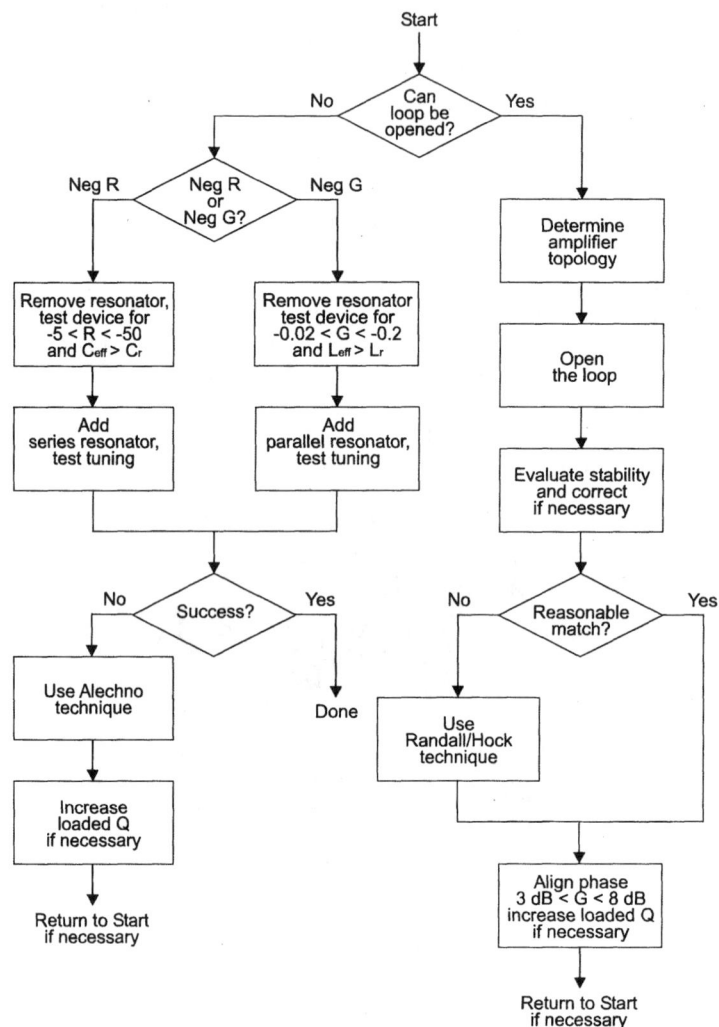

Figure 1.63 Flow diagram with procedures for analyzing an existing oscillator.

A characteristic of oscillators that requires one-port analysis is that an attempt to create two ports by opening a loop results in two ports with no signal path. For example, if the negative-resistance oscillator circuit in Fig. 1.48 is opened at the base, the right port connects to the transistor input and the left port connects to the series resonator to ground. There is no signal path between these two ports and this circuit must be analyzed as a one-port oscillator, or transformed using Alechno's technique.

If analysis is by one port, the next step is determining if the topology operates in a negative resistance or conductance mode. Common-collector and common-drain devices develop negative resistance and common-base and common-gate devices develop negative conductance. This may be confirmed using the test described in Section 1.7.2.4. The resonator on a negative-resistance oscillator should be series mode and the resonator for a negative-conductance oscillator should be parallel mode.

The next step is removal of the resonator and examination of the input characteristics of the device. Negative-resistance oscillator coupling capacitors are adjusted using techniques described in Section 1.7.1.7. Negative-conductance oscillator coupling inductors are adjusted using techniques described in Section 1.7.2.3. The resonator is returned to the circuit and the tuning characteristics are evaluated. If a software simulation is successful but the physical circuit does not operate properly, these steps may be repeated using a vector network analyzer rather than simulation. If additional characterization is required, such as determination of the loaded Q, then Alechno's technique is employed as described in Section 1.7.1.5.

If the loop can be opened this is the preferred analysis method. For example, the Colpitts topology can generally be analyzed by either the one-port or two-port method. The open-loop two-port method is preferred because it provides better insight into potential corrective measures and provides characterization of the loaded Q.

With the open-loop method, the first step is determination of the best point in the circuit to open the loop. The loop should not be opened within the resonator. While the Randall/Hock technique allows this mathematically, there is little point in doing so. To facilitate determining the break point, the amplifier topology should be determined. This is accomplished by determining the grounded terminal of the device. This is typically obvious, however, the ground path may have some resistance to ground.

Next, the stability of the device is determined. Feedback added to common-emitter amplifiers not only improves stability but reduces the sensitivity to device, supply, and temperature variation. Resistance in series with the base may be required to stabilize CC, CD, CB, and CG topologies.

The best place to open the loop of with a common-collector and common-drain device is typically at the emitter or source. The emitter port becomes the loop output and the opposite direction is the loop input. This is a good choice because the emitter output of a common-collector amplifier is relatively low and may be near 50 ohms. The best place to open the loop with a common-base and common-gate device is also the emitter or source. The emitter port becomes the loop input and the opposite direction is the loop output. The emitter input of a common-base amplifier is relatively low and may be near 50 ohms.

If the input or output impedance is near 50 ohms, the design may be confirmed with a vector network analyzer. When one-port of the open-loop cascade is near 50 ohms, a goal of the optimization of the open-loop cascade may be to adjust resonator coupling to match the alternate port to 50 ohms. If this results in excessive gain margin, an attempt to achieve a match may be abandoned in preference to achieving a high loaded Q with moderate gain margin. If no loop break point results in impedances near 50 ohms, the reference impedance of the simulation is adjusted for a best match. If input and output S-parameters better than -10 dB are not achieved, or if the gain margin is low, the Randall/Hock technique described in Section 1.2.1.5 is used. Unfortunately, in this case, design confirmation with a vector network analyzer is less straightforward.

1.9 Optimizing the Design

Fig. 1.64 shows a 40 MHz Colpitts oscillator using a J309 JFET transistor. For analysis, the loop is opened between the source and the resonator capacitive tap. Output power into a 50-ohm load is taken through a 270-pF capacitor at port 3. Resistor R_2 develops a source voltage to reduce the drain current to 10 mA, well below the device I_{dss}. Resistor R_1 is used to stabilize the common-drain JFET amplifier. An inductor value of 270 nH and a ratio of C_2 to C_1 of 4 are selected. C_1 is then tuned to set ϕ_0 at 40 MHz. What is the quality of this design and are the component values optimum?

The solid traces in Fig. 1.65 are the responses of the initial Colpitts oscillator. The output S_{22} is approximately -11.4 dB but S_{11} is only -5.1 db so the true Randall/Hock open-loop gain is used. The maximum phase slope (square symbols) is not aligned precisely with ϕ_0 and the loaded Q (triangular symbols) is only 10.5. The gain margin (circular symbols) is appropriate at 6.3 dB.

The resonator inductor, resonator capacitors, and the base stabilizing resistor are then optimized using the GENESYS linear circuit simulator in an attempt to improve the open-loop characteristics of the oscillator. The optimization goals were set at 40 MHz to G equal to 6 dB, the angle of G to 0, the loaded Q of G to 30, S_{11} to less than −16 dB, and S_{22} to less than −16 dB.

The responses after optimization, setting values on the nearest standard values and tuning the resonator inductor to align ϕ_0 at 40 MHz are shown in Fig. 1.65 as dashed traces.

Linear Techniques 77

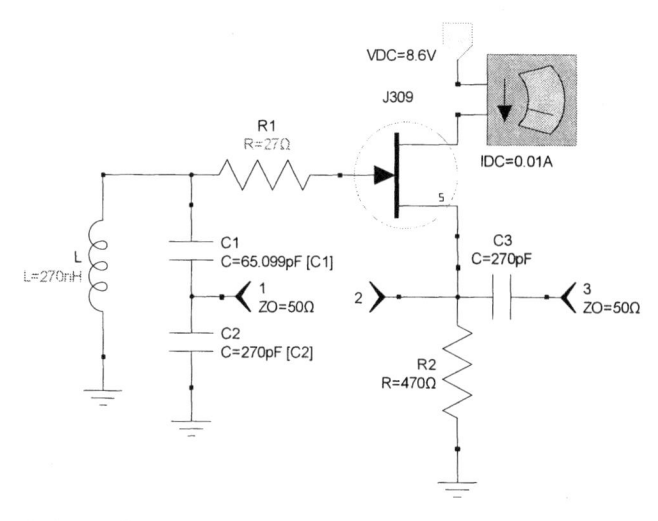

Figure 1.64 Initial 40-MHz Colpitts oscillator using a JFET transistor.

Because G is used, optimizing S_{11} and S_{22} is not necessary, but it simplifies confirmation of the design by vector network analyzer. The maximum phase slope is aligned with ϕ_0 and the loaded Q is improved to the desired 30, largely by decreasing the reactance of the resonator inductor. S_{11} is not significantly changed because the optimized parameters influence it only via the reverse transmission characteristics of the JFET. S_{22} is significantly improved, largely by increasing the ratio of C_2 to C_1. The gate stabilizing resistor was optimized to 0 ohms and may be removed from the circuit. Optimized values are given in Fig. 1.66.

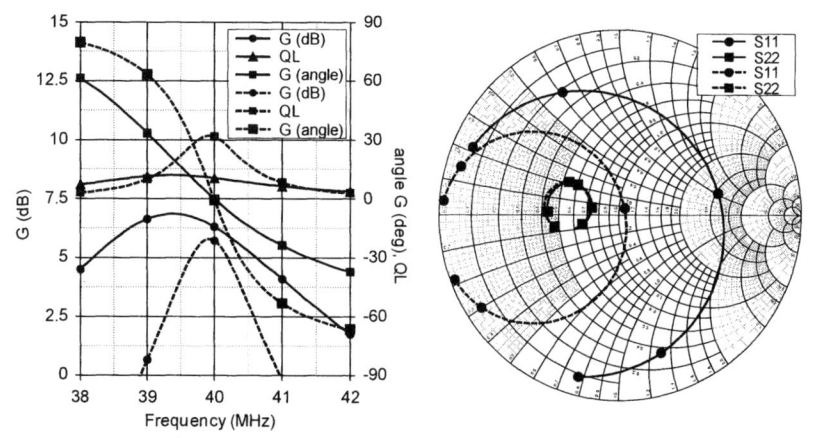

Figure 1.65 The Randall/Hock open-loop gain, G, and loaded Q on the left and S_{11} and S_{22} on the right for the initial Colpitts (solid traces) and after optimization (dashed traces).

The J308, J309, and J310 JFETs are unique in that they don't require the stabilization that many JFET and GaAsFET devices require. These JFETs make excellent medium power oscillators and mixers up to UHF.

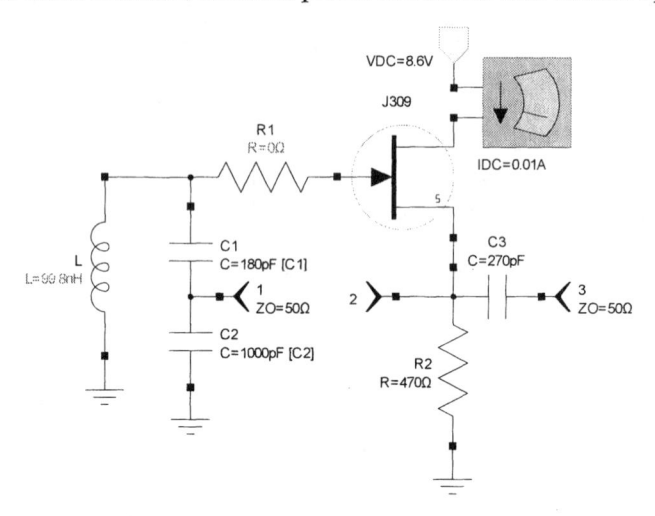

Figure 1.66 Final 40-MHz JFET oscillator with optimized component values.

1.10 Statistical Analysis

In this section statistical analysis is used to investigate the characteristics of a simple negative-resistance oscillator with and without coupling capacitors. Fig. 1.67 is the schematic of the circuit to be evaluated.

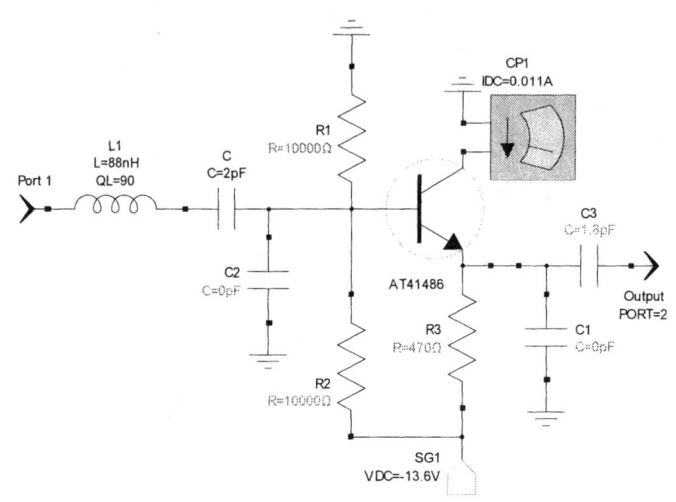

Figure 1.67 Negative-resistance 1000-MHz oscillator used to illustrate statistical analysis.

Parameters of the built-in GENESYS bipolar NPN model (BIPNPN) are optimized so the characteristics match measured data from Avago Technologies for an AT41486 microwave transistor. This is done so parameters of the transistor can be varied to simulate device-to-device tolerances. Other circuit parameters are varied as well. The device and circuit parameters used in the statistical analysis are listed in Table 1.7. A uniform distribution of components is used with the specified tolerance.

Table 1.7 Parameters varied during statistical analysis of the negative-resistance oscillator

Parameter	Symbol	Nominal	Units	% dwn	% up
Ideal maximum forward beta	Bf	68		50	100
B-E zero bias pn capacitance	Cje	3.5	pF	20	20
B-C zero bias pn capacitance	Cjc	0.37	pF	20	20
Ideal forward transit time	Tf	2E-3	nS	20	20
Base inductance	Lb	1.5	nH	20	20
Collector inductance	Lc	0.6	nH	20	20
Emitter inductance	Le	0.1	nH	20	20
Base to ground bias resistor	R1	10000	ohms	5	5
Base to supply bias resistor	R2	10000	ohms	5	5
Emitter to supply bias resistor	R3	470	ohms	5	5
Emitter coupling capacitor	C1	0 or 2	pF	5	5
Base coupling capacitor	C2	0 or 1	pF	5	5
Output capacitor	C3	1.8	pF	5	5
Power supply voltage	VDC	-13.6	Volts	5	5

The first seven parameters listed in Table 1.6 are device parameters. Certain RF critical device parameters are selected. Dozens of other parameters remain fixed. The remaining parameters in Table 1.6 are circuit parameters including the power supply voltage.

The purpose of this statistical analysis is to evaluate the influence of device to device, power supply and general circuit parameter variation on the oscillation frequency. With an ideal oscillator, these parameters have no influence and only the resonator determines the oscillation frequency. Therefore, in this analysis, the resonator inductor, L, and the resonator capacitor, C, remain fixed.

The schematic includes coupling capacitors C_1 and C_2 discussed in Section 1.7.1.6. This statistical analysis reveals the critical role these components play in negative-resistance oscillator performance. On the left in Fig. 1.68 is a 21-sample statistical run without capacitors C_1 and C_2. This represents the most basic form of the negative-resistance oscillator. The value of resonator inductance required to resonate at 1000 MHz is 88 nH. A large value is selected to reduce the dependence of the oscillation frequency on the device. Unfortunately, this large value is difficult at 1000 MHz.

The negative-resistance plots have the lower slopes. The negative resistance is somewhat excessive, which degrades the noise performance. The large variation in negative resistance results in less predictable output power and phase-noise performance. The oscillation frequency, as determined by the frequency at which the reactance is zero, varies from approximately 960 to 1060 MHz.

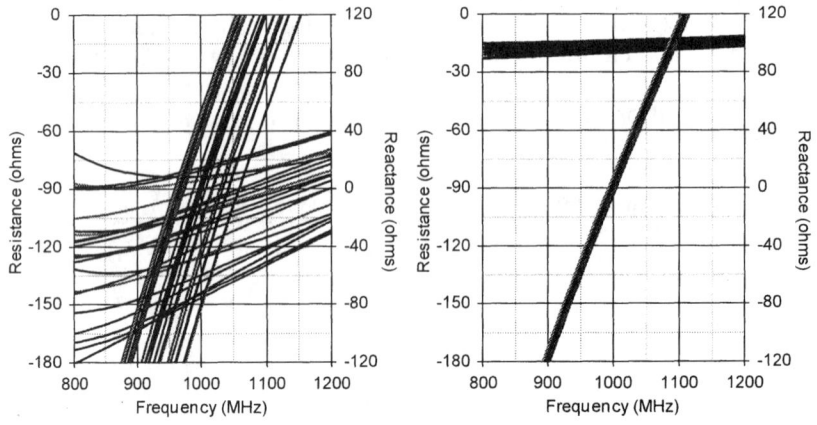

Figure 1.68 Negative resistance and reactance of the negative-resistance oscillator with a statistical analysis for no coupling capacitors (left) and C_1=2 pF and C_2=1 pF (right).

On the right in Fig. 1.68 is another statistical run with identical component tolerances but with C_1 equal to 2 pF and C_2 equal to 1 pF. This increases the effective input capacitance of the device so the resonator capacitor is decreased to return the nominal zero reactance frequency at 1000 MHz. Now both the resonator inductor and capacitor are difficult values. Smaller inductors with larger capacitors degrade performance.

The coupling capacitors moderate and stabilize the negative resistance at -15 to -18 ohms. The oscillation frequency is significantly stabilized with a variation of 995 to 1004 MHz. This example illustrates the effectiveness of linear statistical analysis to evaluate oscillator performance and the critical nature of coupling capacitors.

1.11 Summary

Linear theory is the preferred starting point for the design of oscillators. It is used to establish, evaluate, and optimize the gain margin, oscillation frequency, loaded Q, degree of isolation of the active device, and conditions conducive to spurious free operation.

Additional oscillator performance parameters such as the coupling of power into a load, pulling of the oscillation frequency by load impedance variation, and pushing of the oscillation frequency by power supply voltage change could be studied in additional detail using linear theory. However, these topics are also influenced by nonlinear theory so they will be covered in the following chapter.

Other oscillator performance parameters such as output level, output spectrum, and starting can only be determined using nonlinear theory. These are also covered in the following chapter.

References

[1] Agilent Technologies, GENESYS 2008.07 Documentation Set, www.agilent.com.

[2] K. Harada, "An S-parameter Transmission Model Approach to VCO Analysis," *RF Design*," March 1999, pp. 32-42.

[3] M. Randall and T. Hock, "General Oscillator Characterization Using Linear Open-Loop S-Parameters," *IEEE Trans. MTT*, Vol. 49, June 2001, pp. 1094-1100.

[4] www.avagotech.com.

[5] www.cel.com.

[6] T. Grosch, *Small Signal Microwave Amplifier Design*, SciTech Publishing, Raleigh, NC, 1999, pp. 140-152.

[7] www.specwave.com.

[8] www.minicircuits.com.

[9] www.sirenza.com.

[10] M. Odyniec, "Oscillator Stability Analysis," *Microwave Journal*, June 1999, pp. 66-76.

[11] K. Kurokawa, "Noise in Synchronized Oscillators," *IEEE Trans. MTT*, Vol. 16, April 1968, pp. 234-240.

[12] K. Kurokawa, "Some Basic Characteristics of Broadband Negative Resistance Oscillator Circuits," *Bell System Technical J.*, July-August 1969, pp. 1937-1955.

[13] K. Kurokawa, *An Introduction to the Theory of Microwave Circuits*, Academic Press, New York," 1969, Chap.9.

[14] S. Hamilton, "FM and AM Noise in Microwave Oscillators," *Microwave Journal*, June 1978, pp. 105-109.

[15] S. Alechno, "The Virtual Ground in Oscillator Analysis - A Practical Example," *Applied Microwave & Wireless*, July 1999, pp. 44-53.

2 Nonlinear Techniques

Linear techniques cannot predict the oscillator fundamental output level, the harmonic content of the output spectrum, or internal voltage and current waveforms. Furthermore, nonlinear action modifies the device match and transmission amplitude and phase, thus bringing into question the accuracy of linear analysis. These issues are considered in this chapter.

2.1 Introduction

As the initial signal builds in an oscillator, nonlinear action absorbs the open-loop small-signal gain margin to establish a steady-state operating point. The gain margin is a function of not only device amplification but also the phase zero crossing frequency and the open-loop port impedances. Therefore, phase as well as amplitude compression must be considered. The steady-state operating point is achieved as the gain, phase, and port impedances shift by nonlinear action.

Nonlinear action is a characteristic of the active device, so the sustaining stage's nonlinear characteristics are considered first in this chapter. Then, the characteristics of an open-loop cascade are considered using nonlinear techniques. A method of steady-state nonlinear oscillator analysis is then presented. Finally, these concepts are applied to the analysis of oscillator performance characteristics such as output power coupling, load pulling, and power supply pushing.

2.2 Harmonic Balance Overview

In Chapter 3 the transient behavior of the oscillator is considered. It is helpful to first consider the steady-state nonlinear behavior of the oscillator. In this chapter, harmonic-balance simulation is used to study the oscillator. Specifically, examples and results are illustrated using the Harbec harmonic-balance simulation engine of the GENESYS software suite by Agilent Technologies [1].

In Harbec, a static DC analysis is first automatically performed to linearize the device about the operating point. Any AC sources are turned off for this initial step to simulate the small-signal quiescent bias.

Harmonic-balance simulation assumes that the circuit can be accurately modeled using a finite number of spectral tones. In general, the user specifies the analysis of a sufficient number of harmonics so that the energy in the higher harmonics is significantly less than the total energy. With signal stimulus, Harbec then solves Kirchoff's current law (KCL) in the frequency domain, searching iteratively for voltages that satisfy KCL at each node and each frequency. If a steady-state solution is found, the simulation is said to have converged. Convergence is defined by user-specified limits in the residual current erros in satisfying KCL. For all simulations in this book, the Harbec default limits and simulation parameters are used.

The time required for each iteration would be roughly proportional to the cube of the number of specified spectral components and the number of nonlinear circuit nodes, except that sparse-matrix techniques are used which improve scaling with size. The number of iterations required for convergence depends on many factors. Convergence typically occurs in seconds. If convergence requires more than 60 seconds for a single-device circuit, the design and user options should be evaluated.

2.3 Nonlinear Amplifiers

Consider the 300-MHz amplifier test circuit given in Fig. 2.1. The gain is moderated by shunt feedback resistor R_1 and series feedback resistor R_2. Power is supplied through inductive choke L_2. Dual supplies independently control the device quiescent voltage and current. The emitter current is

$$I_e = -\frac{V_{be} + V_{ee}}{R_2 + R_3} \qquad (55)$$

For example, with V_{be} equal to 0.815 volts, V_{ee} equal to -5 volts, R_2 equal to 8.2 ohms, and R_3 equal to 405 ohms, I_e equals 10.1 mA. The emitter current is controlled by adjusting R3.

Large-signal S-parameters ($LSnm$) are a generalization of small-signal S-parameters where $LSnm$ are level-dependant and the voltages and currents that define the waves are Fourier coefficients rather than vectors. When measured in a 50-ohm system, harmonics are generally terminated differently than they are in a circuit with reactive components. Also, LS22 has questionable meaning. Nevertheless, $LS11$ and $LS21$ are useful concepts, which offer insight into nonlinear oscillation phenomena.

Plotted in Fig. 2.2 with circular symbols is the Harbec simulated amplitude of the 300-MHz fundamental forward large-signal S-parameter ($LS21$) versus the input power of the test circuit in Fig. 2.1. The fundamental and harmonics to the fifth were specified for the simulation. With low input power the large-signal S-parameter equals the small-signal S-parameters, about 10.6 dB.

Nonlinear Techniques

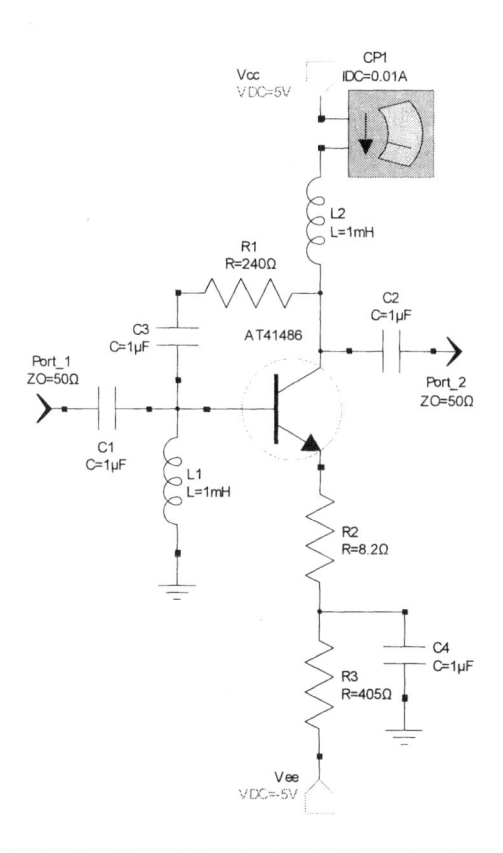

Figure 2.1 CE test circuit with resistive feedback (R_1 and R_2) and dual supplies for independent control of the quiescent device bias voltage and current.

At -6.5 dBm input drive the gain has dropped by approximately 1 dB to 9.6 dB. This is referred to as the 1-dB compression point. Plotted in Fig. 2.2 with a dashed trace and square symbols is the compression. Provided the cascade transmission phase and port impedances are constant with compression, when oscillations build in the closed loop, an open-loop cascade with 6-dB gain margin compresses by 6 dB to reach a steady-state operating level. For this test circuit, the input power is approximately 1 dBm at this compression level. The output power at the fundamental frequency of 300 MHz is plotted in Fig. 2.2 with triangular symbols. At 1-dBm input power, *LS21* is approximately 4.5 dB so the output power is approximately 5.5 dBm. This data and the output coupling method may be used to estimate the output power of the oscillator.

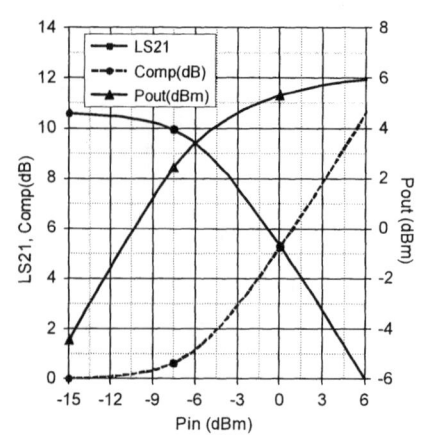

Figure 2.2 LS21 (circular symbols), gain compression (square symbols) and output power (triangular symbols) versus the input power for the CE test circuit in Fig. 2.1.

2.3.1 Quiescent Current and Compression

Essentially, the oscillator output power is the output capability of the sustaining stage amplifier at a compression level that satisfies oscillation conditions, with consideration of the output coupling. Output coupling is discussed later. Next, amplifier output capability is considered.

Plotted in Fig. 2.3 are the gain ($LS21$) and compression at 300 MHz of the CE amplifier test circuit, with 5-, 10-, 20- and 40-mA collector current, versus the input power. For this device, at low input power the gain increases with increasing collector current. With a current of 20 mA or more the gain increase is slight.

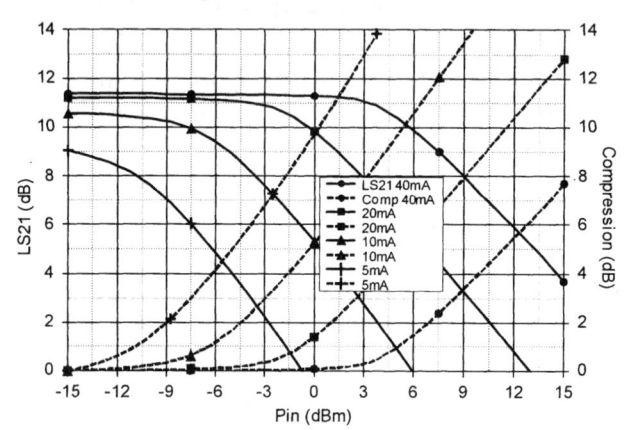

Figure 2.3 LS21 (solid traces) and compression (dashed traces) at 5-, 10-, 20-, and 40-mA collector current versus the input power for the CE amplifier test circuit in Fig. 2.1.

Nonlinear Techniques

At 5 mA, the 1-dB compression point occurs at an input level of approximately -11 dBm with 8 dB of gain. The output power at 1-dB compression is therefore -3 dBm. At 40-mA collector current, the 1-dB compression point occurs at an input level of approximately 5.1 dBm with 10.4 dB of gain. The output power at 1-dB compression is therefore 16.5 dBm. The 1 dB compression point is somewhat independent of the collector-emitter voltage. However, the 1-db compression point is a strong function of collector current. The output power at the 1-dB compression point for a class-A 50 ohm bipolar amplifier of this type is found empirically to be approximately

$$P_{1dB}(dBm) = -17 + 20\log(I_{c-mA}) \qquad (56)$$

The output power of an oscillator is roughly proportional to the device current squared. Therefore, the output power level is limited for oscillators in battery-operated systems with a restricted current budget.

2.3.2 Impedance Shift

The cascade gain margin is at ϕ_o and is a function of the impedance match. When nonlinear behavior modifies these parameters the gain margin is affected. Therefore, reduction of the gain margin to the steady-state level of 0 dB by nonlinear behavior is partly the result of cascade match shifting.

The solid trace in Fig. 2.4 is the small-signal input S-parameter versus frequency for the CE amplifier test circuit biased at 10 mA collector current. The sweep frequency range is 100 MHz to 1000 MHz. The small-signal input S-parameter magnitude at 100 MHz is approximately -30 dB.

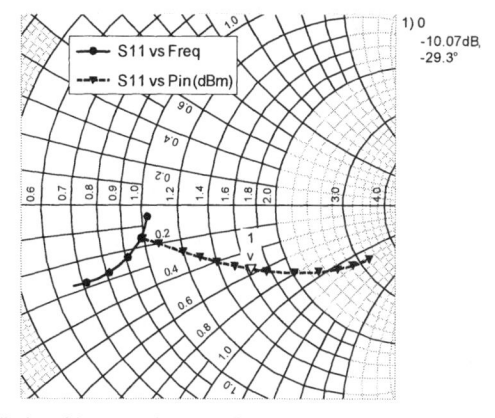

Figure 2.4 Small-signal input and output S-parameters versus frequency for the CE amplifier test circuit biased at 10 mA collector current (solid traces) and large-signal input and output S-parameters versus drive level (dashed traces). The marker represents the large signal input S-parameter at 0-dBm drive.

The second round symbol is at 300 MHz, the frequency used for large-signal tests. The dashed traces in Fig. 2.4 are the large-signal input S-parameters versus drive level. At the lowest test drive level of -15 dBm the large-signal input impedance is equal to the small-signal impedance. The marker represents the large signal input S-parameter at 0-dBm drive. This corresponds to approximately 5.3 dB of gain compression. At this compression level, the impedance shift is significant but not severe.

It is often assumed that nonlinear behavior degrades resonator loaded Q. For this amplifier, nonlinear behavior shifts both the input and output to higher impedance toward the left on the Smith chart. When used with a parallel resonator, this reduces the resonator loading and increases the loaded Q. Surprisingly, not every aspect of nonlinear action is deleterious! However, when used with a series resonator, the increased impedance would decrease the loaded Q. Each cascade must be analyzed.

2.3.3 Phase Shift

A shift in the transmission phase during compression could shift the cascade gain off the amplitude transmission peak and contribute to gain absorption. Given in Fig. 2.5 are the small-signal S_{21} versus frequency (round symbols, solid trace) and the large-signal-S_{21} versus input power (square symbols, dashed trace) plotted on a polar chart for the test circuit biased at 10 mA. At a low drive level of -15 dBm the transmission phase shift is 168°. At approximately 6 dB of compression the phase shift is 166.5° and at 10.6 dB of compression the phase shift is 164.5°. This small shift in transmission phase with compression is evidenced by a nearly straight locus of points on the polar chart, and it results in a small shift in the frequency of ϕ_0. A shift in ϕ_0 may result in either an increase or decrease in gain depending on whether the shift moves ϕ_0 toward or away from the transmission amplitude peak.

The shift in transmission phase shifts the frequency of the cascade ϕ_0 and therefore shifts the oscillation frequency. Rearranging Eq. 32 and converting to degrees we have

$$\frac{\Delta f}{f_0} = -\frac{\pi \Delta \phi^o}{360 Q_L} \qquad (57)$$

where $\Delta f/f_0$ is the fractional frequency shift associated with a transmission phase-shift delta. For example, with a loaded Q of 40 and a 1° phase shift in the CE amplifier test circuit at 6 dB of compression, the fractional frequency shift is 0.0218% or 218 ppm.

Nonlinear Techniques 89

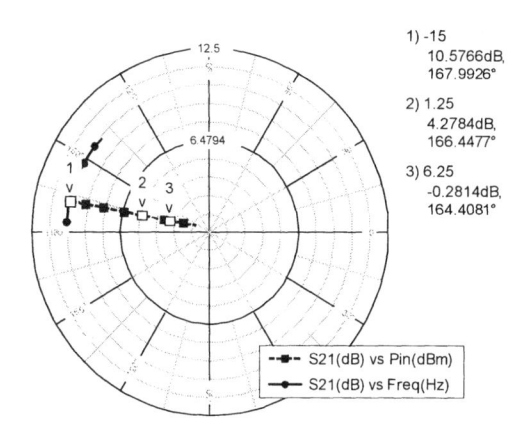

Figure 2.5 Small-signal S_{21} versus frequency (circular symbols, solid trace) and large-signal S_{21} versus input power (square symbols, dashed trace) for the test amplifier biased at 10 mA.

The previous sections illustrate that for this sustaining stage, neither impedance shift nor phase shift delta contribute significantly to absorbing the gain margin. Not all amplifiers are this well behaved and prudent design begins with evaluation of the nonlinear characteristics of the sustaining stage. Nevertheless, nonlinear behavior in an oscillator with moderate gain margin is often not as onerous as might be feared.

2.3.4 Output Spectrum

Repeated in Fig. 2.6 with circular symbols are the output power versus input drive power for the CE amplifier test circuit biased at 10 mA at the 300 MHz fundamental frequency and the second, third, fourth, and fifth harmonics are plotted with cross, triangular, square, and diamond symbols, respectively.

The previous nonlinear simulations of the CE test circuit are performed by balancing five harmonics in the simulator. Given in Table 2.1 are the predicted output levels of the third and fifth harmonics versus the highest harmonic balanced with an input drive level of 0 dBm at 300 MHz. A 0-dBm drive results in approximately 5.3 dB of compression. For this amplifier at 0-dBm drive level, seven harmonics are sufficient for reasonably accurate analysis. Fig. 2.6 is computed balancing harmonics through the seventh.

The frequency-domain output spectrum through the seventh harmonic for the CE amplifier test circuit with a 300 MHz, 0-dBm sinusoidal drive signal is shown in Fig. 2.7. This is approximately 5.2 dB of compression. At 1-dB compression the second harmonic is 18 dB below the fundamental and the third harmonic is 26 dB below the fundamental. At 6-dB compression, the second harmonic is 6 dB below the fundamental and the third harmonic happens to be near a null. As expected, in general higher compression leads to higher harmonic levels.

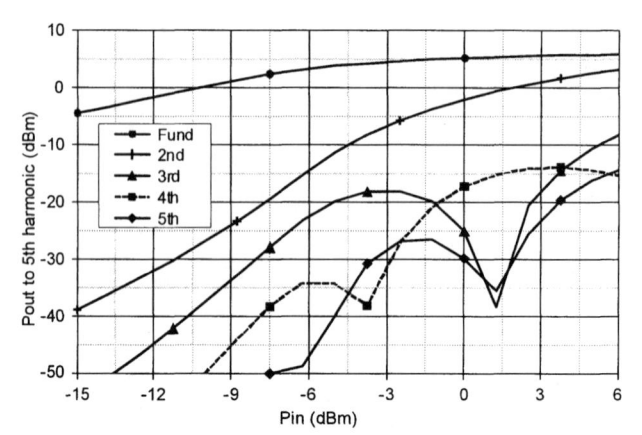

Figure 2.6 Output power versus input drive power for the CE amplifier test circuit at the 300-MHz fundamental (circular symbols) and the second, third, fourth, and fifth harmonics (cross, triangular, square, and diamond symbols).

Table 2.1 Output power at the third and fifth harmonics of the 300-MHz test signal versus the highest harmonic balanced by the simulator for the CE test amplifier biased at 10 mA

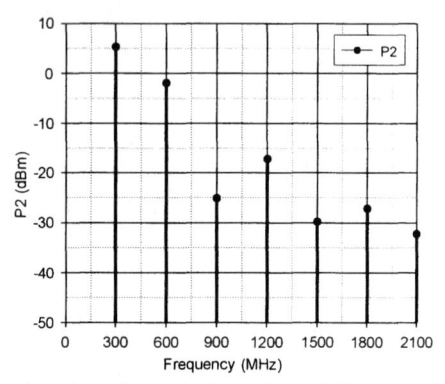

Highest Harmonic Balanced	3rd Harm(dBm)	5th Harm(dBm)
3rd	-17.0	
4th	-23.3	
5th	-24.4	-26.1
7th	-25.0	-29.7
9th	-25.1	-30.1
11th	-25.2	-30.1

Figure 2.7 Frequency domain output spectrum through the seventh harmonic for the CE amplifier test circuit with a 300-MHz, 0-dBm sinusoidal drive signal.

2.3.5 Time Domain Waveform

Fig. 2.8 shows a 15-dB gain type B general-purpose resistive-feedback amplifier with a AT41486 bipolar transistor biased at 10 mA designed using Eqs. 17-24.

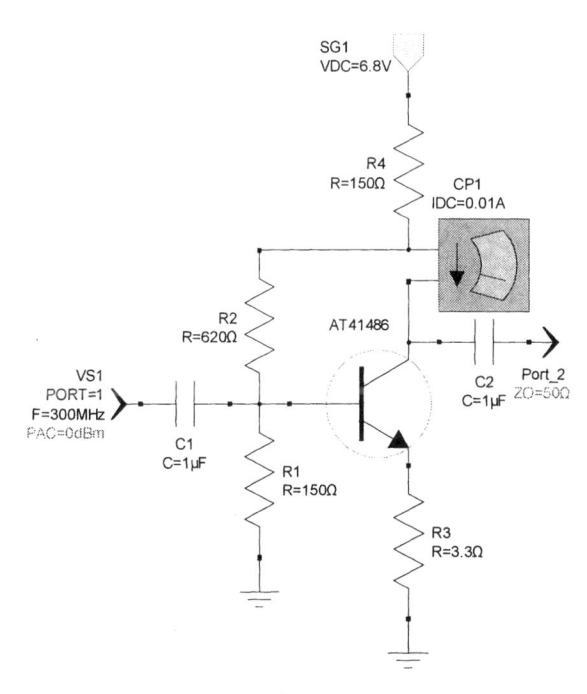

Figure 2.8 A 12-dB gain type B general-purpose resistive-feedback amplifier with a AT41486 bipolar transistor biased at 10 mA.

Fig. 2.9 shows the time-domain voltage waveform at the collector with an sinusoidal input drive level of 0 dBm at 300 MHz. The highest harmonic balanced in the simulation is the ninth.

Clipping is occurring with a maximum voltage of approximately 4.3 volts. This clipping occurs when the transistor is driven into cutoff.

The cutoff voltage where clipping begins is given approximately by

$$V_{cutoff} = V_c - I_c R_p \qquad (58)$$

where V_c is the quiescent collector voltage and I_c is the collector current. R_p is the equivalent parallel collector resistance given by

$$R_p = \frac{1}{1/R_4 + 1/Z_0 + 1/(R_1 + R_2)} \qquad (59)$$

For large device beta, V_c is given by

$$V_c = \frac{V_b(R_1 + R_2)}{R_1} \tag{60}$$

where

$$V_b = 0.79 + 0.0038 I_e(mA) + V_e \tag{61}$$

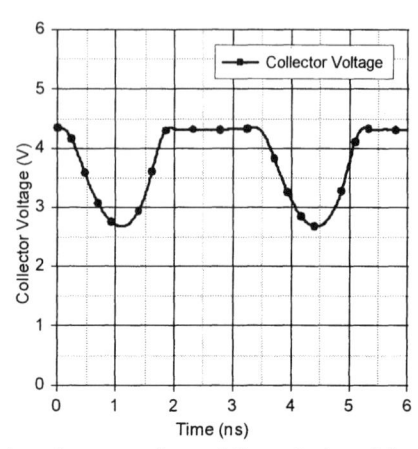

Figure 2.9 Time-domain voltage waveform at the collector of the general-purpose bipolar amplifier with 300-MHz input signal at 0 dBm.

Unless the load impedance is very high, cutoff generally occurs before device saturation. The solid trace in Fig. 2.10 is with a 50-ohm load impedance but the 300-MHz drive signal is increased to 6 dBm. Notice that the maximum voltage at cutoff actually decreased to 3.7 volts from 4.3 volts.

The dash trace depicts the output with a load resistance of 1K ohm. The minimum voltage decreases to the saturation voltage of the device.

2.3.6 Conversion Efficiency

The performance specifications of oscillators typically involve stability, phase noise, and tuning characteristics. With battery-operated devices, limited supply power also dictates a good DC to output power conversion efficiency. DC power to output power conversion efficiency is defined by:

$$\eta(\%) = 100 \frac{P_{out}}{I_{DC} V_{DC}} \tag{62}$$

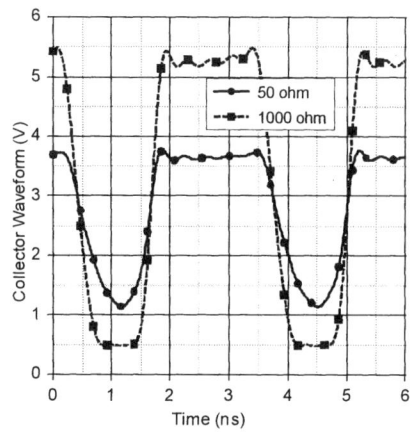

Figure 2.10 Time-domain voltage waveform at the collector of the general purpose bipolar amplifier with a 300-MHz, 6-dBm drive and a 50-ohm load impedance (solid trace) and 1K ohm load (dashed trace).

Typical conversion efficiencies are 2% to 15%. Oscillators may be designed for much higher conversion efficiency. It is important to note that the supply current can change with closing of the loop and the onset of oscillation. Relevant to the discussion of oscillator efficiency is the operating class of the sustaining stage amplifier.

2.3.7 Operating Class

Class identification is a function of the conduction angle, how the amplifier is biased and driven and the topology of the circuit, particularly the output network. An excellent reference on the design of amplifiers of all classes is given by Albulet [2]. Power amplifier is a term commonly used when consideration is given to the operating class, output power, or efficiency an amplifier. A basic circuit for the analysis of power amplifiers is shown in Fig. 2.11.

2.3.7.1 Class-A Amplifiers

The fraction of the RF cycle that the amplifier is in the active region is defined as the conduction angle, $2\theta_c$. For the class-A amplifier the conduction angle is 360° and so the device operates in the active region over the entire RF cycle. Ideally, the amplifier is linear and the output waveform is therefore sinusoidal. For the ideal class-A amplifier the parallel resonator consisting of L_3 and C_3 in Fig. 2.11 is unnecessary. The maximum available output power is

$$P_{out,\max} = \frac{V_{DC}^2}{2R_L} \qquad (63)$$

where R_L is the output load resistance. The circuit is biased for maximum output power by setting the quiescent collector current at

$$I_{DC,\max} = \frac{V_{DC}}{R_L} \qquad (64)$$

Under these conditions the peak output voltage is

$$V_{out} = \sqrt{2P_{out}R_L} \qquad (65)$$

The maximum available efficiency for a class-A amplifier is then

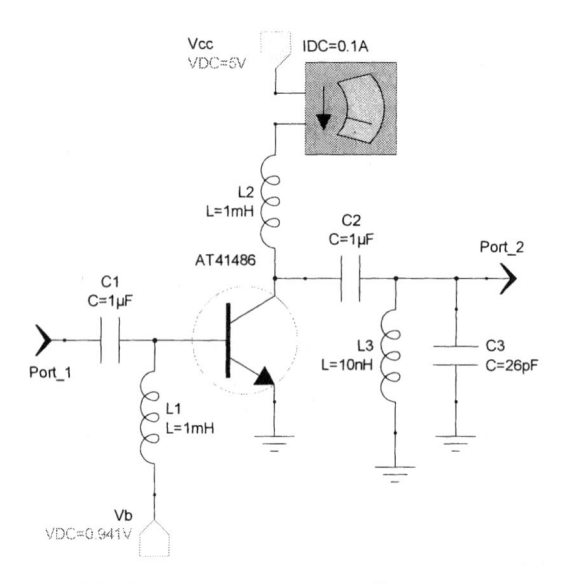

$$\eta_{A,\max}(\%) = 100\frac{P_{out}}{P_{DC}} = 100\frac{V_{out}^2}{2V_{DC}^2} = 50\% \qquad (66)$$

Figure 2.11 Basic circuit for the analysis of power amplifiers.

Because devices are not ideal, and particularly because of nonzero saturation voltage, typical maximum efficiencies for bipolar class-A amplifiers is 40% to 45%. Interestingly, heavy compression increases the output level but may degrade efficiency. Table 2.2 lists the measured characteristics versus drive level at 300 MHz of the basic power amplifier circuit in Fig. 2.11 operated class-A. The small signal gain is 21.84 dB at

300 MHz. At a drive of 10 dBm the compression is 6.78 dB and the conversion efficiency is 46.2%, relatively close to the 50% theoretical maximum. At 20-dBm drive the compression is 16.24 dB and the efficiency decreases to 39.1%.

Table 2.2 Measured characteristics versus input level at 300 MHz of the basic power amplifier in Fig. 2.11 operated class-A

Drive (dBm)	Pout (dBm)	Gain (dB)	Pdc (W)	η_A (%)
-10	11.84	21.84	0.500	30.6
0	21.70	21.70	0.525	28.2
10	25.06	15.06	0.695	46.2
20	25.60	5.60	0.930	39.1

As discussed in Section 1.5.7, power may be applied to the collector using an inductive choke, as in Fig. 2.11, or using a resistor. A resistor is more economical than an inductor and it avoids potential additional resonant modes. However, the use of a resistor destroys the conversion efficiency. The maximum conversion efficiency of an ideal class-A amplifier with a load resistance R_L and a collector supply resistor R_C is

$$\eta_{A,\max,R}(\%) = 100 \frac{R_{Par}^2}{2R_L(R_{Par} + R_C)} \quad (67)$$

where R_{Par} is the parallel equivalent resistance of R_L and R_C. The maximum efficiency peaks at 8 1/3% when R_L equals R_C. When R_L is 3 times R_C, as suggested for the general-purpose feedback amplifier in Section 1.5.7, the ideal maximum conversion efficiency is 7.5%. For nonideal devices, the efficiency is considerably less, partly due to a nonzero saturation voltage. A nonzero saturation voltage reduces the efficiency by a factor $K_{\eta, Vsat}$ given by

$$K_{\eta,Vsat} = 1 - \frac{V_{sat}}{V_{DC}} \quad (68)$$

Conversion efficiencies of 2% to 6% are typical for amplifiers with resistive connections to the supply.

2.3.7.2 Class-B Amplifiers

For the class-B amplifier the conduction angle is 180° and so the device operates in the active region over ½ of the RF cycle. Transistors do not transition abruptly from cutoff to the active region. There is an offset voltage and the transfer characteristic is nonlinear. If the base is biased at exactly 0 volts, at least 0.7 volts of peak drive signal is required to begin conduction. When used as an amplifier, the drive signal can be used to overcome the offset if distortion in the crossover region is acceptable. However, an oscillator sustaining stage must build a signal starting only with noise. (This is not strictly true as will be discussed in Chapter 3.) Therefore, a small quiescent voltage and current are used and the conduction angle is somewhat greater than 180°. This is actually class-AB operation. When the quiescent bias current is small, class-B theory is useful in predicting class-AB amplifier behavior.

Class-B amplifiers are often operated in a push-pull configuration. The maximum output power available with an ideal push-pull class-B amplifier is

$$P_{out,max} = \frac{V_{DC}^2}{2R_L} \qquad (69)$$

The supply DC current is

$$I_{DC} = \frac{2V_{DC}}{\pi R_L} \qquad (70)$$

and

$$P_{DC} = V_{DC} I_{DC} = \frac{2V_{DC}^2}{\pi R_L} \qquad (71)$$

therefore,

$$\eta_{B,max}(\%) = 100 \frac{P_{out}}{P_{DC}} = 100 \frac{\pi}{4} = 78.5\% \qquad (72)$$

Listed in Table 2.3 are the measured characteristics of the basic power amplifier in Fig. 2.11 operated class-AB versus the signal frequency. The base bias voltage is adjusted for a quiescent collector current of 10 mA. The drive level at each frequency is 10 dBm.

Table 2.3 Measured characteristics versus the signal frequency of the power amplifier in Fig. 2.11 operated class-AB

Freq(MHz)	Pout(dBm)	Gain(dB)	Pdc(W)	η_A (%)
3	25.41	15.41	0.606	57.3
30	25.41	15.41	0.607	57.3
100	25.26	15.26	0.612	54.9
300	24.92	14.92	0.602	51.6
1000	22.25	12.25	0.414	40.6

From this data it is evident that for real devices the efficiency falls far short of the theoretical maximum and it degrades significantly with increasing frequency. Although the F_t of this device is 8000 MHz, the efficiency begins to degrade at only 100 MHz. The gain is also decreasing so the output power is recovered by increasing the drive level, but this improves the efficiency only marginally. In terms of efficiency, at higher

Nonlinear Techniques

frequencies the advantage in operating class-AB rather than class-A is marginal.

Listed in Table 2.4 are the measured characteristics versus drive level at 300 MHz of the basic power amplifier in Fig. 2.11 operated class-AB. The first group in the table is with a quiescent collector current of 1 mA and the second group is with 10 mA. Notice with a quiescent collector current of 1 mA that the small-signal gain is lower than the large signal gain. In effect this is the opposite of gain compression with increased drive level. Gain expansion increases the gain from 13.6 dB at -20-dBm drive to approximately 20.3 dB at 0-dBm drive and then compression begins to decrease the gain. The dynamics of this behavior could lead to instability during oscillator starting.

Table 2.4 Measured characteristics versus input drive level at 300 MHz of the power amplifier in Fig. 2.11 operated class-AB at 1-mA and 10-mA quiescent current

Drive(dBm)	Pout(dBm)	Gain(dB)	Pdc(W)	η_A (%)
Ic quiescent =1 mA				
-20	-6.43	13.57	0.011	2.0
-10	9.40	19.40	0.063	13.8
0	20.32	20.32	0.264	40.8
10	24.78	14.78	0.560	53.7
20	25.67	5.67	0.819	45.1
Ic quiescent =10 mA				
-20	5.00	25.00	0.063	5.1
-10	13.81	23.80	0.130	18.6
0	21.28	21.28	0.332	40.5
10	24.92	14.90	0.602	51.6
20	25.67	5.65	0.835	44.2

At a quiescent collector of 10 mA this behavior is absent. Although the maximum conversion efficiency is slightly better with the low quiescent current, the risk of instability may not warrant the improved efficiency associated with a low quiescent collector current.

2.3.7.3 Class-C Amplifiers

Class-C amplifiers have a conduction angle less than 180°. The device quiescent bias is off and the small-signal gain is zero. Therefore, oscillators with class-C amplifiers have similar starting issues as those with class-B amplifiers. The advantage of class-C amplifiers is theoretically higher efficiency. The maximum theoretical efficiency is

$$\eta_{C,\max} = \frac{\theta_C - \sin\theta_C \cos\theta_C}{2(\sin\theta_C - \theta_C \cos\theta_C)} \qquad (73)$$

where θ_c is half the conduction angle in radians. This expression is plotted in Fig. 2.12. Nonideal behavior in real devices significantly reduces the efficiency achieved in practical amplifiers.

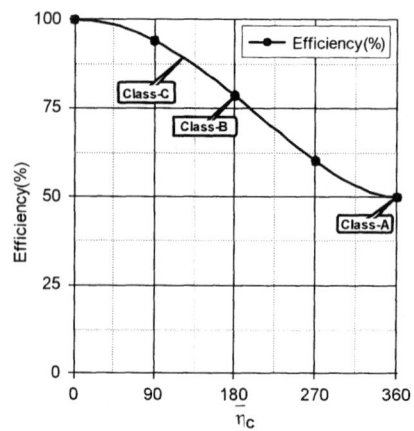

Figure 2.12 Maximum theoretical efficiency(%) of classic power amplifiers versus the conduction angle(degrees).

The efficiencies for class-A and class-B amplifiers correspond with the previously specified values. The efficiency of class-C amplifiers increases with decreasing conduction angle up to a maximum of 100%. As will be illustrated, efficiencies achieved in practical circuits are well below these values.

The design and operation of class-C amplifiers with real devices is more complex than might be intuitive. If the device does not saturate during the RF cycle, operation is referred to as current-source class-C. If the device saturates for a portion of the RF cycle, operation is sometimes referred to as saturated or overdriven class-C. Class-C theory is applied with success to vacuum tube amplifiers but simple theory is unlikely to be successful with solid-state amplifiers. Practical solid-state amplifiers operate in a complex mode referred to as class-C mixed mode. Albulet [2] is an excellent reference on the design of solid-state class-C amplifiers.

Class-C amplifiers require sufficient drive level to bias the device in the active region. At zero drive level, the class-C amplifier has no gain. Given the classic theory for oscillator starting, an oscillator using a class-C operating mode for the active device would never start. Chapter 3 illustrates the noise theory for oscillator starting is flawed and therefore class-C operation may be used for the oscillator sustaining stage.

2.3.7.4 Other Amplifier Classes

Other power amplifier operating modes include D, E, F, and S. Class-D amplifiers use two devices operated in a switched mode. This complexity is seldom justified for oscillators.

Class-E amplifiers are single-ended switch-mode amplifiers. Class E sustaining stages may be worthy of consideration when high conversion efficiency is of the utmost importance.

Class-F refers to a number of topologies that utilize higher-order, multiple-section output networks. The additional harmonic filtering offered by class-F amplifiers is seldom justified.

Class-S amplifiers are switch-mode amplifiers that require the conversion of the signal to a pulse-width modulated signal. These amplifiers offer extremely high efficiency and are typical used with audio-frequency signals to modulate RF power amplifiers. Class-S amplifiers are typically used in conjunction with AM modulation of high-power transmitters.

2.3.8 Power Amplifier Case Study

In this section, the characteristics of example common-emitter Class-A, AB, C, and E power amplifiers are investigated using a silicon bipolar 2N5109 TO-39 case transistor with a typical F_t of 1200 MHz at 50 mA.

2.3.8.1 Model for the 2N5109

A Gummel-Poon BJT Spice model with package and bond-wire parasitics [3] is used for the following computer simulations. The subcircuit is given in Fig. 2.13 and the Gummel-Poon parameters are given in Table 2.5.

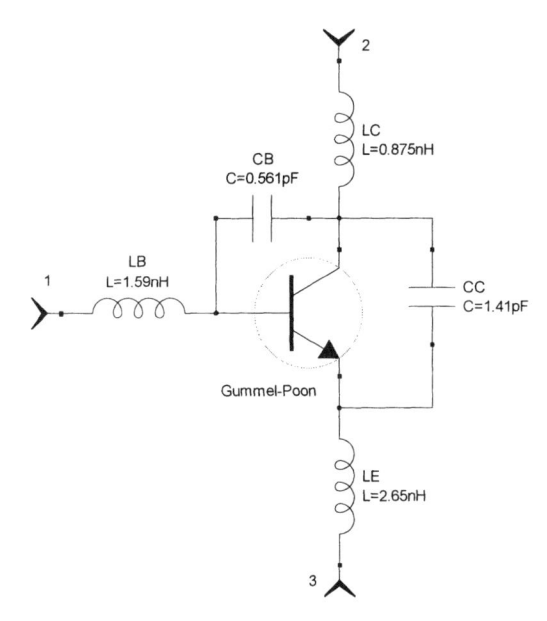

Figure 2.13 Model of the 2N5109 silicon NPN bipolar transistor including package and bond-wire parasitics.

Table 2.5 Nondefault parameters parameters of the Gummel-Poon model of the 2N5109 intrinsic device

Parameter	Symbol	Value
Area scaling factor	Area	1
Transport saturation	Is	4E-15
Ideal max forward beta	Bf	44
Forward early voltage	Vaf	160
Forward beta high-current roll-off	Ikf	0.28
B-E leakage current	Ise	36E-15
B-E p-n leakage emission coefficient	Ne	1.5
Ideal max reverse beta	Br	1.14
Forward Early voltage	Var	16
Reverse beta high-current roll-off	Ikr	0.28
B-C leakage current	Isc	12E-15
B-C p-n leakage emission coefficient	Nc	1.7
Zero-bias maximum base resistance	Rb	1.57
Current where Rb falls halfway	Irb	0.04
Zero-bias minimum base resistance	Rbm	1.1
Emitter ohmic resistance	Re	2.75
Collector ohmic resistance	Rc	0.69
B-E zero-bias p-n capacitance	Cje	1.822E-12
B-E built-in potential	Vje	0.75
B-E p-n grading factor	Mje	0.33
B-C zero-bias p-n capacitance	Cjc	2.758E-12
B-C built-in potential	Vjc	0.75
B-C p-n grading factor	Mjc	0.33
Fraction of Cjc connected to Rb	Xcjc	0.5
Forward bias depletion capacitance coefficient	Fc	0.5
Transit time bias coefficient	Xtf	4.0
Ideal forward transit time	Tf	0.111E-9
Transit time dependence on Vbe	Vtf	6.6
Transit time dependence on Ic	Itf	0.082
Ideal reverse transit time	Tr	8E-9
Flicker noise coefficient	Kf	1E-15
Band-gap voltage	Eg	1.11
Forward and reverse beta temperature coefficient	Xtb	1.5
IS temperature effect exponent	Xti	3

This model is used in a basic simulation test circuit to bias the device at 15 volts V_{ce} and 35 mA collector current. The simulator then linearizes the device and computes the small signal S-parameter forward, reverse, input, and output characteristics which are then compared to S-parameter data originally published for this device by Motorola [4]. The results are given in Fig. 2.14. Agreement is only fair. This is not particularly surprising given device to device variation and that a variety of manufacturers produce this device. Also, the Gummel-Poon model has significant limitations. The manufacturer that supplied the device to which the model was designed fit was not specified by the model developer. The addition of 10° long lines at the input and output of the model improves the agreement significantly, suggesting that the reference plane for the measured data and the model development were not equal. The following amplifier simulations were based on the model.

Nonlinear Techniques 101

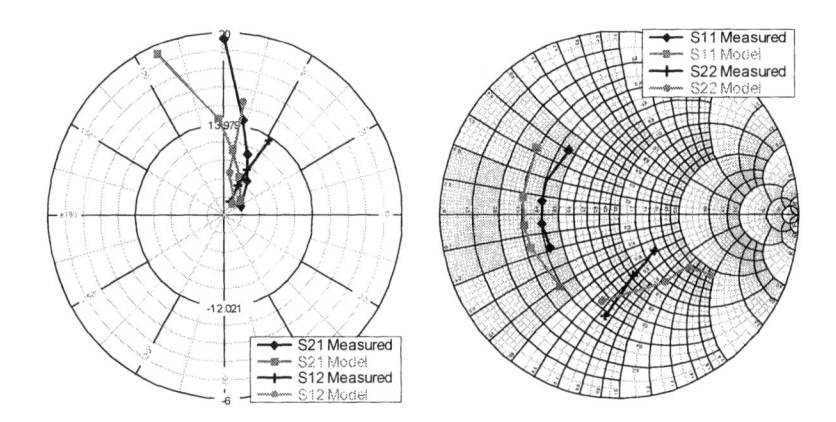

Figure 2.14 Forward, reverse, input, and output S-parameters for the 2N5109 computed from the model (gray traces) and for measured data provided by Motorola (dark traces).

2.3.8.2 Class-A Amplifier Example

Fig. 2.15 is the schematic for a 42 MHz class-A amplifier using this device. Although the model and measured data for this device are compared at 15 volts V_{ce} and 35 mA I_c, the supply voltage for the following amplifiers was set at 5 volts. This is because at the higher current levels required for maximum power output into a 50-ohm load, device dissipation would be excessive with a 15-volt supply. Even with a 5-volt supply, these amplifiers require device heat-sinking.

Eq. 64 specifies that the quiescent collector current is 100 mA for maximum available output power for a class-A amplifier with a 5-volt supply. The reactance of the 1560-nH collector choke is 411 ohms at 42 MHz, significantly higher than the 50-ohm load impedance. The series resonator at the output reduces harmonic current in the load.

The simulated and measured output power, gain, DC current, total supply power, and DC to RF conversion efficiency are tabulated versus input drive level in Table 2.6. For each of the following amplifiers, the output resonator tuning capacitor is adjusted for each input drive level to compensate for changing device reactance at different drive levels.

Notice that at high drive levels, the efficiency exceeds the theoretical maximum of 50% for a class-A amplifier. This is simply because the formula for maximum class-A efficiency assumes the amplifier is linear and is not driven into cutoff or saturation. Even though the amplifier quiescent bias is set up for class-A operation, at high drive levels the circuit becomes nonlinear.

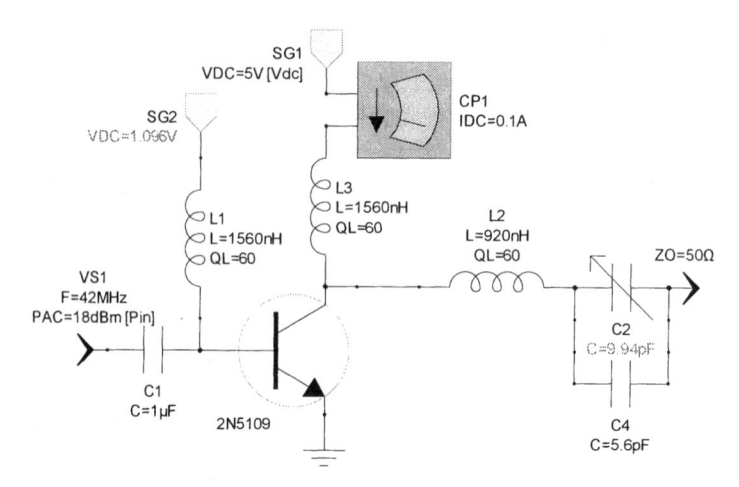

Figure 2.15 Schematic of the 42-MHz 2N5109 class-A amplifier with a supply voltage of 5 volts.

With no input drive, the base bias voltage is adjusted to 1.096 volts to set the collector current at 100 mA. Notice that at 0-dBm drive, the measured DC collector current falls to 92 mA but this behavior is not predicted by simulation. At moderate drive level, the simulated gain and efficiency are significantly higher than the measured values. The simulated and measured saturated output levels are similar.

Table 2.6 Simulated and measured characteristics of the 42-MHz 2N5109 class-A power amplifier

SIMULATED Pin(dBm)	Pout(dBm)	Gain(dB)	Idc(mA)	Pdc(W)	η_A (%)
0	21.1	21.1	101	0.51	25.2
8	26.4	18.4	137	0.68	64.6
10	27.0	17.0	141	0.70	69.6
12	27.1	15.1	141	0.70	72.0
14	27.1	13.1	140	0.70	73.0
16	27.1	11.1	139	0.69	73.6
18	27.1	9.1	138	0.69	73.7

MEASURED Pin(dBm)	Pout(dBm)	Gain(dB)	Idc(mA)	Pdc(W)	η_A (%)
0	22.0	22.0	92	0.46	34.5
8	25.2	17.2	117	0.58	56.6
10	25.7	15.7	125	0.62	59.4
12	26.2	14.2	132	0.66	63.2
14	26.6	12.6	139	0.70	65.8
16	26.9	10.9	149	0.74	65.7
18	27.1	9.1	153	0.76	67.0

It is important to note that in each of the following amplifier types that the bias is supplied by voltage sources. If L_1 is replaced with a resistor and the base bias voltage is increased to provide the necessary base drive, the bias is current sourced and the output power and efficiency of these test circuits is significantly reduced. This phenomenon is described by Lee and Dunleavy [5].

2.3.8.3 Class-AB Amplifier Example

Next, the base bias voltage is decreased to 0.681 volts. This reduces the quiescent collector current to 1 mA and the operating class becomes AB. The simulated and measured results are listed in Table 2.7. Although the theoretical maximum efficiency for class-AB is near that of class-B or 78.5%, notice for this amplifier that both the simulated and measured efficiency are similar to the class-A amplifier.

Table 2.7 Simulated and measured characteristics of the 42-MHz 2N5109 class-AB power amplifier

SIMULATED Pin(dBm)	Pout(dBm)	Gain(dB)	Idc(mA)	Pdc(W)	η_A (%)
0	19.4	19.4	59	0.30	29.2
8	25.1	17.1	111	0.55	58.3
10	25.9	15.9	122	0.61	63.9
12	26.4	14.4	129	0.64	67.4
14	26.7	12.7	134	0.67	70.0
16	26.9	10.9	136	0.68	71.4
18	27.0	9.0	138	0.69	72.1
MEASURED Pin(dBm)	Pout(dBm)	Gain(dB)	Idc(mA)	Pdc(W)	η_A (%)
0	20.2	20.2	57	0.28	36.7
8	24.6	16.6	103	0.52	56.0
10	25.4	15.4	113	0.56	61.4
12	26.0	14.0	122	0.61	65.3
14	26.5	12.5	131	0.66	68.2
16	26.8	10.6	139	0.70	68.9
18	27.1	9.1	145	0.72	70.7

2.3.8.4 Class-C Amplifier Example

Next, the base voltage is set to 0 volts to establish class-C operation and the quiescent collector voltage is near 0. The input signal must overcome the base-emitter offset voltage to turn the 2N5109 on. Therefore, the conduction angle is less than 180°, thus establishing class-C operation. Low drive levels are insufficient to bias the device effectively in the active region, so the gain is reduced. The simulated and measured results are tabulated in Table 2.8. Again, for this amplifier design, there is little advantage to class-C over class-A operation.

Table 2.8 Measured characteristics of the 42-MHz 2N5109 class-C power amplifier

SIMULATED

Pin(dBm)	Pout(dBm)	Gain(dB)	Idc(mA)	Pdc(W)	η_A (%)
0	-24.0	-24.0	0	0	0
8	19.0	11.0	46	0.23	34.4
10	22.2	12.2	67	0.34	49.4
12	23.6	11.6	82	0.41	56.0
14	24.9	10.9	98	0.49	63.1
16	25.6	9.6	110	0.55	66.2
18	26.2	8.2	120	0.60	69.4

MEASURED

Pin(dBm)	Pout(dBm)	Gain(dB)	Idc(mA)	Pdc(W)	η_A (%)
0	-21.0	-21.0	0	0	0
8	11.3	3.3	15	0.08	20.0
10	19.0	9.0	42	0.21	37.8
12	22.6	10.6	68	0.34	53.5
14	24.4	10.4	91	0.46	60.5
16	25.6	9.6	108	0.54	67.2
18	26.3	8.3	125	0.62	68.2

2.3.8.5 Class-E Amplifier Example

The schematic of a 42-MHz class-E power amplifier using a 2N5109 is shown in Fig. 2.16. Broadly speaking, a class-E amplifier can be defined as a single-pole single-throw switch driving a tuned load network [6]. As such, a specific topology is not required. However, a shunt capacitor at the output of the device, such as C_3 in Fig. 2.16, is conducive to class-E operation. This class-E amplifier is similar to the previous class-C amplifier with the addition of this shunt energy storage element.

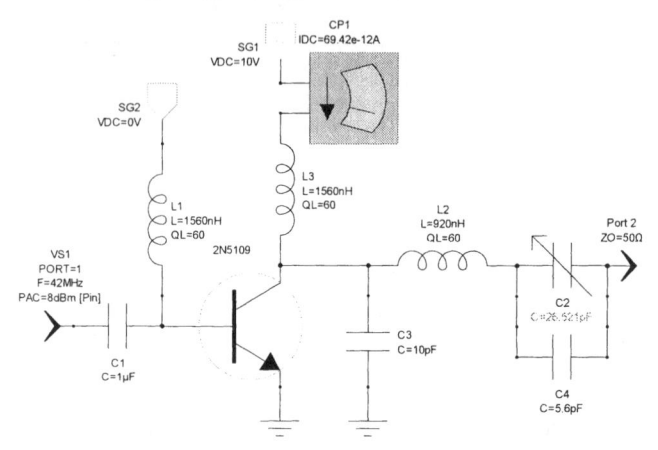

Figure 2.16 Schematic of the 42-MHz 2N5109 class-E power amplifier.

The simulated and measured performance parameters for this amplifier are given in Table 2.9. The output power is somewhat less than the class-A amplifier but the conversion efficiency is the highest of the power amplifiers studied here.

Table 2.9 Simulated and measured characteristics of the 42-MHz 2N5109 class-E power amplifier

SIMULATED Pin(dBm)	Pout(dBm)	Gain(dB)	Idc(mA)	Pdc(W)	η_A (%)
0	-24.0	-24	0	0	0
8	21.6	13.6	45	0.23	63.6
10	23.5	13.5	61	0.30	74.4
12	24.4	12.4	72	0.36	76.8
14	25.0	11.0	81	0.41	77.5
16	25.6	9.6	93	0.47	77.1
18	26.2	8.2	110	0.55	75.3
MEASURED Pin(dBm)	Pout(dBm)	Gain(dB)	Idc(mA)	Pdc(W)	η_A (%)
0	-19.5	-19.5	0	0	0
8	18.3	10.3	26	0.13	52.0
10	21.5	11.5	45	0.22	62.8
12	23.3	11.3	64	0.32	66.8
14	24.6	10.6	86	0.43	67.1
16	25.6	9.6	102	0.51	71.2
18	26.1	8.1	114	0.57	71.4

2.3.8.6 Efficiency Versus Collector-Emitter Path Resistance

The amplifier shown in Fig 2.17 is used to illustrate the importance of low resistance in the collector-emitter path when efficiency is required. This circuit is similar to the class-A amplifier but degeneration resistance, R_1, is included in the emitter and shunt feedback resistance, R_2, is included from the collector to base. When R_1 is 0 and R_2 is infinite, this circuit is identical to the class-A amplifier considered previously.

Table 2.10 gives the simulated and measured performance characteristics of this circuit versus the emitter-degeneration resistance. The drive level is constant at 16 dBm. As the emitter resistance is changed the base bias voltage is adjusted to maintain a collector current of 100 mA.

The efficiency degrades rapidly with increasing emitter resistance. Because the resistance is AC bypassed, the small signal gain is unaffected by the resistance and the resistor serves only as a bias resistor. However, at large signal levels, with increasing resistance, the maximum output level is reduced, compression occurs at lower level and therefore the gain is reduced.

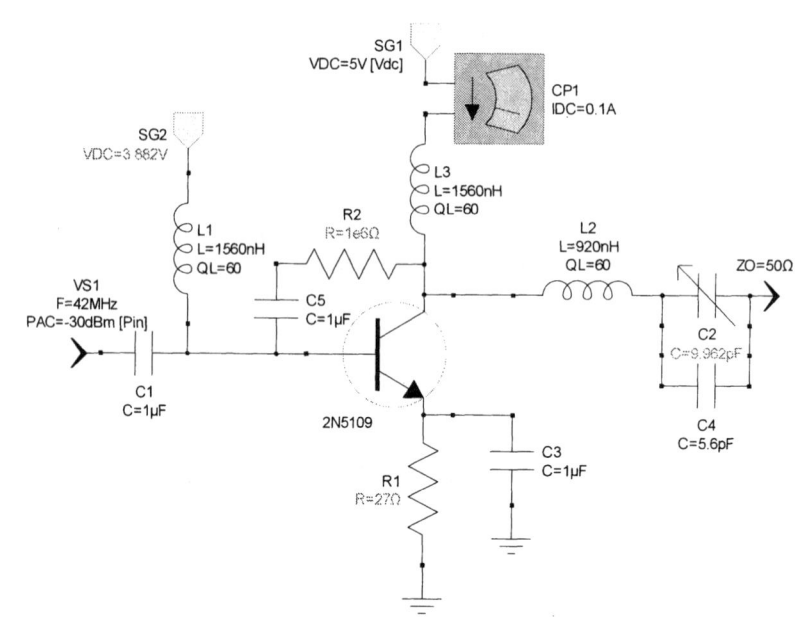

Figure 2.17 Schematic of the 42-MHz 2N5109 Class-A amplifier with emitter series feedback resistance and collector to base shunt feedback resistance.

Table 2.10 Simulated and measured characteristics of the 42-MHz 2N5109 class-A power amplifier with 16-dBm drive versus DC resistance in series with the emitter

SIMULATED DC Resistance	Pout(dBm)	Gain(dB)	Idc(mA)	Pdc(W)	η_A (%)
0	27.1	11.1	139	0.69	73.6
4.7	25.8	9.8	120	0.60	63.9
10	24.6	8.6	103	0.51	55.3
27	19.6	3.6	78	0.39	23.5
MEASURED Pin(dBm)	Pout(dBm)	Gain(dB)	Idc(mA)	Pdc(W)	η_A (%)
0	26.9	10.9	149	0.74	65.7
4.7	25.8	9.8	123	0.62	61.8
10	24.9	8.9	111	0.56	55.3
27	20.7	4.7	102	0.51	23.0

The output power is degraded only moderately with shunt feedback resistance. A resistance of 560 ohms reduces the output power of this circuit only about 1 dBm. With shunt feedback resistance of 10K, 4700, 1800, and 560 ohms, the simulated conversion efficiencies are 73.6, 71.2, 67.3, and 56.6%, respectively.

2.3.8.7 Common-Collector Class-A Amplifier Example

Fig. 2.18 is the schematic of a common-collector 42-MHz class-A power amplifier using the 2N5109 transistor. Resistors R_1 and R_2 are necessary to stabilize the device. They also help stabilize the quiescent bias point against temperature and device variation. Unfortunately, these resistors degrade the conversion efficiency of the class-A amplifier approximately 10%. The base bias voltage is adjusted to set the quiescent collector current at 100 mA.

The tuned circuit at the input is used to step the 50 ohm input impedance up to match the high input impedance of the common-collector amplifier. The low output impedance of this topology is effective for directly driving a 50-ohm load.

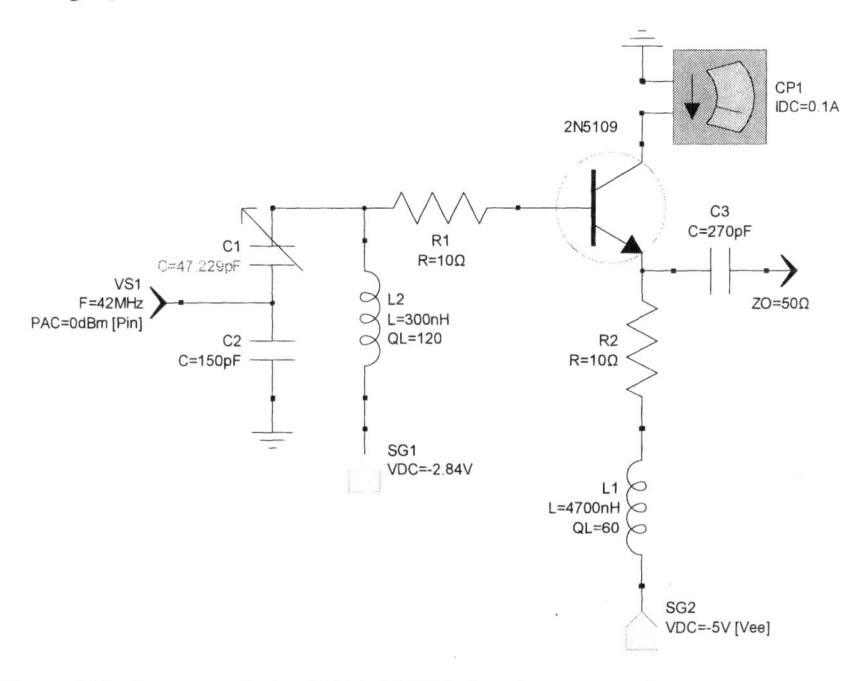

Figure 2.18 Common-collector 42-MHz 2N5109 class-A power amplifier.

Given in Table 2.11 are the simulated and measured performance characteristics of the common-collector class-A power amplifier versus drive level. The measured output power and gain are somewhat higher than simulated. As with the common-emitter topologies, the resonator tuning capacitor is adjusted for maximum output at each drive level. Notice that the conversion efficiency of this common-collector class-A topology is significantly less than the common-emitter topology.

Table 2.11 Simulated and measured characteristics of the CC 42-MHz 2N5109 class-AB power amplifier

SIMULATED Pin(dBm)	Pout(dBm)	Gain(dB)	Idc(mA)	Pdc(W)	η_A (%)
0	13.4	13.4	100	0.50	4.4
8	20.9	12.9	100	0.50	24.9
10	21.7	11.7	99	0.49	30.3
12	22.4	10.4	98	0.49	35.4
14	22.9	8.9	96	0.48	40.4
16	23.2	7.2	98	0.49	42.6
18	23.4	5.4	101	0.50	43.2
MEASURED Pin(dBm)	**Pout(dBm)**	**Gain(dB)**	**Idc(mA)**	**Pdc(W)**	η_A (%)
0	16.5	16.5	100	0.50	8.9
8	21.1	13.1	100	0.50	25.8
10	21.7	11.7	100	0.50	29.6
12	22.3	10.3	100	0.50	34.0
14	22.6	8.6	102	0.51	35.7
16	22.8	6.8	102	0.51	37.4
18	23.0	5.0	105	0.52	38.0

2.3.8.8 Common-Collector Class-C Amplifier Example

Class-C operation of the common-collector topology was tested by adjusting the base bias voltage to equal the emitter supply voltage. As with the common-emitter class-C topology, the input signal must overcome the base-emitter offset voltage to turn the 2N5109 on.

However, unlike the common-emitter amplifiers, the class-C common-collector topology offers significantly improved conversion efficiency over class-A operation. As stated previously, despite the fact that class-C amplifiers require sufficient drive level to bias the device in the active region, triggered class-C oscillators start.

2.3.8.9 Power Amplifier Summary

The output power level of oscillators is generally determined by the saturated output power level of the sustaining stage, which is a strong function of the device current. The conversion efficiency is of little concern for the majority of applications powered by mains and a class-A operating mode is standard. For battery operated applications, when conversion efficiency is important, then:

1) resistance in the collector-emitter path should be avoided;
2) higher class operating modes such as AB, C or E may be helpful;
3) idealized amplifier theory is not adequate to predict performance;
4) simulation of amplifier performance using device models is helpful but precise results require advanced and accurate models;
5) low resistance and use of inductive chokes require greater attention to stability issues.

Table 2.12 Simulated and measured characteristics of the CC 42-MHz 2N5109 class-C power amplifier

SIMULATED Pin(dBm)	Pout(dBm)	Gain(dB)	Idc(mA)	Pdc(W)	η_A (%)
0	6.9	6.9	8.2	0.04	12.0
8	18.2	10.2	33	0.18	39.6
10	20.4	10.4	44	0.22	50.2
12	21.5	9.5	51	0.26	55.6
14	22.2	8.2	57	0.28	58.7
16	22.7	6.7	62	0.31	60.5
18	23.1	5.1	67	0.34	61.3
20	23.4	3.4	72	0.36	61.3

MEASURED Pin(dBm)	Pout(dBm)	Gain(dB)	Idc(mA)	Pdc(W)	η_A (%)
0	3.5	3.5	4	0.02	11.1
8	19.9	11.9	39	0.20	50.1
10	21.2	11.2	48	0.24	54.9
12	22.0	10.0	54	0.27	58.7
14	22.5	8.5	60	0.30	59.3
16	23.1	7.1	66	0.33	61.8
18	23.3	5.3	70	0.35	61.1
20	23.5	3.5	74	0.37	60.5

2.4 Nonlinear Open-Loop Cascade

Nonlinear characteristics of the sustaining stage have been described. Next, the nonlinear behavior of the open-loop cascade is considered. When the sustaining stage and resonator are well matched to the reference impedance, and when the sustaining stage is operating in a linear mode, the open-loop gain is merely the sum of the sustaining stage gain and the resonator loss, and the transmission phase is the sum of the phase shifts. Previously the nonlinear characteristics of the sustaining stage were investigated assuming a 50-ohm source and load. This condition is approximately but not exactly realized in the real cascade. Furthermore, examining the nonlinear behavior of the cascade brings us intuitively one step closer to the final oscillator configuration.

2.4.1 Nonlinear Open-Loop Cascade Example 1

The schematic of Fig. 2.19 cascades a type A common-emitter resistive feedback amplifier with a top-C coupled 300-MHz parallel resonator. The device is an Avago Technologies AT41486 microwave bipolar transistor with an F_t of 8 GHz. To form the oscillator, port 2 is connected to port 1. The supply choke L_2 is not high impedance but provides shunt reactance to compensate the phase shift of the cascade. The oscillator output load termination is represented by resistor R_4 connected to the collector of the device through the 47 pF capacitor C_4. The feedback amplifier gain without the resonator, but including the load resistor R_4, is approximately 10.1 db.

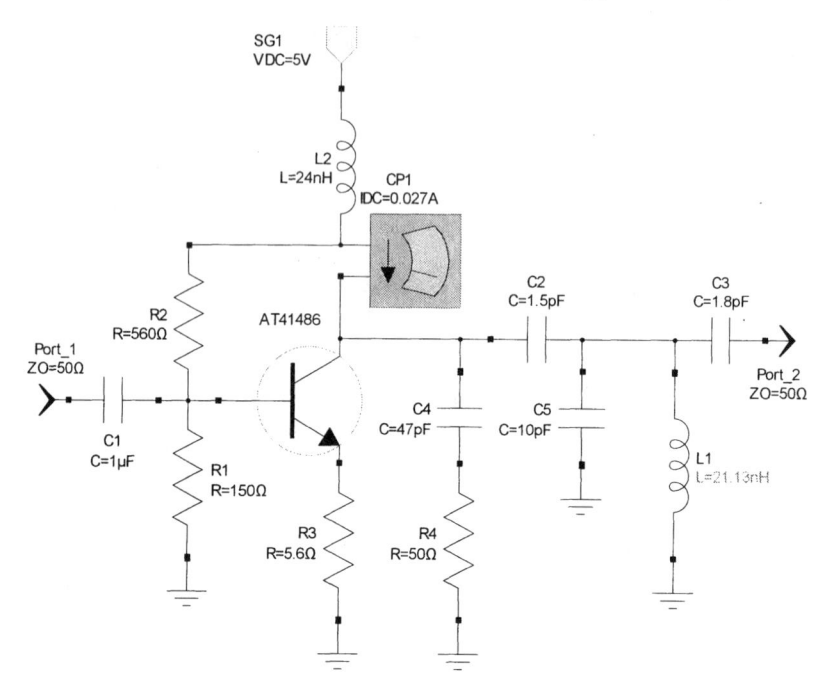

Figure 2.19 Open-loop cascade of a type-A general-purpose amplifier cascaded with a top-C coupled parallel resonator.

The solid traces in Fig. 2.20 are the LS21 amplitude and phase responses and the input and output matches of the cascade with an input drive level of -10 dBm. At this drive level the amplifier mode is nearly linear. The unloaded Q of resonator inductor L_1 is 60. The resulting resonator loss reduces the cascade gain to 6.2 dB at the gain peak. Capacitors C_2 and C_3 have slightly dissimilar values to compensate for dissimilar input and output impedances of the amplifier.

The drive level is then increased to 12.5 dBm where the cascade gain drops to 0 dB at ϕ_o as given by the dashed traces in Fig. 2.20. This represents the final steady-state operating level of the oscillator where the net loop gain is 0 dB. Notice that the transmission phase shifted and the frequency of ϕ_o dropped approximately 1 MHz. This is because nonlinear action not only causes gain compression but also modifies device reactance and shifts the phase. Nevertheless, for this amplifier the phase shift is not excessive. The frequency shift is approximately 0.3%, thus validating the usefulness of linear analysis.

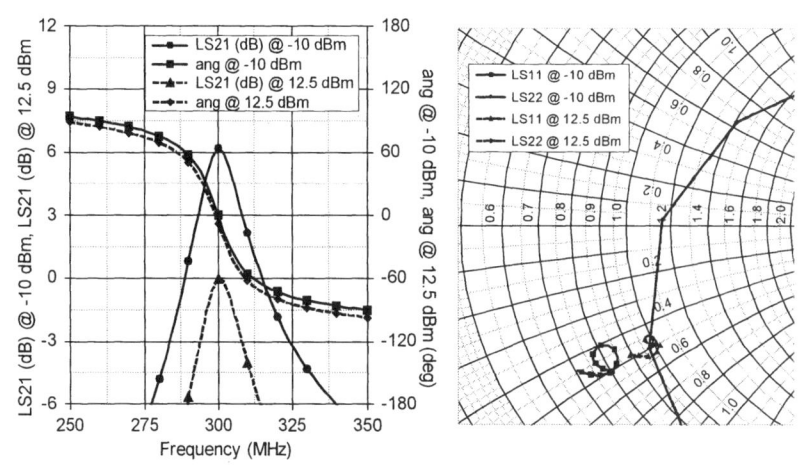

Figure 2.20 LS21 amplitude and phase (left) for the cascade in Fig. 2.19 with -10-dBm drive (solid) and 12.5-dBm drive (dashed) and the LS11 and LS22 (right) with -10-dBm drive (solid) and 12.5-dBm drive (dashed).

2.4.2 Nonlinear Open-Loop Cascade Example 2

A second nonlinear open-loop cascade example schematic is shown in Fig. 2.21. A Colpitts resonator is cascaded with a common-collector amplifier using a BFT92 PNP bipolar transistor with an F_t of 5 GHz [7]. Capacitors C_1 and C_2 step the low output impedance of the common-collector stage up to the higher impedance of the base. Economy and stability are favored over conversion efficiency by the use of entirely resistive bias circuitry. Power is coupled from the emitter using the 270-pF capacitor C_4. Capacitor C_3 is required to avoid the 50-ohm simulator resistance at port 2 from disturbing the bias. When the loop is closed to form the oscillator, C_3 may be eliminated.

The solid traces in Fig. 2.22 are the LS21 amplitude and phase responses and the input and output matches of the cascade with an input drive level of -20 dBm. At this drive level the amplifier mode is nearly linear.

The drive level is then increased to 8.1 dBm where the cascade gain drops to 0 dB at ϕ_o as shown by the dashed traces in Fig. 2.22. This represents the final steady-state operating level of the oscillator. The frequency of ϕ_o this time increased approximately 3 MHz. Again, the shift is not excessive, approximately 1%.

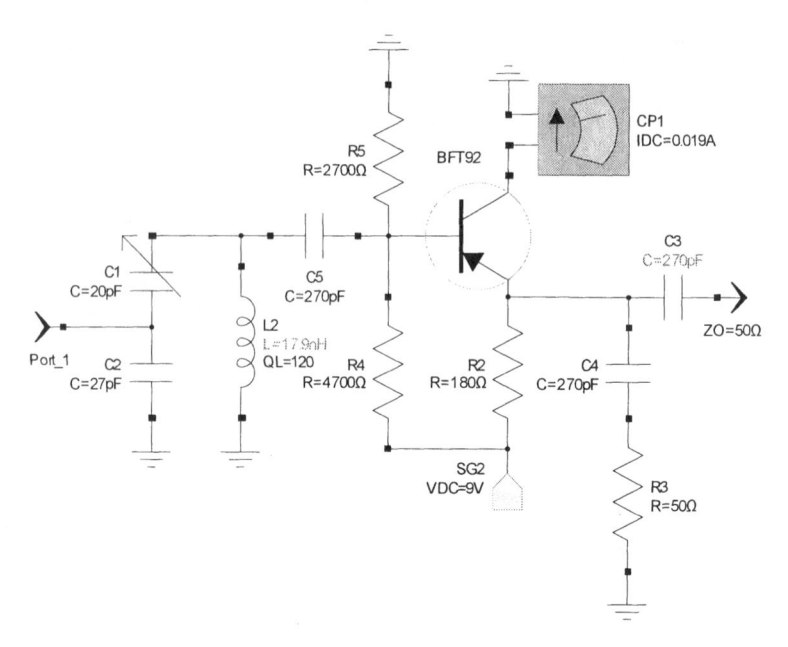

Figure 2.21 Open-loop cascade of a Colpitts type resonator and a PNP common-collector amplifier.

Notice that the phase slope, and therefore the loaded Q, becomes slightly steeper for this circuit as the gain compresses, thus dispelling the myth that nonlinear action in the device reduces the oscillator loaded Q. The magnitude and direction of frequency shift and the impact of nonlinear action on the loaded Q are specific to each circuit and assessment requires a detailed analysis.

2.5 Nonlinear HB Colpitts Example

To this point, both the linear and nonlinear HB analysis involved only a resonator and amplifier cascade. Next, the loop is closed and an HB simulation of the oscillating circuit is performed.

2.5.1 Closing the Loop and Excitation

In Fig. 2.23, the 3-port block labeled Open_Loop duplicates the open-loop cascade of the Colpitts resonator and sustaining stage shown in Fig. 2.21. Ports 1 and 2 are shorted together, thus closing the loop. The load termination depicted by R_3 in Fig. 2.21 is connected to the output port 3 in Fig. 2.21, which becomes the output of the oscillator, or port 1 in Fig. 2.23.

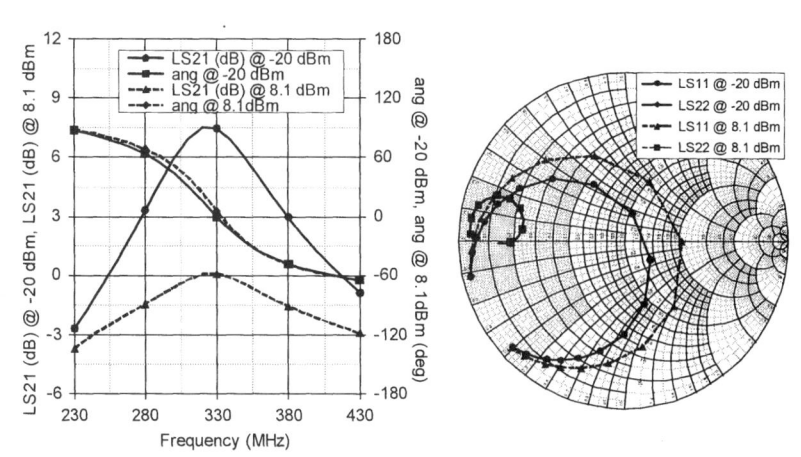

Figure 2.22 LS21 amplitude and phase (left) for the cascade in Fig. 2.21 with -20-dBm drive (solid) and 8.1-dBm drive (dashed) and LS11 and LS22 (right) with -20-dBm drive (solid) and 8.1-dBm drive (dashed).

Because HB simulation is a steady-state analysis, some form excitation is required to establish the operating point of the oscillator. This excitation network is composed of VS_1 and the labeled SFC in Fig. 2.23. SFC is a very high-loaded Q series tuned L-C network specified by the series resonant frequency f_0 and a small capacitor, in this case 0.1 pF. When the oscillation frequency and the voltage at the closed input/output node of the cascade exactly match the amplitude and frequency of the voltage source VS_1, no current would flow through the excitation circuit. The voltage of VS_1 is adjusted and the tuning inductor L_2 of the oscillator is adjusted until this condition is met. This is accomplished by either manual tuning of these two parameters or by optimization. To avoid loading the harmonic currents of the oscillator waveform, the high, loaded Q SFC network provides a high impedance at all frequencies other than the fundamental.

There is a 1-port model *Oscport* in the GENESYS simulator that represents a grounded AC voltage source with an amplitude *Vprobe* and frequency *Fosc* for the oscillation spectral component, and nothing for any other spectral components of the HB solution spectrum. This model does not accept any user input. Its parameters are filled by the simulator. Initially when *Oscport* is launched, *Fosc* is equal to the small signal oscillation frequency. After the nonlinear analysis is complete, *Fosc* is equal to the steady-state oscillation frequency, and *Vprobe* is equal to the amplitude of the first harmonic of voltage at the node at which the *Oscport* is connected.

Because knowledge of the characteristics of the open-loop cascade is the first step and is critical to the management of the design, it is advisable to create a 3-port open-loop cascade schematic and display the results of that simulation, either by linear or nonlinear HB analysis, and then include an additional schematic to close the loop. Either the manual excitation circuit depicted in Fig. 2.23 or the *Oscport* model may be used with this closed circuit to find the steady-state oscillation solution.

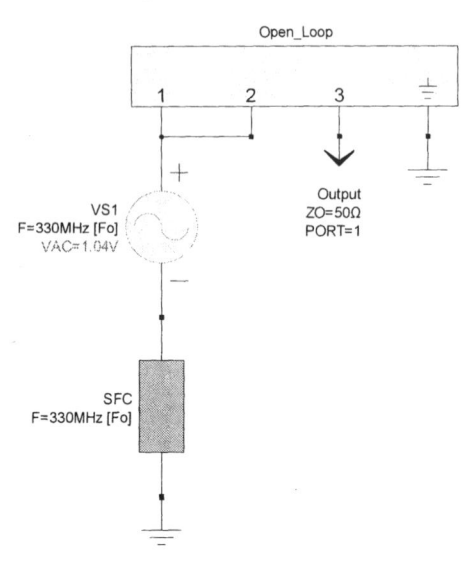

Figure 2.23 Closing the open-loop cascade and adding a steady-state excitation circuit for a harmonic balance simulation of the oscillator.

2.5.2 Harmonic Balance Colpitts Output Spectrum

Fig. 2.24 is the frequency spectrum through the seventh harmonic of the oscillator output computed by the HB simulation. The fundamental output power is 10.24 dBm. The second and third harmonicas are -11.1 and -19.9 dBc, respectively. The seventh harmonic is -39.6 dBc. The peak current in VS_1 at the fundamental frequency is optimized to a negligible current, in this case 32.7 uA.

The results of a HB simulation versus the number of balanced harmonics is given in Table 2.13. Although this oscillator is a highly nonlinear circuit requiring approximately 7.5 dB of gain compression to reach steady-state, Table 2.13 suggests that five or seven harmonics are sufficient to achieve reasonable accuracy. It is also evident that the number of balanced harmonics should exceed the highest harmonic for which accurate results are required. Nevertheless, an excessive number of harmonics is not required and it increases simulation time. Before performing extensive design refinement or optimization, sample runs with a varying number of harmonics are helpful.

Nonlinear Techniques 115

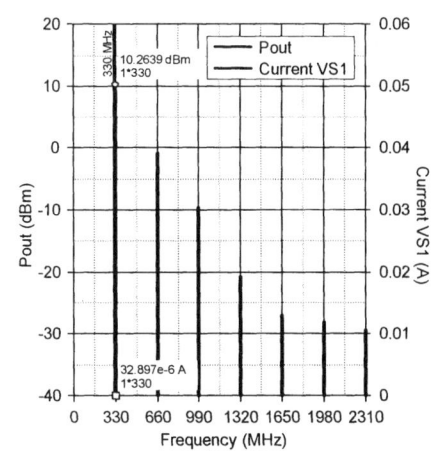

Figure 2.24 Output spectrum of the example Colpitts oscillator computed by a harmonic-balance simulation through the seventh harmonic.

Table 2.13 HB simulation results versus the number of balanced harmonics for the 330-MHz Colpitts oscillator example

Number of Harmonics	Pout(dBm) Fund.	2nd Har. (dBc)	3rd Har. (dBc)	4th Har. (dBc)	5th Har. (dBc)
2	12.12	-6.1			
3	10.82	-11.1	-15.9		
5	10.21	-10.4	-18.3	-28.9	-30.1
7	10.24	-11.1	-19.9	-31.1	-37.7
9	10.17	-11.2	-20.1	-30.7	-39.0
11	10.12	-11.2	-19.9	-30.3	-40.5
21	10.16	-11.1	-19.9	-30.2	-40.8

2.5.3 Excitation Current Versus Oscillator Parameters

Given in Fig. 2.25 are the peak currents at 330 MHz in the excitation source versus the excitation voltage for resonator inductor values of 17.65 nH (circular symbols), 19.65 nH (square symbols), and 21.65 nH (triangular symbols). When the inductor value is correct, the current in the excitation source dips sharply at the correct excitation voltage. When the inductor value is incorrect, the dip is not as pronounced. If the inductor is sufficiently incorrect, there is no dip in the source current. With an incorrect estimate of the inductor value, when the excitation source voltage is optimized, the voltage will be incorrectly driven to 0 volts.

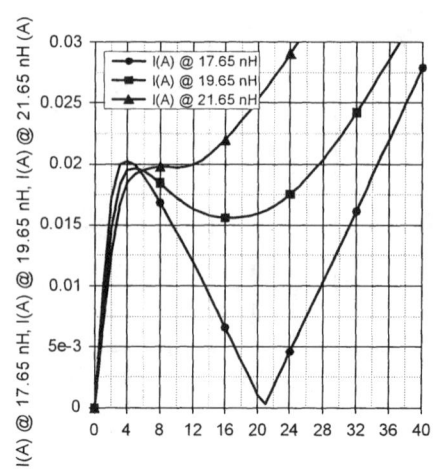

Figure 2.25 Current in the excitation voltage source, VS_1, versus the voltage of VS_1 with the resonator inductor equal to 17.65 nH (circular symbols), 19.65 nH (square symbols), and 21.65 nH (triangular symbols).

These response curves are representative of all oscillator circuits. Successful strategies include placing a lower optimization limit on the oscillator output power, but it is better to estimate the initial inductor value as closely as possible and even to manually tune the inductor for minimum excitation source current before launching optimization of the excitation current.

Oscillator HB analysis may also be performed by adjusting the resonator capacitor, or by using fixed resonator values and determining the frequency that minimizes the excitation source current.

2.6 Nonlinear Negative-Resistance Oscillator

In this section, the closed-loop HB analysis is performed for a negative-resistance oscillator. A schematic for a 900-MHz oscillator using an Avago Technologies AT41486 transistor is shown in Fig. 2.26. The device is biased at 23.5 mA using a -9-volt supply.

Port 1 is the small-signal analysis port and output power is taken by tapping the resonator inductor at port 2. To form the oscillator port 1 is grounded. The output tap is close to the ground point. In fact, the inductance of a via used to ground the inductor L_2 provides most of the inductance of L_1. Tuning is accomplished by C_1.

The small-signal linear input resistance and reactance are given in Fig. 2.27. The reactance is zero ohms at approximately 940 MHz and the resistance at that frequency is -11.8 ohms. When the tuning capacitor C_1 is adjusted the dip the in the input resistance shifts in frequency and the resistance remains negative across the entire tuning range of 700 to 1100 MHz. Tuning this range requires capacitance from 6 to 1 pF.

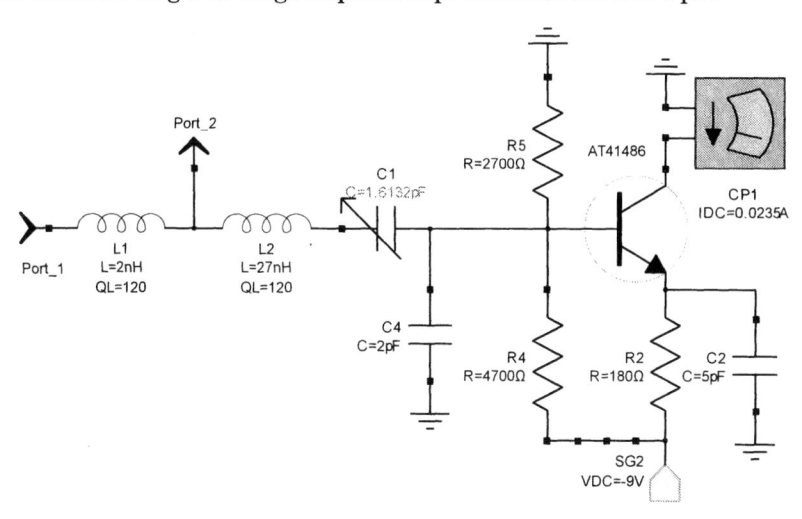

Figure 2.26 Common-collector 900 MHz negative-resistance oscillator.

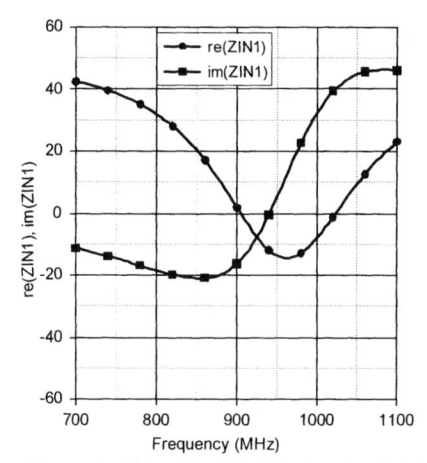

Figure 2.27 Small-signal linear input resistance (circular symbols) and reactance (square symbols) for the negative-resistance oscillator given in Fig. 2.26.

Fig. 2.28 illustrates grounding the input test port 1 and adding an excitation source for the HB analysis. The block *Sch_Open* represents the 2-port ungrounded schematic. This 2-port has a ground reference connected to terminal 3 of *Sch_Open*. The input port 1 is grounded to form the oscillator and the output port 2 of *Sch_Open* becomes the oscillator output and the only port of the oscillator. The excitation source is connected to this oscillator output port.

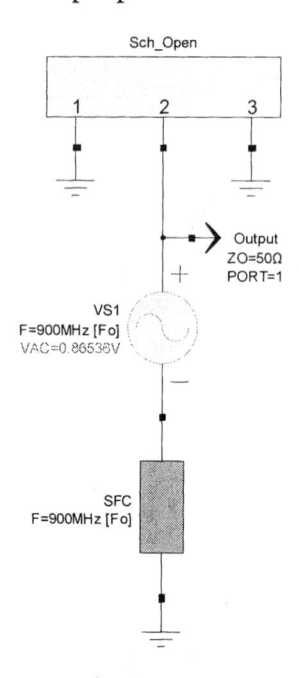

Figure 2.28 Grounded input test port 1 and excitation source for the HB analysis of the negative-resistance oscillator.

Fig. 2.29 shows the output spectrum and excitation source current of the grounded negative-resistance oscillator computed by HB analysis. The tuning capacitor C_t and the excitation source voltage are adjusted to minimize the excitation source current at the desired center frequency of 900 MHz.

The second harmonic is -24 dBc and the third harmonic is -29.9 dBc. This is significantly better harmonic suppression than the previous Colpitts oscillator. The superior harmonic suppression is the result of coupling the output power from the resonator. Output coupling is covered in detail in the next section.

Fig. 2.30 shows the excitation source current versus excitation source voltage. The oscillator tuning capacitor is adjusted to set the oscillation frequency at the excitation source frequency of 900 MHz. Notice the general shape corresponds to that of the 330-MHz oscillator analyzed by the open-loop method.

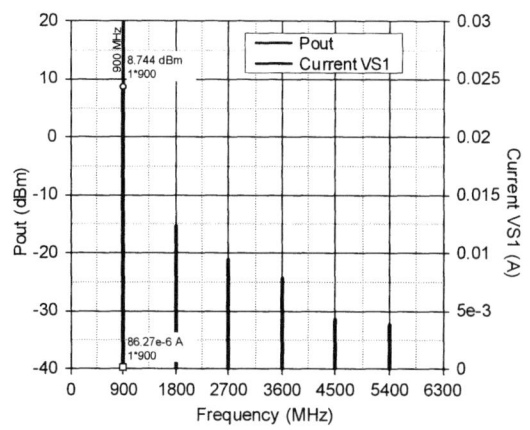

Figure 2.29 Output spectrum and excitation source current computed by HB analysis for the grounded negative-resistance oscillator.

Notice that a tuning capacitance value of 1.613 pF results in an oscillation frequency of 900 MHz. However, the small-signal linear analysis indicates a reactance zero-crossing at 940 MHz. This discrepancy is the result of reactance shift as device operation becomes nonlinear, driving the operating frequency downward in this case by 4.4%. The resonator is coupled directly to the sustaining stage so the operating frequency is more susceptible to nonlinear behavior. It is important to note that the frequency is also more susceptible to bias, device, and temperature effects.

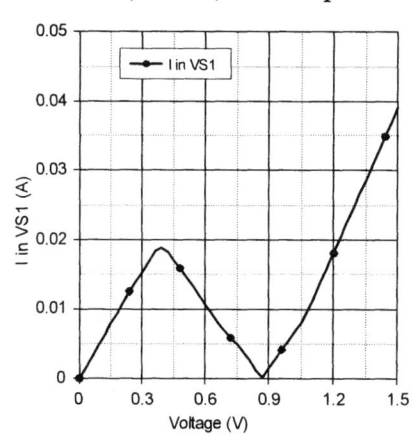

Figure 2.30 Current in the excitation source versus excitation source voltage with the source frequency aligned with the oscillator frequency.

2.7 Output Coupling

In this section, issues associated with coupling output power from the oscillator are cover in detail. Many of these topics can be explored using linear simulation techniques, but other topics require nonlinear analysis. For consistency, all output coupling issues are covered in this chapter.

The process of loading the oscillator with an output termination may disturb the gain margin, transmission phase, match, and loaded Q of the cascade. The impact on the phase, match, and loaded Q can be entirely avoided by using a matched directional coupler. This is useful for oscillator study, but the additional expense is generally not justified for production oscillators. Rather, power is extracted from some node of the oscillator using a capacitor. This offers the additional benefit of not impacting the device bias with the load resistance.

2.7.1 Coupling Node

Fig. 2.31 shows the 330 MHz Colpitts oscillator with potential output power coupling points at the emitter, collector, and the base.

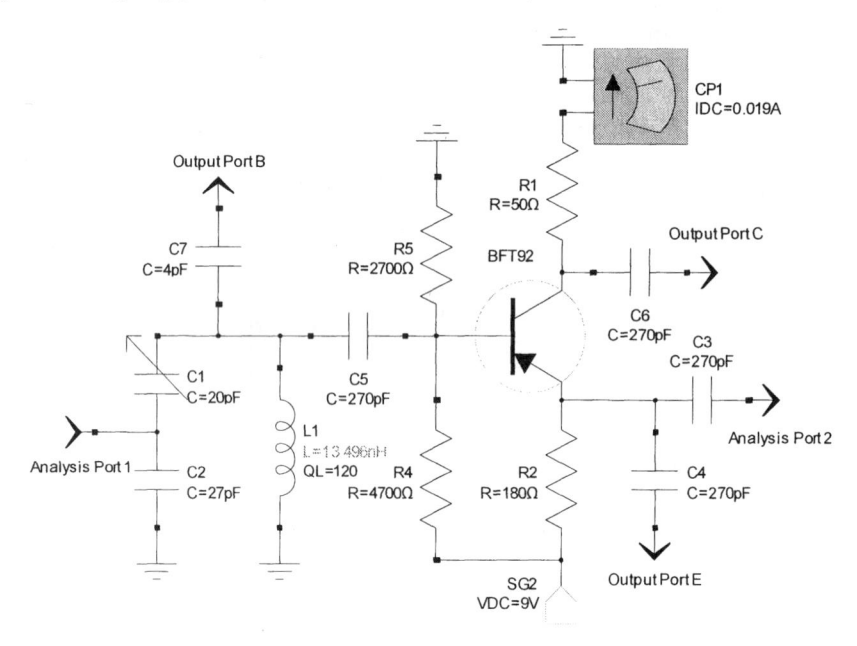

Figure 2.31 Colpitts oscillator showing potential output extraction points.

Shown on the left in Fig. 2.32 is the output spectrum computed by HB simulation and taken at the emitter at Output Port E. Also plotted is the excitation source current indicating that the frequency and voltage of the source are adjusted for low source current. During each spectrum sampling, the alternate output ports are terminated in high impedance so they do not load the oscillator.

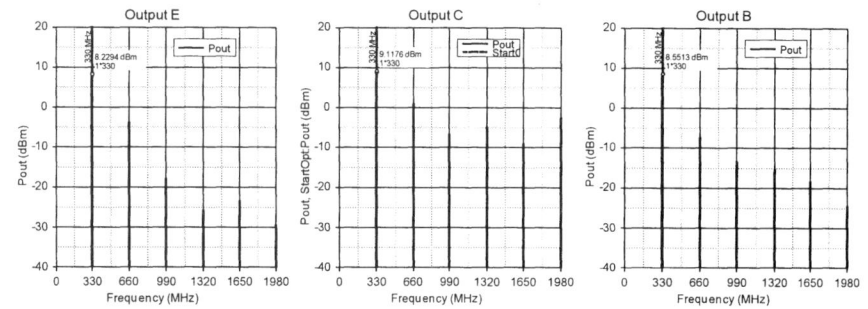

Figure 2.32 Spectrums of the oscillator in Fig. 2.30 with outputs taken at the emitter (Output E), the collector (Output C) and the base (Output B).

Resistor R_1 is added to facilitate taking power at the collector. The impedance of the collector to ground path is not critical in most common-collector oscillators, so R_1 has a minimal effect on operation. With R_1 and a 50-ohm load termination the collector to ground AC resistance is 25 ohms. Alternatively, the entire collector to ground path can be through the load by using a high-impedance inductive choke in place of R_1. The output spectrum at the collector is shown in the middle of Fig. 2.32.

Shown on the right in Fig. 2.32 is the output spectrum taken at Output Port B. Notice in Fig. 2.31 that the coupling capacitor is 4 pF. This is a reactance of 120 ohms at 330 MHz. Eq. 43 reveals that the effective load resistance in parallel with the Colpitts resonator is therefore 340 ohms rather than 50 ohms. A 50-ohm termination directly across the parallel resonator is excessive loading. The coupling capacitor loads the resonator not only with resistance but capacitance as well. The resonator inductor is reduced to 13.495 to compensate for this additional capacitance and re-center the oscillating frequency at 330 MHz.

Notice from Fig. 2.32 that power is successfully extracted from all three nodes in this oscillator. In general, the greatest output power and the highest levels of harmonics result from output coupling at the collector. In general, the best harmonic performance is obtained by extracting power from the resonator.

2.7.2 Load Pulling

The oscillator load termination is often not constant impedance. Load impedance variation can pull the oscillation frequency and in extreme cases may cause the gain margin to fall below 0 dB and stop oscillation. When load variation influences the oscillation frequency it is called load pulling. Load pulling is often specified as a frequency shift resulting from a 2:1 load

VSWR with any phase angle. Load pulling is conveniently analyzed with linear simulation by terminating the oscillator through a long length of transmission line in a 2:1 VSWR load, such as 25 ohms in a 50 ohm.

Shown in Fig. 2.33 is an open-loop linear simulation of the transmission amplitude and phase and matches for the Colpitts oscillator example with a 25-ohm load connected to Output Port E through a transmission line that is 24,000 degrees long at 330 MHz. As the simulation frequency is swept, the long length of line rotates the load termination through a 2:1 VSWR circle of virtually all phase lengths, creating the patterns in Fig. 2.33.

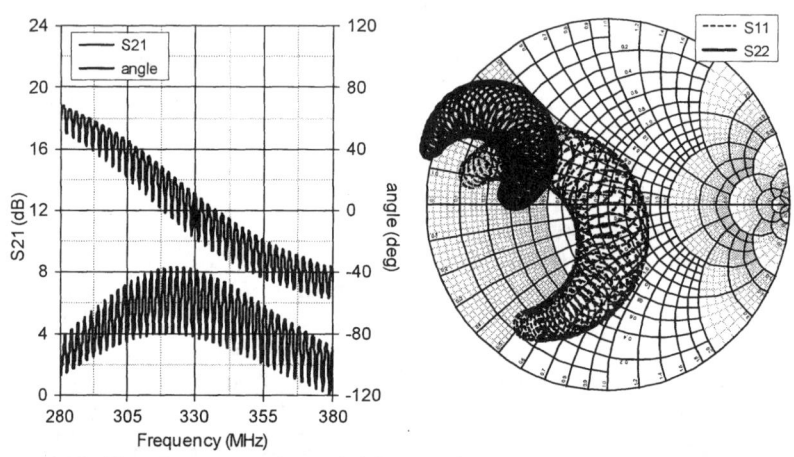

Figure 2.33 The linear open-loop Colpitts load pulling characteristics generated by terminating the emitter output in a 2:1 *VSWR* load through a long transmission line.

An examination of the transmission phase reveals that this 2:1 VSWR pulls the phase-zero crossing and therefore the oscillation frequency approximately 1.6 divisions or 20 MHz. The analysis also reveals that the gain margin ranges from approximately 4 to 8 dB so oscillation is not threatened. From the input and output impedance plots it is evident that the open-loop cascade match is also effected but not to a degree to invalidate the analysis. This technique can be combined with the Randall/Hock equation to improve analysis accuracy for poorly matched cascades.

In Fig. 2.34, the load-pulling characteristics of the Colpitts example are given with power extracted at Output Port C and Output Port B. Notice that extracting power at the collector significantly reduces load pulling.

2.7.3 Loaded Q and Load Pulling

As outlined in Section 1.6.12, loaded Q is critical to oscillator phase noise and immunity to perturbations introduced by the active device.

Nonlinear Techniques 123

Higher loaded Q also reduces load pulling. Fig. 2.35 shows a load pull analysis for the oscillator in Fig. 2.31 with power extracted from Output Port E and the loaded Q approximately doubled by doubling capacitors C_1 and C_2 and halving inductor L_1. The 2:1 VSWR load pull is reduced to approximately 11 MHz from the initial 20 MHz. Other factors being equal, load pull is proportional to the inverse of the loaded Q.

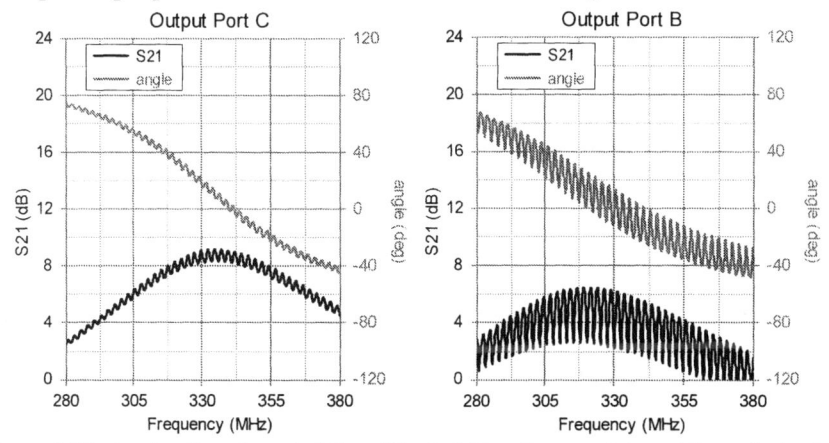

Figure 2.34 Load pulling characteristics of the Colpitts with power taken at the collector (left) and the base (right).

Load pull is a function of factors beside the loaded Q. Specifically, the output coupling location and the degree of coupling play significant roles. The degree of coupling is considered next. A typical specification for 2:1 VSWR load pull for an oscillator with an L-C resonator is 1 part per thousand or 0.1%.

2.7.4 Degree of Coupling

The 4-pF output coupling capacitor C_7 in Fig. 2.31 has a reactance of 120.6 ohms at 330 MHz. Eq. 43 reveals that the parallel load resistance presented to the resonator is not 50 ohms but rather 340 ohms. With low coupling capacitor reactance the load is coupled directly to the resonator but with high coupling capacitor reactance the load coupling is loose.

Given in Table 2.14 are characteristics of the closed-loop Colpitts oscillator determined by HB simulation versus values of the coupling capacitor. The coupling capacitor also loads the resonator with capacitive reactance as determined by Eq. 44. Therefore, for each value of coupling capacitor the resonator inductor, L_1, is optimized to tune the frequency to 300 MHz by minimizing the excitation source current.

The output power versus the coupling capacitance is given in column 3. Initially, as the coupling capacitance increases the output power increases. Notice that with higher values of coupling capacitance the heavy resonator loading reduces the open-loop gain margin (column 6) and the internal signal level drops, as evidenced by the voltage at the closed input/output node (column 4). As the gain margin drops, lower amplifier compression reduces the internal oscillator power available for delivery to the output and the output power drops. This behavior is common in other oscillators. Full load coupling may kill oscillation. For this oscillator, the maximum output power occurs over a broad range of capacitor values of 4 to 6 pF.

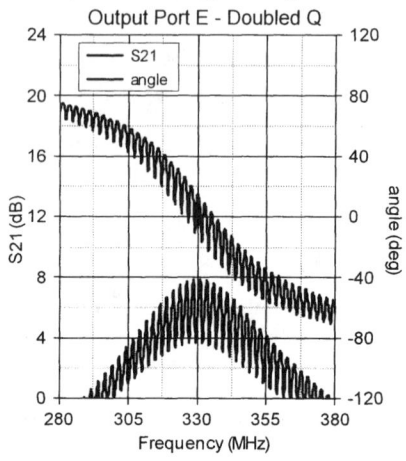

Figure 2.35 Load-pulling test with twice the loaded Q.

Table 2.14 HB simulation results versus the value of coupling capacitor for the oscillator in Fig. 2.31 with power extraction at Output Port B

C7(pF)	L1(nH)	Pout (dBm)	VS1 (Volts)	2nd Har. (dBc)	Margin (dB)	Pull (%)	QL
0.5	15.74	-4.36	1.86	-11.1	7.06	0.13	4.9
1	15.26	1.29	1.78	-11.7	6.98	0.52	5.0
2	14.48	6.10	1.55	-13.3	6.91	1.98	4.7
4	13.50	8.57	1.06	-15.9	6.36	6.35	3.7
6	13.02	8.21	0.78	-15.2	5.58	12.3	2.9
12	14.09	6.16	0.55	-13.8	3.71	27.7	1.7
18	18.51	4.12	0.49	-14.9	2.23	39.1	1.3

2.7.5 Loaded Q and Coupling

The open-loop loaded Q versus the coupling capacitance is given in column 8 of Table 2.14. With loose coupling, the loaded Q is approximately 5. With heavier coupling, the resistance loading the resonator significantly degrades the loaded Q. With a coupling capacitance of 2 pF the output power is within 2.5 dB of the maximum output and the loaded Q is only moderately degraded.

The second harmonic level relative to the fundamental is given in column 5 in Table 2.14. Initial increase in the coupling capacitance reduces the harmonic level because the heavier loading reduces the gain margin and compression in the amplifier. Still heavier loading begins degrading the loaded Q. The resonator becomes less effective at filtering harmonics and harmonic levels stop improving.

2.7.6 Coupling Reactance and Load Pulling

The 2:1 VSWR load pull (%) versus the coupling capacitance is given in column 7 of Table 2.14. The load pull for this oscillator is rather poor. Only with very loose coupling is a typical load pull performance of 0.1% approached. At this coupling, the output power is 13 dB below maximum. With a coupling capacitance of 2 pF, the load pull is almost 2%.

Not shown in Table 2.14 is the fact that the gain margin with 4-pF coupling capacitance is less than 2 dB with the worst case 2:1 $VSWR$ load. A 2:1 $VSWR$ load can result in oscillation failure with coupling capacitance of 6 pF or greater. These are risks of coupling output power directly from the resonator. The advantage is generally improved harmonic performance.

2.7.7 Coupling Reactance and Harmonics

Common practice couples power from the oscillator using a capacitor as previously illustrated. Capacitors are small, economical, and DC bias voltages are not perturbed by the load. However, the use of a coupling capacitor has a disadvantage. If finite reactance is used to control the coupling level, the lower reactance at higher frequency couples harmonics to the output more efficiently than the fundamental. The harmonic performance is degraded.

On the left in Fig. 2.36 is the output spectrum of the Colpitts oscillator in Fig. 2.31 with a 2-pF coupling capacitor at output port B. The second harmonic is -13.3 dB below the output fundamental level. On the right, the spectrum is given with the coupling capacitor replaced by a 116.3 nH coupling inductor with the same reactance magnitude, 241.1 ohms. The output fundamental power is similar but the second harmonic -23.6 dBc. Inductive coupling acts as a single pole lowpass filter improving the second harmonic performance by 10 dB. The resonator inductor is increased to compensate for the shift of load reactance from capacitive to inductive.

It is fortunate that the coupling node of this particular oscillator is at DC ground potential and the inductor does not disturb the bias. If this were not the case, a blocking capacitor is required in series with the coupling inductor. For oscillators with a narrow output frequency range this blocking capacitor can be of high reactance tuned with a higher inductive reactance to improve harmonic performance further.

2.7.8 Output Coupling Example 2

This section illustrates a rigorous method of coupling power from the oscillator. The input and through ports of a directional coupler close the cascade loop and power is extracted via the coupled port. Because the coupler is matched at each port, the load impedance does not shunt the coupling node and the coupler value precisely controls the level of signal extracted from the oscillator system. A schematic is given in Fig. 2.37. Design information on wire-wound couplers is given in Appendix A.

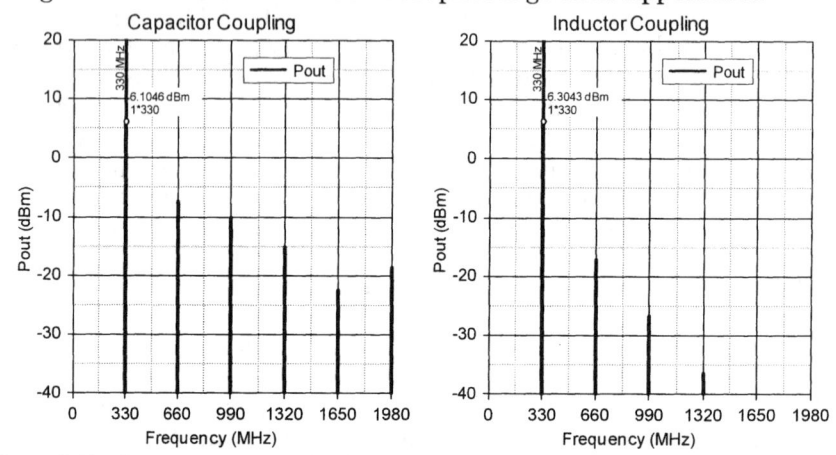

Figure 2.36 Output spectrum of the Colpitts oscillator with a 2-pF output coupling capacitor (left) and a 106.3-nH output coupling inductor (right).

The oscillator is a 100-MHz Colpitts similar to the design in Fig. 2.31. This oscillator was analyzed using harmonic balance simulation.

Table 2.15 shows the results of the harmonic balance simulation for coupling values of -2 to -15 dB. For this coupler, the approximate through port loss ranges from -5.4 dB for -2 dB coupling, -1.3 dB for -6.02 dB coupling to -0.14 dB for -15 dB coupling. For each case, L_1 and the source excitation voltage are optimized to minimize the source excitation current.

For tight coupling at -2 dB, the through loss of the coupler is highest and the loop gain margin is only 2.0 dB. The resulting internal voltage at the closed-loop node as indicated in column 3 is only 0.68 volts. The harmonic performance as indicated in columns 6 and 7 is relatively good because the low gain margin results in minimal nonlinear action in the amplifier. Although a maximum amount of power is extracted via the coupler, the low amplifier compression results in lower available power for the output, as indicated in column 4. With loser power coupling values of -3 and -4.5 dB, the output power is actually greater because of the heavier amplifier compression and available power. At even loser coupling values, the extracted power declines and the higher levels of retained internal power increase nonlinear action and the harmonic performance degrades.

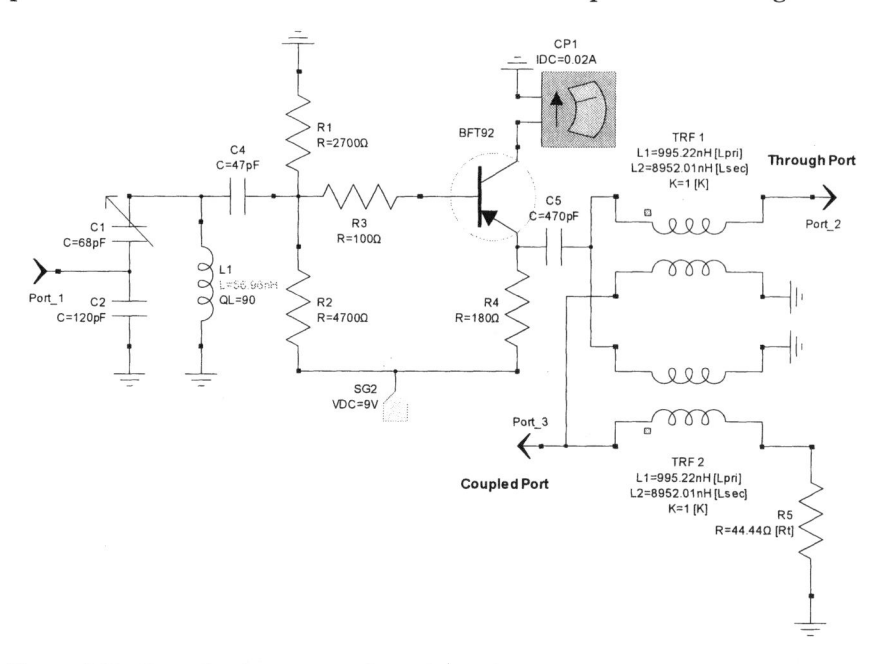

Figure 2.37 Example of output coupling using a directional coupler inserted in the closed-loop. Power is extracted at the coupled port and the through port is used to close the loop.

Table 2.15 HB simulation results versus the directional coupling value for the oscillator in Fig. 2.37 with power extracted at the coupled port

Coupling (dB)	L1 (nH)	VS1 (volts)	Pout (dBm)	Margin (dB)	2nd Har (dBc)	3rd Har (dBc)	QL
-2.00	59.56	0.68	8.86	2.0	-19.9	-25.0	6.1
-3.01	58.20	1.13	10.2	3.8	-10.8	-43.4	6.2
-4.50	57.58	1.57	10.0	5.4	-8.81	-19.6	6.2
-6.02	57.28	1.87	9.16	6.3	-8.15	-15.8	6.2
-9.54	56.96	2.24	6.27	7.4	-8.76	-13.6	6.2
-15.0	56.75	2.44	1.14	7.9	-7.87	-13.7	6.1

Note that the loaded Q of oscillator open-loop cascade is not a function of the coupler value. This is because the impedance loading the resonator is unaffected by the coupling value. There is a common misconception that output loading disturbs the resonator and degrades the loaded Q. This example illustrates this is not necessarily true. Output loading did degrade the loaded Q in the previous example because the output termination shunted the parallel resonator. This was a design choice and not inherent to all oscillators. In fact, if the output termination shunts either the input or output of a series resonator, then the loading actually increases the loaded Q!

The cost and complexity of directional couplers is typically not justified for oscillators and power is more often extracted via a simple coupling capacitor or a tap point on an inductor or transmission line.

2.7.9 Coupling Summary

Power may be extracted from the oscillator at any node. An appropriate choice is made by considering the following factors.

1) The coupling circuitry is included during the open-loop cascade simulation to assess the impact of coupling.
2) The degree of coupling is controlled by the reactance of the coupling capacitor.
3) Generally, coupling at the collector provides the highest output and worst harmonics, coupling across the resonator provides the best harmonic performance, and worst load-pulling and coupling at the emitter is a common compromise.
4) Maximum output power may or may not occur with the tightest coupling.
5) Load coupling may decrease, increase, or have little effect on the loaded Q.
6) Load pulling is easily simulated using a 25-ohm termination at the end of a long transmission line.

2.8 Passive Level Control

Nonlinear behavior is not fundamental to the oscillator operation. Limiting caused by nonlinear behavior in the active device of a typical oscillator is merely a convenient means of establishing the steady-state operating point.

Bill Hewlett's prototype oscillator shown in Fig. 2.38 eventually became the HP 200A, the first product of Hewlett-Packard. It was Hewlett's electrical engineering thesis subject at Stanford University in 1938.

Hewlett's unique contribution was using a small incandescent lamp for level control, placed in series with the cathode of a 6J7 cascaded with a 6F6 vacuum tube in a Wien bridge oscillator. Two additional tubes served as an output buffer. The signal level partially excited the 3-watt incandescent bulb. An increase in the signal level would increase the filament temperature and increase the resistance. This was an elegant solution to two problems. It leveled the output signal versus frequency and the output level was not limited by nonlinear action in the active device. The limiting behavior of the lamp is a thermal process with a long time constant,

Nonlinear Techniques

typically 0.1 to 1 second for small lamps. For a given RF signal level, the V-I transfer characteristic is linear. The oscillator harmonic distortion was improved because the oscillator was operating in a nearly linear mode.

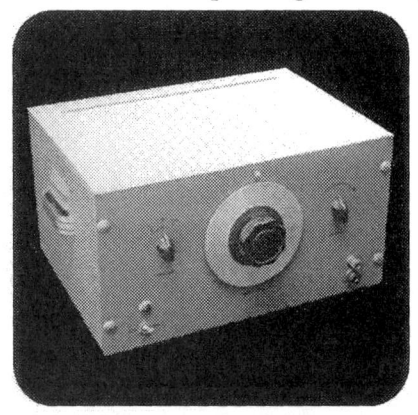

Figure 2.38 Prototype of the HP 200A audio oscillator designed by Bill Hewlett in 1938. (Photo courtesy of Hewlett-Packard Development Company.)

Fig. 2.39 shows the schematic for a 100-MHz oscillator using a 2N5109 bipolar transistor. L_3 models the inductance of the emitter path to ground. L_1 is selected to adjust the cascade transmission phase. R_1 decouples the supply and stabilizes the quiescent bias. R_2 moderates the low frequency gain and biases the device. The amplifier is a type A resistive feedback amplifier described in Section 1.5.7.

The resonator is a top-C coupled parallel resonator with a loaded Q of approximately 18. Power is extracted with C_4 from the resonator to reduce harmonics in the output. The loop is closed by connecting port 2 to the input port 1. C_6 is required for simulation but is not required in the final oscillator.

The results of a HB simulation of this oscillator are given in Fig. 2.40. Harmonics to the seventh are balanced and the circuit is optimized to minimize the excitation current. The simulated output power is 22.27 dBm and the second harmonic is approximately -24 dBc.

The relatively high output power is the expected result of a quiescent bias of approximately ½ watt. The current measured by the current probe CP_1 in Fig. 2.39 is for the open-loop circuit before the onset of oscillation. The simulated steady-state DC current during oscillation is 0.11 amps, a power supply level of 1.32 watts. The total output power to DC power efficiency of 13%.

Given in Fig. 2.41 is the output spectrum measured with an HP8568A spectrum analyzer. Reasonable agreement with the simulated spectrum is achieved through approximately the fifth harmonic. The measured spectrum reveals sixth and seventh harmonics are less than -40 dBm while the simulated spectrum predicts significant sixth and seventh harmonics. The difference is attributed to device model error and circuit parasitics.

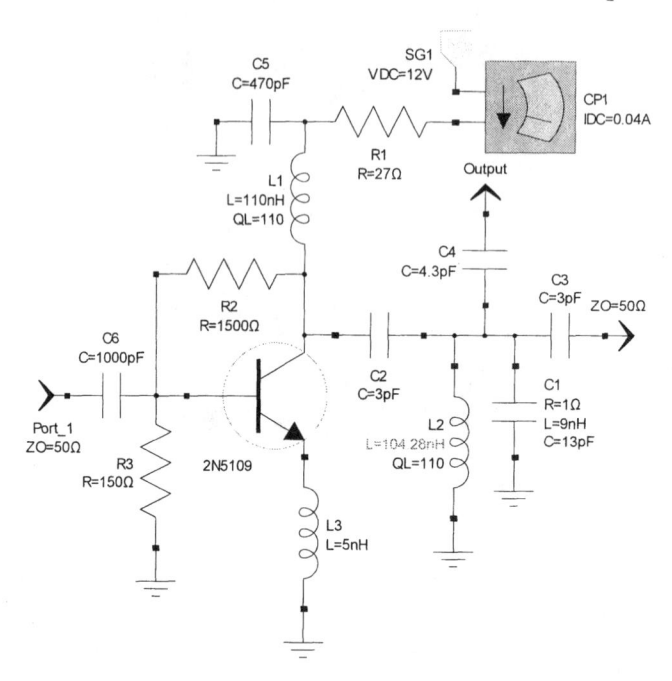

Figure 2.39 Open-loop cascade for a 100-MHz oscillator using a 2N5109 bipolar transistor.

Next, the cascade loop is closed using a 1.5 volt, 25 mA, incandescent lamp. The RF signal excites the lamp to a dim glow well below the rated 37.5 mW operating point. Nevertheless, excitation is sufficient to raise the filament temperature and resistance to stabilizes the output level at approximately 15.3 dBm. The measured lamp current and large-signal resistance versus DC voltage is given in Table 2.16. Even at an excitation voltage as low as 0.1 volts the lamp resistance is elevated.

Nonlinear Techniques 131

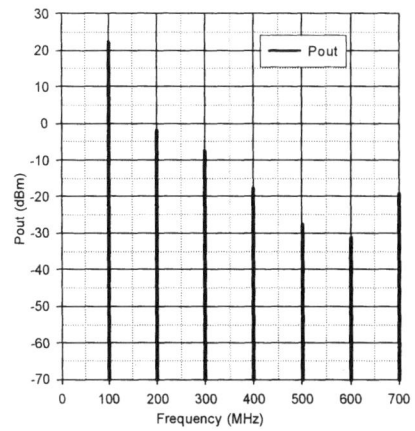

Figure 2.40 Output spectrum of the 100-MHz oscillator computed by HB simulation.

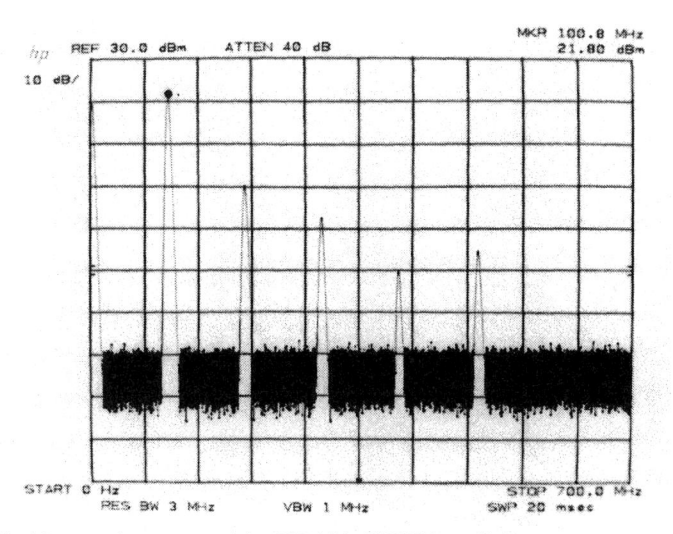

Figure 2.41 Measured spectrum of the 100-MHz 2N5109 oscillator.

The measured spectrum is shown in Fig. 2.42. Notice that the second harmonic is reduced to -27 dBc.

Finally, a 47-ohm resistor is placed in series with the 1.5-volt, 25-mA lamp closing the cascade loop. The increased resistance decreases the loop gain and further reduces the output level and harmonics. The lamp filament is still sufficiently heated to stabilize the output level. The results are given in Fig. 2.43. The output level is 7.5 dBm and the second harmonic is -40 dBc and higher harmonics are below the spectrum analyzer noise level in 3-MHz bandwidth. At this point, the oscillator is nearly linear.

Table 2.16 Measured current of the 1.5-volt, 25-mA incandescent lamp versus DC voltage and the computed resistance

Voltage (volts, DC)	Current (mA)	Resistance (ohms)
0.1	8.0	12.5
0.2	13.2	15.2
0.4	16.6	24.1
0.6	17.9	33.5
0.8	19.3	41.4
1.0	20.9	47.8
1.2	22.6	53.1
1.5	25.1	59.8
1.8	27.3	65.9

In theory, the oscillator could be operated with sufficient padding in the closed loop to operate the oscillator at this level. However, the required gain margin would be very low. Circuit tolerances and supply, temperature, and device-to-device variation could result in a negative gain margin and no oscillation.

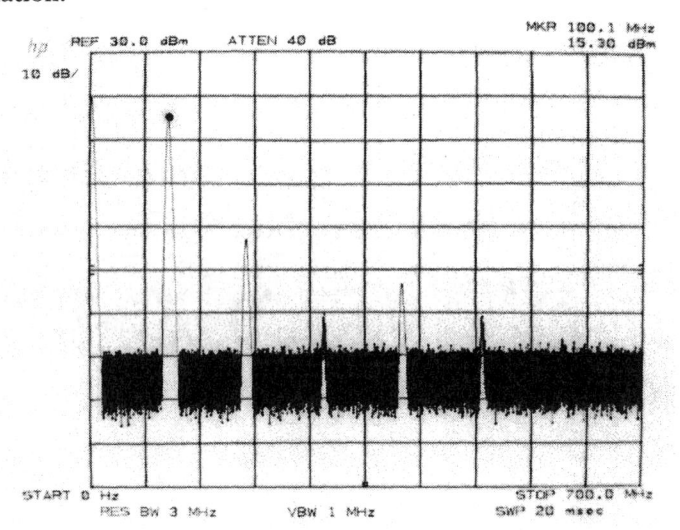

Figure 2.42 Measured spectrum of the 100-MHz oscillator with an incandescent lamp closing the loop.

2.9 Supply Pushing

The sensitivity of oscillator parameters to the supply voltage is referred to as pushing. A change of the supply voltage changes the quiescent bias and the steady-state operating point. In general, an increase in the supply voltage increases the quiescent bias in the device and the output power of the oscillator. However, a further increase in the supply voltage may bias the device at a higher current and dissipation that may not be optimum and the output power may decline. Of particular concern is the operating frequency as a function of the supply voltage. If supply voltage variation causes excessive frequency shift then supply noise modulates the oscillator frequency, thus potentially degrading the phase-noise performance of the oscillator. This is considered further in Chapter 4.

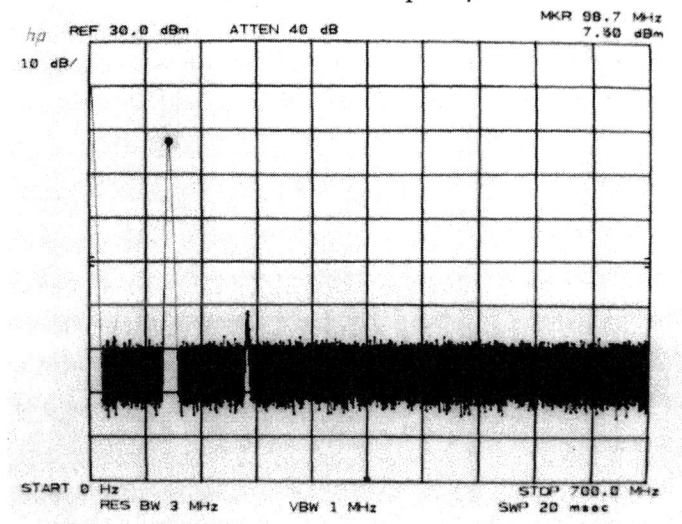

Figure 2.43 Measured output spectrum of the 100-MHz oscillator with a 47-ohm resistor in series with an incandescent lamp closing the loop.

Pushing, K_p, is defined as the ratio of the frequency shift to a supply DC voltage shift.

$$K_p = \frac{\Delta f_{osc}}{\Delta V_{DC}} \qquad (74)$$

The fractional pushing is defined as the ratio of the frequency shift with a 1-volt supply delta to the operating frequency, measured about the nominal operating supply voltage.

$$k_p = \frac{\Delta f_{osc}}{f_{osc}}\bigg|_{\Delta V_{DC}=1v} \qquad (75)$$

A typical fractional pushing specification is

$$k_{p(typical)} = \frac{0.02}{Q_L} \qquad (76)$$

where Q_L is the open-loop cascade loaded Q.

Fig. 2.44 shows the HB simulated output power and frequency of the 100-MHz oscillator in Fig. 2.39 versus the power supply voltage from 8 volts to 15 volts in 1-volt steps. For this oscillator k_p is approximately 0.0005, or 500 ppm. With a loaded Q of 18, the typical fractional pushing estimated by Eq. 79 is 0.0011. This 100-MHz oscillator has good pushing performance. The typical fractional pushing predicted by Eq. 79 is only a guideline, but if an oscillator has significantly worse pushing performance the design should be evaluated.

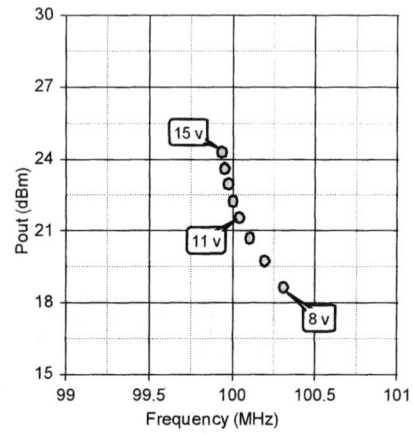

Figure 2.44 Output power and frequency of the 100-MHz oscillator versus the supply voltage.

2.10 Spurious Modes

The ideal oscillator spectrum contains only a component of zero width at the desired output frequency. Harmonics of the fundamental are expected and in many applications are not troublesome. Spectrum widening, noise sidebands, and discrete sideband components from supply ripple and external interferers are also expected and must be managed to meet specifications. Noise and discrete components in the output spectrum are considered in Chapter 4. Other spectrum components such as sub-harmonics of the fundamental and spurious oscillations should not be tolerated. Tips for avoiding these later components are discussed in this section.

2.10.1 Unstable Amplifiers

An oscillator designed using the open-loop cascade method must utilize a stable amplifier. The same techniques used to design an amplifier that is stable are used to design an oscillator sustaining stage. Numerous texts are available that cover this subject in detail [8,9]. Resistive or transformer negative feedback generally improves device stability. A small amount of inductance in series with the emitter of common-emitter amplifiers is often helpful.

A successful strategy involves designing the sustaining stage amplifier and plotting stability circles. If the amplifier is not unconditionally stable, the amplifier is not terminated in impedance spaces that are unstable. This analysis must be conducted at all frequencies where the device has gain and not just near the oscillation frequency.

Resistive shunt and series feedback as described in Section 1.5.6 not only helps stabilize the amplifier, it reduces the high gain present in common-emitter bipolar amplifiers at low frequency.

Common-collector and common-base topologies are generally less stable than the common-emitter topology. Nevertheless, the CC and CB topologies are used in popular Colpitts oscillators. The CB topology is destabilized by inductance in the base path to ground. The CC topology is destabilized by a capacitive load at the emitter. Unfortunately the CC Colpitts oscillator places a capacitive load at the emitter. Both of these topologies are stabilized by adding resistance in series with the base. Resistance values from 15 to 100 ohms is often effective.

2.10.2 Multiple Phase Zero Crossings

The open-loop amplifier-resonator cascade should have only one phase zero crossing (ϕ_o) with gain margin. Again, this analysis is performed not just near the frequency of oscillation but rather broadband at all frequencies where the gain potentially exceeds 0 dB. An oscillator will generally oscillate at the lowest frequency ϕ_o with gain. However, good design practice establishes only one ϕ_o.

2.10.3 Bias Relaxation Modes

When oscillation commences, the current drawn by the circuit may increase. The voltage provided by the power supply should not shift when this occurs. The onset of oscillation may require a time period far in excess of the RF period, so the time constant associated with a potential supply voltage shift may be relatively low. A resulting oscillation mode may both FM and AM modulate the oscillator. This oscillation mode is often at ultrasonic or audio frequencies. When the oscillator is used in a communication system an audio buzz or puttering may result. This is referred to as motor boating.

Similarly, relaxation modes can exist in biasing circuitry, particularly with bypassed resistive bias networks. Consider the negative-conductance CB 722-MHz oscillator in Fig. 2.45.

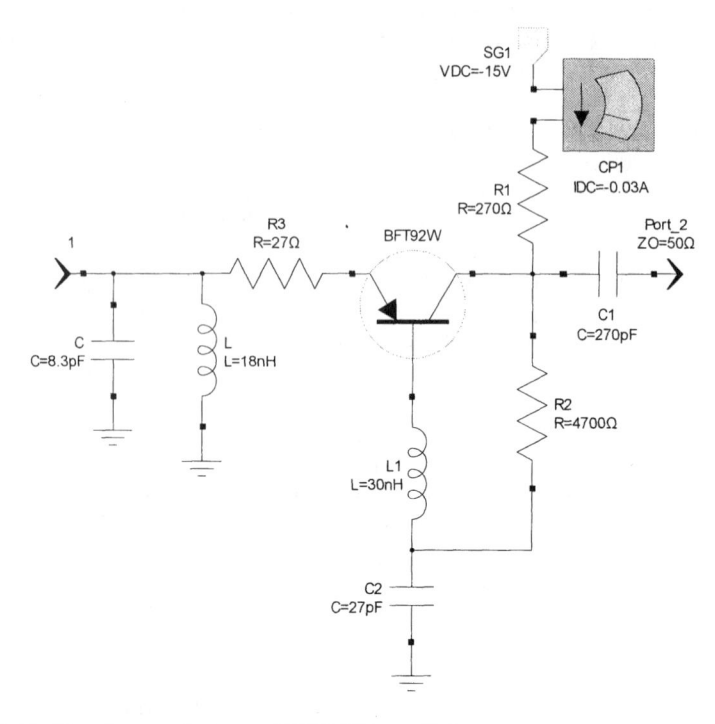

Figure 2.45 Negative-conductance CB 722 MHz oscillator.

C_2 was initially 1000 pF. A strong bias relaxation oscillation mode existed, resulting in the spectrum given in Fig. 2.46. This was resolved by decreasing C_2 to 27 pF.

2.10.4 Parametric Modes

The collector-base junction of bipolar transistors possess capacitance that is a nonlinear function of the collector-base voltage. A high RF voltage present at the collector "pumps" this nonlinear capacitance and produces negative-resistance at subharmonics of the pumping (oscillation) frequency. These are referred to as parametric modes and are the desired operating mode used in parametric amplifiers, popular low-noise microwave amplifiers employed in the 1960s before the development of low-noise microwave transistors.

Nonlinear Techniques 137

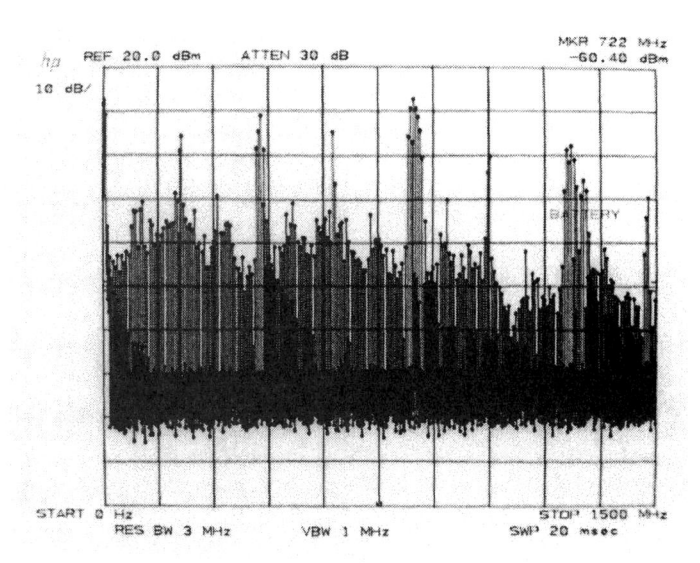

Figure 2.46 Spectrum of the negative-conductance oscillator in Fig. 2.45 with relaxation-mode spurious oscillations.

Fig 2.47 is the output spectrum with subharmonic components of the negative-resistance oscillator in Fig. 2.45. The maker identifies the desired 722-MHz output component of the spectrum at a power level of 8.8 dBm. The normal second harmonic at 1444 MHz on the far right is at approximately 0 dBm. In this circuit, subharmonics exist at ½ the desired operating frequency producing spectral components at 361 and 1083 MHz. Subharmonics can be produced at 1/2, 1/3, and 1/4, and so forth, of the fundamental frequency.

Parametric mode spurious oscillations are often the result of excessive gain margin, excessive negative resistance, or excessive negative conductance. The negative conductance oscillator is particularly susceptible to parametric modes, but they can exist in any oscillator with excessive gain. The spectrum given in Fig. 2.47 was produced in the oscillator shown in Fig. 2.45 with R_3 absent. Including R_3 at 27 ohms stopped this parametric mode.

2.10.5 Multiple Resonance Modes

Analogous to multiple phase zero crossings that threaten the stability of an oscillator are multiple resonances in negative-resistance or negative conductance oscillators. Consider the 1-GHz negative-resistance oscillator in Fig. 2.48.

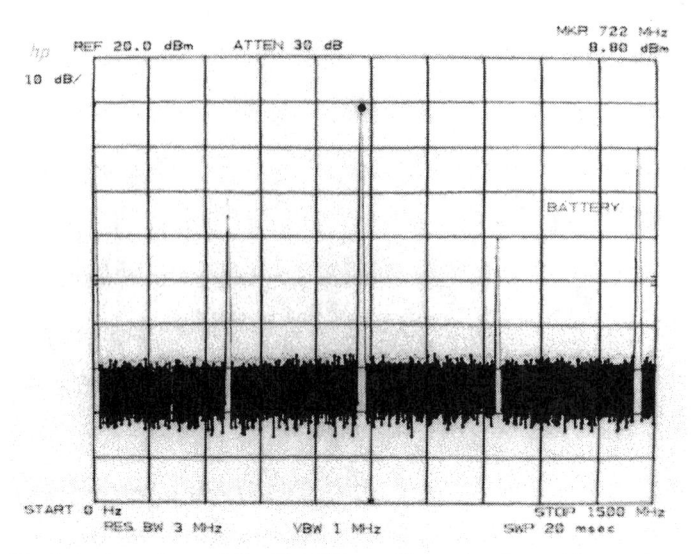

Figure 2.47 Output spectrum of the 722-MHz oscillator showing subharmonic components at ½ the operating frequency.

Resistor R_2 and R_3 form a voltage divider to provide base bias. This bias voltage is bypassed with capacitor C_2 and then coupled to the base using the inductive choke L_1. Seemingly innocent, this introduces a second series resonant circuit to ground. Proper design insures that the resistance looking into the base of the *MRF5812* is not negative at the resonant frequency of L_1, C_2, and the effective input capacitance of the *MRF5812*. An even better design practice is to eliminate L_1 and C_2 entirely as depicted in the negative-resistance oscillator of Fig. 1.47. I have a passion for the elimination of unnecessary components. Besides introducing unnecessary expense, they are potentially harmful to design quality. Many oscillator designs include unnecessary components.

2.10.6 Spurious Mode Summary

The following important design concepts will generally avoid spurious modes.
1) Design the sustaining stage amplifier using well-known techniques for amplifier design. Evaluate stability over a broadband frequency range.
2) Use resistive feedback to stabilize the amplifier and eliminate excessive gain at low frequencies.
3) Use resistance in series with the base of CC and CB amplifiers.
4) Ensure there is one phase zero crossing at the desired frequency.

5) Ensure there is only one resonance mode in negative-resistance and negative-conductance oscillators.
6) Design for moderate gain margin, negative resistance, or negative conductance. Do not design for maximum gain, negative resistance, or negative conductance as suggested in many publications.
7) Ensure the supply voltage is constant at currents drawn by the circuit both before and after commencement of oscillation.
8) Avoid bias network time constants at frequencies for which the amplifier has gain.

In extreme cases, if the cause of spurious modes can not be identified and rectified, it may be necessary to use an alternate device or oscillator topology.

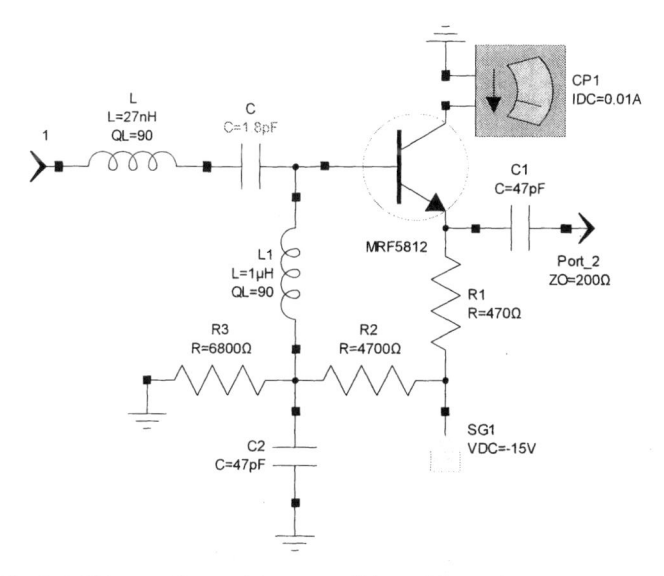

Figure 2.48 A 1-GHz negative-resistance oscillator with spurious resonance mode in the base circuit.

2.11 Ultimate Test

A benefit of careful open-loop cascade gain, phase, match, and loaded Q analysis is that the design objectives are verifiable open-loop using a vector network analyzer. Assessing the design quality when the loop is closed and oscillation is underway is difficult.

However, there is a simple but powerful qualitative test of design quality that should be performed on every oscillator. If you employ no simulation or analytical technique described in this book, you should at least utilize this test.

The oscillator is connected to a variable power supply and the output is viewed on a spectrum analyzer. The supply voltage is slowly increased from 0 volts and the following observations are made.

1) Oscillation should start at a voltage well below the desired operating voltage, perhaps 50% to 70% of the operating voltage. This suggests adequate gain margin. Some bias schemes may preclude this, but you should at least understand why.
2) As the voltage is increased, the output power generally increases because of increasing output capability of the sustaining stage. At voltages higher than the operating voltage the output power may decline due to thermal stress or device operation above the current for optimum F_t. If the output power declines at voltages below the operating voltage the design should be evaluated for device thermal stress or excessive device current.
3) The device should not fail at voltages moderately higher than the operating voltage. This insures breakdown voltages and dissipation limits are not exceeded.
4) Output power changes with supply voltage should be smooth with no sudden jumps. Sudden jumps are indicative of spurious modes.
5) The output frequency should change smoothly with the supply voltage. It may first rise or fall and then change direction, but sudden jumps in frequency are indicative of spurious modes.
6) The output frequency change should be as expected by the pushing specification. Larger than expected shifts are indicative of loaded Q lower than expected or high bias sensitivity to the supply voltage. Fig. 2.44 is representative of well-behaved power/frequency points versus supply voltage.
7) A final refinement involves testing the oscillator with variations in the load impedance using a sliding transmission-line tuner.

The above "ultimate" tests, temperature testing and finally "in system" testing provide a high degree of performance confidence.

References

[1] Agilent Technologies, GENESYS 2008.07 Documentation Set, www.agilent.com.

[2] M. Albulet, *RF Power Amplifiers*, Noble/Scitech Publishing, Raleigh, NC, 2001.

[3] "New Models Simulate RF Circuits," *Intusoft Newsletter*, November 1991, Gardena, CA.

[4] *RF Device Data*, Motorola Semiconductor Products, Phoenix, 1983.

[5] B. Lee and L. Dunleavy, "Understanding Base Biasing Influence on Large Signal Behavior of HBTs," *High Frequency Electronics*, May 2007, pp. 66, 68-70, 72-73.

[6] F.H. Raab, "Idealized Operation of the Class-E Tuned Power Amplifier," *IEEE Trans. Circuits and Systems*, December 1977, pp. 725-735.

[7] www.nxp.com.

[8] T. Grosh, *Small Signal Microwave Amplifier Design*, Noble/Scitech Publishing, Raleigh, NC, 1999, pp. 140-145.

[9] R. Gilmore and L. Besser, *Practical RF Circuit Design for Modern Wireless Systems*, Artech House, Norwood, MA, 2003, pp. 15-59.

3 Transient Techniques

Chapters 1 and 2 characterize the oscillator at steady-state. Harmonic balance characterization provides steady-state time-domain waveform through the Fourier transform. This chapter considers transient behavior, specifically starting. While certain time-domain issues, such as the steady-state RF voltage across a tuning varactor, could have been considered in Chapter 2, most waveforms and time-domain characterizations are considered in this chapter.

3.1 Introduction

In many applications, the oscillator remains on during system operation and starting characteristics are unimportant. For gated oscillators and oscillators employing very high loaded Q resonators, where starting is slow, the starting time may be important.

Randall and Hock, in their essential paper [1], solved not only the open-loop cascade match problem but they also offered an estimate for the rise time of an oscillator. Assuming the output voltage is an exponentially increasing sine wave, and recognizing the input and output voltages must be equal

$$e^{\alpha t}\sin(\omega_0 t) = G_0 e^{\alpha(t-\tau)}\sin(\omega_0(t-\tau)+\theta) \quad (77)$$

where α is the exponential rise rate, τ is the open-loop cascade time delay and G_o is the open-loop gain at ϕ_o. The assumption that the output is an exponentially increasing sine wave is confirmed impirically in Section 3.2. Again, with equal input and output voltage the delay and phase shift must be such that

$$\sin(\omega_0 t) = \sin(\omega_0(t-\tau)+\theta) \quad (78)$$

therefore

$$e^{\alpha t} = G_0 e^{\alpha(t-\tau)} \quad (79)$$

and so

$$\alpha = \frac{\ln(G_0)}{\tau} \quad (80)$$

Approximating the time delay, τ, with the group delay, t_d, then the rise rate is

$$\alpha = \frac{\ln(G_0)}{t_d} \qquad (81)$$

and from Eq. 37 relating the group delay to the open-loop cascade loaded Q

$$\alpha = \frac{\omega_0 \ln(G_0)}{2Q_L} \qquad (82)$$

The 10% to 90% rise time is then

$$t_r \approx \frac{t_d \ln 9}{\ln(G_0)} \qquad (83)$$

Again using Eq. 34 and with G_o in decibels, the rise time is

$$t_r \approx \frac{38.2 Q_L}{\omega_0 G_{0dB}} \qquad (84)$$

3.2 Starting Modes

Starting is a transient condition initiated by the application of power to the system. Alternatively, oscillation can be initiated by closing the loop of the cascade or grounding the resonator of a one-port oscillator. To understand starting, it is critical to realize that the system is operating in a linear mode as the signal level builds. Ultimately, the level increases until the system becomes nonlinear and the steady-state level is achieved. In other words, initially starting is a linear process and then near the end of the process, the system becomes nonlinear.

Traditional wisdom is that noise provides the initial signal. In fact, resonator band-limited noise can provide a starting signal. However, oscillator starting is also induced by transients associated with the application of power to the oscillator. In fact, with many oscillators this is the predominant starting mode. Both starting modes are considered in the following sections.

Fig. 3.1 diagrams a matched closed-loop system consisting of a noninverting amplifier cascaded with a lossless series $L\text{-}C$ resonator. The backward wave coupler is mathematical with the unreal properties of 0-dB insertion loss. It injects a test signal in the forward direction at port 1 and samples the test signal at port 2, again with no insertion loss.

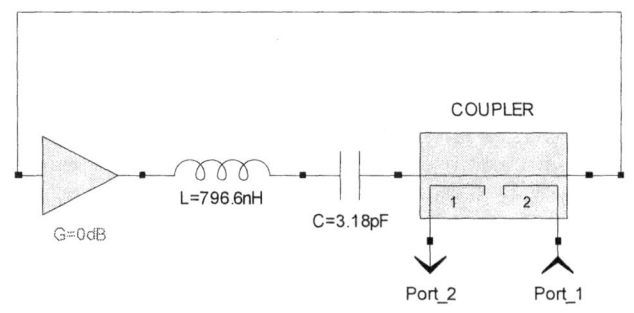

Figure 3.1 Closed-loop system with an amplifier, a series L-C resonator, and a coupler.

Fig. 3.2 shows the closed-loop gain versus the open-loop gain measured from port 1 to port 2. It is interesting to note that the closed-loop gain is greater than 0 dB even with an open-loop gain less than 0 dB. For example, the closed-loop gain is 4.66 dB with an open-loop gain of -4 dB.

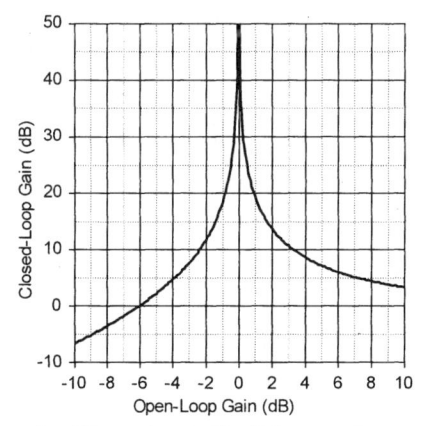

Figure 3.2 Closed-loop gain of the system in Fig. 3.1 versus the open-loop gain.

The classic closed-loop transfer function is

$$CL(s) = \frac{G(s)}{1 - G(s)H(s)} \qquad (85)$$

where s is the frequency variable $j\omega$, $G(s)$ is the amplifier transfer function and $H(s)$ is the feedback transfer function. With an inverting summing node at the input to the amplifier, as used in some references, the negative sign in the denominator is positive. $G(s)H(s)$ is the open-loop transfer function. When it is 1, or 0 dB, the denominator vanishes and the closed-loop gain is infinite. This is the oscillator steady-state operating point.

3.2.1 Noise Mode of Starting

From Eq. 85 and Fig. 3.2, it is clear that with 0-dB open-loop gain margin that the closed-loop gain is infinite. Goldberg [2] writes:
"Many designers assume that oscillation occurs because the device must be initially powered. This powering effect must create a transient....The major point here is that the oscillator is a device with immense gain (and Q) that continuously amplifies noise spectra, not a transient after effect..."

The difficulty with this view from the standpoint of oscillator starting is that at $t=0+$ the system is linear with excess gain. The open-loop gain does not stabilize to 0 dB, resulting in an immense closed-loop gain, until after the signal has built to nonlinear levels. For example, consider a cascade with an initial gain margin of 6 dB. Notice from Fig. 3.2 that in this case the closed-loop gain is only 6 dB. If in fact the gain was immense at $t=0+$, starting would be nearly instantaneous, clearly at odds with the facts.

Noise, however, does provide an initial signal, and starting proceeds as described by Randall and Hock. Deriving an equivalent RMS noise voltage from the effective input amplifier noise filtered by the resonator, the initial starting peak voltage is

$$v_{init} = \sqrt{\frac{2kTBF}{Z_0}} \qquad (86)$$

where k is Boltzmann's constant, T is the temperature in degrees Kelvin, B is the 3-dB bandwidth of the resonator in Hertz, and F is the amplifier noise factor. At 300 K, kT equals 4.14×10^{-21}. For example, in a 50-ohm system, with a resonator bandwidth of 10 MHz and an amplifier noise factor of 3 (4.77 dB), the effective starting voltage is 70.48 nV.

Since the final steady-state voltage is given approximately by

$$v_{ss} \approx v_{ini} e^{\alpha t_{start}} \qquad (87)$$

then

$$t_{start} = \frac{\ln(v_{ss}/v_{ini})}{\alpha} \qquad (88)$$

and using Eq. 85 for α then

$$t_{start} = \frac{2Q_L \ln(v_{ss}/v_{init})}{\omega_0 \ln G_0} \qquad (89)$$

For a 100-MHz oscillator with a loaded Q of 10 (B = 10 MHz), an initial small-signal voltage gain of 2 (6 dB), an effective input noise starting voltage of 70.48 nV and a final steady-state level at the input of 1-volt peak, the starting time is 756 nS. Noise is a statistical process and the start time is only known approximately.

Notice from Eq. 89 that the starting time increases with the loaded Q and decreases logarithmically with increasing voltage gain. Increased loaded Q decreases the resonator bandwidth, reducing the initial starting voltage and thereby further increasing the starting time.

3.2.2 Transient Mode of Starting

The application of power to the oscillator excites the resonator which then rings at its natural frequency. Since the resonator natural frequency approximately equals the oscillation frequency, this provides an initial starting signal. Fig. 3.3 depicts a series L-C resonator and load resistance R_L driven by a voltage step with source resistance R_S.

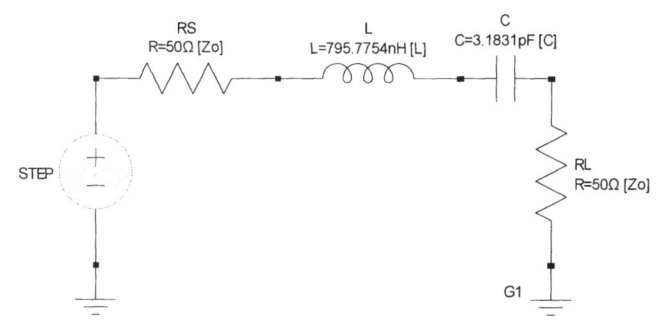

Figure 3.3 Series L-C resonator excited by a step voltage source.

With $Q_L >> 1$, the step generates a voltage across R_L given approximately [3] by

$$v_o(t) = \frac{V_{step} e^{-\kappa t}}{2Q_L \left(1 - \frac{\kappa^2}{\omega_o^2}\right)} \sin \omega_d t \qquad (90)$$

where

$$\kappa = \frac{R_{total}}{2L} \qquad (91)$$

$$\omega_0 = \frac{1}{\sqrt{LC}} \qquad (92)$$

$$\omega_d = \sqrt{\omega_o^2 - \kappa^2} \qquad (93)$$

The first peak of the damped sinewave is the initial starting voltage driving the input of the sustaining stage.

$$v_{init} = \frac{V_{step} e^{-\kappa t}}{2Q_L \left(1 - \frac{\kappa^2}{\omega_o^2}\right)} \qquad (94)$$

where

$$t = \frac{1}{4} \frac{2\pi}{\omega_d} \qquad (95)$$

This is a damped sinewave. From Eq. 94, with a loaded Q of 10 and a supply step of 9 volts, the initial peak starting voltage at 100 MHz is 0.38 volts, or nearly seven orders of magnitude greater than the noise starting voltage of the previous example.

In general, the supply voltage step does not couple directly to the resonator and the starting transient is significantly less than predicted by Eq. 94. This effect must be considered on a case-by-case basis for each oscillator. Once the initial transient voltage is estimated, Eq. 89 may be used to predict oscillator starting time.

3.2.3 Time Constant of the Supply Step

The previous analysis assumes a step rise time of 0. A finite rise time reduces the peak value and increases the time to the first peak, thus effectively decreasing the effective ring frequency. Both effects decrease the efficiency of starting by the step. When the step is filtered by an R-C network, the time constant is given by $\tau = RC$.

Shown as the dashed trace in Fig. 3.4 are the time domain responses of a supply voltage step filtered by an R-C network with a time contsant equal to the oscillation period. Shown as a solid trace is the response of a high loaded Q 100-Hz series resonator to this filtered step.

If the oscillation period is t, the voltage at $t/4$ versus τ/t is given in Table 3.1. The voltage is relative to 1 volt at $\tau/t=0$. Also given in Table 3.1 is the time required to reach 90% of the steady-state filtered supply voltage. The required time is given as a percentage of the oscillation time period.

From Fig. 3.4 and Table 3.1 it is clear that as the power supply rise time increases, the resonator response to the step produces a smaller initial starting voltage. However, even with supply time constants orders of magnitude greater than the oscillation time period, the step response voltage typically exceeds the noise voltage.

Power supplies are typically low-frequency circuits with rise times of 1 uS and longer. Even if the power supply rise time is several orders of magnitude longer than the oscillation period, the supply transient is sufficient to provide an initial starting voltage. If the supply has an extremely long rise time, noise provides the initial starting signal.

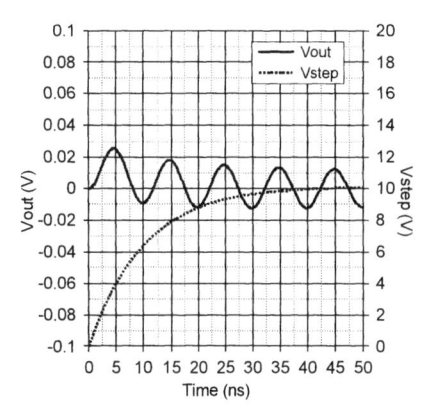

Figure 3.4 R-C filtered supply voltage step (dashed trace) and output ringing voltage response for a very high Q series resonator.

Table 3.1 The voltage $V_{0.25t}$ at $t/4$ and the time required to reach 90% of the steady-state step voltage versus the supply R-C time constant

τ/t	$V_{0.25t}$ (rel. to $\tau/t = 0$)	$t_{90\%}$ % of t
0	1.000	0%
0.01	0.997	2.3%
0.1	0.756	23%
0.3	0.401	68%
1	0.146	226%
3	0.052	680%
10	0.016	2260%

If the supply rise time is less than 10% of the RF period, starting is unaffected by the supply rise time. This is unlikely to be the case except for low-frequency oscillators. Essentially, then the starting time of a typical oscillator is the power supply rise time plus a small additional time associated with signal buildup. A potential exception is crystal oscillators which typically have a very high Q and may have a starting time significantly longer than the supply rise time. An example is covered later in this chapter.

3.3 Starting Basic Example

Fig. 3.5 diagrams the cascade of a voltage controlled voltage source (VCVS), a series 100-MHz L-C resonator and a resistive splitter to couple output power. Each component of the cascade has a phase shift of 0° at 100 MHz for a total cascade transmission phase shift of 0°. R_3 couples the power supply to the system. Each component in the cascade is matched to 50 ohms since the Thevenin equivalent resistance of R_2 and R_3 is 50 ohms. To form an oscillator, ports 1 and 2 are connected and power is extracted at port 3.

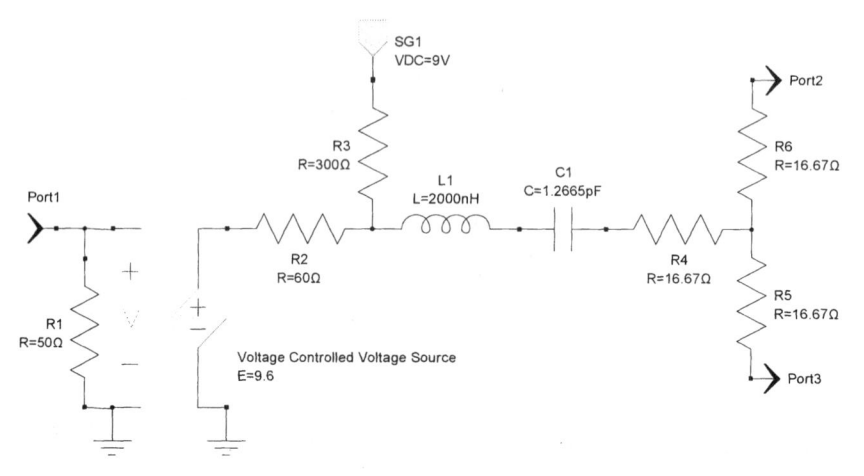

Figure 3.5 Open-loop cascade of a voltage controlled current source driving a series *L-C* resonator with a resistive splitter for coupling output power.

The voltage gain of the VCVS, R_2/R_3, the VCVS driving the resonator, the resonator and the splitter are 9.6, 0.833, 0.5, 1.0, and 0.5, respectively, for a total voltage gain of 2.0 or 6.02 dB. From Eq. 33, the loaded Q of the resonator, with a 2000-nH inductor and terminated at each end in 50 ohms, is 12.57. The gain, phase, and loaded Q are confirmed by a linear simulation of this cascade with results given in Fig. 3.6.

The loop is closed and 9 volts is applied at the supply at $t=0$. A time-step simulation of the closed-loop system is given in Fig. 3.7. The dashed response is the absolute value of the resulting output voltage at port 3. The absolute value is plotted so that the waveform can be plotted using a log vertical scale. The absolute function full wave rectifies the waveform producing twice the number of peaks and halving the period. Notice that the output voltage rises linearly on a log vertical scale, confirming that the output waveform rises exponentially.

The output waveform would rise indefinitely if the VCVS were capable of supplying the necessary output voltage. Of course this is not feasible. The solid trace in Fig. 3.7 is with exponential limiting of the VCVS output with increasing input voltage. The nonlinear parameters of the VCVS were adjusted to limit the maximum output voltage at 1.0 volts.

Using Eq. 84, the estimated 10-90 rise time is 126.9 nS, a useful estimate of the starting time predicted by the time-step simulation. Starting is investigated further after the time-step simulation process is described.

Transient Techniques 151

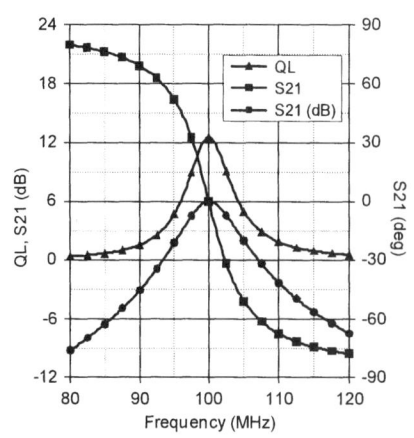

Figure 3.6 Open-loop gain (round symbols), phase (square symbols) and loaded Q (triangular symbols) of the cascade in Fig. 3.1.

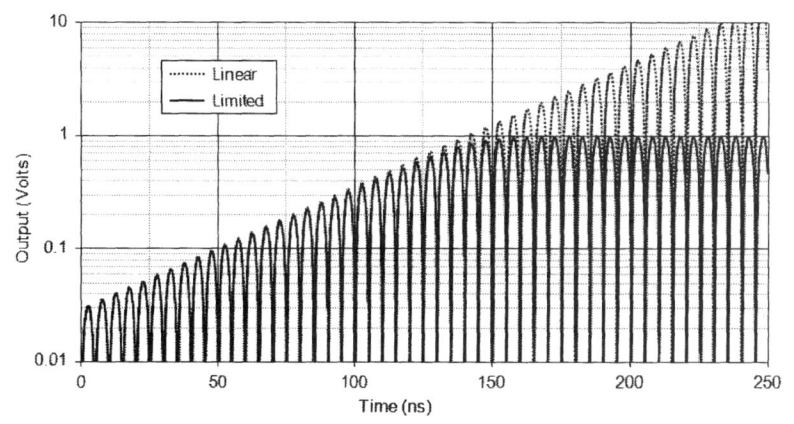

Figure 3.7 Absolute value of the output voltage versus time for the cascade in Fig. 3.1 with a linear sustaining source (dashed trace) and with a sustaining source with limited output capability (solid trace).

3.4 Simulation Techniques

Realistic evaluation of starting in practical circuits involves device models, parasitics and a plethora of detailed circuit topology issues. Accurate design requires nonlinear, time-variant, analysis. The only realistic approach is digital computer simulation with good hardware and sophisticated software.

3.4.1 SPICE

Since the 1970s, SPICE has been essentially synonymous with time-domain circuit simulation. SPICE1 (Simulation Program with Integrated Circuit Emphasis) was developed at the Electronics Research Laboratory of the University of California and released in 1973. SPICE2 released in 1975 included additional models, solved some practical issues and popularized this program. It was available in the public domain and a number of commercial variations have been released.

SPICE2 uses a variable timestep transient analysis using either trapezoidal or Gear integration of equations formulated with a modified nodal analysis. If SPICE is set up with supply voltages transitioning from 0 volts at turn on, the simulation process replicates what happens when an oscillator is turned on. The resulting simulation predicts both starting characteristics and the final output time-domain waveform. Kundert [4] provides additional information on the theory and use of SPICE-based simulators.

3.4.2 Cayenne

The time-domain simulator used for example simulations in this book is Cayenne from Agilent Technologies [5]. Cayenne is not derived from SPICE code but the techniques used are advancements of similar concepts. It is a time-stepped transient simulator.

3.4.2.1 Cayenne Basic Operation

The following description is abbreviated from Cayenne documentation. The simulation flow follows.

1) Solve the circuit at $t=0$. If Skip Bias is selected, the initial solution at all nodes is zero volts. Otherwise, the initial solution is the DC solution.
2) Advance the time by an amount determined by the simulator.
3) Solve the circuit at the new time step, replacing charge-based elements (like capacitors) with equivalent current sources and resistances. If the circuit cannot be solved or the error exceeds specified limits, then return to the last good time point and go back to step 2 using a different step size.
4) Save required output data to the dataset.
5) Repeat steps 2, 3, and 4 until the stop time has passed.

The time points that are output to the dataset are not the same time points calculated by the simulation. To ensure accuracy, Cayenne simulates more time points than are saved, using a number of rules to set the internal simulation time steps. For details refer to the Cayenne documentation.

Transient Techniques 153

3.4.2.2 Cayenne Numerical Precision

Cayenne is different from traditional SPICE simulation in that it tracks and controls the current error at each node. SPICE tracks charge, which is not normally observed by the user. This difference is important when a circuit has a capacitance that is large relative to the time step. For a small time step and large capacitance, the voltage differential during the step is undetectable given the limits of double precision numbers in digital computers and convergence is compromised. Cayenne detects this condition and increases the time step.

SPICE does not directly track current error and does not warn the user if the current is not within a reasonable tolerance. For more details on this subject, see Kundert [4].

When Cayenne increases the time step, the increased time step may cause accuracy errors. Cayenne displays an error message and the user has the following options.

1) Increase the current tolerance. This prevents the error and forces Cayenne to give an answer similar to SPICE.
2) Increase the minimum and maximum time-step values. This may cause a decrease in the precision of the current.
3) Change the time-step method to fixed, thus always using the max step. If very small time steps were specified for the output, this may cause an excessive quantity of output data points.
4) Reduce the value of the capacitors or inductors. Using 1 uF capacitors for coupling or bypass in a microwave circuit is often pointless and undesirable.

3.4.2.3 Cayenne Frequency-Dependant Models

Frequency-dependent models can cause difficulties for all time domain simulators. Traditional SPICE simulation does not allow use of frequency dependencies. Later versions and derivations of SPICE added the capability to simulate s-domain devices defined by rational polynomials, or they use other methods for dealing with frequency domain models. While traditional models work with simple structures like filters or for first-order frequency roll-off effects, they are ineffective with more complicated models typically found in RF and microwave simulation, such as dispersive and coupled transmission lines, measured S-parameter data, lumped elements with frequency dependent unloaded Q, and elements with frequency dependent loss and skin effect.

Cayenne has several different strategies for simulating frequency dependent models. The two basic methods are approximate models and convolution. For each component Cayenne determines whether the model is nonlinear, frequency-dependent, and/or time-dependent.

Frequency-dependent models recognized by Harbec include the following types.

1) Models that use the FREQ variable in equations that define their parameters.
2) Internal models that use frequency in their definitions, such as transmission lines and INDQ/CAPQ, inductor, and capacitor models with frequency-dependent Q.
3) Aggregate or user models that include either of the above.
4) Frequency domain sources, such as IAC and PAC, are not considered frequency-dependent models. These sources all have direct time domain equivalents such as $v=\sin(\omega t)$.
5) Charge-dependent elements, such as ideal inductors and capacitors, are not considered frequency-dependent. Cayenne directly simulates models with an impedance or admittance of the form $R + j\omega X$, where R and X are constant and ω is the radian frequency.

Time-dependent models recognized by Harbec include the following types.

1) Models that use the TIME variable in equations that define their parameters.
2) Internal models that use time or delay in their definitions. Currently, the only models in this category are nonlinear transistor or Verilog-A models that contain delay.
3) Aggregate or user models that include either of the above.
4) Time domain sources, such as IPWL and VPULSE, are not considered time-dependent models. These sources all have direct frequency domain equivalents using Fourier transforms.

Nonlinear models include elements like diodes and nonlinear transistors such as Gummel-Poon and BSIM. If a model is frequency-dependent, but is also either nonlinear or time dependent, Cayenne gives a warning and ignores the frequency dependency.

Otherwise, if a model is frequency-dependent, Cayenne checks the response as set in the Convolution tab in the Accuracy Testing section. By default, this does two things.

1) If the only frequency dependency is due to loss, and the "Always Use Constant Loss" box is checked, then Cayenne uses the impedance or admittance at the "Most Accurate Frequency," specified on the general tab, to calculate equivalents of the form $R + j\omega X$ for all matrix entries for that element. This allows elements like INDQ, CAPQ, and WIRE, which are based on RLOSS or GLOSS models, to avoid convolution.
2) Otherwise, the response for the element is calculated over a range of frequencies as specified in the Accuracy Testing section. For additional details, please refer to the Cayenne documentation.

Transient Techniques 155

3.4.2.4 Cayenne and SPICE Oscillator Simulation Setup

If a Cayenne or SPICE simulation is set up biased steady-state it will not start. Many solutions to this problem have been promulgated such as injecting a step, injecting a pulsed sinusoid, or changing the charge state of a capacitor. Cayenne in the General tab of the Transient Analysis Properties dialog includes an option to "Help Oscillators Start."

Such techniques are unnecessary if the simulation schematic supply voltage is set up as a step (or pulse) rather than as a DC voltage. This causes the simulation to start naturally as it does when power is initially applied to the circuit. Furthermore, this simulates the actual starting time of the oscillator. If the simulation includes the supply rise time and supply line filtering, the effects of these parameters on the starting time are simulated.

3.5 Second Starting Example

Fig. 3.8 shows the schematic and photograph of a 10-MHz Rhea type oscillator [6, 7]. The distinguishing characteristics of this oscillator are an inverting amplifier and a highpass resonator, here consisting of C_2, L_2 and C_3. The primary feature of this oscillator is economy since the resonator capacitors also serve as the bias decoupling elements. The frequency application range is limited because a rather large value of resonator inductor is required to achieve good loaded Q.

Although the resonator is highpass, at higher loaded Q the frequency response is pseudobandpass. This implementation of the oscillator extracts power through a large-valued coupling capacitor in series with the resonator inductor. Extracting power from the resonator results in excellent harmonic performance. The measured and harmonic-balance simulated output spectrum for this oscillator is given in Fig. 3.9.

Fig. 3.10 shows measured starting waveforms for the 10-MHz oscillator with supply pulses of 1.85 volts (top left), 2.9 volts (top right), and 4 volts (bottom). The oscillator is powered by a pulsed supply so the waveforms depict both starting (turn-on) and turn-off characteristics. The waveforms were captured using an Agilent 54855A oscilloscope using the TESTLINK module of GENESYS.

The vertical scale of the oscilloscope trace of the output at the top left is 15 mV/division, so the initial ringing voltage is approximately 50 mV peak. With a supply step of 1.85 volts, the open-loop gain of the sustaining stage is less than unity and the ringing voltage dies. The supply pulse trace is also shown with a vertical scale of 500 mV/division.

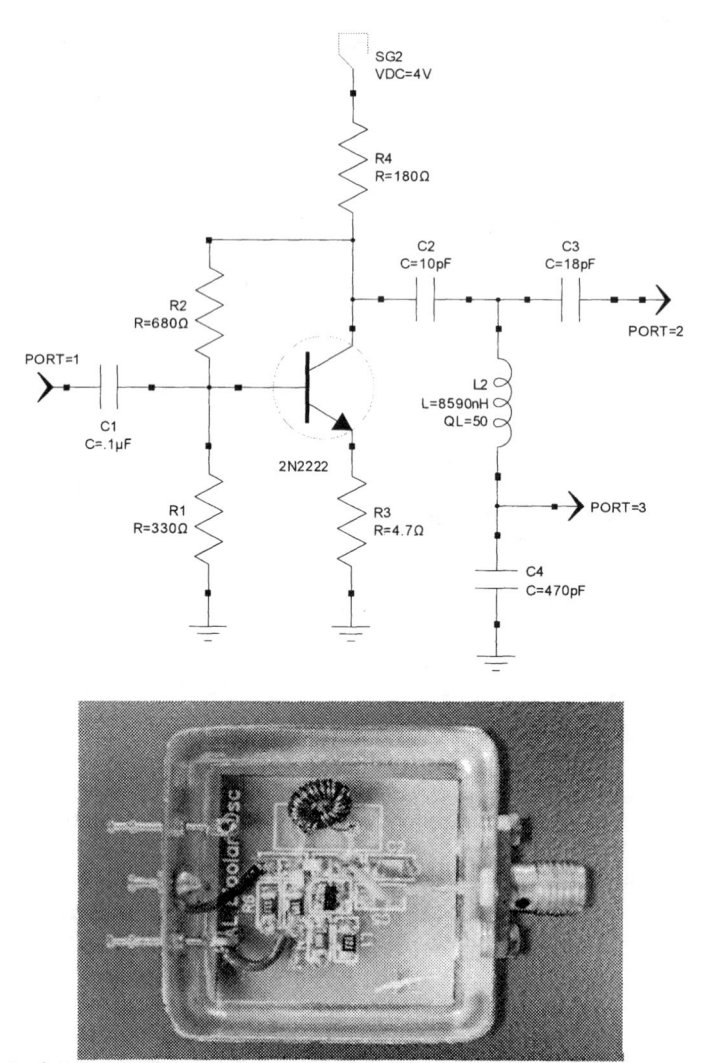

Figure 3.8 Schematic (top) and photograph (bottom) of a 10 MHz Rhea type oscillator.

The vertical scale of the oscilloscope trace of the output at the top right is 20 mV/division. With a supply pulse of 2.9 volts the open-loop gain of the sustaining stage is near unity and the ringing voltage remains near the initial starting voltage. This is an unrealistic oscillator operating point because any decrease in the gain caused by temperature or other changes would threaten oscillation.

Transient Techniques 157

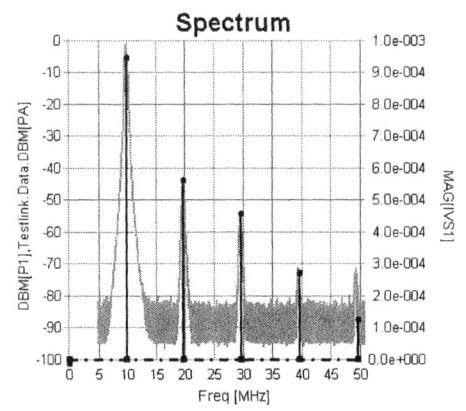

Figure 3.9 Measured and simulated output spectrum of the 10-MHz oscillator of Fig. 3.8.

The vertical scale of the oscilloscope trace of the output at the bottom is 50 mV/division. With a supply pulse of 4 volts the open-loop gain of the cascade is approximately 2.5 dB and the initial ringing waveform builds to the nonlinear steady-state operating level of 145-mV peak. The horizontal scale is 400 nS/division.

The waveforms in Fig. 3.10 were captured with 500 mS of persistence with an applied supply pulse period of 5 uS. The data thus depicts approximately 100,000 repeated and superimposed starting waveforms triggered off the leading edge of the pulse. With no supply bypass capacitors, the amplifier becomes active as soon as power is applied. Notice there is no observable starting jitter within the resolution of the figure. The oscillator not only starts reliably but it starts at the same phase each time. Clearly, starting in this oscillator is deterministic and not a noisy process.

3.6 Starting Case Study

Fig. 3.11 is the schematic of a typical 100-MHz Colpitts oscillator using an MRF901 bipolar transistor. To perform a linear analysis, the loop is opened between the emitter and the resonator capacitive tap. The open-loop input and output matches are rather poor so the Randall/Hock equation is used to analyze the open-loop gain, approximately 8.5 dB at ϕ_0. The loaded Q is approximately 9. The loop is closed by connecting ports 1 and 2 and power is extracted at port 3.

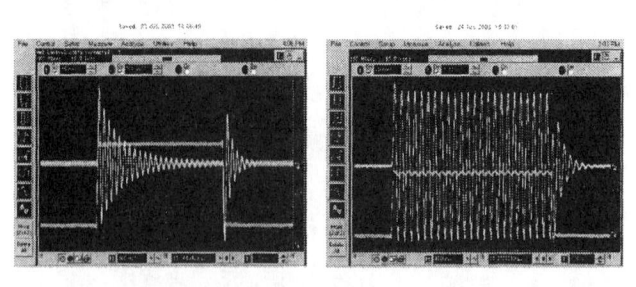

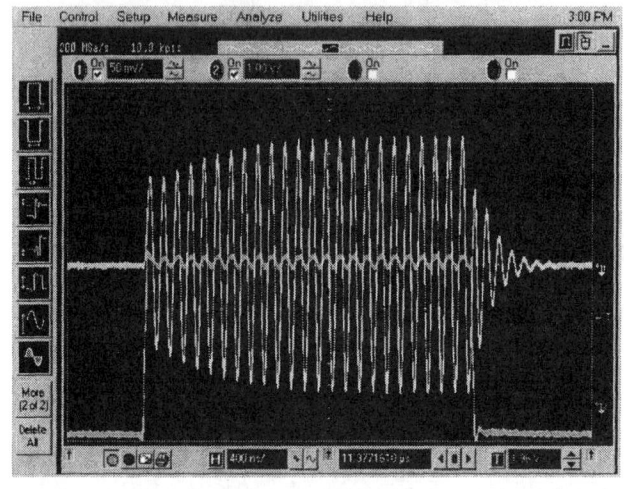

Figure 3.10 Starting and turn-off waveforms for the 10-MHz oscillator with supply pulses of 1.85 volts (top left), 2.9 volts (top right), and 4 volts (bottom).

Capacitor C_6 is not required in the oscillator; it prevents shorting the emitter bias voltage by the port 2 simulation termination. The resistors R_1 and R_2 bias the base at approximately 3.8 volts. After a base-emitter drop of 0.8 volts, the 3 volts across emitter resistor R_3 establishes the emitter current at 11 mA. Capacitor C_5 bypasses the base bias resistors and choke L_2 prevents shorting the resonator signal. Capacitor C_3 prevents the resonator inductor from shorting the base bias voltage. Resistor R_4 assists in stabilizing the CC amplifier.

Shown on the left in Fig. 3.12 is the output starting waveform computed by the Cayenne simulator. The initial starting time is approximately 3.7 uS. The output voltage continues to increase and has not stabilized by 10 uS.

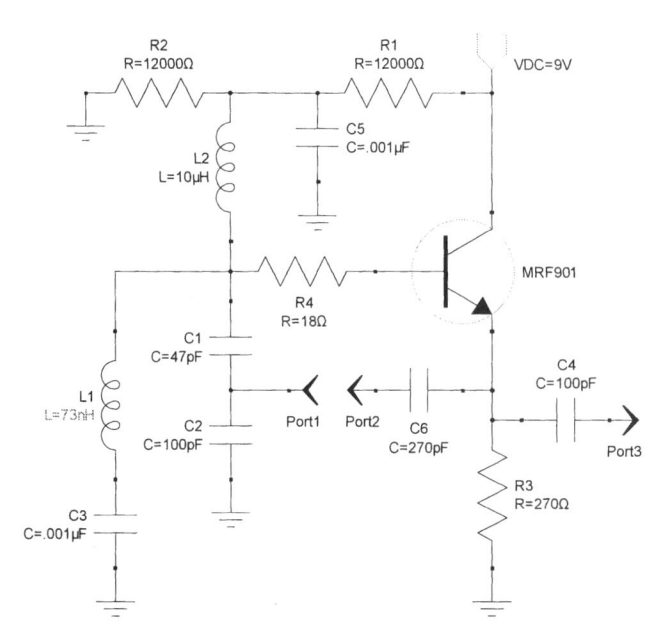

Figure 3.11 Typical 100-MHz oscillator used for the starting case study.

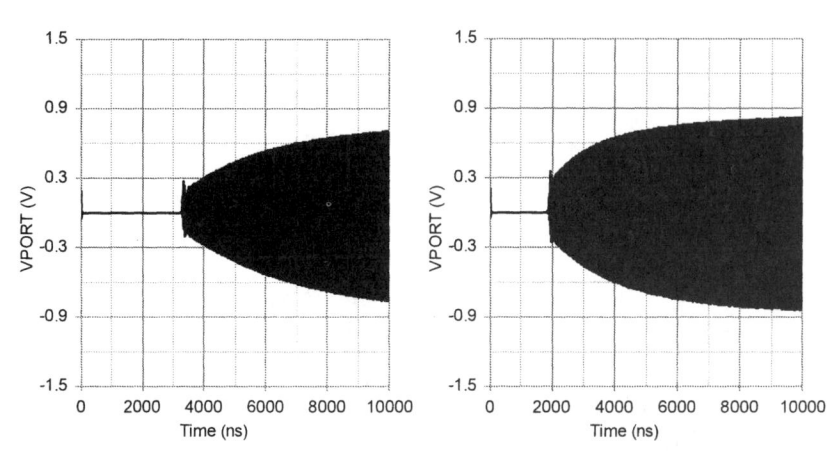

Figure 3.12 Output starting waveform of the 100-MHz case study oscillator given in Fig. 3.11 (left) and with C_5 and L_2 removed (right).

On the right in Fig. 3.12 the bypass capacitor C_5 has been removed. The common node of resistors R_1 and R_2 is now at high impedance and choke L_2 is not required. The starting waveform with these components removed is given on the right in Fig. 3.12. Notice the initial starting time is decreased to 2 uS and the waveform is stabilizing sooner. The time required to charge C_5

through resistors R_1 and R_2 prevents the amplifier from immediately biasing active. The base voltage and therefore the emitter current continue to rise as C_5 charges.

Next, consider capacitor C_3 that also must be charged through the bias resistors to allow the base voltage to rise. If the value of C_3 is significantly reduced, the value of L_1 must be increased to achieve resonance. For example, if the reactance of C_3 is set equal to the initial reactance of the inductor at ω_0 then the new value of inductor is approximately doubled. This modified Colpitts oscillator is referred to as a Clapp [8]. The Clapp has the advantage that the larger inductor value typically has a higher unloaded Q. The starting waveform of the Clapp is given in Fig. 3.13. Notice the change of the independent time axis to 500 nS. Steady-state operation is achieved in approximately 400 nS.

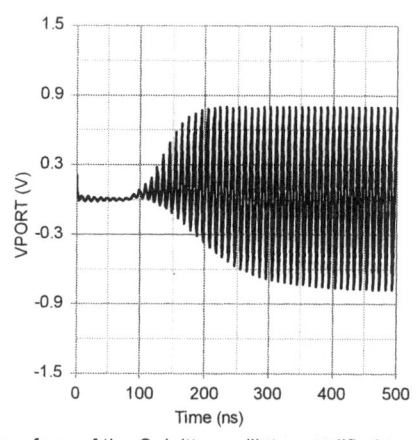

Figure 3.13 Starting waveform of the Colpitts oscillator modified to a Clapp.

3.7 Triggering

Next, consider the 100-MHz Clapp oscillator with capacitor C_2 returned to the supply rather than ground as shown in Fig. 3.14. The supply is at RF ground and this modification has no effect on steady-state operation. However, this modification applies the supply step directly to the resonator rather than through the high-impedance bias resistor R_1. Resonator ringing then provides a strong initial starting signal for the oscillator.

Transient Techniques 161

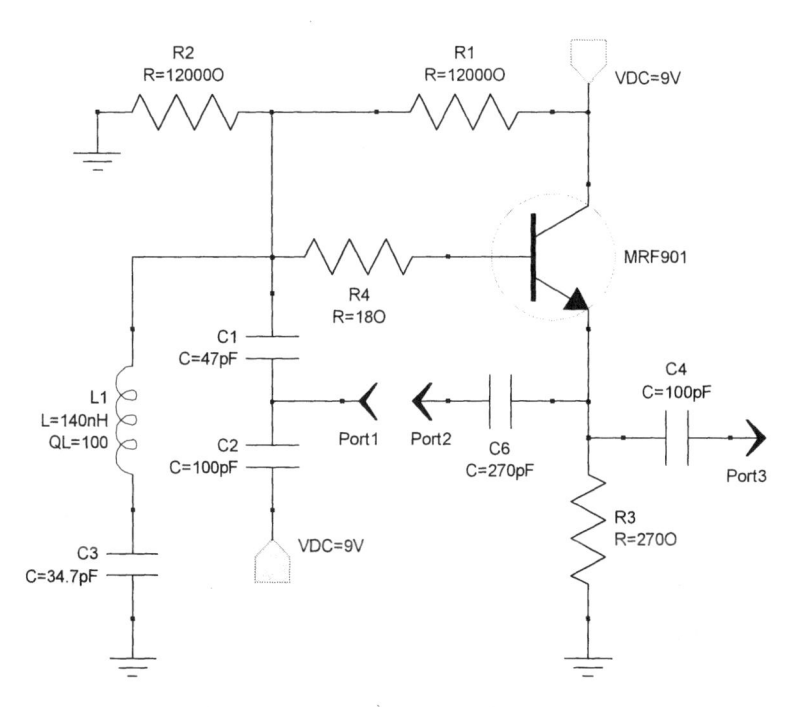

Figure 3.14 A 100-MHz Clapp oscillator with capacitor C_2 returned to the supply.

The resulting starting waveform is shown in Fig. 3.15. Starting is almost immediate, occurring within a few RF cycles. Steady-state operation is achieved in approximately 150 nS. This "triggering" effect is applicable for other topologies and has been applied to crystal oscillators [9].

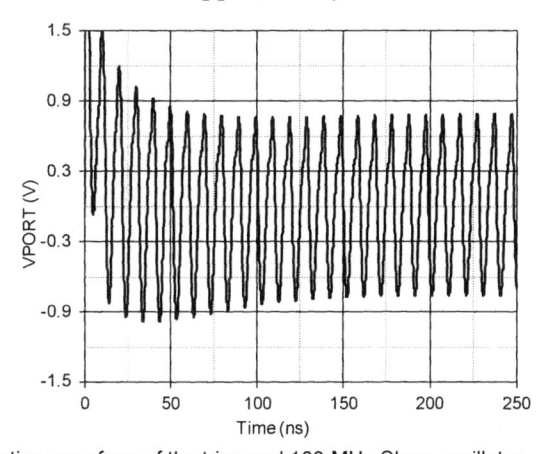

Figure 3.15 Starting waveform of the triggered 100-MHz Clapp oscillator.

3.8 Simulation Techniques for High Loaded Q

Eq. 82 reveals that the rate of signal build is inversely proportional to the loaded Q. Eq. 94 reveals that the magnitude of the initial signal is also inversely proportional to the loaded Q. The combined effect is that oscillators with high loaded Q require lengthy start times. The manifestation can be extreme for crystal oscillators. High-stability oscillators are covered in detail in Chapter 8. In summary, oscillators employing bulk-mode quartz resonators commonly have loaded Qs of 20,000, and 500,000 is feasible. Simulation of the start time for very high loaded Q oscillators requires a large number of time steps and large data storage for simulators such as SPICE and Cayenne.

Consider the JFET 10-MHz bulk-mode quartz crystal oscillator shown in Fig. 3.16. It is a Colpitts oscillator with C_1 and C_2 forming the capacitive tap. The operating frequency is just above the crystal series-resonant frequency, thus presenting inductance in parallel with C_1 and C_2. Bias resistor R_1 holds the gate near DC ground potential. Resistor R_2 biases the source positive with respect to the gate in order to stabilize the drain current below the device I_{dss}. The crystal resonator is returned to RF ground potential through the supply in order to trigger oscillation. The linear analysis ports are 1 and 2 and power is extracted from port 3. Capacitor C_3 is not required when the loop is closed to form the oscillator.

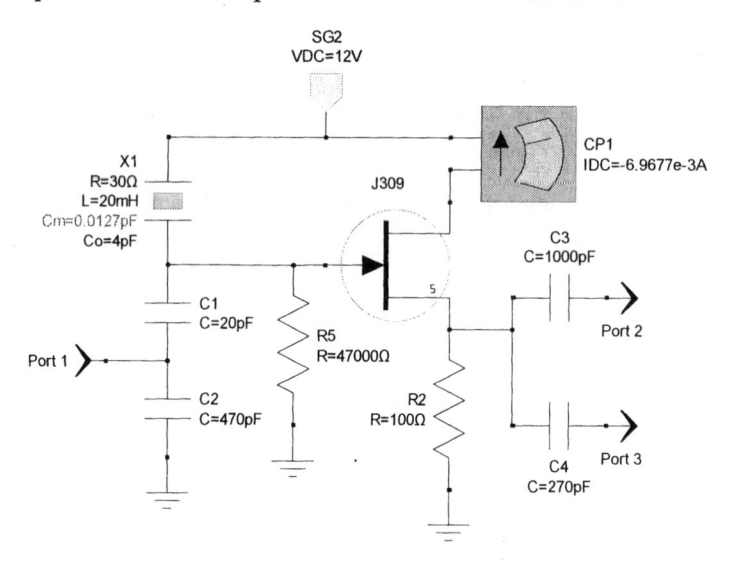

Figure 3.16 A JFET 10-MHz bulk-mode crystal oscillator.

The Randall/Hock corrected gain and transmission phase for the JFET 10 MHz crystal oscillator are shown in Fig. 3.17. The gain margin at ϕ_0 is

Transient Techniques 163

7.72 dB, a voltage gain of 2.43. The unloaded Q of the crystal resonator as determined from the series resistance of 30 ohms and motional inductance of 20 mH is 41,890. The loaded Q at ϕ_o is 25,450, approximately 61% of the unloaded Q. The peak loaded Q of 32,000 occurs slightly below ϕ_o. It is typically difficult aligning ϕ_o with the peak loaded Q in a Colpitts crystal oscillator. Extremely high device input impedance is required, which is why a JFET rather than a bipolar device is used in this example.

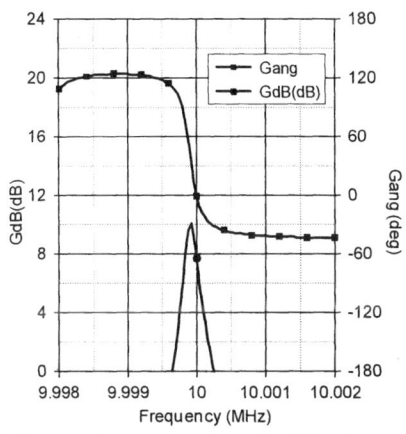

Figure 3.17 Randall/Hock open-loop cascade gain (circular symbols) and transmission phase (square symbols) for the JFET 10-MHz crystal oscillator.

A Cayenne time-step starting waveform simulation of this oscillator is shown in Fig. 3.18. A maximum time step of 4 nS is selected, 25 samples per 10-MHz period. With a stop time of 8 mS, 2 million steps are required. This requires storage of a large quantity of simulation data. The Cayenne option to store only port voltages and not node voltages is selected. Cayenne also supports storing only a portion of the output data. In this case, data only every 40 nS is stored. While 40-nS sampling is insufficient for accurate time-step simulation, it is adequate for displaying the output waveform.

With a voltage gain of 2.43 and loaded Q of 25,450, the rate of signal build predicted by Eq. 85 is 1096 rad/sec. The initial ring signal for this oscillator is approximately 1-mV peak and the steady-state signal is approximately 4-volts peak. From Eq. 89, the estimated starting time is 5.47 mS, in excellent agreement with the simulated waveform.

It is interesting to note that a state-of-the-art crystal 10-MHz oscillator with a loaded Q of 500,000 requires nearly a tenth of a second to start!

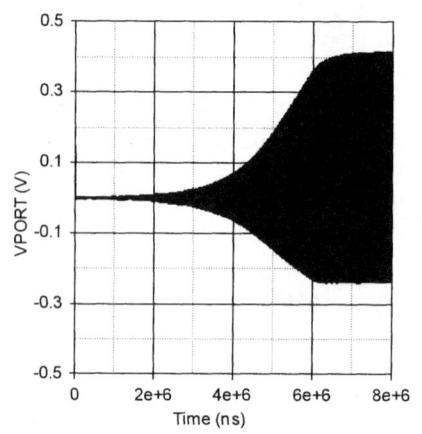

Figure 3.18 Starting waveform for the JFET 10-MHz crystal oscillator.

3.9 Steady-State Oscillator Waveforms

The steady-state oscillator spectrum may be computed using nonlinear harmonic balance techniques described in the previous chapter and then applying a built-in Fourier transform function to compute the voltage and current waveforms. However, in this chapter, displayed waveforms are computed using time-step simulation.

3.9.1 Clapp Oscillator Waveforms

Consider the Clapp oscillator in Fig. 3.19. This oscillator example is used to compare the time-domain waveforms of an oscillator at the available extraction nodes. This oscillator is similar to the oscillator used for the starting case study except resistor R_5 is added to transform the collector current to a voltage to extract power at the collector. This resistance has only a minor impact on the gain of the common-collector amplifier.

The output waveform on the left in Fig. 3.20 is extracted from the resonator using a small value of capacitance. This capacitor effectively increases the load resistance to reduce resonator loading. The capacitor value is adjusted to couple maximum output power into the 50-ohm load. This capacitor also loads the resonator with capacitance so the value of the Clapp capacitor C_3 is reduced to adjust the frequency back to 100 MHz.

Transient Techniques

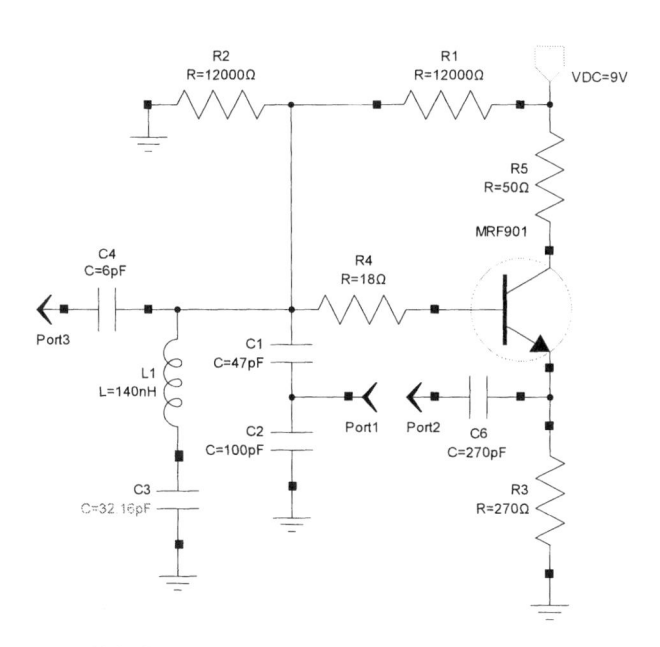

Figure 3.19 A 100-MHz Clapp oscillator similar to the starting case study oscillator but with power extracted at the resonator.

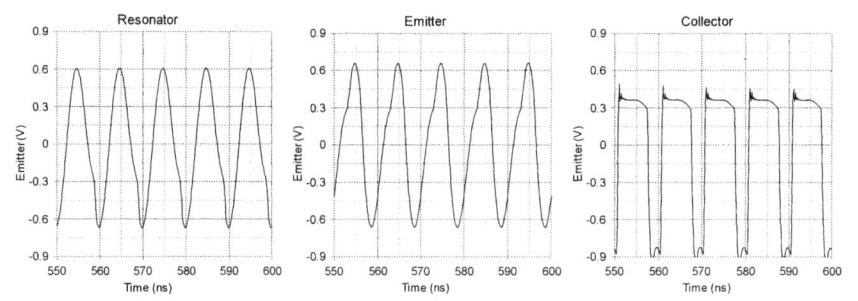

Figure 3.20 Steady-state time-domain waveforms of the 100-MHz Clapp oscillator with power extracted at the emitter (left), the resonator (center) and the collector (right).

Notice that the waveform is reasonably sinusoidal, indicating only moderate distortion and low harmonic levels. A smaller value of capacitor further improves the harmonic performance at the expense of output power. As mentioned in Chapter 1, coupling power from the resonator typically provides the lowest harmonic levels at the expense of stability and load-pulling performance.

Next, the resonator load is removed and power is extracted from the emitter with full coupling using a low-reactance, 100-pF capacitor, into a 50-ohm load. The time-domain waveform is given at the center in Fig. 3.20.

In this oscillator the waveform distortion is only slightly greater than with resonator coupling. The emitter of Colpitts and Clapp oscillators is coupled directly to the resonator through the capacitive tap. This is not the case with all oscillator topologies and the waveform distortion at the emitter is then significantly worse.

Next, the emitter load is removed and power is extracted from the collector with full coupling using a low-reactance, 100 pF capacitor, into a 50-ohm load. The time-domain waveform is shown on the right in Fig. 3.20. The waveform is nearly square going positive and is impulsive going negative. To accurately model the amplitude of the rising-edge ringing requires a simulation time-step significantly smaller than $1/F_t$ of the active device. For this simulation, the time step is 0.1 nS.

The collector current of CC oscillator topologies can be extremely impulsive. This is used to advantage for oscillators also serving as frequency multipliers. The effect is enhanced by using active devices with F_t as high as possible. High gain margin also enhances harmonics. Conversely, harmonic performance is improved by using devices with F_t only moderately higher than the fundamental oscillation frequency and using lower gain margins.

3.9.2 The Resonator Voltage

The desirable attributes of high loaded Q are described earlier. However, one disadvantage is that high loaded Q produces high circuit voltage and current. When varactors are used to tune a resonator, forward conduction may result. High power oscillators may also exceed the voltage breakdown or current limit of fixed capacitors.

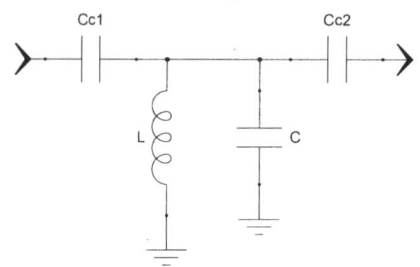

Figure 3.21 Top-C coupled parallel resonator.

Consider the top-C coupled parallel resonator depicted in Fig. 3.21. The total resistance effectively in parallel with the resonator is the parallel combination of the effective parallel resistance from the unloaded Q of the resonator, Q_R, and the two termination resistances stepped up by the transformer action of the coupling capacitors. The total resistance is

$$R_{total} = \left(\frac{1}{Q_R X_L} + \frac{2R_0}{R_0^2 + X_{cc}^2} \right)^{-1} \tag{96}$$

Then the peak voltage across the resonator is

$$V_{R,pk} = \frac{\sqrt{2} R_{total} V_{source-rms}}{\sqrt{R_0^2 + X_{cc}^2}} \tag{97}$$

The loaded Q is

$$Q_L = \frac{R_{total}}{X_{ind}} \tag{98}$$

which may be used to compute the insertion loss. Shown in Fig. 3.22 are the peak resonator voltage (solid trace), the loaded Q (dotted trace), and insertion loss (dashed trace) versus the reactance of the coupling capacitors. The assumed inductor reactance is 100 ohms, the unloaded Q is 200, the termination resistances are 50 ohms each, and the drive level is 2.82 volts rms (22 dBm). The maximum resonator voltage occurs with a loaded Q at 50% of the unloaded Q. The IL at this point is 6.02 dB. With a drive of only 2.83 volts rms, the peak resonator voltage is 28.3 volts!

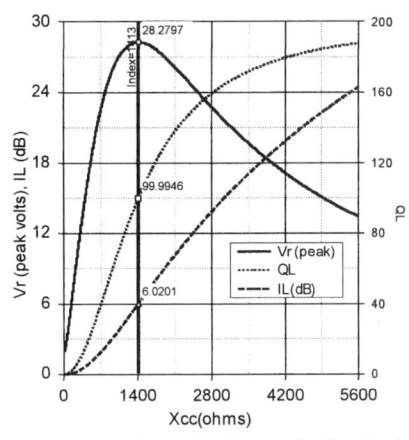

Figure 3.22 The peak resonator voltage (solid trace), the loaded Q (dotted trace), and insertion loss (dashed trace) versus the reactance of the coupling capacitors for the resonator in Fig. 3.16.

A series L-C resonator with shunt coupling elements at the input and output is the dual form of the parallel resonator. Interestingly, the voltage at the center node of this series L-C resonator is equal to the voltage across the coupled parallel resonator.

3.9.3 Varactor Coupling

For this resonator, a parallel tuning varactor is driven into forward conduction at a drive voltage of only 70 mV rms, or -10 dBm. For even a low power oscillator, the nonlinear device is the resonator rather than the sustaining stage. This is disastrous for any oscillator! The output power is far less than expected, the noise performance destroyed, and spurious modes are prevalent.

The resonator voltage of the parallel resonator is reduced by decreasing the inductive reactance. The voltage is reduced by the square root of the reactance. However, reducing the resonator voltage by decreasing the inductance then requires a large-value tuning capacitance. Varactors with large capacitance have lower unloaded Q.

Fig. 3.23 shows three methods for coupling a tuning varactor to a parallel resonator. Schematic A directly couples the varactor to the resonator using a low-reactance fixed capacitor. The resonator RF voltage can be high enough to cause varactor reverse breakdown. But the greater threat is forward conduction, which occurs at the PN junction forward potential. Coupling method A is generally used only with a combination of low oscillator power, low inductor reactance, and a high minimum tuning voltage. The RF voltage across the varactor should be kept at 1 volt peak-to-peak or lower.

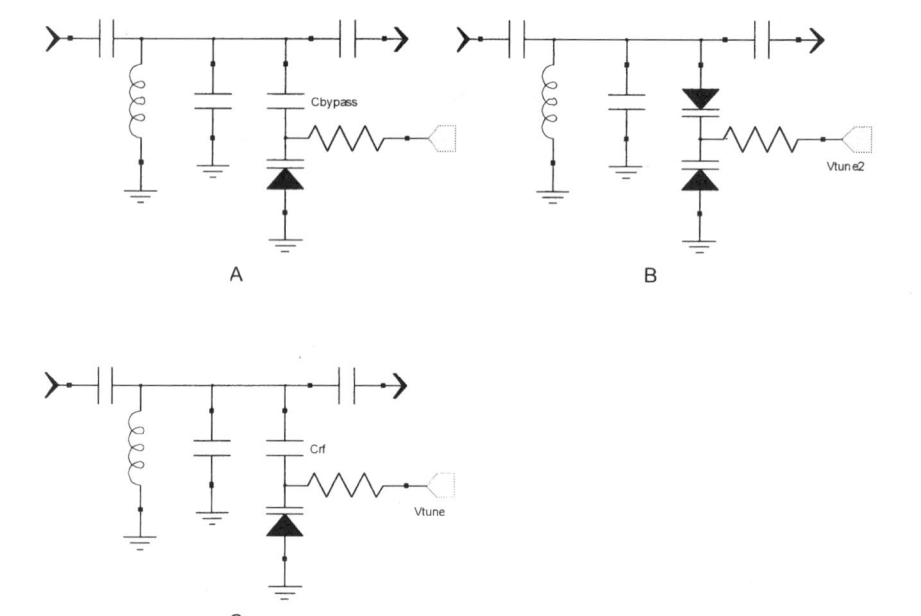

Figure 3.23 Three methods for coupling a tuning varactor to a parallel resonator.

The coupling method labeled B is much preferred. Two back-to-back varactors insure that one varactor is reverse-biased and high impedance at any RF voltage. An RF voltage up to 2 volts peak-to-peak is acceptable.

The method shown in Fig. 3.23(c) uses a small value of capacitance in series with the varactor. The RF voltage across the varactor is reduced by the factor

$$K_r = \frac{C_{rf}}{C_{rf} + C_{var}} \qquad (99)$$

The greatest threat for forward conduction occurs with tuning voltages close to 0 volts reverse bias. The value used in the equation for C_{var} is this lowest tuning voltage. The difficulty with method C is that it reduces the available tuning range because C_{rf} reduces the change of the effective total capacitance for a given change in C_{var}. The design objective is not the minimization of the varactor RF voltage because this restricts the loaded Q. The design objective is as high a loaded Q as possible while ensuring that the RF voltage on the varactor is kept below 1 volt peak-to-peak for a single diode and 2 volts peak-to-peak for back to back diodes.

Tuning varactors are typically PN junction devices and not Schottky diodes. At high frequency, PN junction varactors don't conduct as effectively as Schottky diodes, so the above-mentioned voltage limits are conservative. Nevertheless, control of the RF voltage across tuning varactors must be considered. Indications of varactor overdrive include lower than expected output power, oscillation failure at low tuning voltage, and erratic or sudden steps in frequency with tuning. It is imperative that these conditions be avoided.

3.10 Waveform Derived Output Spectrum

Harmonic-balance is the nonlinear tool of choice when steady-state spectrum simulation is desired and time-step tools such as SPICE or Cayenne are a natural choice when time-domain data is desired. However, because harmonic-balance analysis requires the careful setup of a starting circuit, and because of the proliferation of time-step simulators, there is some motivation for using time-step simulation to obtain the output waveform and then post-processing the output data by Fourier transform to obtain the output spectrum.

Consider the example 300-MHz negative-resistance oscillator shown in Fig. 3.24. A common-collector 2N5109 NPN bipolar transistor is biased at approximately 13 mA using a negative supply. Output is taken at the emitter through a 2-pF coupling capacitor. An initial linear analysis is performed by examining at port 1 the input resistance and reactance through the series resonator. The inductor is tuned to set the input reactance at 0 ohms at 300 MHz.

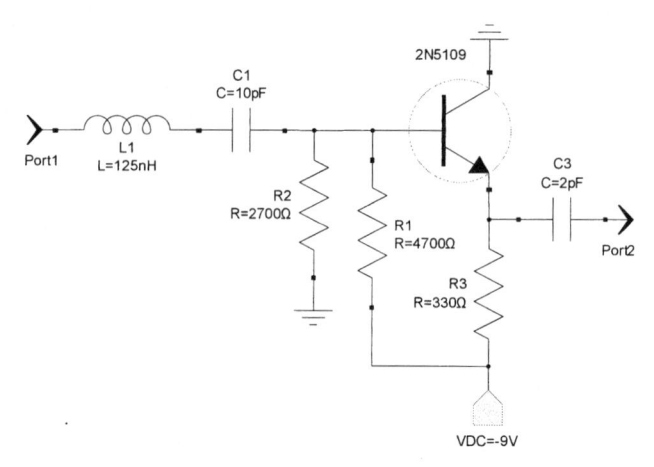

Figure 3.24 A 300-MHz negative-resistance oscillator using a 2N5109 bipolar transistor.

A second schematic reuses the linear network but grounds port 1 and then a Cayenne time-step simulation is performed to obtain the starting waveform. The results are given on the left in Fig. 3.25. Steady-state levels are achieved in approximately 40 nS.

Cayenne postprocessing is then used to perform a Fourier transform over four short term time slices of the output data. The time slice duration chosen is 12.5 nS and slices begin at 1, 18.75, 31.25, and 43.75 nS. The slices are shown on the waveform on the left in Fig. 3.25.

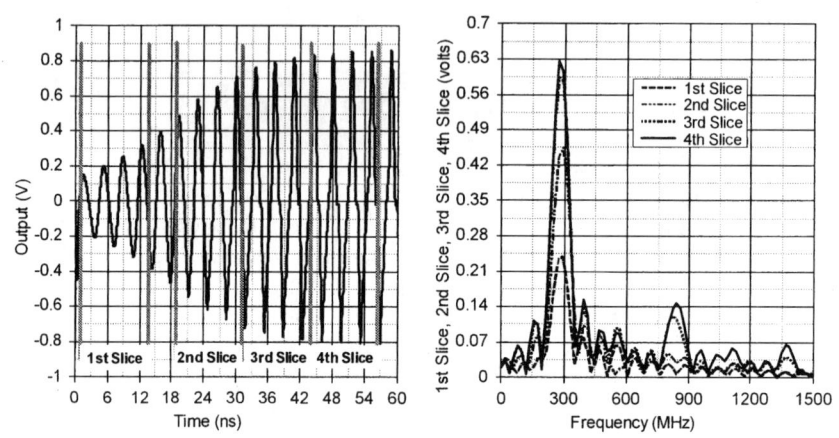

Figure 3.25 Starting waveform of the 300-MHz negative-resistance oscillator (left) and the discrete Fourier transform of four short-term time slices of the waveform (right).

On the right in Fig. 3.25 are the frequency-domain results of the Fourier transforms for the four slices. The oscillation frequency is just below 300 MHz. The width and sidebands of each spectral component are artifacts of the properties of the Fourier transform.

The spectrum of the first slice is the dashed trace with a peak amplitude of approximately 0.24 volts near 300 MHz. Notice the low harmonic voltages for this slice. This is because as the signal is initially building the signal level is insufficient to drive the device to nonlinear operation. This is also evident from the sinusoidal shape of the waveform in the time-domain plot on the left.

In the second slice, the level has increased to 0.46 volts but harmonic levels are still low. Note the 900-MHz component of the Fourier transform is evident in the third slice and it continues to build in the fourth slice. Due to limitations of the Fourier transform on discrete data, the frequency resolution of oscillator time-step simulation is not as good as harmonic-balance simulation. However, the Fourier transform of time-step simulation data provides additional insight into the behavior of the starting process in oscillators.

References

[1] M. Randall and T. Hock, "General Oscillator Characterization Using Linear Open-Loop S-Parameters," *IEEE Trans. MTT,* Vol. 49, June 2001, pp. 1094-1100.

[2] B. Goldberg, "Oscillator Phase Noise Revisited, a Heuristic Review," *RF Design,* January 2002, pp. 52, 56, 58, 60, 62-64.

[3] C. Close, *The Analysis of Linear Circuits,* Harcourt, Brace & World, NY, 1966.

[4] K. Kundert's, *The Designer's Guide to SPICE and Spectre,* Kluwer Academic Publishers (Springer), 1995.

[5] Agilent Technologies, GENESYS 2008.07 Documentation Set, www.agilent.com.

[6] R. Rhea and B. Clausen, "Recent Trends in Oscillator Design," *Microwave Journal,* January 2004, pp. 22-24, 26, 28, 30, 32, 34.

[7] R. Rhea, "A New Class of Oscillators," *IEEE Microwave Magazine,* June 2004, pp. 72, 74, 76, 78, 80, 82-83.

[8] J.K. Clapp, "An Inductance-Capacitance Oscillator of Unusual Frequency Stability," *Proc. Of the IRE,* Vol. 36, 1948, pp. 356-358.

[9] Y. Tsuzuki, T. Adachi and J.W. Zhang, "Fast Start-Up Crystal Oscillator Circuits," *1995 IEEE Frequency Control Sym.*, pp. 565-568.

4 Noise

The ideal oscillator has constant amplitude and linearly advancing phase with time. The output waveform zero crossings are perfectly periodic. In practice, the amplitude and advancing phase include noisy components. When considered over a time period generally less than a second, these nonideal variations are referred to as noise. When considered over a longer time period, the resulting frequency variation is referred to as long-term instability or frequency drift. In digital systems, nonperiodicity in the zero crossing is referred to as jitter. In many system applications, these issues drive the oscillator design process.

Noise terminology in this book generally complies with the excellent reference *Phase Noise Characterization of Microwave Oscillators: Frequency Discriminator Method* [1]. Complimenting this is a reference subtitled *Phase Detector Method*. Both briefly review phase noise concepts and then measurement in detail using Agilent instrumentation.

Oscillator noise is increasingly significant because communications channels are more heavily loaded and closer spaced, data systems utilize higher bit to bandwidth efficiencies, EW and CCC systems are increasingly complex, and system frequencies reach ever higher. These factors place stricter demands on noise performance.

4.1 Definitions

The ideal oscillator output voltage is given by

$$V(t) = V_0 \sin(\omega_0 t) = V_0 \sin(2\pi f_0 t) \qquad (100)$$

However, in practice the output voltage is

$$V(t) = [V_0 + \varepsilon(t)]\sin[2\pi f_0 t + \phi(t)] \qquad (101)$$

where $\varepsilon(t)$ is a zero-mean random process that introduces amplitude noise on the signal and $\phi(t)$ is a zero-mean random process that introduces phase noise on the signal. The practical output voltage is visualized using Fig. 4.1. At an instant in time, the amplitude of the oscillator output voltage is represented by the length of the vector and the phase is represented as the angle, in this case 45°, relative to the positive real axis to the right. With time, the vector rotates counterclockwise with a rotation rate determined by the frequency.

4.1.1 Vector Representation of the Oscillator Output

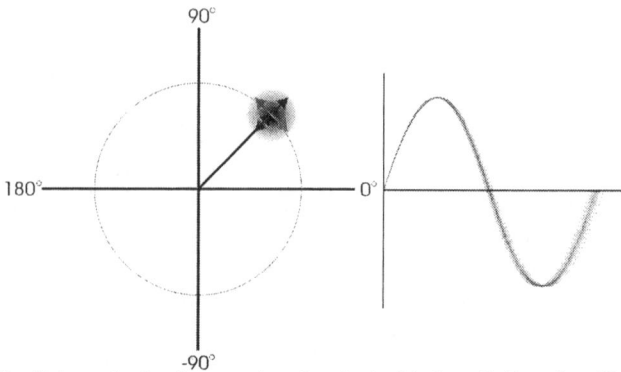

Figure 4.1 Oscillator output voltage vector at an instant in time (left) and multiple samples of the output waveform, each triggered on a rising edge (right).

The ideal output may be represented as a vector with constant amplitude and rotational rate. At an instant in time this is depicted as the black vector on the right in Fig. 4.1. The real output signal vector is uncertain, but likely to be within the region depicted as gray. Amplitude uncertainty is represented by the smaller gray vectors parallel to the desired signal vector. Phase uncertainty is represented by the smaller gray vectors perpendicular to the desired signal vector.

The literature often attributes limiting in the sustaining stage to a reduction of amplitude noise and bases analysis on the predominance of phase noise [2, 3]. While this assumption is useful for analysis, the assumption is not general. This assumption is addressed by Lee and Hajimiri [4], Odyniec [5], and Kurokawa [6].

4.1.2 Jitter

When multiple samples are viewed with a storage oscilloscope, with each sample triggered at the leading edge, the waveform on the right in Fig. 4.1 results. In this example, little amplitude variation is observed. The variation in the voltage zero crossing is referred to as jitter. Jitter is commonly used as a noise performance specification in digital systems. Jitter is a nondeterministic, statistical measure.

4.1.3 The Output in the Frequency Domain

Fig. 4.2 is the output of a HP8640B analog free-running signal generator tuned to approximately 1 GHz at a level of 0 dBm observed with an HP8568A spectrum analyzer. The span width is narrow: 50 Hz per horizontal division. The sweep time is 10 sec or 1 sec per horizontal division. As the sweep approached the center frequency, during approximately a 0.5-sec time, the instantaneous frequency peaked multiple times over a frequency range of approximately 25 Hz. A subsequent sweep revealed a different but again noisy pattern at the signal peak. The spreading of the peak in the frequency-domain is referred to as residual FM modulation. Residual FM is a nondeterministic, statistical measure. Residual FM, residual PM, and jitter are mathematically related.

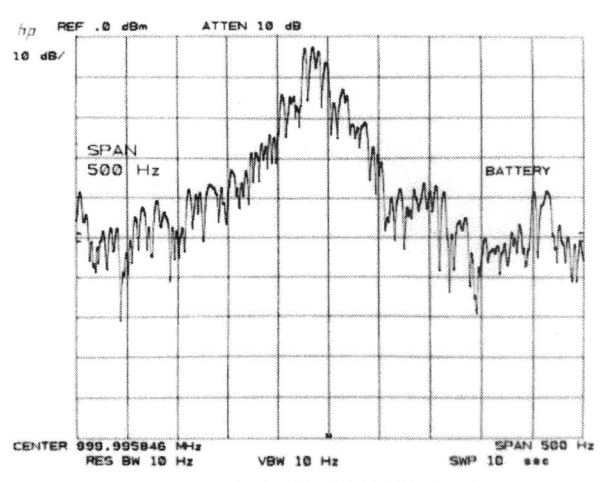

Figure 4.2 Narrow spectrum sweep of a 1-GHz HP8640B signal.

The noisy amplitude and phase modulation of the carrier introduce noise sidebands in the frequency domain representation of the output. Fig. 4.3 shows a display of the 1-GHz HP8640B signal with a total frequency span of 5 kHz. Residual FM of the signal is less observable on this wider frequency span, but the generally symmetrical nature of sideband noise becomes evident.

If the spectrum analyzer video bandwidth is reduced, the "grassy" nondeterministic noise sidebands are smoothed. The discrete components at 240, 360, and 480 Hz offset above and below the carrier are probably harmonics of the 60-Hz line frequency. These components are not evident on the spectrum analyzer when using its 20-MHz front-panel calibrate signal and they are probably from the signal generator. There are also discrete components at approximately 1560-Hz offset. All of these discrete spectrum components are caused by repeatable disturbances, perhaps power supply ripple or synthesizer reference sidebands. These components are deterministic.

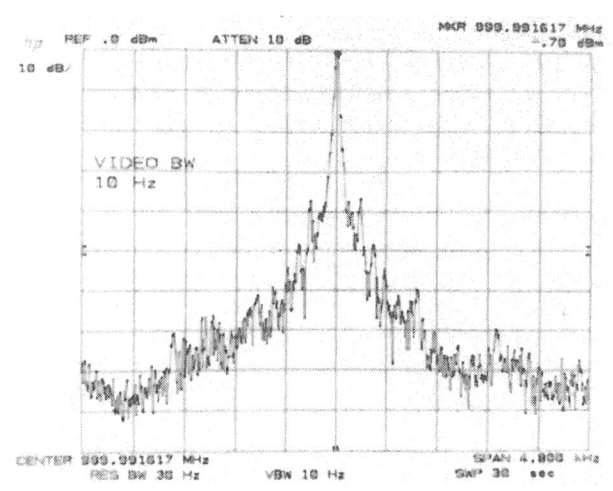

Figure 4.3 Wider spectrum sweep of the 1-GHz HP8640B signal.

4.1.4 SSB Phase Noise

An important definition of oscillator phase instability is the spectral density of phase fluctuations. It is often specified per hertz.

$$S_\phi(f_m) = \frac{\Delta\phi_{rms}^2(f_m)}{BW_{\Delta\phi_{rms}}} \quad (rad^2 / Hz) \tag{102}$$

More commonly used for specification purposes is the ratio of the power in one phase-modulated sideband to the total signal power. It is specified in a single sideband at a specified offset from the carrier and 1-Hz bandwidth and is referred to as the single-sideband (SSB) phase noise, generally expressed in decibel format. Provided that the total phase deviations are

$$\Delta\phi_{peak} \ll 1\, rad \tag{103}$$

then the SSB phase noise is

$$L(f_m) = \frac{1}{2} S_\phi(f_m) \tag{104}$$

The condition stipulated by Eq. 103 is an important one. If $\Delta\phi_{peak}$ is not small, then $\mathscr{L}(f_m)$ in dB is potentially positive, an impossibility.

The total frequency span of the phase noise plot in Fig. 4.3 is 4 kHz, or 400 Hz per horizontal division. At 800-Hz offset below the carrier, if the noise is averaged, the sideband level is approximately 72 dB below the carrier, or -72 dBc. The resolution bandwidth is 30 Hz. To reference the noise to 1-Hz bandwidth a correction factor is applied.

$$\mathscr{L}(f_m)(dBc/Hz) = \mathscr{L}(f_m)(dBc/BW) - 10\log BW \qquad (105)$$

In this case, 10 log (BW) is 14.77 and the SSB phase noise at 800-Hz offset below the carrier is -86.8 dBc/Hz. As is described in Section 4.3.1, the characteristics of noise power measurement with Agilent spectrum analyzers requires a 1.7-dB correction factor so this SSB phase noise is actually -85.1 dBc/Hz. Notice that the SSB phase noise at 800-Hz offset above the carrier in Fig. 4.3 is approximately 4 dB lower. The asymmetry here is probably the result of inadequate averaging and long-term instability during the slow sweep time. Noise measurement is a statistical process. A second sweep will give slightly different results. A faster sweep time and additional averaging using a lower video bandwidth removes much of the asymmetry. Unfortunately, lower video bandwidth requires longer sweep time and the measurement is more likely to be affected by long-term stability. With higher resolution bandwidth, faster sweep times are practical and the HP 8640A noise sidebands become more symmetrical.

Another definition of oscillator instability is the spectral density of frequency fluctuations. It is often specified per hertz.

$$S_{\Delta f}(f_m) = \frac{\Delta f_{rms}^2(f_m)}{BW_{\Delta f_{rms}}} \quad (rad^2/Hz) \qquad (106)$$

The SSB phase noise, spectral density of phase fluctuations, and spectral density of frequency fluctuations are mathematically related.

$$S_\phi(f_m) = \frac{S_{\Delta f}(f_m)}{f_m^2} \quad (rad^2/Hz) \qquad (107)$$

$$\mathscr{L}(f_m) = \frac{S_{\Delta f}(f_m)}{2 f_m^2} \qquad (108)$$

4.1.5 Residual FM and Residual PM

A narrow spectrum sweep of a real oscillator reveals spreading of the carrier. This is referred to as residual frequency modulation (FM) or incidental FM. If the oscillator is detected with a frequency discriminator, a baseband voltage results. Because frequency and phase are related, residual PM is also definable.

Residual PM is related to the SSB phase noise by

$$\Delta\phi^2 = 2\int_{f_a}^{f_b} \mathscr{L}(f_m) df_m \quad (rms) \qquad (109)$$

where f_a and f_b are the lower and upper baseband frequencies of interest in the system that correspond to offset frequencies at RF. For example, in stereo FM broadcast the baseband frequencies of interest are 30 to 15,000 Hz. Residual FM is related to the SSB phase noise by

$$\Delta f^2 = 2 \int_{f_a}^{f_b} f_m^2 \mathscr{L}(f_m) df_m \; (rms) \qquad (110)$$

Residual PM and residual FM are directly related to system performance. For example, in a system utilizing FM modulation, the ultimate detected baseband S/N ratio with high input C/N is given by

$$\left. \frac{S}{N} \right|_{ultimate} = 10 \log \frac{\Delta f_{sig,rms}^2}{\Delta f^2} \qquad (111)$$

where $\Delta f_{sig,rms}$ is the desired signal deviation. For example, with a desired signal deviation of 53-kHz rms, to achieve 60 dB ultimate S/N requires a local oscillator residual FM of 53 Hz or better. This value is approximate and dependent on the baseband frequency because preemphasis and deemphasis are used in FM broadcast to improve the S/N ratio. Nevertheless, an advantage of residual FM and PM as noise specifications is the direct relationship to system performance.

Additionally, residual FM and PM are measurable with less expensive test equipment than is SSB phase noise. The reason SSB phase noise is generally preferred for specification is that integration of the SSB phase noise reveals the residual noise, while determination of the SSB phase noise from the residual FM or residual PM data requires assuming the shape of the SSB phase noise.

4.1.6 Two-Port Noise

The noisy disturbance of the phase of the oscillator output signal considered above is referred to as absolute phase noise. When a signal transfers through a noisy two-port device then AM and PM noise is induced into the signal, thus again producing sidebands. The device noise is referred to as two-port or additive noise. Typically, two-port noise is less than absolute source noise. Also, unlike absolute phase noise, the AM component of two-port noise is typically comparable in magnitude to PM noise. Unless stated otherwise, in the remainder of this book the term phase noise refers to absolute rather than two-port noise.

Given in Fig. 4.4 is the two-port noise figure in decibels of a bipolar amplifier circuit with resistive shunt and series feedback. Here, white noise from 1 to 100 MHz results in a flat noise figure of 2.5 dB. Above 100-MHz high-frequency effects cause the noise figure to increase. Below 1 MHz, device flicker noise causes the noise figure to increase. The intersection of the flicker noise and the flat noise figure is referred to as the flicker corner frequency. In this case the flicker corner is 6 kHz.

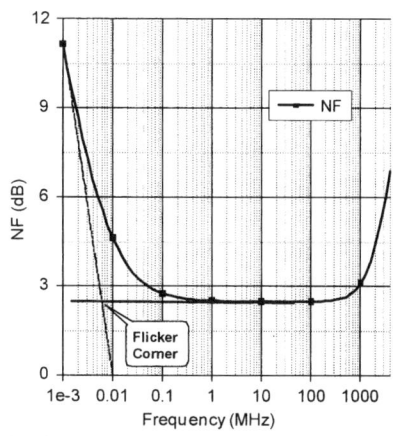

Figure 4.4 Two-port noise figure of a bipolar amplifier circuit with resistive series and shunt feedback. Flicker noise effects are evident on the left and high frequency effects on the right.

4.1.7 Acoustic Disturbances

Vibration and other acceleration forces on the oscillator can disturb the oscillation phase and frequency. Fig. 4.5 shows the spectrum of the 1GHz HP8640B signal with a small mechanical vibrator placed on the same laboratory bench as the HP8640B. Touching the bench reveals only moderate vibration. At offset frequencies of 400 to 1000 Hz the sideband levels are increased by 10 to 15 dB. In this case, the sidebands are multiple discrete frequencies. These are referred to as acoustic sidebands. Acoustic sidebands are deterministic for sinusoidal excitation or are partially noisy, non-deterministic in the case of noisy bearings. For aeronautical and other mobile environments, acoustic disturbances often dominate sideband levels. In some applications, bearing noise from cooling fans limit performance.

4.2 Predicting Phase Noise

This section introduces techniques for estimating the SSB phase noise of oscillators. The classic method described by D.B Leeson [7] is straightforward and insightful for minimizing oscillator phase noise. This method is reviewed by Lee and Hajimiri [4] who refer to Leeson's method as the linear time invariant (LTI) theory. They point out certain limitations to this theory that are alleviated using the linear time variant (LTV) theory.

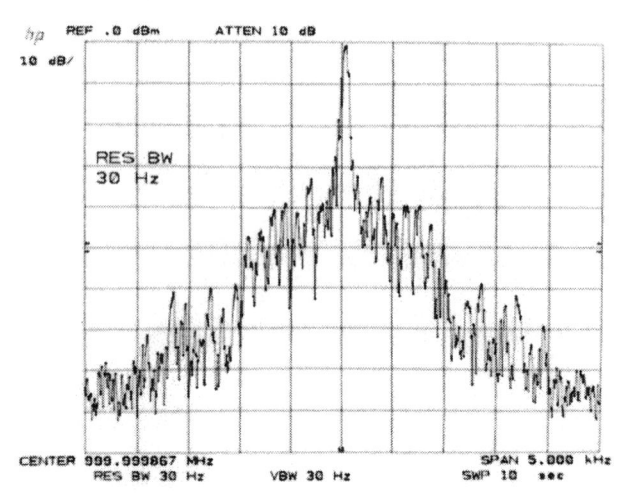

Figure 4.5 A 1-GHz HP8640B spectrum illustrating acoustic induced sidebands.

4.2.1 Linear Time Invariant Theory

D.B. Leeson gave "a heuristic derivation, presented without formal proof" of the phase noise of the output of an oscillator. A formal proof was offered by Sauvage [8]. Leeson stipulated that the oscillator is linear with "minor corrections to the results that are necessary to account for nonlinear effects." He also stipulated that AM components << FM components except at large offset. He also implied time invariance. Assuming a simple series or parallel resonator in the feedback network, his expression for the power spectral density of the phase fluctuations is

$$S_\phi(\omega_0) = S_{\Delta\phi}\left[1 + \left(\frac{\omega_0}{2Q\omega_m}\right)^2\right] \qquad (112)$$

where $S_{\Delta\phi}$ is the additive voltage noise component, which is flat and equal to $2FkT/P_s$ for white noise and rises toward lower frequency at 6dB/octave for flicker noise. F is an empirical factor loosely correlated to device noise, k is Boltzmann's constant, T is the operating temperature and P_s is the output power. Q is the open-loop unloaded Q of the feedback oscillator. With a term added for flicker noise, the decibel form of Leeson's equation is

$$\mathscr{L}(f_m) = 10\log\left[\frac{1}{2}\left(1 + \frac{f_c}{f_m}\right)\left(1 + \left(\frac{f_0}{2f_m Q_L}\right)^2\right)\left(\frac{FkT}{P_s}\right)\right] \quad (dBc/Hz) \quad (113)$$

where

f_c is the flicker corner frequency
f_m is the offset, modulation or baseband frequency
f_o is the carrier frequency
Q_L is the open-loop loaded Q
F is an empirical factor correlated to device noise
k is Boltzmann's constant
T is the operating temperature, and
P_s is the output power

Because F is an empirical factor and because f_c is only indirectly related to the device baseband flicker corner, the quantitative predictive power of Eq. 114 is limited. Nevertheless, the phase noise predicted by Leeson's equation correlates relatively well with measured data for oscillators I have tested. Fig. 4.6 is a plot of SSB phase noise as predicted by Eq. 114 with a carrier frequency of 100 MHz, a flicker corner of 6 kHz, a loaded Q of 100, an amplifier noise factor of 1.77 (2.5 dB) and 0-dBm output power. At offset frequency higher than the resonator half bandwidth, $f_o/2Q$, the SSB phase noise is flat with frequency as determined by the third term in Eq. 113. Generally, AM noise dominates in this frequency region. At offset frequency lower than $f_o/2Q$, the SSB phase noise increases proportional to f_m^{-2} as determined by the second term in Eq. 113. At offset frequency lower than the flicker corner, the SSB phase noise increases proportional to f_m^{-3} with the first term providing the additional factor of f_m^{-1}. At offset frequencies very close to the carrier, a random walk term results in a total "noise" slope of $1/f^4$.

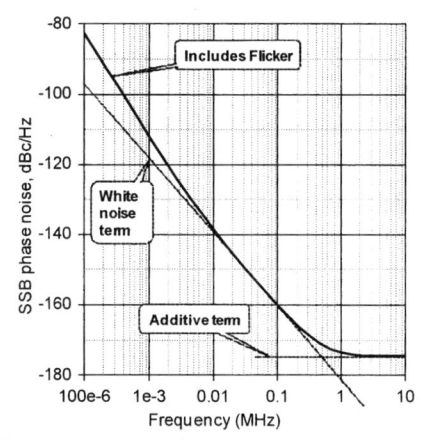

Figure 4.6 Typical SSB phase noise versus carrier offset.

4.2.2 Extensions to LTI-Based Theory

For convenience, the phase noise predicted by Leeson's equation is referred to in the remainder of this book as Leeson noise. However, oscillator phase noise, both deterministic and nondeterministic, is generated by mechanisms other than phase perturbations in the sustaining stage as predicted by Leeson. These contributors are considered next.

4.2.2.1 Pushing Induced Noise

Recall that a change in the supply voltage changes the bias and transfer characteristic of an amplifier. A change in the transmission phase results in an oscillator frequency shift. Therefore, noise on the power supply is translated into phase noise on the oscillator output spectrum.

This noise is predictable, based on the static pushing sensitivity of the oscillator, determined by measuring the oscillation frequency for supply voltages above and below the nominal voltage. The phase deviation of the carrier is given by

$$\theta_d = \frac{2K_p V_{ns}}{f_m} \qquad (114)$$

where K_p is the frequency sensitivity to the supply voltage in Hz/volt and V_{ns} is the noise voltage of the supply. The resulting SSB level is then

$$\mathscr{L}(f_m) = 20\log\frac{\theta_d}{2} \qquad (115)$$

If V_{ns} is a discrete signal the resulting sideband pair is deterministic. If V_{ns} is noise with units of volts/Hz then the sidebands are nondeterministic.

4.2.2.2 Varactor Modulation Noise

Oscillators are often tuned by a voltage-controlled variable capacitance called a varactor. Noise on the tuning voltage then modulates the carrier. However, even if the tuning voltage is noiseless, an internal noise source is present in all varactors. The effective noise resistance, R_{enr}, for abrupt and hyperabrupt silicon varactors is 300 ohms to 10 kohms. This resistance is not the effective series resistance used to model the unloaded Q of varactors. The effective series resistance used to model varactor unloaded Q is typically less than 10 ohms. The effective noise resistance is poorly understood and not specified by varactor manufacturers. R_{enr} is often determined empirically by measuring oscillator phase noise and deducing R_{enr}. R_{enr} may then be used to assess design trade-offs.

The noise voltage produced by the effective noise resistance is

$$V_{nv} = \sqrt{4kTR_{enr}} \qquad (116)$$

and the peak phase deviation in 1-Hz bandwidth is

$$\theta_d = \frac{\sqrt{2}K_v V_{nv}}{f_m} \qquad (117)$$

where K_v is the VCO gain constant in hertz/volt. The resulting SSB phase noise is again given by Eq. 115. The total SSB phase noise of a varactor-controlled oscillator is the power sum of the phase noise due to Leeson, pushing-induced noise, and varactor-modulation noise. From Eq. 117 it is evident that degradation from varactor-modulation noise is most severe for oscillators with wide absolute tuning bandwidth. It is typically the limiting factor in oscillators with gain constants of 50 MHz/volt and higher.

4.2.2.3 Buffer Noise

Buffer amplifiers are used to isolate the load and to avoid load variation from influencing the oscillator amplitude, phase, or frequency. A buffer amplifier degrades phase noise performance whenever its thermal noise, referenced to its input, exceeds the oscillator output noise. Buffer noise referenced to its input, then referenced to the oscillator carrier is given by

$$\mathscr{L}(f_m) = 10\log\frac{F_{buf}kT}{P_s} \qquad (118)$$

4.2.2.4 Master Noise Equation

The above terms are uncorrelated and add on a power basis. Therefore, the combined oscillator phase noise is

$$L(f_m) = 10\log\left[\frac{1}{2}\left(1+\frac{f_c}{f_m}\right)\left(1+\left(\frac{f_0}{2f_mQ_L}\right)^2\right)\left(\frac{FkT}{P_s}\right) + \frac{2kTR_{enr}K_v^2}{f_m^2} + \left(\frac{K_pV_{ns}}{f_m}\right)^2 + \frac{F_{buf}kT}{P_s}\right] \quad (dBc/Hz) \qquad (119)$$

where the variables are defined in previous sections. This master expression for SSB noise includes terms for all sources of noise in well designed oscillators without excessive open-loop gain margin and with passive components possessing only thermal noise. As discussed in Section 4.2.3, this expression can be pessimistic (conservative) for oscillators using active devices with a flicker corner frequency higher than the offset frequencies of interest.

4.2.2.5 Frequency Multiplication

Frequency multiplication of the output of an oscillator degrades the SSB noise performance at the increased output frequency. With a frequency multiplication factor of N,

$$\mathscr{L}_{xN}(f_m) = \mathscr{L}(f_m) + 10\log N^2 + A \quad (dBc/Hz) \qquad (120)$$

where A is an additive factor that depends on the type of frequency multiplier used. Frequency multipliers operate by two principles: nonlinear reactance such as a varactor, or nonlinear conductance such as diode doublers. Schottky barrier diode doublers have a low additive noise factor A of a decibel or less. Reactive varactor and snap diode multipliers are generally noisier. Alternatively, multiplier noise may be specified as a limiting noise floor. For example, the specified noise floor of Hittite GaAs PHEMT active ×2 microwave frequency multipliers is -140 dBc/Hz at 100 kHz offset [9].

4.2.3 Linear Time Variant Theory

The typical oscillator achieves a steady-state output level by nonlinear behavior that adjusts the gain, phase, and match for unity open-loop gain. Typical oscillator operation is therefore nonlinear. Various noise theories attempt to describe noise phenomena as the result of nonlinear behavior. Indeed, excessive loop gain can contribute to additional sources of noise in an oscillator. However, oscillator noise is not, as a first principle, the result of nonlinear behavior. Recall that through either active or passive AGC, linear oscillators are practical, such as with Hewlett's HP200A. If noise were the result of nonlinear action these oscillators would be essentially noiseless, which of course is not the case. If noise sidebands are not attributed to nonlinear action, what is their source?

Lee and Hajimiri [4] suggest that the linear assumption of Leeson's theory is valid but that time invariance is not. They suggest that considering time variance leads to a more accurate oscillator noise theory. Indeed, LTV theory offers new approaches to oscillator noise analysis. LTV theory validates, to a degree, the role Leeson's parameters play in oscillator performance but LTV offers insight leading to improved performance.

Consider the voltage responses of a parallel LC resonator to small current impulses. If the impulse occurs at the peak of an existing oscillating voltage, as illustrated at the top of Fig. 4.7, the effect is to temporarily increase the amplitude of the waveform. However, if the impulse occurs at any other time, both the amplitude and the zero crossing time are affected. This phase perturbation manifests as noise.

The response to an impulse at time τ is

$$h_\phi(t,\tau) = \frac{\Gamma(\omega_0 \tau)}{q_{max}} u(t-\tau) \qquad (121)$$

where q_{max} is the maximum charge displacement across the resonator capacitance, $u(t-\tau)$ is the unit step function and $\Gamma(\omega_0 \tau)$ is the impulse sensitivity function (ISF). Dividing $\Gamma(\omega_0 \tau)$ by q_{max} makes the function independent of the signal amplitude. From this, Lee and Hajimiri derive an expression for the SSB phase noise in the $1/f^2$ region

$$\mathscr{L}(\omega_m) = 10\log\left(\frac{\overline{\dfrac{i_n^2}{\Delta f}}\Gamma_{rms}^2}{2q_{max}^2\omega_m^2}\right) \qquad (122)$$

where $\overline{i_n^2}/\Delta f$ is the mean squared spectral density of the white noise current and Γ_{rms} is the rms value of the ISF. The SSB phase noise in the $1/f^3$ region is then

$$\mathscr{L}(\omega_m) = 10\log\left(\frac{\overline{\dfrac{i_n^2}{\Delta f}}c_0^2}{8q_{max}^2\omega_m^2}\frac{\omega_{1/f}}{\omega_m}\right) \qquad (123)$$

where $c_0/2$ is the DC term of a Fourier series expansion of $\Gamma(\omega_0\tau)$ and $\omega_{1/f}$ is the device $1/f$ corner frequency. The $1/f^3$ corner frequency is

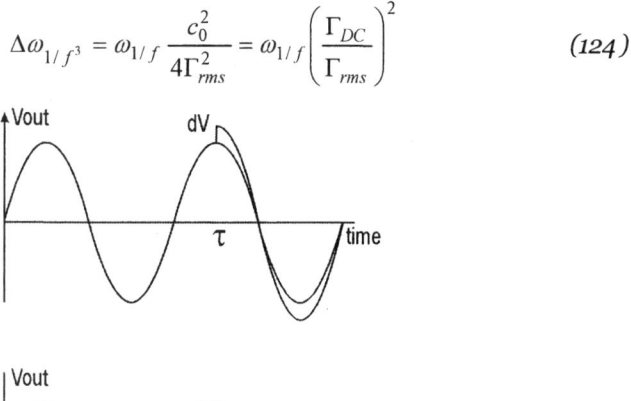

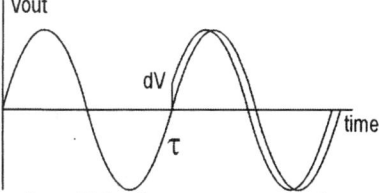

$$\Delta\omega_{1/f^3} = \omega_{1/f}\frac{c_0^2}{4\Gamma_{rms}^2} = \omega_{1/f}\left(\frac{\Gamma_{DC}}{\Gamma_{rms}}\right)^2 \qquad (124)$$

Figure 4.7 Response of a sinusoidal voltage across a simple resonator to a current impulse occurring at the peak of the voltage (top) and at a zero crossing (bottom).

Notice that the frequency of transition from the $1/f^3$ region to the $1/f^2$ region is not necessarily equal to the device $1/f$ corner frequency, f_c. This is a primary significance of the LTV theory. The outcome is that with certain design techniques, the $1/f^3$ transition frequency can be closer to the carrier than f_c, thus offering hope for oscillator design utilizing device technologies with notoriously poor $1/f$ performance, such as CMOS and GaAs.

The difficulty with LTV is in determining the ISF. Analytical methods for finding the ISF exist for only special cases and then only approximately. The ISF is generally found by simulation. The advantage of LTV theory is that it reveals methods for suppressing flicker noise by using symmetry. LTV theory is worthy of exploration when processes with high flicker noise must be used, such as GaAs FET and CMOS devices. The reader is referred to Lee and Hajimiri's paper [4] for a more detailed description of the LTV theory and for additional references.

4.3 Measuring Phase Noise

The measurement of oscillator phase noise below -100 dBc/Hz is challenging. Various methods are used with each offering specific advantages and disadvantages. The following sections review available techniques and the suitability of each for different applications.

4.3.1 Direct Method with a Spectrum Analyzer

Perhaps the simplest method for measuring SSB noise is direct measurement by a spectrum analyzer. This technique has the added advantage that the spectrum analyzer may be used for other measurements and it is often already included in the laboratory instrumentation. Discrete frequency components are also easily identified. Spectrum analyzer displays of signals are given earlier.

The sidebands and their "grassy" appearance is the result of noisy perturbation of the ideal oscillator. The width of the central carrier component of the spectrum is the response of analyzer IF filter. The IF filter bandwidth is referred to as the resolution bandwidth. Filtering of the detected analyzer signal at baseband is referred to as video filtering. Narrow resolution bandwidth and video filtering limit the sweep speed. The sweep time in Fig. 4.3 is 30 seconds. During this time, random walk occurred in the signal, which is responsible for the observed asymmetry in the sidebands. Narrower video filtering smoothes the grassy sidebands and assists in "reading" the desired average value, but this requires a slower sweep speed.

The average level of a sideband at a given offset relative to the carrier is directly the SSB noise. However, three correction factors are required [10]. First, the measurement resolution bandwidth is seldom 1 Hz as this would require excessive sweep time. In Fig. 4.3 the panel resolution bandwidth is 30 Hz. The noise in 1-Hz bandwidth is corrected by

$$BW_{corr} = 10\log\frac{1Hz}{BW_{resolution}} \qquad (125)$$

In this case, the SSB noise in 1-Hz bandwidth is -14.77 dB below the displayed value. The second correction factor is required because the analyzer's Gaussian 3-dB resolution bandwidth is not equal to the effective noise bandwidth. The noise bandwidth is approximately 1.2 times the panel resolution BW. Therefore

$$NBW_{corr} = 10\log(1/1.2) = -0.79 dB \qquad (126)$$

Finally, the analyzer detector is calibrated for discrete sinusoids. Most analyzers utilize a logarithmic IF amplifier followed by a peak detector. The peak detector reads lower than the true rms value and the logarithmic amplifier amplifies noise peaks less than the remainder of the signal. Consequently, the detection correction factor is

$$SD_{corr} = +2.5 dB \qquad (127)$$

The total correction for direct spectrum analyzer SSB noise measurement is then

$$N_{corr} = 10\log\frac{1Hz}{BW_{corr}} - 0.79 + 2.5 \cong 10\log\frac{1Hz}{BW_{resolution}} + 1.7 \qquad (128)$$

The above correction is easily applied. However, the sensitivity of the spectrum analyzer is severely limited due to the demands placed on the analyzer LO for use as a general-purpose instrument. Given in Table 4.1 is the noise floor versus offset frequency for the HP8568A 100-Hz to 1500-MHz spectrum analyzer used in the author's laboratory. This data was obtained using the 20-MHz calibrate signal available at the front panel of the analyzer. The noise floor of the HP8568A is a weak function of RF frequency. At higher RF frequency, the noise floor below 3-kHz offset degrades by 5 to 10 dB [11].

Table 4.1 Noise floor of a HP8568A, 100-Hz to 1500-MHz spectrum analyzer relative to the carrier (column 2) and SSB phase noise of a 100 MHz oscillator (column 3)

Offset(kHz)	SSB noise(dBc/Hz)	Example 100 MHz Osc
0.3	-103.3	-85
1	-107.3	-100
3	-110.3	-114
10	-112.3	-128
30	-111.8	-139
100	-115.3	-150
300	-125.8	-160
1000	-128.3	-170

The last column in Table 4.1 is the SSB phase noise of an example 100 MHz L-C oscillator with a loaded Q of 20 and output level of 7 dBm. It is evident that in this case the direct method is suitable for measuring SSB phase noise only close to the carrier.

An additional issue with the direct method is that measurement close to the carrier requires narrow resolution bandwidth and therefore long sweep time, resulting in oscillator drift during the measurement.

When the noise of the measurement system is comparable to the noise of the DUT, and when the noise of the measurement system is known, the noise power of the measurement system may be power subtracted from the measured data to correct the measurement.

4.3.2 Selective Receiver Method

The selective receiver method is diagrammed in Fig. 4.8. The method requires an LO that is quieter than the DUT to mix the DUT signal down to an IF frequency where a narrow bandwidth filter such as a crystal filter can be realized. The IF filter bandwidth must be less than the required minimum offset frequency. A power measurement is made at the carrier frequency and the desired offset frequency by tuning the low-noise LO. The level difference, when corrected for the bandwidth of the IF filter, is the SSB noise. Alternatively, if the DUT is tunable, the LO may be a low-noise, fixed tuned oscillator. To compensate for the wide range in measured level, a step attenuator is used prior to the mixer.

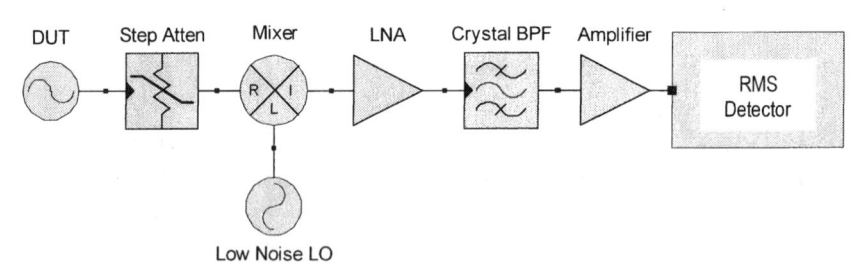

Figure 4.8 System for measuring phase noise using a selective receiver.

When the LO converts a desired sideband frequency into the IF filter, power from the image is also converted. Fig. 4.9 illustrates this for a desired measurement centered at 300 kHz offset, an IF filter centered at 300 kHz, and the image centered at 900 kHz offset. In this case, the average level of the image power is approximately 4 dB below the desired level. The image correction factor is

$$C_i = -10\log\left(10^{N_i/10} + 1\right) \qquad (129)$$

where

$$N_i = P_{image}(dB) - P_{desired}(dB) \qquad (130)$$

In this case, N_i is -4 dB and C_i is -1.46 dB. The actual SSB noise is -1.46 dB lower than the detected power. The image correction factor is largest for low IF filter center frequency and for measurement at larger offset.

The selective receiver method is straightforward, it has a low noise floor, it is easily automated and it is relatively inexpensive. It is poorly suited for offsets close to the carrier because the required IF filter bandwidth becomes excessively narrow.

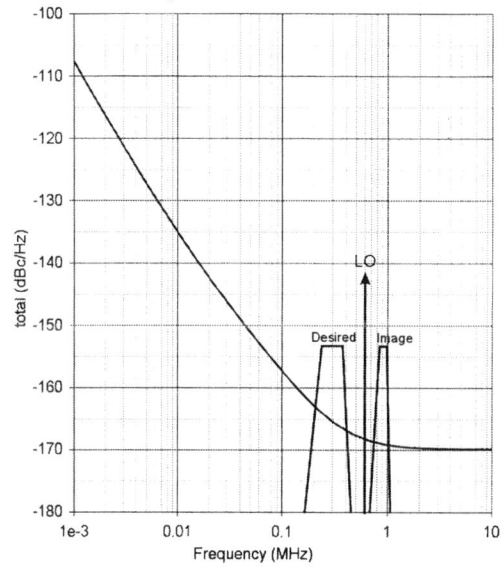

Figure 4.9 Diagram illustrating the desired and image signals in a selective receiver SSB noise measurement system.

4.3.3 Heterodyne/Counter Method

In this method, a low-noise reference LO downconverts the DUT signal to a low frequency beat note, which is then counted multiple times using a constant time between measurements. From this data the Allen variance is computed. The process is repeated with differing constant times between measurement. The Allen variance and phase-noise are related. For additional information on the Allen variance, refer to [12].

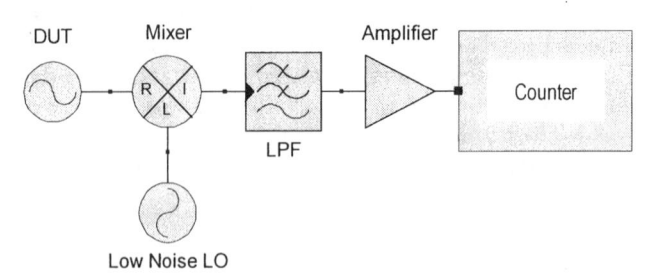

Figure 4.10 System for the measurement of multiple frequency samples for computing the Allen variance and ultimately the phase noise.

The heterodyne to a low beat note frequency makes the method very broadband, limited primarily to the quality and availability of the low-noise LO. The method is suitable for measurement to very low offset frequency, easily below 1 Hz. There are a number of caveats. The beat note is low frequency so the long-term stability of both the DUT and low-noise LO must be excellent. As with other methods employing a low-noise LO, the LO phase noise must be better than the DUT. The maximum measurable offset frequency is limited to about 10 kHz by the sampling speed of the counter. Finally, the counter sampling is essentially a digital filtering process that also has responses at $3f$, $5f$, $7f$, and so forth, and noise is integrated at these higher frequencies. Unless the noise is falling at f^2 or higher, the higher frequency components introduce significant error. Therefore, the method is not suitable for SSB noise plots that are flat or shallow with frequency such as within the loop bandwidth of synthesized sources. This method is primarily suited for close in measurement of frequency and time standards.

4.3.4 Reference Oscillator Method

The reference oscillator method, also referred to as the phase detector method, is sensitive over a wide range of offset frequencies. The method is the basis of commercially available oscillator phase-noise test sets. It is diagrammed in Fig. 4.11.

A double balanced mixer is typically used as the phase detector. The reference low-noise oscillator is locked and kept in phase quadrature with the DUT. The lock bandwidth must be less than the lowest offset frequency of interest so that the loop does not correct the DUT phase noise to that of the reference. The phase detector output, with the $2f_0$ image filtered by a lowpass filter, is the combined noise of the reference and DUT at baseband frequency.

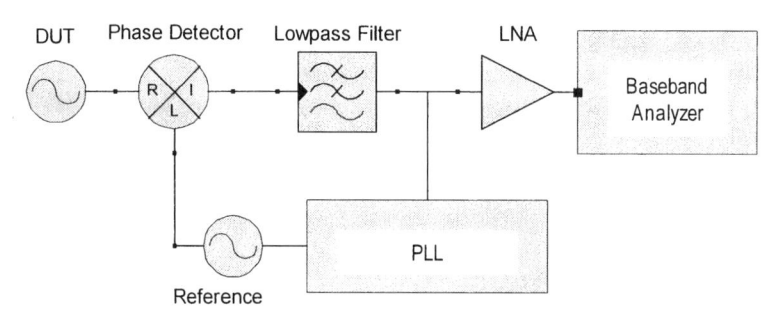

Figure 4.11 System for measuring phase noise using the reference oscillator (phase detector) method.

The reference is operated at the normal LO signal level, for example, 7 dBm for a typical Schottky diode-ring double-balanced mixer. Maximum sensitivity occurs if the RF drive is near the mixer compression point, typically 3 dBm. If the reference and DUT are unlocked, the frequency difference produces a beat note. When locked in quadrature, the mixer output voltage change for a phase change is

$$\Delta v(t) = v_{peak} \sin[\Delta\phi(t)] \qquad (131)$$

where v_{peak} is the beat note peak voltage and $\phi(t)$ is the phase error. At a 3 dBm input beat note level, 7 dB of loss and a 50 ohm system, v_{peak} is 0.2 volts and the phase detector gain, K_ϕ, is 0.2 volts/radian.

The baseband analyzer can be a spectrum analyzer to measure the noise power as a function of offset frequency, $\Delta P_{rms}(f_m)$. Because the carrier is suppressed, an LNA can precede the spectrum analyzer to significantly increase its sensitivity and therefore mitigate the poor sensitivity of the direct spectrum measurement method. The noise voltage as a function of the offset frequency, $\Delta v_{rms}(f_m)$, is computed from the measured noise power.

$$\Delta v_{rms}(f_m) = \sqrt{\Delta P_{rms}(f_m) Z_0} \qquad (132)$$

where Z_0 is the characteristic impedance of the spectrum analyzer. The spectral density of phase fluctuations is

$$S_\phi(f_m) = \left(\frac{\Delta v_{rms}(f_m)}{K_\phi}\right)^2 \qquad (133)$$

and the SSB phase noise is then

$$\mathscr{L}(f_m) = 10\log\left[\frac{1}{2}\frac{\Delta P_{rms}(f_m)Z_0}{K_\phi^2}\right] \quad (134)$$

The noise floor of the measurement system with an ideal reference oscillator is estimated by replacing the signal noise power in Eq. 134 with the noise floor of the baseband analyzer. If the noise figure of the baseband analyzer is 3 dB, then $\Delta P_{analyzer}$ in 1-Hz bandwidth is -171 dBm. In this case, with a phase detector gain of 0.2 volts/radian, the system noise floor is -173 dBm. However, the phase noise of the reference oscillator is generally the limiting factor.

4.3.5 Frequency Discriminator Method

The frequency discriminator method using a mixer and delay line for measuring phase noise is diagrammed in Fig. 4.12. Short-term frequency fluctuations are converted into voltage fluctuations as is done in the reference oscillator method. An advantage of the frequency discriminator method is that a reference oscillator is not required. Frequency fluctuations are transformed into phase fluctuations in the delay line. A phase shifter sets the transformed and direct signal in phase quadrature. The transformed signal and the direct signal drive a mixer that serves as a phase fluctuation to voltage converter. A baseband analyzer processes the voltage as with the reference oscillator method.

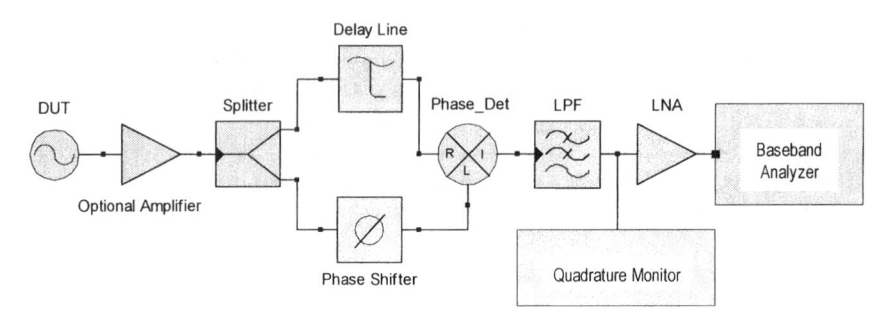

Figure 4.12 System for measuring phase noise using the discriminator method.

The transfer response of the system [1] is

$$\Delta v(f_m) = K_\phi 2\pi\tau_d \Delta f(f_m)\frac{\sin(\pi f_m \tau_d)}{\pi f_m \tau_d} \quad (135)$$

where τ_d is the delay of the delay line and other variables were defined previously. Ideally, the delay is large to maximize the baseband voltage for a given frequency fluctuation. Unfortunately the $sin(x)/x$ term causes a null in the system sensitivity at f_m equal to $1/\tau_d$. To avoid a necessary correction for the $sin(x)/x$ term, generally

$$f_m \leq \frac{1}{2\pi\tau_d} \qquad (136)$$

or for a given maximum offset frequency of interest

$$\tau_d = \frac{1}{2\pi f_m} \qquad (137)$$

For example, if the measurement system must operate to 1-MHz offset, then the chosen delay is 159 nS.

The LO port of the mixer is driven by the DUT through the splitter and phase shifter. If a standard level Schottky diode double balanced mixer is used, the optimum LO drive level is 7 dBm. The maximum K_ϕ occurs with a signal level near mixer compression, or 3 dBm for a standard level mixer.

Eq. 135 reveals that system sensitivity is directly proportional to both K_ϕ and τ_d. K_ϕ increases with increased signal level, but because the delay line attenuation increases with length, the signal level decreases with increasing τ_d. There is an optimum delay line attenuation for maximum system sensitivity and it is 8.7 dB [1]. With this delay line attenuation, the optimum DUT signal at the input to the delay line is the drive level for a standard level mixer, 3 dBm, plus 8.7 dB, or +11.7 dBm. The optimum frequency discriminator system therefore has a delay line length dictated by Eq. 137 and a type and diameter that result in 8.7 dB of attenuation at the carrier frequency of interest. Improved system sensitivity results with the use of a higher-level mixer for the phase detector and a higher-level signal.

The frequency discriminator method output is the spectral density of frequency fluctuations. For the frequency discriminator

$$S_{\Delta f}(f_m) = \frac{\Delta v_{rms}^2(f_m)}{(K_\phi 2\pi\tau_d)^2} \qquad (138)$$

and therefore

$$\mathscr{L}(f_m) = 10\log\left[\frac{1}{2}\frac{\Delta v_{rms}^2(f_m)}{(K_\phi 2\pi\tau_d f_m)^2}\right] = 10\log\left[\frac{1}{2}\frac{\Delta P_{rms}(f_m)Z_0}{(K_\phi 2\pi\tau_d f_m)^2}\right] \qquad (139)$$

The sensitivity of the frequency discriminator method is then

$$\mathscr{L}(f_m) = 10\log\left[\frac{1}{2}\frac{\Delta P_{analyzer}(f_m)Z_0}{(K_\phi 2\pi\tau_d f_m)^2}\right] \qquad (140)$$

For example, If the noise figure of the baseband analyzer is 3 dB, then $\Delta P_{analyzer}$ in 1 Hz bandwidth is -171 dBm. In this case, with a phase detector gain of 0.2 volts/radian and τ_d equal to 300 nS, the maximum offset frequency for uncorrected measurement is 530 kHz and the system noise floor at offsets of 1, 10, and 100 kHz are -118.5, -138.5 and -158.5 dBc/Hz, respectively.

Notice that when f_m equals $2\pi\tau_d$ the sensitivity of the frequency discriminator method equals the sensitivity of the reference oscillator method (assuming the noise of the reference oscillator is less than the noise of the DUT). Closer to the carrier the sensitivity of the frequency discriminator degrades as f^{-2}. The frequency discriminator is best suited for noise measurement at higher offsets, although it is more sensitive than the direct method at a wide range of offset frequencies.

4.3.6 Example Phase-Noise Measurement System

Two methods are used for measuring SSB phase-noise in the author's lab. An HP8568A, 100-Hz to 1.5-GHz spectrum analyzer (SA) is used to directly measure the SSB phase-noise in VHF and UHF VCOs. Varactor modulation noise or varactor RF voltage limitations make it difficult to achieve low SSB phase noise and the sensitivity of the SA direct method is often sufficient for these oscillator types.

When higher sensitivity is required, the frequency discriminator method is used and the HP8568 spectrum analyzer is used as the baseband noise detector. The noise figure of the HP8568A is approximately 43 dB. Therefore, an SA preamplifier with a low noise figure and gain of 50 dB or more is required to maximize the system sensitivity.

Fig. 4.13 shows the phase detector and preamplifier system used by the author. It is derived from a system described by Wenzel [13]. A Mini Circuits SBL-1 double-balanced mixer is used as the phase detector. L_1 and C_1 filter the RF components of the phase detector output. Lossless reactors result in a response peak at the natural frequency. To avoid this, R_7 is manually selected to combine with the loss resistance of L_1 and flatten the response. The Toshiba 2SK369 preamplifier has a noise figure of almost 0 dB from 100 Hz through 300 kHz. This preamplified signal is AC coupled to a pair of OPA604 opamps. With +7 dBm signal at the phase-shifter port, the overall system gain from the phase detector output to the SA 50-ohm input is 57.2 dB with a 3-dB cutoff bandwidth of 30 Hz to 700 kHz. The gain and bandwidth were measured using a fixed 20-MHz, -43-dBm signal at the delay line port (SBL-1 mixer RF port) and sweeping the frequency of a +7-dBm signal at the phase-shifter port. The gain is defined by taking 20 times the log of the ratio of the peak voltage of the beat note at the SA port and at the uA741 port. The SA port is terminated in 50 ohms. It is important that the output waveform is observed with an oscilloscope to insure the beat-note level does not saturate the OPA604 amplifiers.

Noise

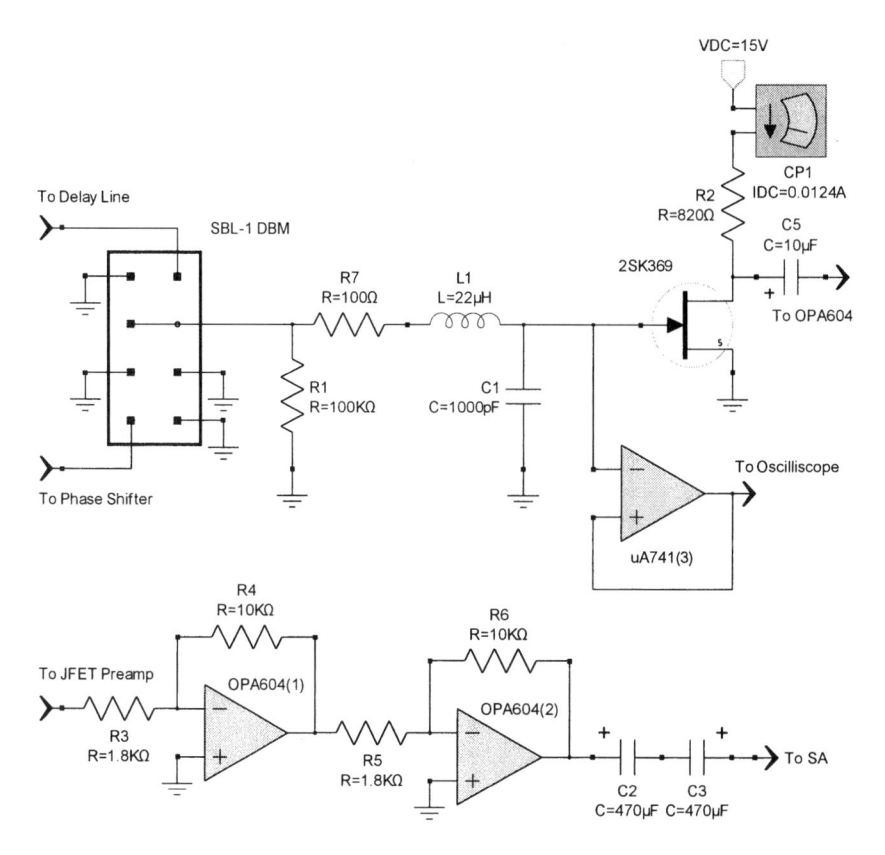

Figure 4.13 Phase detector and preamplifier for a frequency discriminator SSB phase-noise measurement system.

The delay line is 152 meters of foam RG-8U type foam dielectric coaxial cable. The measured delay of the cable with 0.5 meter lengths of RG-58 flexible connecting cables at each end is 583 nS. The oscillator-under-test drives a Mini Circuits ZFSC-2-2500 splitter. One port of the splitter drives the phase shifter port of the phase detector either directly or through a phase-shifter. The phase shifter is adjusted for phase quadrature, resulting in 0 volts DC at the uA741 output. Alternatively, with 583 nS of delay, phase quadrature occurs at multiples of 858 kHz and a tunable DUT oscillator is adjusted to the nearest frequency resulting in phase quadrature. The phase-detector gain constant, K_ϕ, is measured by tuning the oscillator frequency or adjusting the phase-shifter to observe the peak-to-peak swing at the uA741 port. K_ϕ in volts/radian equals the peak voltage swing. At 400 MHz and a +10-dBm signal to the splitter, the phase-detector gain constant of this system is 0.182 volts/radian.

The uA741 operational amplifier (opamp) drives an oscilloscope that serves as a quadrature detector. The phase shifter is adjusted until the DC component of the baseband signal is 0 volts. The 10-Hz minimum resolution bandwidth of the HP8568A spectrum analyzer limits the minimum practical offset measurement frequency to approximately 200 Hz. Therefore, this system is capable of measuring the SSB phase-noise at offsets from 200 Hz to 300 kHz with a system noise floor listed in Table 4.2.

Table 4.2 SSB phase-noise sensitivity of the preamplifier in Fig. 4.13, a 500 foot foam RG-8U delay line and an HP8568A spectrum analyzer

Offset (Hz)	SSB PN sensitivity (dBc/Hz)
300	-116
1000	-126
3000	-136
10K	-146
30K	-156
100K	-166
300K	-176

Measured using this system and given in Table 4.3 is the SSB phase noise at 400 MHz and +10-dBm output level of the HP8640B signal generator used in the author's laboratory.

Table 4.3 Measured SSB phase-noise of an HP8640B signal generator at 400 MHz and +10 dBm output level

Offset (Hz)	SSB phase noise (dBc/Hz)
300	-75
1000	-91
3000	-108
10K	-124
30K	-138
100K	-149
300K	-155

4.4 Designing for Low Phase Noise

As described in Section 4.2, numerous mechanisms may contribute to the SSB phase noise of an oscillator. Improving the phase noise performance requires determining the predominant sources of noise. For example, if the predominant source is varactor modulation noise, then increasing the cascade loaded Q offers little benefit for the phase noise performance.

4.4.1 Estimating the Predominant Noise Source

Provided the necessary parameter values are known, the master noise equation, Eq. 119, is used to estimate whether Leeson noise, varactor modulation noise, pushing-induced noise or buffer noise predominates.

If parameter values are not known, or for confirmation, varactor modulation noise is temporarily removed by replacing the varactor with a fixed capacitor. If phase-locking is required during phase-noise measurement, the tuning range is significantly reduced by using fixed capacitance in series and/or in parallel with the varactor.

Pushing-induced phase noise is temporarily removed by utilizing a battery for the power supply.

Unless oscillator output power is lower than 0 dBm or the buffer amplifier noise figure is high, buffer noise generally does not limit the oscillator system noise performance, except perhaps at the highest offset frequencies. Since a buffer is often used for load isolation, for testing purposes it too may be temporarily removed.

4.4.2 Reducing Leeson Noise

If the predominant source of noise is Leeson, then the following parameters are managed during the design process. They are listed roughly in order of significance. It is assumed the operating frequency, f_o, and the offset frequencies, f_m, are set by system requirements.

4.4.2.1 Loaded Q

Eq. 113 reveals that the SSB phase-noise decreases with the square of the cascade loaded Q. Phase-noise performance is affected by other parameters only linearly. Therefore, the most important design parameter for phase-noise performance is the cascade loaded Q. Increased loaded Q also reduces pushing so it reduces pushing-induced noise. Loaded Q deserves special attention during the oscillator design process.

Eq. 38 reveals that as the loaded Q approaches the resonator unloaded Q, the resonator insertion loss approaches infinity. Therefore, the resonator unloaded Q sets an upper limit on the loaded Q. Using a specific oscillator configuration with a simple uncoupled series resonator and with a high-efficiency amplifier with a switching and low-impedance output stage, Everard [14] demonstrated that the optimum ratio of loaded to unloaded Q is 2/3 with an open-loop gain margin of 3. The optimum Q ratio is broad and values from 0.4 to 0.9 produce phase-noise performance within 3 dB of optimum. A loaded to unloaded Q ratio of 2/3 produces an insertion loss of 9.54 dB in the resonator. While the optimum ratio does not strictly apply to configurations other than Everard's, his ratio offers a guideline. However, a ratio of 2/3 results in excessive loss if the resonator unloaded Q is not as high as expected. A ratio of 1/2 offers a compromise with excellent phase-noise performance and reduced resonator insertion loss. With simple series or parallel resonators, high loaded Q may require extreme values of resonator source and load resistance. This problem is readily solved using resonator coupling as described in Section 1.6.8.

Since the loaded Q is limited by the resonator unloaded Q, state-of-the-art phase-noise performance is achieved using a resonator technology with the highest possible unloaded Q. The available unloaded Q of various resonator technologies is discussed in Section 6.1. However, it is important to recognize that merely increasing the resonator unloaded Q without increasing the loaded Q is of limited benefit.

4.4.2.2 Output Power

The second most useful parameter for improving phase-noise performance is the output power. SSB phase-noise is specified relative to the carrier. Noise is a fixed, low-level disturbance. Increasing the output power reduces the noise relative to the carrier. If high power seems inconsistent with lownoise, consider the example oscillator and measured data given in Section 7.4.6.

4.4.2.3 Output Power Coupling

Leeson's equation for the SSB phase noise of an oscillator assumes tight coupling of the output load to the oscillator. Load pulling is improved by loosely coupling the load. This reduces output power but may not degrade the SSB phase noise at offsets lower than resonator half bandwidth, $f_o/2Q_L$. However, the reduced output power will degrade the SSB amplitude noise at offsets higher than the resonator half bandwidth. Therefore, the best noise performance is achieved using full coupling.

4.4.2.4 Device Noise Figure

It would seem that good active device noise figure is critical to good oscillator noise performance. However, modern active devices have excellent spot (white) noise figure. The noise figure difference between an economical or state-of-the-art device may be only a few decibels. Only marginally better oscillator noise performance is achieved using the best devices. It is unwise to choose a device based on noise figure if it has excess gain that results in a gain margin over 6 to 8 dB.

If the loop gain margin and resulting amplifier compression are moderate, linear amplifier noise theory provides design guidelines. The operating noise figure of an amplifier is a function of the source impedance. The optimum oscillator noise performance may not be achieved by matching the cascade input and output impedance but rather by designing the match for the appropriate source impedance.

4.4.2.5 Device Flicker Noise

While the device spot noise figure at the output frequency is only moderately important, achieving good SSB phase noise close to the carrier requires the selection of a device process with good flicker noise performance.

Flicker noise is associated with contamination and crystal defects that capture and release carriers in a random fashion. The associated time constants result in increasing noise with decreasing frequency. A $1/f$ distribution in the amplifier becomes a $1/f^3$ oscillator noise distribution inside the resonator half bandwidth.

Silicon and silicon-germanium, bipolar, and junction field effect devices generally have a flicker corner below 10 kHz. Bi CMOS also typically has low flicker noise. However, gallium arsenide, MES FET, and PHEMT devices can have flicker corners well over 10 MHz. Some RF CMOS processes have flicker corners of a few megahertz. A high flicker corner frequency destroys the SSB phase noise of an oscillator. LTV theory introduced in Section 4.2.3 is helpful in dealing with devices with a high flicker corner.

The parameters in device models for flicker noise are the flicker noise constant, *KF*, and the flicker noise exponent, *AF*. In SPICE they define the noise power spectral density equal to *KF* times the current raised to the power of *AF* and divided by the offset frequency. While this SPICE technique has limitations, it is commonly used. Device flicker noise is often not characterized by manufacturers. Exceptions are the NE68519 and NE68819 devices by NEC/California Eastern Laboratories [15]. Given in Table 4.4 are the flicker parameters and corner frequencies of these devices versus DC collector current and the collector-emitter voltage.

Table 4.4 Flicker model parameters AF and KF and the flicker corner of NE68819 and NE68519 devices versus collector current

	Vce = 1 volt			Vce = 3 volts		
Ic (mA)	AF	KF	Fc (kHz)	AF	KF	Fc (kHz)
NE68819						
5	1.90	5E-15	5.6	1.18	10E-15	5.1
10	1.39	72E-15	5.9	1.22	15E-15	5.8
20	1.19	14E-15	7.7	1.23	16E-15	6.7
30	1.16	10E-15	8.3	1.26	26E-15	7.3
NE68519						
5	1.56	454E-15	5.8	1.35	80E-15	7.3
10	1.48	231E-15	8.8	1.19	18E-15	9.9
20	1.89	7600E-15	12.4	1.12	13E-15	11.2
30	1.60	774E-15	18.6	1.25	30E-15	11.9

Flicker noise rises proportionally with collector current. However, the white noise floor increases with increased device dissipation, which tends to lower the effective flicker corner. Nevertheless, with the measured bias conditions, the flicker corner increases with increased collector current.

Unfortunately, this data reveals that the flicker parameters are a function of both the quiescent DC current and voltage, thus raising concern about the validity of noise modeling under large-signal, nonlinear operating conditions.

Unbypassed resistance in series with the emitter of bipolar amplifiers has a strong influence on flicker noise [16], as do the specific bias point and component values. Ferre-Pikal et.al. offer an excellent article on the design of amplifiers for low flicker noise [17]. In general, using high F_t devices and designing the sustaining stage with high collector-base voltage, wide bandwidth, unbypassed emitter resistance and low gain are helpful. With the exception of the high collector-base voltage, these characteristics are typical to certain commercial BJT and HBT MMIC gain blocks. Indeed, these amplifiers have exceptionally low flicker corner frequencies. Arambruro et al [18] investigated the flicker noise of three families of commercial MMIC amplifiers. Results are summarized in Table 4.5.

Table 4.5 Parameters and flicker corner of certain wideband MMIC amplifiers

Model	Typ Gain @100MHz (dB)	Bandwidth (GHz)	Supply (volts)	Supply (mA)	Flicker corner (kHz)
E001 HBT	11.8	DC to 8	3.8	50	0.20
E003 HBT	22.1	DC to 8	3.8	35	0.10
E004 HBT	13.8	DC to 8	5.0	80	0.30
M001 BJT	18.5	DC to 1.0	5.0	17	0.20
M002 BJT	13.0	DC to 2.7	5.0	25	0.20
M006 BJT	20.0	DC to 0.8	3.5	16	0.20
M011 BJT	12.7	DC to 1.0	5.5	60	0.80
Q-101 FF	11.3	0.002 to 0.07	15.0	91	0.02

The reference does not specify the device manufacturer of each series. However, the E001, 003, and 004 devices are Mini-Circuits ERA series, the M001, 002, 006, and 011 devices are Mini-Circuits MAR series (originally Agilent MSA series) devices, and the Q-101 is an amplifier with transformer feedback manufactured by Spectrum Microwave (originally Q-Bit). The flicker corner frequency is extracted from graphs in the reference with a carrier frequency of 10 MHz with nonsaturating signal levels. Saturation affected flicker noise variously from device to device and with carrier frequency, in some cases improving and in others degrading flicker noise. Nevertheless, the flicker performance of these broadband MMIC devices is excellent and at offset frequencies above 100 Hz, device flicker noise is insignificant.

4.4.2.6 Operating Temperature

With oscillators not utilizing a quartz-crystal resonator, the dominant source of noise originates within the active device. At higher power levels, device dissipation increases the operating temperature and thus degrades the oscillator noise performance. However, semiconductor device reliability degrades by 2 for each 10°C increase in junction temperature. A typical absolute maximum junction temperature for silicon devices is 150°C, or 423K. Since room temperature is approximately 300K, the maximum allowed operating junction temperature represents only a 1.5-dB decrease in SSB phase noise performance. Therefore, because noise performance improves linearly with output power, the optimum noise performance is achieved with highest safe operating quiescent operating bias. Consideration should be given to the impact of device current on gain, noise figure, and flicker noise. Nevertheless, the best noise performance requires high power operation. Other practical issues, such as battery life, may dictate design trade-offs.

4.4.3 Reducing Pushing Induced Noise

As described in Section 4.2.2.1, supply noise modulates the carrier through pushing. Differential oscillators with controlled current sources typically have low pushing and therefore good immunity to pushing induced noise. When a change in the supply voltage alters the collector-base voltage and capacitance in a bipolar transistor, the transmission phase-shift is altered, which modulates the frequency. Bias schemes with low sensitivity to the supply voltage also result in good pushing performance. A steep phase slope reduces this frequency shift and reduces pushing. Therefore, loaded Q is again critical when oscillator noise performance is limited by supply noise.

To avoid pushing induced noise, extremely low supply noise is required. From Eq. 115, with a typical pushing sensitivity of 900 kHz for a 900 MHz oscillator and a noise voltage of 35 nV/Hz peak, the contribution to single sideband phase noise at 1-kHz offset is -90 dBc/Hz. Only high-quality commercial regulators achieve this level of noise performance. For better phase noise, it is necessary to follow voltage regulators with passive, or well-designed discrete regulators. To achieve less than 0.1 radians peak deviation, a discrete 60-Hz ripple voltage must be less than 3-uV peak!

It is the author's experience that monolithic IC voltage regulators may have erratic output noise. For low-current oscillators, R-C filtering is less expensive and more reliable. For high-current oscillators, discrete active supply filtering is recommended. If an IC voltage regulator is required, at a minimum, a device with specified noise performance should be used.

Chemical batteries offer noise voltages well below those of IC voltage regulators and high-quality power supplies [19]. Over a broad frequency range, the noise voltage is approximately equal to the Johnson noise of the battery internal resistance. High capacity and Ni-Cd batteries with low internal resistance are particularly quiet. This suggests that very low-noise power supplies might be feasible using constantly charged chemical batteries.

4.4.4 Reducing Buffer Noise

The impedance of certain loads vary during operation. For example, the terminal impedance of an antenna varies when objects move in the near field. In receivers, the isolation of a mixer or RF preamplifier reduces oscillator pulling, however, in critical applications a buffer between the oscillator and the system is required.

Because a buffer amplifies both the carrier and the sideband noise, provided the sideband-noise power being amplified is above the buffer noise referenced to its input, the buffer does not degrade the oscillator noise performance. Eq. 118 quantifies the effect of buffer noise.

Active buffer amplifiers are commonly used. However, there is a more reliable and less costly method for isolation that improves oscillator sideband noise performance. Shown in Fig. 4.14 are the active buffer load isolation method (left) and pad isolation method (right). Both systems are designed to deliver 7 dBm to the load. Both systems offer load isolation.

With the active buffer method on the left, an oscillator with 0-dBm output drives a buffer with 7 dB of gain. The oscillator has an SSB phase noise of -140 dBc/Hz at 100 kHz offset. This noise level is significantly higher than the -171 dBm/Hz input noise level of the buffer so the buffer does not measurably degrade the SSB phase noise.

In the system on the right, the oscillator is operated at 13-dBm output. A 6-dB pad reduces the output level to 7 dBm. However, with other oscillator parameters being constant, the higher oscillator output level improves the SSB phase noise to -153 dBc/Hz. The pad attenuates both the carrier and the sideband noise. Thus, the pad isolation system has a 13-dB better phase noise performance. The pad system replaces an active circuit with a simple pad and offers significantly improved phase-noise performance. Broadband gain blocks with feedback have relatively poor reverse isolation. In this case, the pad also offers better isolation. The higher power oscillator requires additional oscillator circuit current but the current of the active buffer is avoided.

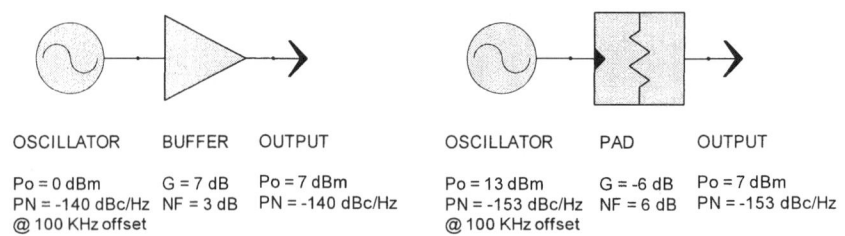

OSCILLATOR	BUFFER	OUTPUT		OSCILLATOR	PAD	OUTPUT
Po = 0 dBm	G = 7 dB	Po = 7 dBm		Po = 13 dBm	G = -6 dB	Po = 7 dBm
PN = -140 dBc/Hz @ 100 KHz offset	NF = 3 dB	PN = -140 dBc/Hz		PN = -153 dBc/Hz @ 100 KHz offset	NF = 6 dB	PN = -153 dBc/Hz

Figure 4.14 Active buffer method (left) and pad method (right) for reducing oscillator system load pulling.

4.4.5 Reducing Varactor Modulation Noise

Varactor modulation noise is often the dominant source of noise in varactor-controlled oscillators with a wide tuning range. Varactor modulation noise is quantified in Section 4.2.2.2. Ignoring flicker, buffer, and pushing noise, at offset frequencies lower than the resonator half bandwidth, varactor modulation noise dominates when

$$K_v > \frac{f_0}{4Q_L} \sqrt{\frac{F}{R_{enr} P_s}} \quad (141)$$

For example, for a 100-MHz oscillator with a loaded Q of 40, a sustaining stage noise factor of 2 and a power of 7 dBm and R_{enr} of 3KΩ, varactor modulation noise dominates if the tuning sensitivity exceeds 228 kHz/volt. For a similar 900-MHz VCO with a loaded Q of 10, varactor modulation dominates if the tuning sensitivity exceeds 8 MHz/volt. There is no advantage of the loaded Q exceeding a specific value when varactor modulation noise dominates.

$$Q_L < \frac{\Delta v_t}{4bw} \sqrt{\frac{F}{R_{enr} P_s}} \quad (142)$$

where bw is the fractional tuning bandwidth and Δv_t is the voltage delta required to tune the frequency range. For example, with a noise factor of 2, a power of 7 dBm, R_{enr} of 3 KΩ, a fractional tuning range of 0.2 (20%) and a tuning voltage delta of 5 volts, there is little advantage of a loaded Q much higher than 2.3 from the perspective of phase noise! Varactor modulation noise dominates except for narrow tuning VCOs.

Few desirable options exist for wide-tuning varactor VCOs. To avoid varactor modulation noise entirely, YIG-tuned oscillators are employed. Historically this was the case for high-performance instrumentation. YIG oscillators offer the added benefits of high loaded Q, excellent phase-noise performance, and broad, linear tuning bandwidth. Unfortunately, YIG oscillators are expensive and require a high control current.

Varactor selection for low R_{enr} is a first step. Unfortunately, varactor manufacturers are of little help. Testing a wide-tuning oscillator with varactor samples is the typical solution. However, even the best varactors provide only marginally better phase noise performance than typical varactors.

Because phase-noise performance is proportional to the tuning sensitivity squared, the best option is reducing the tuning sensitivity. It might seem prudent to design a VCO with a wide margin on the tuning range. However, doubling the tuning range degrades the varactor modulation noise by 6 dB. A large tuning voltage delta is extremely beneficial. A 15-volt delta provides 10-dB better phase-noise performance

.... a 5-volt delta! While this option is typically not available in battery-powered applications, it is easily accomplished in fixed station operation. High reverse-breakdown tuning varactors typically have a higher series resistance and lower unloaded Q. Therefore, using a tuning voltage near the varactor specified reverse-bias breakdown voltage is a small advantage. However, wide tuning VCOs typically have a lower loaded Q and resonator unloaded Q is less critical.

The final step in reducing tuning sensitivity is combining switched inductors to cover the tuning range while the varactor tunes a smaller bandwidth. Ignoring the required tuning overlap margin, a 2-range VCO improves varactor-modulation noise by 6 dB. Using a DC-controlled course tuning varactor and a fine tuning varactor for phase-locking offers no varactor-modulation noise advantage since the course varactor contributes to the noise. A course and fine tuning varactor does reduce cleanliness demand on a phase-lock control line.

4.4.6 Reducing Oscillator Noise Summary

Statements such as "avoid saturation at all costs," "use high-impedance devices," and other similar statements abound in the literature. Such rules of thumb for minimizing the noise are misleading because they presume certain configurations. For example, "use high-impedance devices" presumes the use of a parallel-mode resonator and the absence of coupling elements. Leeson's simple equation is a remarkably accurate predictor of phase-noise performance and it has no term associated with the impedance level of the sustaining stage. The following guidelines are offered for the minimization of oscillator noise (listed approximately in order of importance).

1) Estimate contributions to the total noise performance using the master noise equation.
2) Verify the noise performance using nonlinear simulation.
3) Verify the analytical estimate and simulation by powering a prototype oscillator by battery to eliminate pushing induced noise and temporarily replacing the varactor with a fixed capacitor to eliminate varactor modulation noise.

If performance is limited by Leeson noise, utilize the following steps.

1) Increase the loaded Q. Use a higher unloaded resonator Q and technology if necessary.
2) Increase the oscillator power level. Consider improving DC to RF conversion efficiency if supply power is limited.
3) Maintain reverse DC bias on the tuning varactors and the RF voltage below 1 volt peak. Use back-to-back varactors.
4) Use a moderate gain margin to avoid severe nonlinear action.
5) Couple the output tightly to the load.

6) Avoid semiconductor processes with a high flicker noise corner such as GaAs FETs. If these processes are unavoidable consider LTV theory to minimize flicker noise.
7) Use a modest degree of unbypassed emitter degeneration resistance.
8) Select a device with a good noise figure at the required operating current.
9) Design the cascade for a good noise match.
10) Once a good design foundation is established via the above principles, investigate different bias points using nonlinear simulation and testing. Small bias changes may significantly influence flicker noise.

If performance is limited by varactor modulation noise, then utilize these steps.

1) Select the varactor for low R_{enr}.
2) Increase the tuning voltage delta required to cover the frequency range.
3) Range switch the resonator inductor.

If performance is limited by pushing and supply noise, then utilize these steps.

1) Use a bias scheme insensitive to the supply voltage.
2) Use a high supply voltage and a device active-region voltage well above the knee voltage where the current is somewhat independent of the active-region voltage.
3) Avoid zeners and inexpensive voltage regulators that do not have noise specified.
4) For low current oscillators, R-C filter the supply voltage. For high current oscillators, utilize discrete transistor supply filtering.

4.5 Nonlinear Noise Simulation

In this section, Harbec [20] nonlinear noise simulation of oscillators is described. Harbec mixes small-signal noise with the oscillator carrier signal and its harmonics to create noise sidebands. The noise on each sideband and their correlation is then used to compute the phase noise. Phase noise in dBc/Hz is available as the measurement NPM, amplitude noise in dBc/Hz is available as the measurement NAM, and the relative power of the AM-PM SSB noise is available as the measurement NAPM. Additionally, Harbec supports reporting all individual contributors to the oscillator noise. Harbec nonlinear oscillator noise analysis considers

1) normal frequency conversion noise;
2) AM to PM noise conversion;

3) frequency translated noise caused by nonlinearity in the presence of the large-signal oscillation;

4) bias shift due to oscillation.

Both the upper and lower sideband noise is computed for each harmonic of the carrier and therefore nonlinear noise simulation requires 4 times the memory of a normal harmonic balance simulation. To minimize the required memory, the number of specified harmonics in the General Tab of Harbec should be modest; five or seven harmonics is sufficient for oscillators with an appropriate gain margin. Details of the process and setup are considered next using examples.

4.5.1 Negative Resistance Oscillator Noise Example

Shown in Fig. 4.15 on the left is a silicon 2N5109 medium power bipolar transistor with a specified F_t of 1.2 GHz in a CC configuration biased at approximately 10.6 mA. On the right a negative-resistance oscillator is formed by grounding the base through the series resonator L_1 and C_1. The ungrounded schematic on the left is used to simulate the transistor small-signal input resistance and reactance. At the desired oscillation frequency, the negative resistance is -11 ohms and the reactance is -38.4, an effective series capacitance of 5 pF.

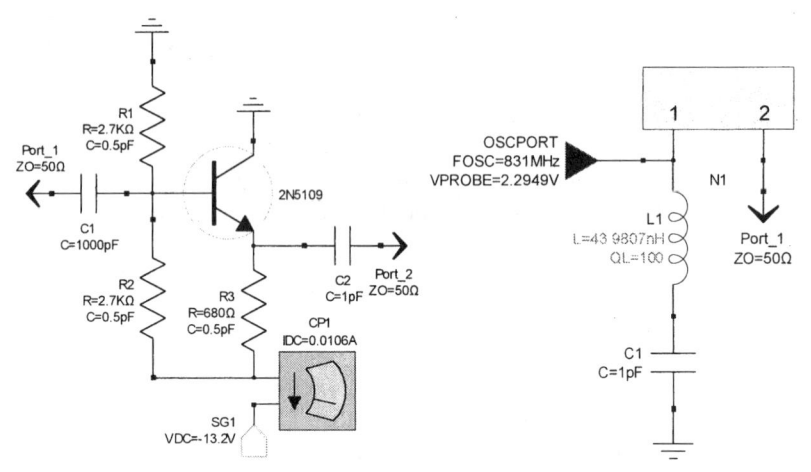

Figure 4.15 CC 2N5109 biased at 10.5 mA (left) and a negative-resistance oscillator formed by grounding the base through a series resonator.

The *Oosport* probe in the right schematic is a Harbec feature that first searches for the natural frequency and then optimizes the large-signal oscillation amplitude and frequency for minimum current in the probe. The frequency range for the search is specified in the Oscillator Tab of the Harbec properties dialog.

Noise

This action is similar to the manual technique first described in Section 2.5.1. An oscillator prototype was constructed with a wire-wound resonator inductor L_1 calculated to be 40 nH. L_1 was then manually tuned in the simulator until the oscillation frequency determined by *Oscport* matched the measured prototype operating frequency, 831 MHz. The simulated vale of L_1 is 43.9807 nH. The calculated resonant frequency of L_1, C_1 and the small-signal effective input capacitance is 831.34 MHz. Therefore, for this circuit, the large-signal and small-signal effective input capacitances are similar.

The simulated output spectrum with five harmonics balanced is shown in Fig. 4.16. The simulated output power is 3 dBm. The measured output power at 831 MHz is 3.4 dBm.

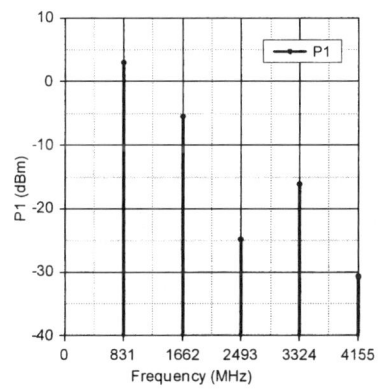

Figure 4.16 Simulated output spectrum of the negative-resistance oscillator shown in Fig. 4.15.

Oscport also implements the nonlinear phase and amplitude noise analysis in Harbec. The Noise Tab of the Harbec dialog shown in Fig. 4.17 is used to specify the output port and the oscillator offset frequency range. In the following analysis the noise is swept from 100 Hz to 10 MHz.

As the offset frequency approaches the carrier the simulator noise conversion matrix becomes ill-behaved. This causes the displayed noise to flatten with decreasing offset rather than continuing to rise. Because it is caused by ill conditioning, small changes in simulation parameters may cause this effect to come and go. Checking the Polishing option invokes a routine that is often successful in correcting the problem. If the problem persists, it is sometimes helpful to slightly modify the bias network or to specify a slightly smaller or larger quantity of harmonics to balance.

The SSB noise may be calculated for all carriers (harmonics) or a specific carrier. The carrier with index 1 is the fundamental, the carrier with index 2 is the second harmonic, and so forth. This option only controls which carrier SSB noise is calculated and displayed and has no impact on the simulated noise of the fundamental.

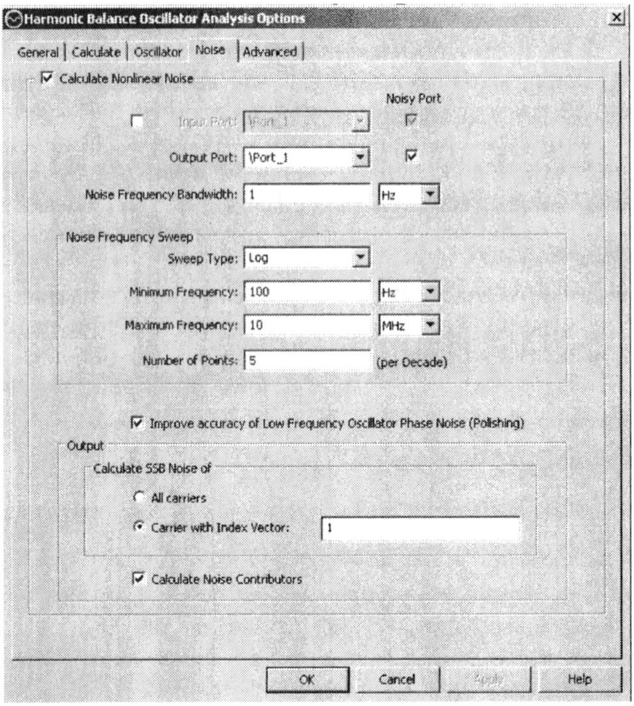

Figure 4.17 Noise Tab of the *Oscport* dialog.

Harbec also reports the contributions to the SSB noise from each source within the oscillator. If the option Calculate Noise Contributors is selected, the contributions are available in the automatically created output variable NCValue[index]. The index is ordered with the most significant contributor at large offset labeled as index 1. The automatically computed output variable NCName[index] holds a text string indicating the source of the contributor within the oscillator. The NCName data for the Harbec SSB noise analysis of the negative-resistance oscillator in Fig. 4.15 is given in Fig. 4.18.

Plotted in Fig. 4.19 is the SSB noise from the five most significant contributors to the 831-MHz negative-resistance oscillator. For example, the most significant source is current shot noise, Irb, in the base resistance of the 2N5109. The "*" indicates this source is generated by oscillator nonlinear mixing. The term nonlinear should not be interpreted too literally. When external resistance is placed in series with the base to reduce the magnitude of the negative resistance until operation is nearly linear and the second harmonic is over 60 dB below the carrier, this "nonlinear" source remains the predominant source of noise.

Noise

Reactors influence noise but they do not generate noise. The noise contribution from the resonator inductor L_1 is from the series resistance associated with the unloaded Q of the inductor. The remaining contributors are below the Port_1 contribution and are not displayed in Fig. 4.19.

Variable	Q	NCIndex	NCName
CNOISE	1	1	\N1\2N5109\Q1:Irb:shot:*:(white)
FNOISE	2	2	\N1\2N5109\Q1:Ic:shot:*:(white)
Freq	3	3	\Port_1:thermal:(white)
FreqID	4	4	\L1:thermal:(white)
FreqIndexIM	5	5	\N1\R2\R1:thermal:(white)
FSPNOISE	6	6	\N1\R1\R1:thermal:(white)
IN1_CP1	7	7	\N1\R3\R1:thermal:(white)
IPORT	8	8	\N1\2N5109\Q1:Irb:flicker:*:(flicker)
LogOutput="Harmonic	9	9	\N1\2N5109\Q1:Rb:*:(white)
NAM	10	10	\N1\2N5109\Q1:Re:(white)
NAPM	11	11	\N1\2N5109\Q1:Rc:(white)
NCIndex	12	12	\N1\R2\C1:thermal:(white)
NCName	13	13	\N1\R1\C1:thermal:(white)
NCValue	14	14	\N1\R3\C1:thermal:(white)
NPM			
Osc_dPhaseZdF			
Osc_Freqs			
Osc_ReZ			
P1=[PPORT[1]]			

Figure 4.18 NCName data of the Harbec oscillator noise analysis.

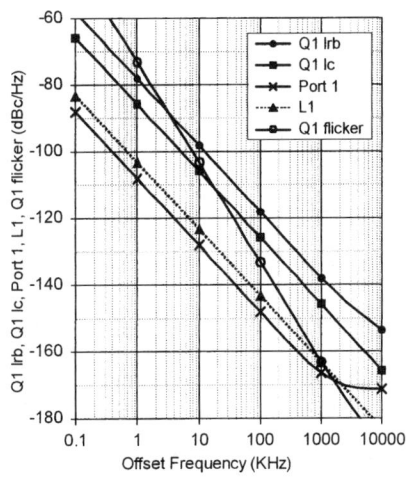

Figure 4.19 Simulated four most significant contributors of SSB noise of the negative-resistance oscillator.

Fig. 4.20 shows simulated, Leeson predicted, and measured SSB noise of the 831-MHz negative-resistance oscillator. The nonlinear simulated AM, PM, and total noise are plotted with triangular, square, and circular symbols, respectively. As is often assumed in SSB noise analysis, for this oscillator the phase noise totally dominates except at high offset frequency. The small-signal, open-loop, loaded Q determined using Alechno's technique as described in Section 1.7.1.5 is 4.3. The flicker corner frequency of the 2N5109 was unknown. A flicker corner of 1 kHz and a device NF of 3 dB are assumed. The power used in Leeson's equation is the measured power, 3.4 dBm. The phase noise predicted by Leeson's equation agrees well with both the nonlinear simulation and the measured data. Phase noise of -78, -88, -98, and -107 dBc/Hz at 1, 3, 10, and 30 kHz, respectively, were measured using an HP8568A spectrum analyzer. The noise floor of the analyzer precluded measurement at offsets higher than 30 kHz. The measured data suggests the flicker corner is lower than assumed and predicted by nonlinear simulation.

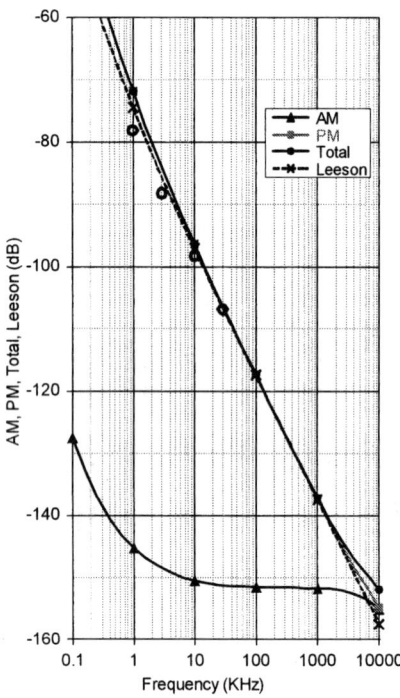

Figure 4.20 SSB noise of the 831-MHz negative resistance oscillator including the nonlinear simulated AM noise (triangular symbols), the simulated PM noise (square symbols), the total simulated noise (solid circular symbols), the phase noise predicted by Leeson (cross symbols), and the noise measured using a spectrum analyzer (open circular symbols).

4.5.2 Linear Oscillator Phase Noise Example

In this example, the SSB phase noise of a 300-MHz negative-resistance oscillator is investigated with the oscillator operating in a conventional nonlinear limiting mode and with a small incandescent lamp to establish a nearly linear operating mode. The open-loop and closed-loop schematics are shown in Fig. 4.21. The open-loop schematic simulates the small-signal negative resistance and reactance at port 1. The simulated negative resistance at 300 MHz is -33.4 ohms and the reactance is -121 ohms, corresponding to an effective series input capacitance of 4.4 pF.

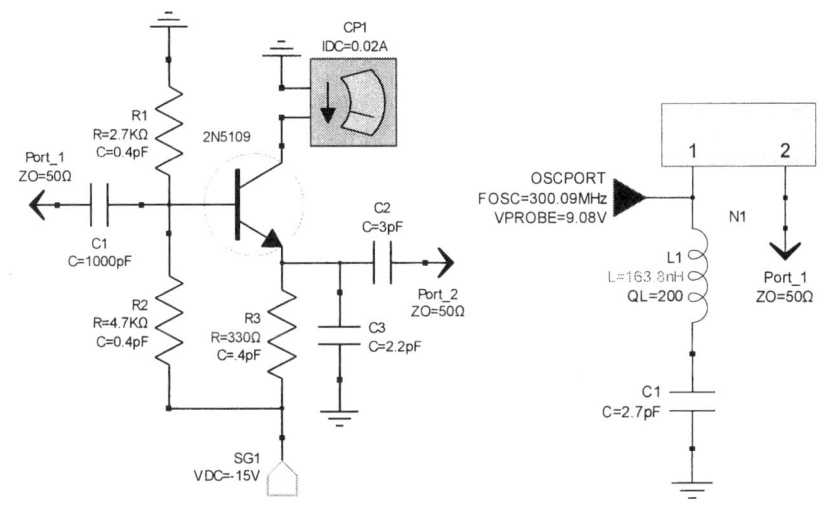

Figure 4.21 Open-loop (left) and closed-loop (right) schematics of the 300 MHz negative-resistance oscillator.

The oscillator is formed by placing a series resonant circuit from port 1 to ground as illustrated on the right in Fig. 4.21. A 2.7-pF capacitor is selected and L_1 is adjusted for a simulated oscillation frequency of 300 MHz. The harmonic-balance simulated output spectrum is shown in Fig. 4.22. The simulated output power is 11.5 dBm and the second and third harmonics are -10 and -15 dBc, respectively. The measured output power is 15.2 dBm and the second and third harmonics are -13.0 and -19.5 dBc, respectively. Closer agreement would be desirable, but the measured DC current is 28 mA, significantly higher than the predicted 20 mA, thus suggesting that the model DC characteristics do not accurately describe this device.

Next, a Micro Lamps 1730ML, 6-volt, 40-mA miniature lamp is placed in series with the resonator. The measured cold resistance of the lamp is 21 ohms and the resistance at 6 volts is 1670 ohms. Although the RF excitation of the lamp is insufficient to cause a visible glow, the oscillator output power is constant with a change in the supply voltage, suggesting the lamp is providing AGC. The measured output power is 0.5 dBm. The second and third harmonics are -29 and -49 dBc, respectively, further suggesting the oscillator is operating in a quasi-linear mode.

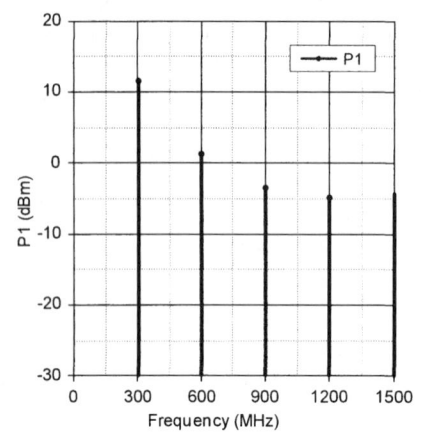

Figure 4.22 Simulated output spectrum of the 300 MHz negative-resistance oscillator operating the device in a nonlinear mode.

Given in column 2 of Table 4.6 is the phase-noise in column 2 of the 300-MHz oscillator operating in a conventional nonlinear limiting mode and given in column 3 is the phase noise with the lamp and operating in quasi-linear mode.

Table 4.6 Measured SSB phase-noise of the 300-MHz oscillator operating in conventional nonlinear limiting mode (column 2) and in a quasi-linear mode (column 3)

Offset (Hz)	Nonlinear Osc Po=+15.2 dBm Phase Noise (dBc/Hz)	Quasi-Linear Osc Po=0.5 dBm Phase Noise (dBc/Hz)
1K	-72	-59
10K	-102	-82
100K	-126	-112

Notice that the phase noise when operating in the quasi-linear mode is significantly worse than with nonlinear operation. The cause for the improved phase-noise performance in the nonlinear limiting mode is the higher operating level. This example suggests that Lee and Hajimiri's assertion [4] that phase noise is not fundamentally a nonlinear affect is valid. The use of AGC to improve phase-noise performance is ill advised. If the oscillator is designed with excessive open-loop gain or excessive negative resistance, then heavy nonlinear behavior can result in degraded performance. When designed with appropriate gain margin or negative resistance, backing off the maximum output power of the oscillator using AGC degrades the phase-noise performance.

Alternatively, a Micro Lamps 1738ML, 2.7-volt, 60-mA miniature lamp placed in series with the collector path to ground provides AGC to limit the output to +10 dBm. In this case, the lamp is well lit but operating below full brightness. Phase-noise data was not taken for this case.

4.6 PLL Noise

The phase-locked loop (PLL) is ubiquitous in modern electronic systems. In addition to providing a method of frequency tuning by digital control, PLLs can improve both long- and short-term frequency stability and reduce the impact of vibration. In the basic single-loop PLL, the frequency of a VCO is divided by a digitally controlled divider chain and the resulting signal is compared to a reference oscillator signal in a phase/frequency detector (PD). The output of the PD is filtered and this signal provides low-frequency control to lock the divided VCO to the reference. The divider number is either integer or fractional.

Within the bandwidth of the control loop, the VCO spectral density of frequency fluctuations assumes the character of the reference oscillator. The reference oscillator is often frequency divided from a high-stability, bulk quartz crystal oscillator. This reference signal has excellent long- and short-term stability.

Given in Table 4.7 is an abbreviated form of an automatically generated report from the Agilent Technologies =PLL= software program originally developed by Eagleware Corporation [20]. This PLL implements a 40- to 70-MHz frequency synthesizer with 250-kHz steps. The Loop Statistics section of the report provides simulated characteristics of the resulting PLL. The Input Parameters section repeats the data entered by the user for each of the functional blocks of the PLL.

Table 4.7 Abbreviated data from an automatically generated report of the =PLL= program by Agilent Technologies

Report created on Sat May 23 16:54:55 2009 by =PLL=, (c) 2003 Eagleware Corporation
PLL Type: Frequency Synthesizer
Loop Statistics Follow
===============
Residuals, RMS:
FM: 0.604006 Hz
PM: 0.000499289 radians
VCO f0 (Hz): 5.5e+07
VCO gain (Hz/Volt): 7.5e+06
PD gain (A/Radian): 0.000254648
Approx. Lock Time (sec): 0.000121951
Input Parameters Follow
================
Sim Tab:
Frequency Plot Control:
Min Freq (Hz): 10
Max Freq (Hz): 1e+6
Pts/Decade: 50
Test Conditions:
Divider N1: 160
Divider N2: 280
Residuals Analysis Control:
Flow (Hz): 100
Fhi (Hz): 2000

Ref Tab:
Reference:
Freq (Hz): 250000
Load Res (ohm): 40000
Pout (dBm): 0
Loaded Q: 50000
Flicker Corner (Hz): 10000
Noise Figure (dB): 3
PD/÷ Tab:
Phase Detector Types: Charge Pump P/F Detector
PFD Current (A): .001600
Noise Floor (dBc/Hz): -157
Logic "0" (V): 0.5
Logic "1" (V): 4.5
Convert Sine to Square: Checked
Div. Noise (dBc/Hz): -150
Divider N: 216
Filter Tab:
Integrator: Passive 3-Pole
Loop BW (Hz): ~8200 Hz
Phase Margin (Deg): ~49
C1 (F): 8.2e-9
Override: Checked
C2 (F): 4.7e-8
R1 (ohms): 1100
VCO Tab:
Fmin (Hz): 4e+7
Fmax (Hz): 7e+7
V for Fmin (V): 0.5
V for Fmax (V): 4.5
Use Tuning Table: Not Checked
$1/(2 \cdot pi \cdot R \cdot C)$ (Hz): 1e+6
Load Res (ohm): 50
Pout (dBm): 7
Loaded Q: 10
Flicker Corner (Hz): 10000
Noise Figure (dB): 3

Entries in the Sim Tab define analysis and display parameters such as the digital divider range and the frequency range for the simulation. The Ref Tab is used to describe the reference oscillator. The phase noise of the reference oscillator is computed from the entered parameters using Leeson's equation and the divider number. Alternatively, the measured reference oscillator phase noise versus frequency is manually entered. The PD/÷ Tab specifies the type and sensitivity of the PD and the estimated noise floor of the digital divider. The characteristics of the loop filter are specified in the Filter Tab. The user enters the desired loop bandwidth and phase margin. =PLL= then computes the values for the loop R_1, C_1, and C_2. In this case, the compute filter values Override box is checked and the nearest standard values to the computed values are entered. Also, for suppression of reference sidebands in the PLL output, the Filter Tab supports an additional loop filter of user specified type and order. No additional filter was used in this example. Finally, the VCO Tab is used to specify the characteristics of the VCO. A constant tuning voltage is used, as in this example, or a VCO frequency versus applied voltage table is entered. The $1/(2\pi R C)$ entry provides for the specification of the time-constant for an *R-C* tuning voltage filter. As with the reference oscillator, the phase noise of the VCO is computed by Leeson's equation or is entered by table.

The Output Tabs of the =PLL= program are then used to display the PLL open and closed-loop responses, the phase noise characteristics, the time-domain response to the initial lock, and a step change in the divider number, the schematic, or the report. The time-domain responses are not generated by analytical formula but rather by a time-step simulation of the loop starting from 0 volts at turn-on. This simulation includes nonlinear behavior so that of the PD that accurately simulates lock behavior.

Fig. 4.23 shows plots of the phase-noise contribution of each loop component to the total phase noise of the PLL. At the bottom, markers 1 through 4 provide the SSB phase noise at 0.1-, 1-, 10-, and 100-kHz offsets, respectively. The Total trace simulates the SSB phase noise of the output of the PLL. As with the time-domain simulation, the phase noise is not determined by analytical expression, but rather the noise of each loop component, including resistors in the loop filter, are propagated through the PLL, resulting in an accurate simulation of the steady-state phase noise of the system. The traces are not the phase noise of the individual loop functional blocks but rather the contribution to the total phase noise by that block as transferred by the closed loop. For example, notice that the phase noise contribution of the VCO peaks near the loop bandwidth at 8 kHz. The phase noise of the VCO continues to rise at 30 dB/decade toward the carrier but the noise correction properties of the loop reduce the phase noise contribution of the VCO inside the loop bandwidth. Similarly, the acoustic noise of the VCO is reduced inside the loop bandwidth and the primary source of acoustic noise becomes the quartz resonator in the reference oscillator.

This type of analysis is insightful for determining the steps that must be taken to improve the PLL phase-noise performance. For example, in this PLL, the primary contributor to phase noise below 50 Hz is noise in the reference oscillator, the primary contributor from 50 Hz to 500 kHz is divider noise and the primary contributor above 500 kHz is the VCO. Differing PLL and functional block parameters may change these relationships significantly. The quantity of interacting parameters that determine PLL performance make intuitive assessment difficult. Achieving the best phase-noise performance requires careful analysis.

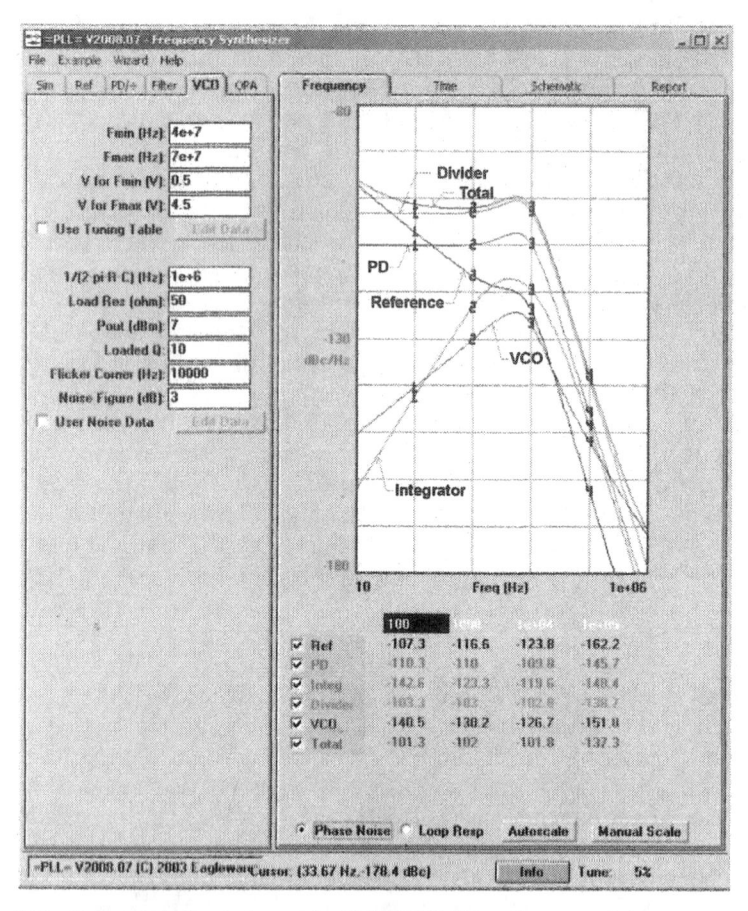

Figure 4.23 Noise analysis window of the GENESYS PLL software module.

References

[1] Phase Noise Characterization of Microwave Oscillators: Frequency Discriminator Method, Product Note 11729C-2, Agilent Technologies.

[2] U. Rohde, *Microwave and Wireless Synthesizers: Theory and Design*, John Wiley, New York, 1997.

[3] R. Rhea, *Oscillator Design and Computer Simulation*, Noble Publishing (Scitech), Raleigh, 1995.

[4] T.H. Lee and A. Hajimiri, "Oscillator Phase Noise: A Tutorial," *IEEE Journal of Solid-State Circuits*, Vol 35. 35, No. 3, March 2000, pp. 326-336.

[5] M. Odyniec, *RF and Microwave Oscillator Design*, Artech House, Norwood, MA, 2002.

[6] K. Kurokawa, *An Introduction to the Theory of Microwave Circuits*, Academic Press, New York, 1969.

[7] D.B. Leeson, "A Simple Model of Feedback Oscillator Noise Spectrum," *Proc. of the IEEE*, February 1966, pp. 329-330.

[8] G. Sauvage, "Phase Noise in Oscillators: A Mathematical Analysis of Leeson's Model," *IEEE Trans. on Instrumentation and Measurement*, December 1977, pp. 408-410.

[9] www.hittite.com.

[10] Spectrum Analysis: Noise Measurements, Application Note 150-4, Hewlett Packard, April 1974.

[11] Automated Noise Sideband Measurements Using the HP 8568A Spectrum Analyzer, Application Note AN 270-2, Hewlett Packard, 1980.

[12] D.W. Allen, "Time and Frequency (Time-Domain) Characterization, Estimation and Prediction of Precision Clocks and Oscillators," *IEEE Trans. Ultrasonics, Ferroelectronics and Frequency Control*, November 1987, pp. 647-654.

[13] C. Wenzel, "A Low Noise Amplifier for Phase Noise Measurement," lowamp.pdf, www.wenzel.com.

[14] J.K.A. Everard, "Minimum Sideband Noise in Oscillators," *IEEE Int. Frequency Control Sym.*, 1986, pp. 336-339.

[15] California Eastern Laboratories, 1/f Noise Characteristics Influencing Phase Noise, Application Note AN1026, February 2003.

[16] J. Li, E.S. Ferre-Pikai, F.L. Walls, D. Forbes and S.H. Talisa, "Predicting 1/f AM and PM Noise in Si BJT Amplifiers: A New Computer Program," *IEEE Int. Frequency Control Sym.*, 1998, pp. 204-209.

[17] E.S. Ferre-Pikal, F.L. Walls and C.W. Nelson, "Guidelines for Designing BJT Amplifiers with Low 1/f AM and PM Noise," *IEEE Trans. Ultrasonics, Ferroelectrics and Frequency Control*, Vol. 44, No. 2, March 1997, pp. 335-343.

[18] M.C.D. Aramburo, E.S. Ferre-Pikal, F.L. Walls and H.D. Ascarrunz, "Comparison of 1/f PM Noise in Commercial Amplifiers," *IEEE Int. Frequency Control Sym.*, 1997, pp. 470-477.

[19] C.K. Boggs, A.D. Doak and F.L. Walls, "Measurement of Voltage Noise in Chemical Batteries," *IEEE Int. Frequency Controls Sym.*, 1995, pp. 367-373.

[20] Agilent Technologies, GENESYS 2008.07 Documentation Set, www.agilent.com.

5 General-Purpose Oscillators

The next four chapters apply previously discussed principles to example oscillators. Each chapter includes oscillators categorized by general type. The categories are somewhat arbitrary. For example, any oscillator can be tuned by control of a reactor in the resonator. However, certain topologies are better suited for VCOs, so there is a chapter devoted to VCOs. Primarily, the categories are based on applications or overriding design principles. Some of the oscillators to be illustrated were covered during the introduction of oscillator concepts. Previously illustrated types are repeated so each chapter is complete.

This chapter includes a variety of general-purpose oscillators. In some cases, these oscillators illustrate underlying principles of later, more specific topologies. These general oscillators typically utilize resonators with simple R, L, or C resonators and exclude distributed, piezoelectric, and newer technology resonators.

5.1 Comments on the Examples

In the majority of the examples in this and later chapters the fundamental form of each oscillator is illustrated. Active devices are modeled by nonlinear Spice-type models provided by the manufacturers. To the degree that these models are accurate, devices represent real-world characteristics. However, lumped element and PWB layout parasitics are ignored since they are strongly dependent on the specific manufacturing process. In certain cases, where parasitics are critical, they are discussed. Unless otherwise noted, the spectrum graphs are simulated by the Harbec harmonic balance engine of the GENESYS suite by Agilent Technologies and the waveform graphs are simulated by the Cayenne engine of GENESYS. A number of examples include measured data and serve to increase confidence in the theory and simulation techniques.

Formulas are provided in Section 1.6.7 to compute element values for Colpitts resonators. However, because active devices are generally non-unilateral, the application of these formulas is often difficult. Formulas were not derived for all resonators in the following examples. To shift the frequency, initially the element values of the examples are scaled. Aside from satisfying the necessary initial criteria, the most important goal is increasing the loaded Q. To increase the loaded Q for parallel resonators, the resonator inductor value is decreased and the capacitor values are increased. To increase the loaded Q for series resonators, the resonator inductor value is increased and the capacitor value is decreased. Then linear simulation is used to tune or optimize the gain, phase, match and loaded Q, using the Randall/Hock correction if necessary. Then the design is finalized by nonlinear analysis using harmonic-balance simulation, time-step simulation, or both.

5.2 Oscillators Without Resonators

The term resonator here signifies a structure that (1) involves resonant properties such as a peak in transmission amplitude, (2) a transmission phase shift of 0° or 180° in the limiting case of high loaded Q, and (3) loaded $Q >> 1$. With a finite loaded Q, the phase shift at the transmission amplitude peak may not be 0° or 180°. Resonator examples include coupled L-C networks, distributed elements, and piezoelectric structures. Oscillators without resonators include R-C structures, ring oscillators, and delay line oscillators. These oscillators have poor loaded Q and therefore have relatively poor stability and phase-noise performance. They are used at lower frequencies where inductors are large and expensive, and because the lower operating frequency naturally results is better phase-noise (see Eq. 113). Oscillators without resonators are also used in IC design since high-Q inductors are difficult to realize on chip.

5.2.1 R-C Oscillators

R-C oscillators cascade an R-C phase-shift network with the active device to realize a net transmission-phase of 0° at the desired oscillation frequency. An R-C network introduces significant attenuation as well as the phase shift. To minimize the loss, an inverting sustaining stage is used to realize 180°, leaving 180° additional shift by cascaded R-C networks. Because the attenuation is infinite at the -90° limiting phase shift of a single R-C network, three or more networks must be cascaded.

The phase-shift of a series resistor, shunt capacitor network in a system with source and load resistance of R_o is

$$\phi = \tan^{-1} \frac{2R_0}{3X_c} \quad (143)$$

where X_c is the capacitor reactance. Solving for the required reactance

$$X_c = \frac{2R_0}{3\tan\phi} \quad (144)$$

General-Purpose Oscillators

For example, for -60° phase shift the required reactance with a 200-ohm resistor is -76.98 ohms. At 100 MHz this is a capacitance of 2067 pF. The insertion loss is given by

$$IL(dB) = -10\log\left(\frac{4X_c^2}{9X_c^2 + 4R_0^2}\right) \qquad (145)$$

The IL for -60° phase shift is 9.54 dB. When multiple sections are cascaded, the termination of the previous section is not purely resistive. Therefore the phase-shift and attenuation of a cascade is only approximately the sum of the above-estimated values. The accuracy is suitable for practical use. For example, given in Table 5.1 are the phase shift per section, IL, capacitance, and loaded Q versus the number of sections, N, for an R-C cascade with 200-ohm source, series, and load resistors. From Eq. 144, the required reactance for a -45° section with 200 ohms series resistance is 133.33 ohms, which is 1193.7 pF. From Table 5.1, the actual capacitance required in a cascade of four sections is 1125.4 pF. In any event, at finite frequency, real amplifiers are only approximately inverting so simulation is advised for accurately estimating the oscillation frequency.

Table 5.1 The phase shift per section, total IL, capacitance, and loaded Q versus the number of sections, N, for a 1 MHz R-C cascade

N	ϕ/section (°)	IL (dB)	Cap (pF)	Q_L
2	-90	∞	∞	0
3	-60	32.36	2250.8	0.55
4	-45	29.25	1125.4	0.73
5	-36	28.58	712.36	0.81
6	-30	28.62	500.17	0.86
7	-25.7	28.92	373.52	0.88
8	-22.5	29.34	290.86	0.90

Notice that increasing the number of sections has only a modest impact on the IL and the loaded Q. Of course, the IL must be overcome by the sustaining stage gain, so a minimum gain of approximately 34 dB is required.

Given in Fig. 5.1 is a 1 MHz R-C phase-shift oscillator. The input and output impedances of the 2N3904 amplifier were determined to be near 200 ohms, which was therefore used as the open loop, reference impedance. To reduce the loss, the amplifier output impedance is used as the first series resistor and capacitor C_2 completes the first R-C section. There are therefore three R-C sections of -60° each. Capacitor C_5 is only required for simulation as a DC block for the termination resistance. When the loop is closed capacitor C_1 is required to block the DC collector voltage from the base. Because the R-C sections present a poor match to the 200 load resistance, the Randall/Hock correction is advised for the initial linear open-loop cascade analysis.

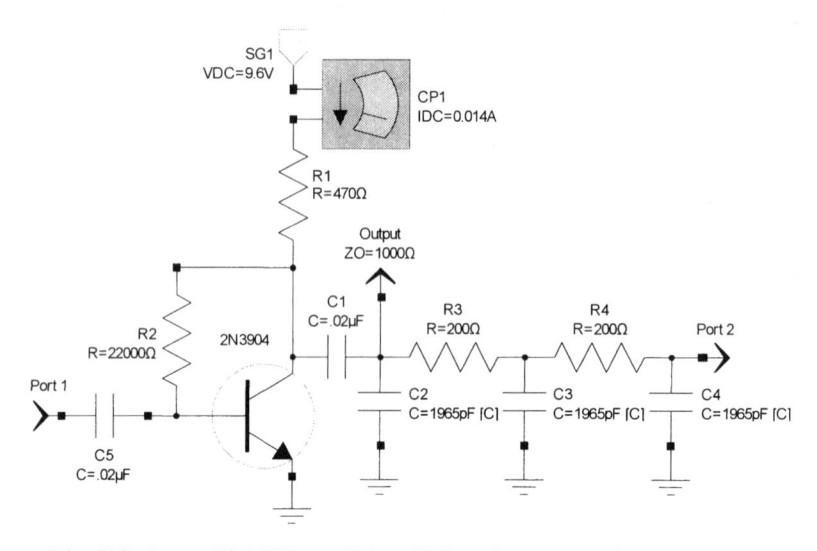

Figure 5.1 R-C phase-shift 1-MHz oscillator with three lowpass sections.

Harmonic balance simulation was used to adjust capacitors C_2 through C_4 for operation at 1 MHz. Given on the left in Fig. 5.2 is the simulated spectrum of the oscillator through seven harmonics.

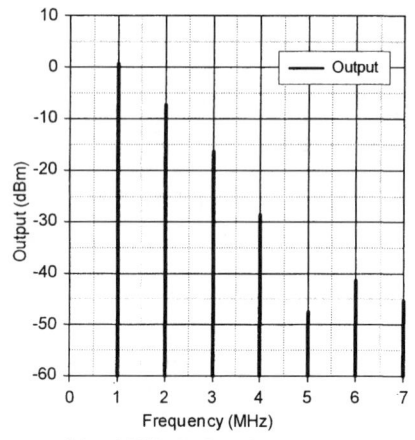

Figure 5.2 Output spectrum of the 1 MHz R-C oscillator.

Fig. 5.3 shows the amplitude and phase of G from Randall/Hock. Notice that with the capacitors adjusted for steady-state nonlinear oscillation at 1 MHz that the linear open-loop phase zero crossing is approximately 1.19 MHz. Nonlinear behavior of the active device shifts the transmission phase of the amplifier. This problem is aggravated by the

General-Purpose Oscillators

shallow phase-slope of the R-C oscillator, which introduces a significant frequency shift with a change in amplifier phase. This is a consequence of low loaded Q.

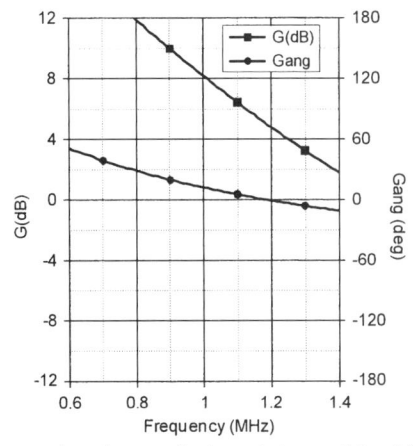

Figure 5.3 Open-loop cascade gain magnitude and phase of the 1 MHz R-C oscillator.

If the starting time is critical, C_1 should be selected with care. A large value slows starting while a small value acts as a high-pass element that leads the phase and requires more phase-shift from the R-C cascade. On the left in Fig. 5.4 is the starting waveform through 20 uS of the 1 MHz R-C oscillator. After approximately 1 cycle the period has settled at 1 MHz. This fast starting is a consequence of the low loaded Q.

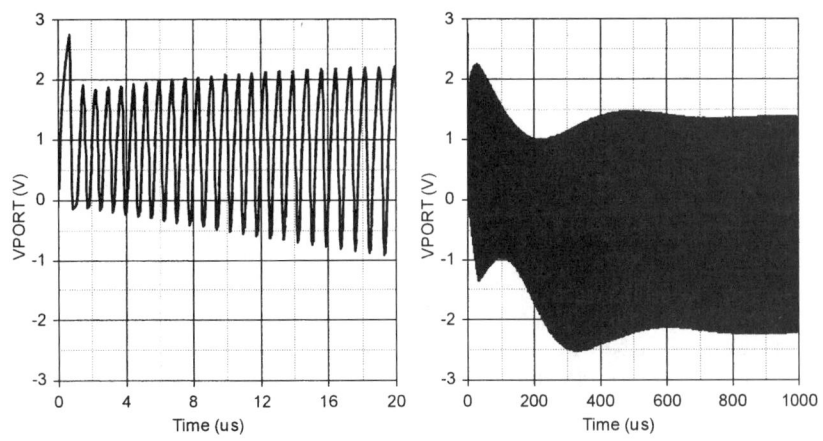

Figure 5.4 Starting waveforms through 20 uS (left) and 1000 uS (right) for the 1 MHz R-C oscillator.

On the right in Fig. 5.4 is the starting waveform through 1000 uS. It takes approximately 800 uS for all transients to settle. These longer term transients are the consequence of C_1 and C_5 charging through R_1. Recall that when the loop is closed that C_5 is no longer required, which will affect the starting waveform. Given in Fig. 5.5 is an *R-C* phase-shift oscillator using four highpass sections.

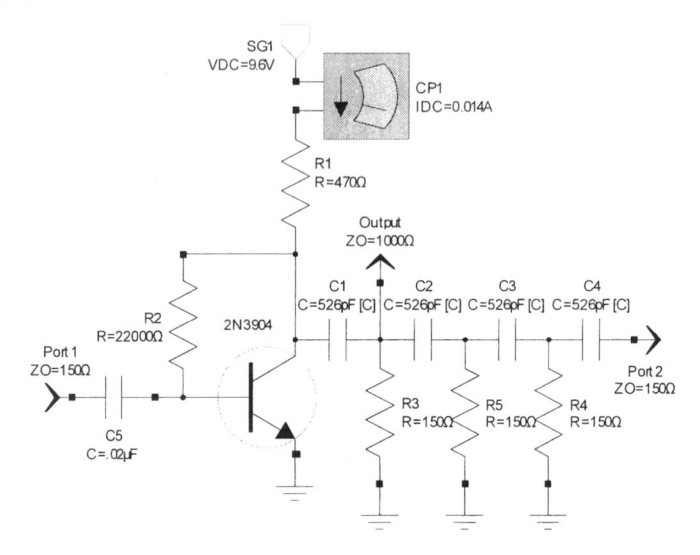

Figure 5.5 *R-C* 1 MHz oscillator with four highpass sections.

The output spectrum and waveforms for this oscillator are given in Fig. 5.6. While it appears that an additional capacitor is required, C_1 replaces a blocking capacitor required in the lowpass *R-C* oscillator. Additionally, C_1 is much smaller than the blocking capacitor thus eliminating secondary transient effects inherent to the lowpass oscillator, as evidenced in Fig. 5.6. This topology provides improved starting characteristics as indicated by the waveform on the right in Fig. 5.6.

Fig. 5.7 shows a 100 kHz *R-C* phase-shift oscillator using an operational amplifier (opamp). To insure the opamp has sufficient gain and the phase is 180°, the gain bandwidth product should be a minimum of 100X the oscillation frequency. Then, provided the network resistors are significantly greater than the operational amplifier output impedance, the oscillation frequency is given approximately by

$$f_0 = \frac{1}{2\pi RC\sqrt{6}} \qquad (146)$$

The feedback resistor is adjusted to provide 3 to 8 dB of gain margin.

General-Purpose Oscillators 225

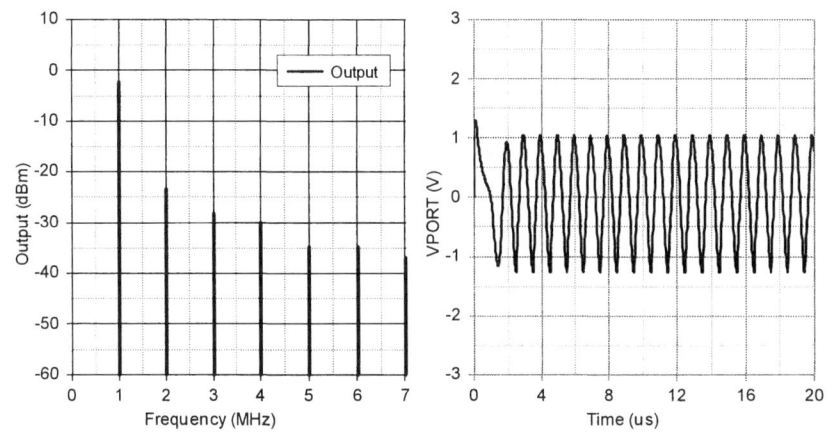

Figure 5.6 Output spectrum (left) and waveform (right) of the highpass R-C 1-MHz oscillator.

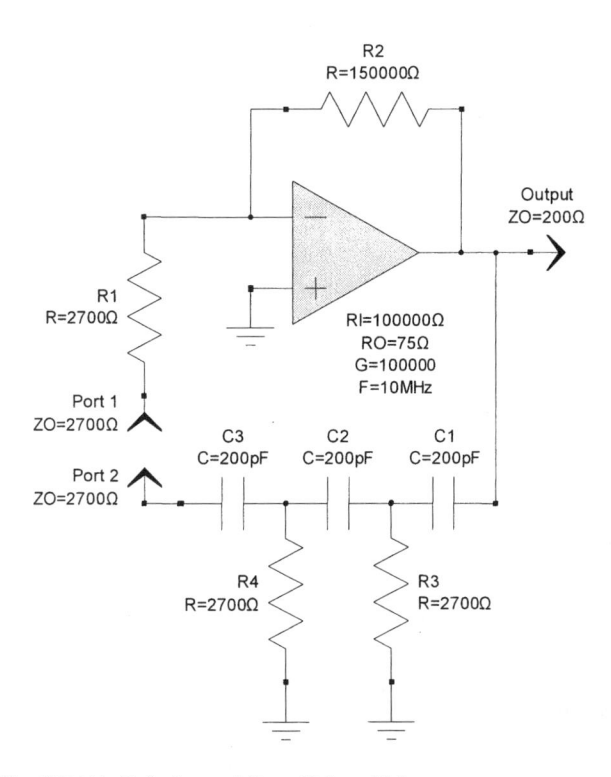

Figure 5.7 The 100-kHz R-C phase-shift oscillator utilizing an opamp.

5.2.2 Wien Bridge

The Wien bridge oscillator was theorized by Max Wien before devices offering gain were available, thus an oscillator was not realized. William Hewlett popularized the concept with his famous vacuum-tube audio oscillator (see Section 2.8). A bipolar differential amplifier implementation is shown in Fig. 5.8. A desirable attribute of this topology is the ability to tune the frequency with two equal-valued resistors or capacitors.

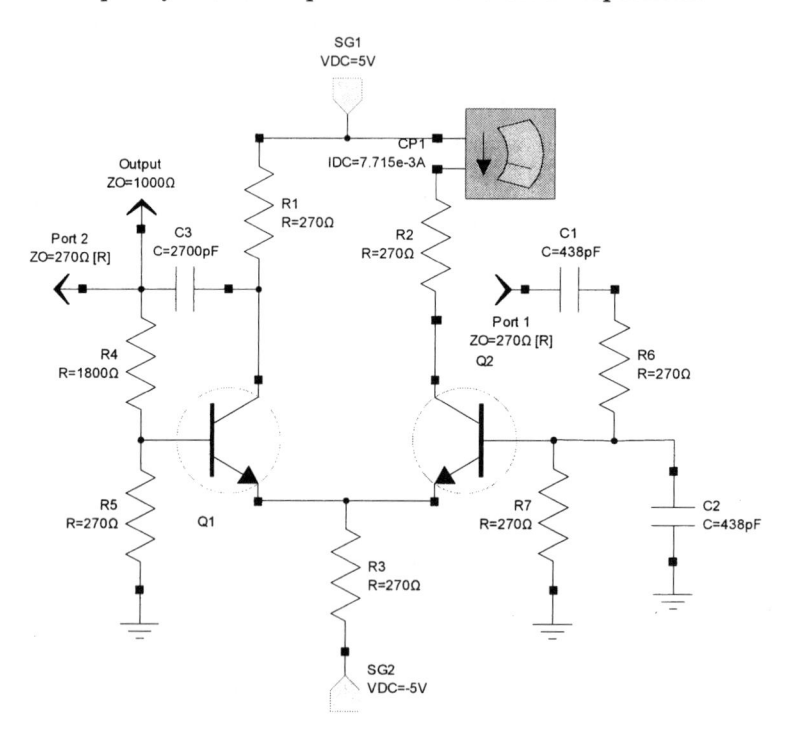

Figure 5.8 Wien bridge 1 MHz oscillator using a differential amplifier.

The frequency controlling elements are C_1, R_6, C_2, and R_7. They form a network with a peak of transmission at -9.54 dB and a transmission phase shift of 0°. The loaded Q is 0.333. The transmission phase-shift from the base of Q_2 to the collector of Q_1 of the differential amplifier using high F_t devices is also 0°, for a net open-loop transmission phase of 0°. Provided the transmission phase-shift of the differential amplifier is near 0°, the oscillation frequency is given by

$$f_0 = \frac{1}{2\pi RC} \qquad (147)$$

General-Purpose Oscillators

where R and C are the Wien bridge resistor and capacitor values.

Resistors R_4 and R_5 limit the gain by providing negative feedback from the collector to the base of Q_1. The classic technique of forcing linear operation by passive control of the level uses an incandescent lamp for resistor R_5.

The output spectrum of the differential Wien bridge 1-MHz oscillator is shown on the left in Fig. 5.9. Notice the predominance of odd harmonics as further evidenced by the semisquare waveform given on the right. Starting is rapid with relatively fast secondary transient settling.

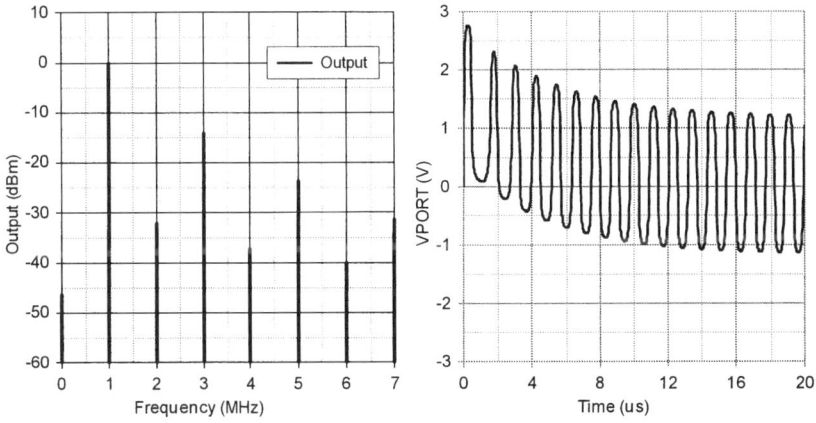

Figure 5.9 Output spectrum (left) and waveform (right) of the differential amplifier Wien bridge 1-MHz oscillator.

If a suitable opamp is available for the desired oscillation frequency, the Wien bridge oscillator is a better choice than the R-C phase-shift opamp oscillator. A 100-kHz Wien-bridge opamp oscillator is shown in Fig. 5.10. The effective attenuation of the bridge is -9.54 dB, much smaller than an R-C cascade, and only moderate gain is required of the opamp. Consequently, operation closer to the unity gain crossover frequency is possible. The opamp has a relatively low output impedance, as enhanced by the feedback of R_1 and R_2, and can therefore drive a 50 ohm or higher impedance load.

5.2.3 Multivibrators

Fig. 5.11 shows a 100 kHz multivibrator using bipolar transistors. The multivibrator is essentially a digital device with one transistor on while the other is off, and operational descriptions of the multivibrator are typically consistent with this viewpoint. It is interesting to note that if the connection of the right side of C_1 and the bottom of R_2 is opened that the linear open-loop transmission properties satisfy the oscillation criteria described in Chapter 1. The cascade input port 1 is the base of Q_2, the cascade output port 2 is through C_1 and power is extracted at either collector. R_2 and R_3 are selected to bias the transistors in the active region. The loaded Q is approximately 0.5.

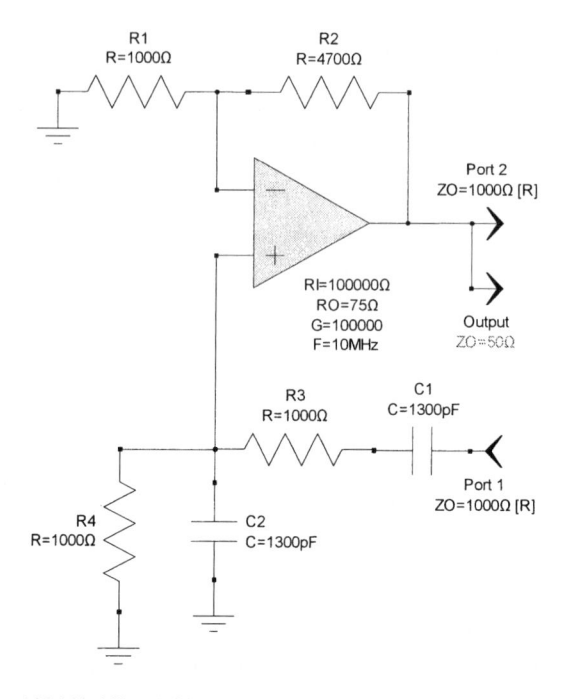

Figure 5.10 The 100-kHz Wien-bridge oscillator utilizing an opamp.

In Fig. 5.11, power is extracted at the collector of Q_2. For balanced circuits, 180° out of phase signals are extracted from both collectors. Transistors Q_1 and Q_2 alternately switch on/off states producing square waves at each collector.

With ideal transistors, $R_1<<R_2$ and $R_4<<R_3$, the operating frequency is given by

$$f_0 = \frac{1}{\tau} = \frac{1}{\ln(2)(R_2C_1 + R_3C_2)} \quad (148)$$

When R_2C_1 and R_3C_2 are equal the output is approximately square wave. With the duty factor (DF) of Q_2 high, the capacitor values are given by

$$C_1 = \frac{DF}{0.693 R_2 f_0} \quad (149)$$

$$C_2 = \frac{1-DF}{0.693 R_3 f_0} \quad (150)$$

General-Purpose Oscillators

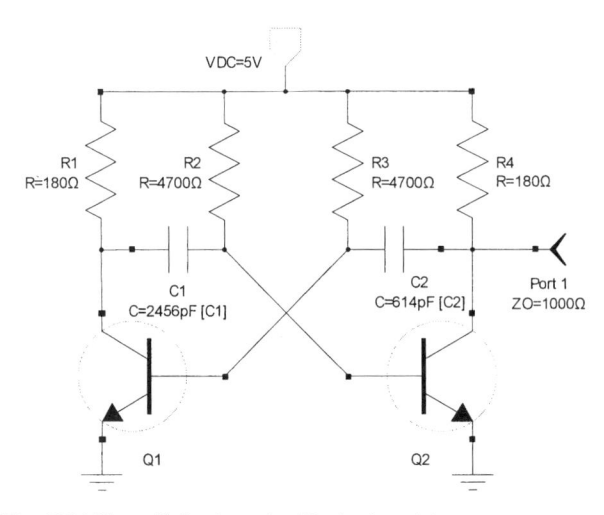

Figure 5.11 The 100-kHz multivibrator using bipolar transistors.

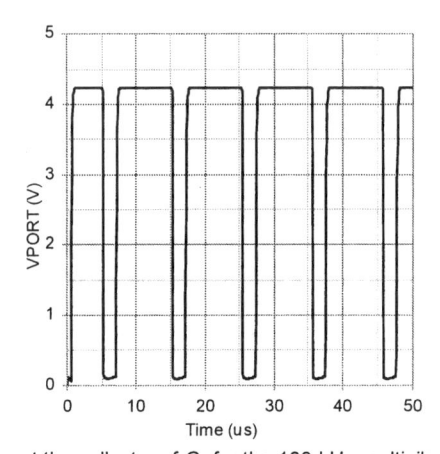

Figure 5.12 Waveform at the collector of Q_2 for the 100-kHz multivibrator given in Fig. 5.11.

For example, with DF=0.8 (high 80% of the period), an oscillation frequency of 100 kHz and equal resistors R_2 and R_3 of 4700 ohms, then C_1 is 2456 pF and C_2 is 614 pF. The resulting waveform is given in Fig. 5.12.

Notice that after settling, the oscillation period is approximately 10.15 uS, or 98.5 kHz. As the desired frequency is increased, real-devices introduce greater error in the operating frequency predicted by the simplified formula and simulation is advised.

When the multivibrator is initially powered, both transistors initially have a high voltage and tend to switch on. Imbalance invariably causes one to switch on first and stable oscillation begins. With small values of R and

C, the bases may remain high for a longer time than is required to charge the capacitors, resulting in latching both transistors on. In practice this is seldom an issue at frequencies below 10 MHz. Therefore, multivibrators are typically not used above a few megahertz so devices with F_t of 100 MHz or greater are suitable.

5.2.4 Ring Oscillators

When the input of the inverter is low, the output is high, and when the input is high, the output is low. Thus when an odd number of inverters are cascaded and the output drives the input, the system toggles at a rate determined by the delay through the system. Ring oscillators are therefore the cascade of an odd number of inverters with the output returned to the input.

The output R-C time constant is primarily responsible for gate delay. The oscillation frequency is given by

$$f_o = \frac{1}{N \times R \times C} \qquad (151)$$

where N is the number of stages, R is the load resistance and C is the load capacitance. In terms of the delay of each stage

$$f_o = \frac{1}{2 t_d N} \qquad (152)$$

A ring of gates is often included on die as a convenient means of measuring the process gate delay based on the frequency of oscillation. Because inverting gates are ubiquitous elements in IC design, ring oscillators are one of the more common forms of oscillators employed in IC designs. Unfortunately, since they are essentially R-C, it is not possible to achieve high loaded Q and therefore good phase noise performance.

The Q of ring oscillators is approximately

$$Q = \frac{N}{2} \sin\left(\frac{\pi}{N}\right) \qquad (153)$$

The Q for three stages is 1.3. With an infinite number of stages the Q increases to only 1.57.

The gate delay is shifted by varying the bias of the buffers, thus facilitating VCOs. Phase-locking the ring oscillator mitigates the poor phase-phase for certain applications.

Fig. 5.13 shows a basic, single-stage, TTL inverter gate model. Buffered inverters are available that cascade three inverting sections. They offer higher gain and somewhat lower power consumption. However, while buffered gates work well for logic circuits, unbuffered gates have more moderate gain and are preferred in oscillator circuits. The delay associated

General-Purpose Oscillators

with a three inverter gate is not 3X the delay of a single inverter because the internal gates are faster.

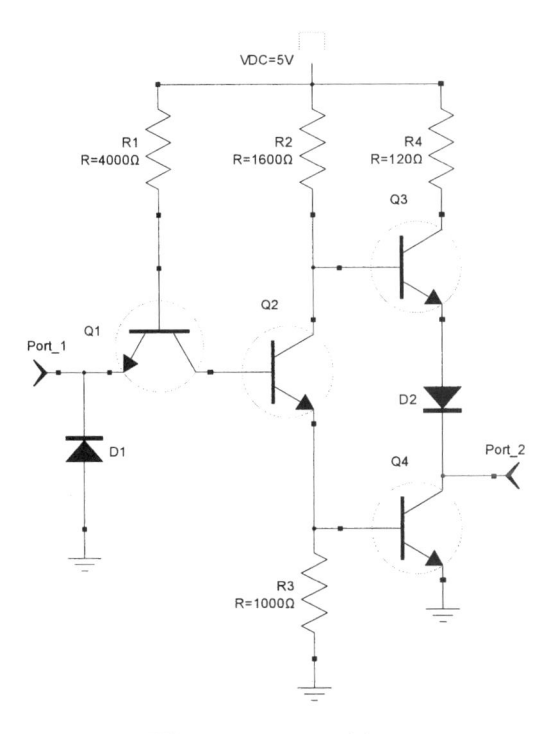

Figure 5.13 Basic single-stage TTL inverter gate model.

Shown in Fig. 5.14 is a ring oscillator consisting of a cascade of seven of the single-stage inverters depicted in Fig. 5.13. The ring oscillator drives a 4700-ohm resistive load.

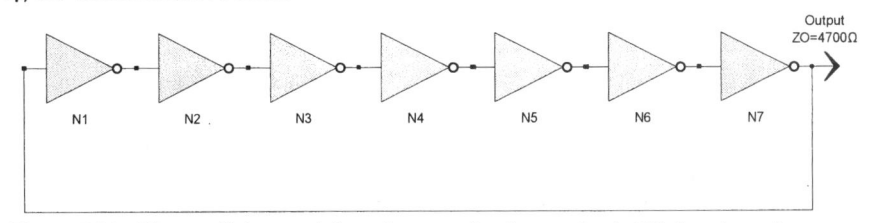

Figure 5.14 Ring oscillator consisting of a cascade of seven basic TTL inverter gates.

The starting waveform of the seven-stage ring oscillator is given in Fig. 5.15. The steady-state period is approximately 94 nS, a frequency of 10.64 MHz. From Eq. 152, the delay associated with each inverter in the cascade is 6.71 nS.

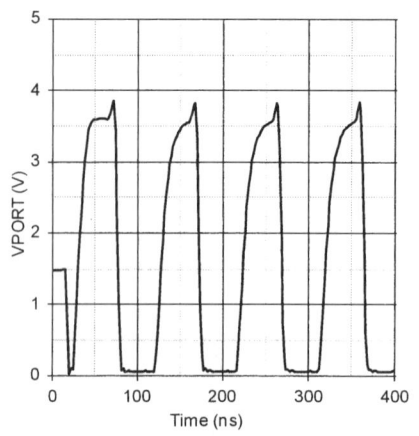

Figure 5.15 Starting waveform of the seven-stage ring oscillator.

5.2.5 Twin-T Oscillators

Fig. 5.16 shows a 43.5-kHz twin-T oscillator using a uA741 opamp. The twin-T notch network connected between the output and the inverting input provides negative feedback reducing the system gain except at the notch frequency. The twin-T notch frequency is given by

$$f_{notch} = \frac{1}{2\pi R_x C_x} \qquad (154)$$

where R_x is the value of the series resistors and C_x is the value of the series capacitors. The shunt resistor is set equal to $R_x/2$ and the shunt capacitor is set equal to $2C_x$.

The operating frequency of this oscillator as calculated by Eq. 154 is 53 kHz while the operating frequency of this circuit is 43.5 kHz. The difference is associated with the finite gain-bandwidth product of the opamp. Increasing the gain-bandwidth of the opamp from 1 to 10 MHz significantly reduces the error. For best simulation accuracy and so the frequency is more device-independant, the gain-bandwidth product of the opamp should be 100X greater than the desired operating frequency.

Fig. 5.17 shows a 19.6 kHz twin-T oscillator using a bipolar transistor. The transistor operates as a simple CE amplifier and the base bias drive is provided through resistors R_2 and R_3. In the classic mode, the twin-T provides a deep transmission notch rather than a peak and this circuit is not valid.

General-Purpose Oscillators

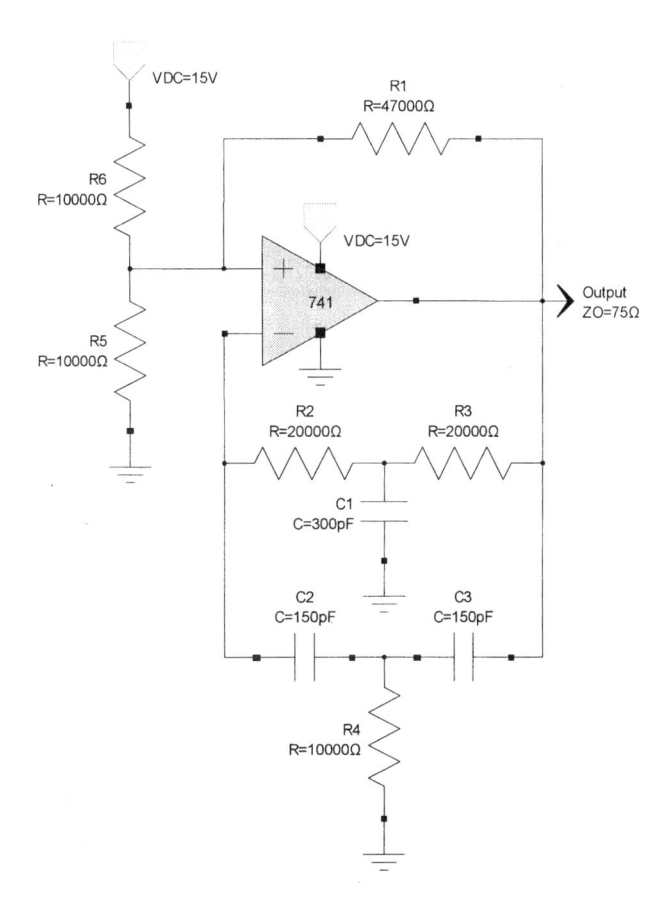

Figure 5.16 Twin-T 43.5 kHz oscillator using a 741 opamp.

In this circuit, resistor R_4 is set at $R_x/10$ rather than $R_x/2$. The transmission characteristics of just the notch with 10-kohm source and load resistances are shown in Fig. 5.18. The solid traces are the open-loop transmission amplitude (round symbols) and the transmission phase (square symbols) for the classic notch values. Component values were chosen to provide a notch at 10 kHz. The dashed traces are the responses with R_4 reduced to $R_x/10$. Notice that the attenuation is reduced to -25 dB and the frequency of the 180° phase reversal is significantly increased. When the notch is cascaded with an inverting amplifier with greater than 25 dB of gain, the oscillation criteria are satisfied.

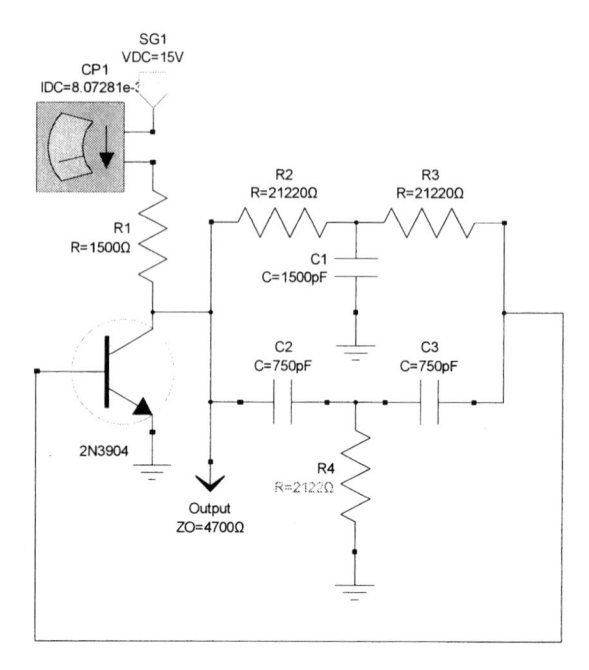

Figure 5.17 The 19.6-kHz twin-T oscillator using a bipolar transistor.

Fig. 5.19 is the starting waveform of the twin-T bipolar oscillator. After settling, the period is approximately 51 nS for an operating frequency of 19.6 kHz. When the shunt resistor is reduced, the termination impedances presented to the twin-T by the transistor impacts the actual operating frequency. Simulation is advised to determine the operating frequency.

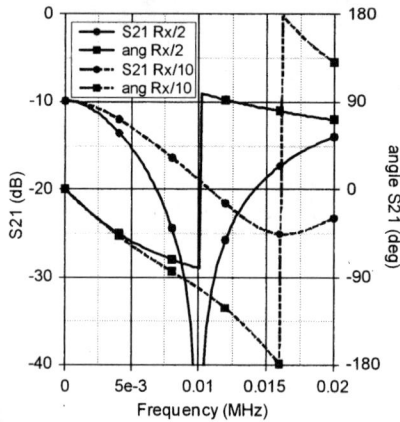

Figure 5.18 Transmission responses of the twin-T notch with classic component values (solid traces) and with the shunt resistor reduced to $R_x/10$ (dashed traces).

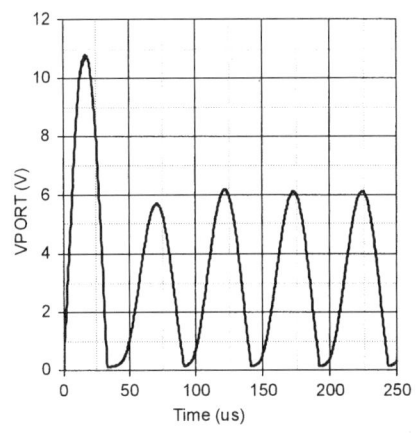

Figure 5.19 Starting waveform of the 19.6-kHz twin-T bipolar oscillator.

5.3 L-C Oscillators

Oscillators using L-C resonators offer versatility, high performance, and applicability to a wide frequency range. The first coherent electronic oscillators were vacuum tube with L-C resonators, so historically L-C resonators represent an early technology, proceeded only by spark-gap transmitters and then motorized mechanical frequency generators. Even spark-gap transmitters were band limited by utilizing resonance of the antenna and its loading elements.

Modern capacitors are available in a wide range of options, they are small and inexpensive, parasitics are moderate, and they provide high unloaded Q. Rather, the inductor typically limits the performance of both discrete and integrated L-C oscillators. The unloaded Q of the benchmark inductor, a single-layer, copper-wire solenoid, ignoring parasitic capacitance and radiation is

$$Q_U = 7.48 r \Psi \sqrt{f} \qquad (155)$$

where r is the mean radius of the solenoid in millimeters, f is the frequency in megahertz, and Ψ is a factor that ranges from 0.5 to 0.9 depending on winding parameters. Increasing the radius increases the unloaded Q. However, at a given frequency, parasitic capacitance limits the useful radius. Therefore, the maximum available unloaded Q for the solenoid is approximately 200 to 250, independent of the frequency. Modern chip inductors often have poor unloaded Q, not because of poor construction, but because they are small. The design of solenoid inductors and more detailed models are given in Appendix A.

5.3.1 Colpitts

The Colpitts is realized using a variety of active devices. However, it is the resonator that defines whether an oscillator is Colpitts, Hartley or other named type. The classification of types by name is primarily a study of resonator topologies.

As was pointed out in Section 1.4, the modern convention for the Colpitts resonator presents to the signal path a shunt capacitor, series capacitor, and shunt inductor. The shunt inductor and series capacitor are highpass and the shunt capacitor is lowpass. Therefore in the stopband, the Colpitts includes two transmission zeros at DC and one transmission zero at infinite frequency. With zeros above and below the passband, the Colpitts is a bandpass resonator.

The primary features of the Colpitts resonator are a transmission phase shift near 0° at resonance and dissimilar input and output impedance. The CE and CS devices are inverting and have relatively equal input and output impedance. Therefore, Colpitts oscillators use CC, CB, CD, and CG devices.

The Colpitts resonator is economic and a properly designed oscillator provides good performance. The Colpitts is arguably the most common oscillator topology.

A caution must be noted for the Colpitts and other oscillators with similar topologies. Examine the first Colpitts example in Fig. 5.20. Notice the collector is grounded for RF and the emitter is terminated in a relatively high resistance. This is the description of a tuned-base CC negative resistance oscillator. Next notice the capacitors C_1 and C_2 in series to ground at the base. All capacitors have parasitic series inductance and so this path represents a series L-C to ground. This circuit is prone to oscillate at UHF frequencies because of this effect. If parasitic oscillations are observed with the Colpitts or other similar oscillators, resistance should be placed in series with the base to absorb the device negative resistance. A small value from 18 to 100 ohms generally suffices.

5.3.1.1 Colpitts CC Bipolar Dual Supply

Fig. 5.20 shows a 100-MHz Colpitts bipolar transistor oscillator biased with dual positive and negative voltage supplies. Again, when ports 1 and 2 are connected to close the loop capacitor C_3 is not required. The gain margin of this circuit is approximately 6 dB and the loaded Q is approximately 10.

The output spectrum is given on the left in Fig. 5.21. The reasonably good harmonic suppression is the result of taking output power at the emitter at the resonator capacitive tap and the use of a transistor with F_t only a few times greater than the operating frequency.

General-Purpose Oscillators 237

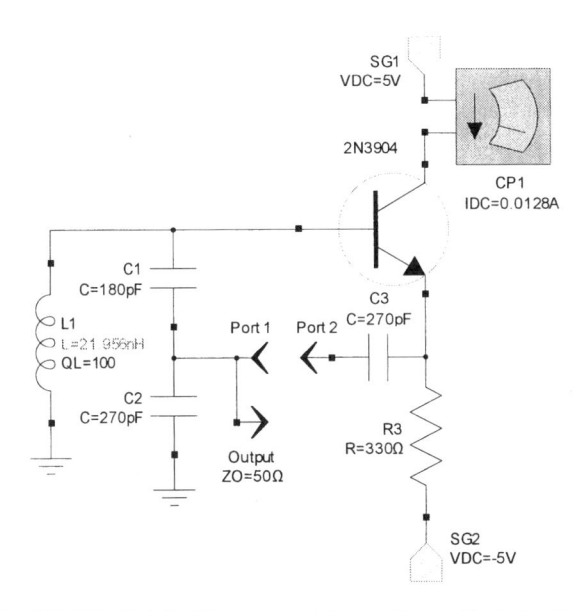

Figure 5.20 The 100-MHz Colpitts bipolar transistor oscillator with dual-voltage supplies.

The output starting waveform is given on the right in Fig. 5.21. The starting time for this oscillator, and most oscillators in general, can be significantly reduced by employing the triggering techniques described in Section 3.7.

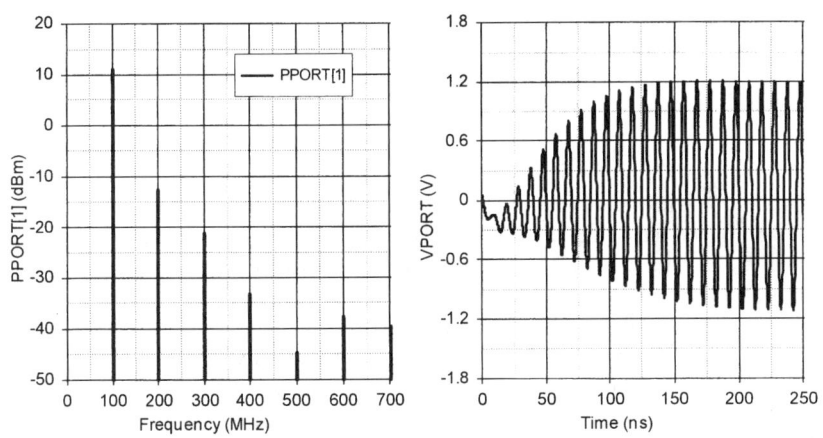

Figure 5.21 Output spectrum (left) and starting waveform (right) for the dual supply bipolar Colpitts.

5.3.1.2 Colpitts CC Bipolar Negative Supply

Fig. 5.22 shows the schematic of a 100-MHz Colpitts CC bipolar oscillator biased with a single negative supply. If a positive supply is required a PNP transistor is used. The circuit is similar to the dual supply Colpitts oscillator except that capacitor C_3 is required to avoid shorting the base bias voltage.

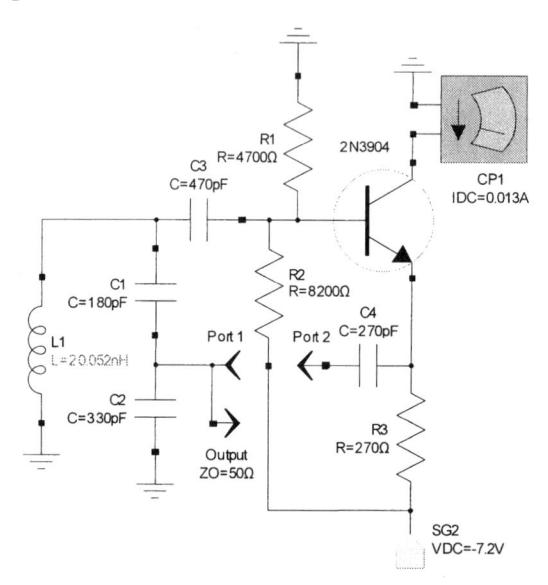

Figure 5.22 The 100-MHz Colpitts bipolar transistor oscillator with a single negative supply.

Notice the small value of the resonator inductor L_1. Improving the loaded Q requires an even smaller value.

5.3.1.3 Colpitts CB Bipolar

Fig. 5.23 shows a 100-MHz CB bipolar Colpitts oscillator. Again, capacitor C_3 is required only for simulation. Capacitor C_4 provides an RF ground for the base. A larger value for C_4, acting with resistors R_1 and R_2, filters power supply noise but requires time to charge and slows starting.

Even though the resonator values are similar to the CC Colpitts, the loaded Q of this CB circuit is approximately 18, significantly better. The impedance of the collector loading the resonator is higher than the impedance of the base that loads the resonator in the CC circuit. However even with this CB circuit, realizing a high loaded Q requires a very small value of inductance. Achieving high unloaded Q with small inductor values is difficult. This issue is addressed in Section 5.3.2, which covers the Clapp oscillator.

General-Purpose Oscillators 239

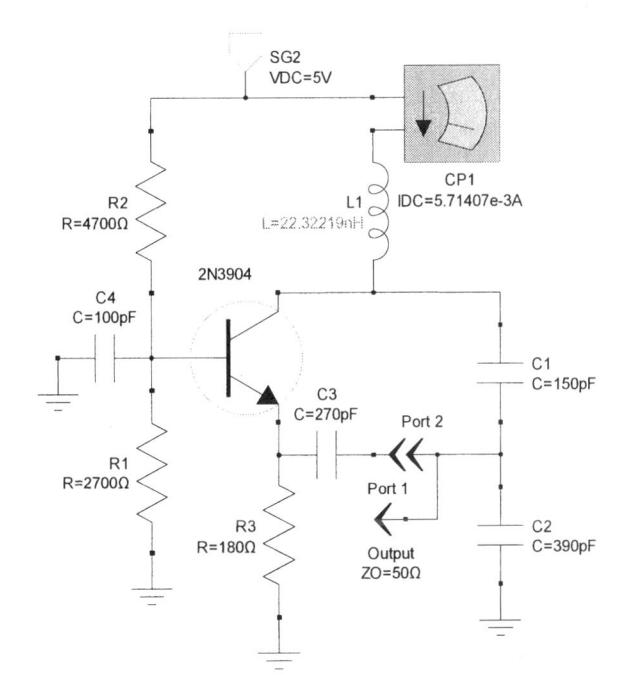

Figure 5.23 The 100-MHz CB bipolar Colpitts oscillator.

On the left in Fig. 5.24 is the output spectrum of this 100-MHz CB Colpitts and the starting waveform is given on the right.

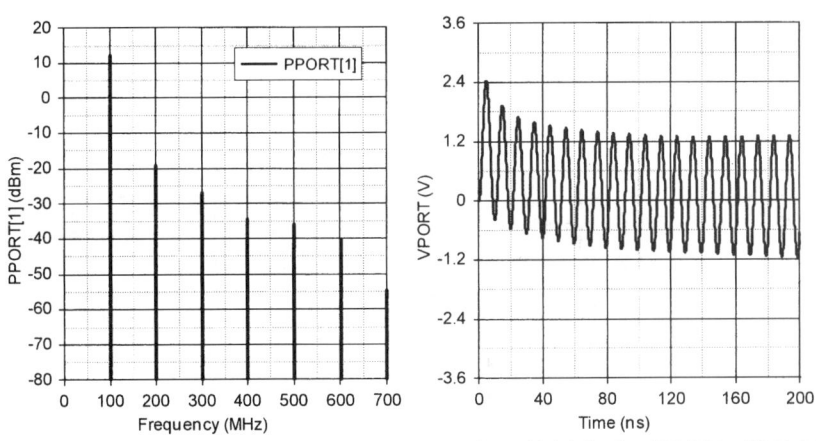

Figure 5.24 Output spectrum (left) and starting waveform (right) for the 100-MHz CB Colpitts oscillator.

5.3.1.4 Colpitts CD FET

Fig. 5.25 shows a 100-MHz CD FET oscillator using a J309 N-type junction field-effect transistor. This FET has a moderately high I_{dss}, 18 mA typical. Resistor R_3 reduces and stabilizes the drain current. The loaded Q of this circuit is approximately 18 even though the resonator inductor is nearly 4X larger than with the bipolar oscillators. The very high input impedance of the FET is an advantage with the Colpitts resonator.

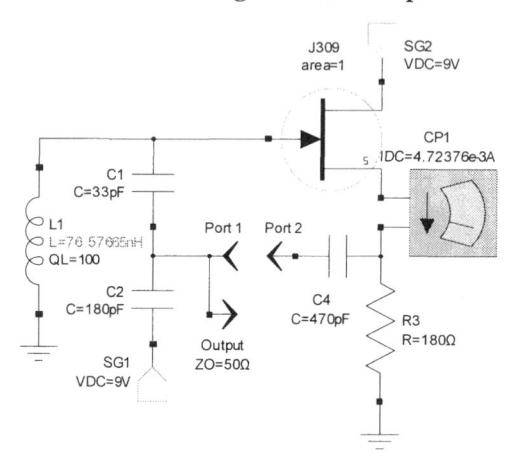

Figure 5.25 The 100-MHz CD FET oscillator.

Fig. 5.26 shows the output spectrum (left) and the starting waveform (right) for this oscillator. The output power is relatively low for this oscillator. Starting is fast, partly because triggering is used by returning C_2 to the supply rather than ground.

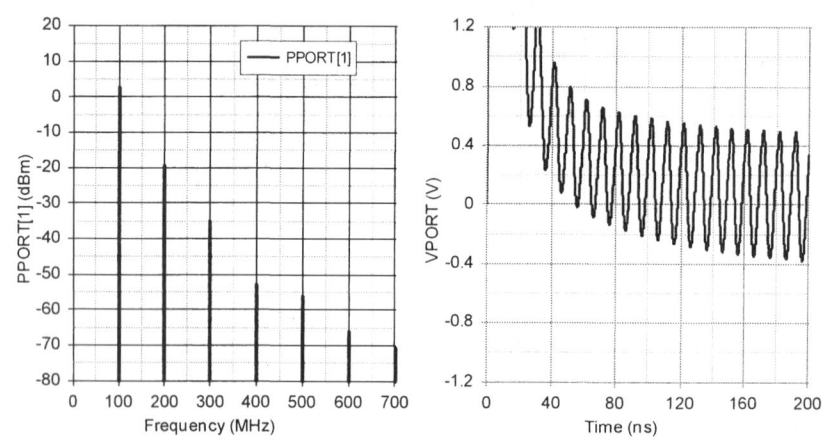

Figure 5.26 Output spectrum (left) and starting waveform (right) for the 100-MHz CD FET oscillator.

5.3.1.5 Colpitts CG FET

Fig. 5.27 shows a 100-MHz CG Colpitts FET oscillator. The gain margin is approximately 4 dB and the loop input and output match are good. The loaded Q with these resonator values is approximately 18. Because the drain current is only a weak function of the supply voltage, this circuit offers good load pulling and power supply noise immunity.

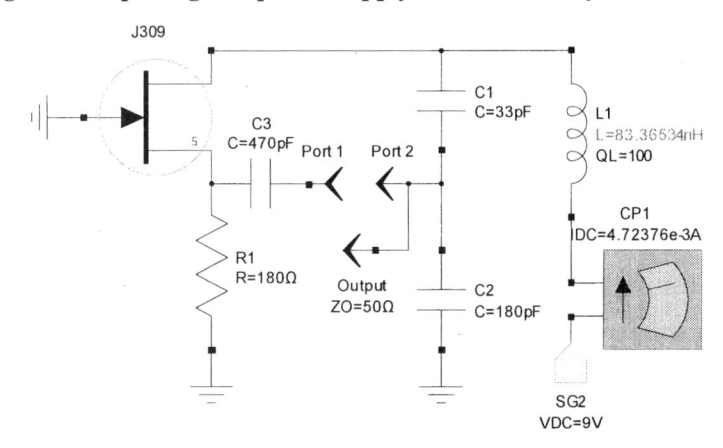

Figure 5.27 The 100-MHz CG Colpitts FET oscillator.

The output spectrum is given on the left in Fig. 5.28. Harmonic suppression is good. The starting waveform is given on the right. Starting is fast, aided by natural triggering because the resonator inductor is driven by the supply. This oscillator is an excellent general-purpose oscillator requiring only the Colpitts resonator, the FET, and one resistor.

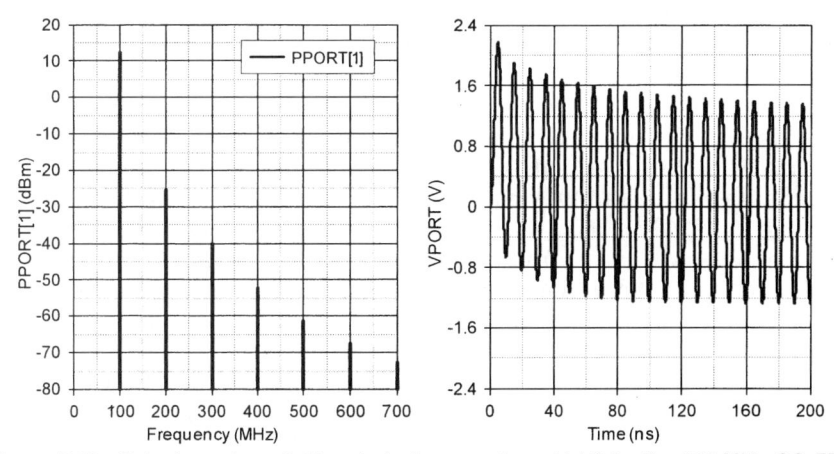

Figure 5.28 Output spectrum (left) and starting waveform (right) for the 100-MHz CG FET oscillator.

5.3.2 Clapp

As indicated, an issue with the bipolar Colpitts oscillator, particularly at higher frequency, is a small value of resonator inductance. Also, for the single-supply bipolar CC Colpitts, a blocking capacitor is required at the base. This additional capacitor is wasted with the Colpitts but used to advantage in the Clapp [1] oscillator to increase the resonator inductor value. An example CC bipolar Clapp oscillator is shown in Fig. 5.29.

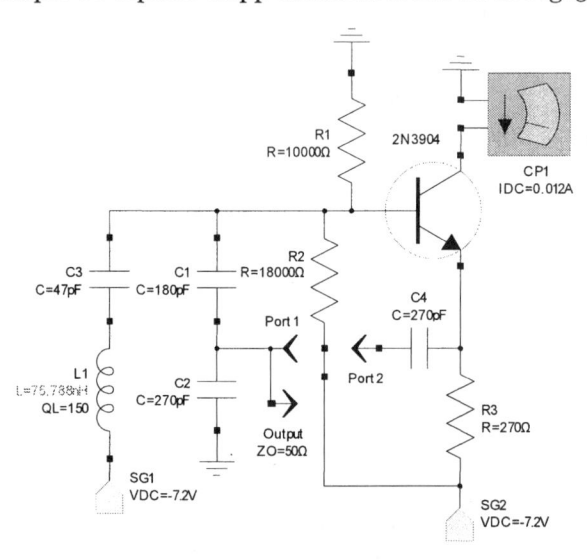

Figure 5.29 The 100-MHz CC bipolar Clapp oscillator.

In this oscillator, C_3 has a reactance of 33 ohms. The value of the resonator inductor is increased from 20 nH for the Colpitts to almost 76 nH, a significant advantage for winding an inductor with good unloaded Q. More significantly, for the same values of C_1 and C_2, the loaded Q for this Clapp is 27, a significant improvement over the loaded Q of 18 for the Colpitts. In fact, the loaded Q is high enough that the unloaded Q of the inductor should be carefully considered. If this circuit is built using small chip inductors, which have poor unloaded Q, the loaded Q must be reduced to avoid excessive resonator loss and therefore low or negative gain margin.

Shown on the left in Fig. 5.30 is the output spectrum of the CC bipolar Clapp oscillator and the output wave form is shown on the right.

Resistance or a ferrite bead in series with the base may be necessary to suppress negative-resistance oscillator modes associated with the series L-C or the capacitive tap and its parasitic inductance. The CB bipolar Clapp requires an additional choke in parallel with the Clapp capacitor to bias the collector. It is therefore not a compelling topology. Also, the Clapp capacitor compromises the simplicity of FET oscillators and the resonator

General-Purpose Oscillators

inductor value is reasonable with the FET Colpitts, so the Colpitts is more compelling than the CB Clapp.

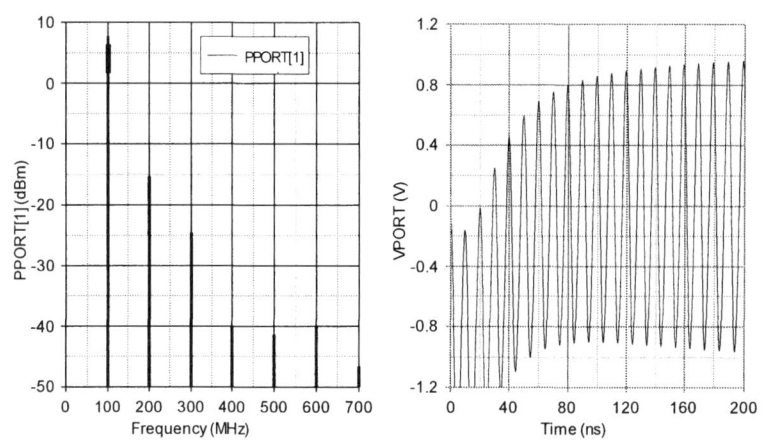

Figure 5.30 Output spectrum (left) and starting waveform (right) for the 100-MHz CC bipolar Clapp oscillator.

5.3.3 Seiler

A variation on the Clapp oscillator is the Seiler [2], which places an additional capacitor in parallel with the resonator inductor. This reduces the value of the inductor somewhat. However, the Seiler capacitor is useful for reducing the inductor value in lower frequency oscillators. The primary use for the Seiler is that back-to-back varactors can be used in place of C_4 for tuning. If the inductor is returned to ground rather than the supply, starting slows, but the varactors are both returned to ground. An example Seiler oscillator is shown in Fig. 5.31, and the simulated spectrum and starting waveform are shown in Fig. 5.32. Other than efficient tuning there is little motivation for the Seiler over the Clapp.

5.3.4 Hartley

The Hartley resonator presents to the signal path a shunt inductor, series inductor, and shunt capacitor. The shunt inductor is highpass and series inductor and shunt capacitor are lowpass. Therefore in the stopband, the Hartley includes one transmission zero at DC and two transmission zeros at infinite frequency. The Hartley is a bandpass resonator.

The primary features of the Hartley resonator are similar to the Colpitts; a transmission phase shift near 0° at resonance and dissimilar input and output impedance. The CE and CS devices are inverting and have relatively equal input and output impedance. Therefore, Hartley oscillators also use CC, CB, CD, and CG devices.

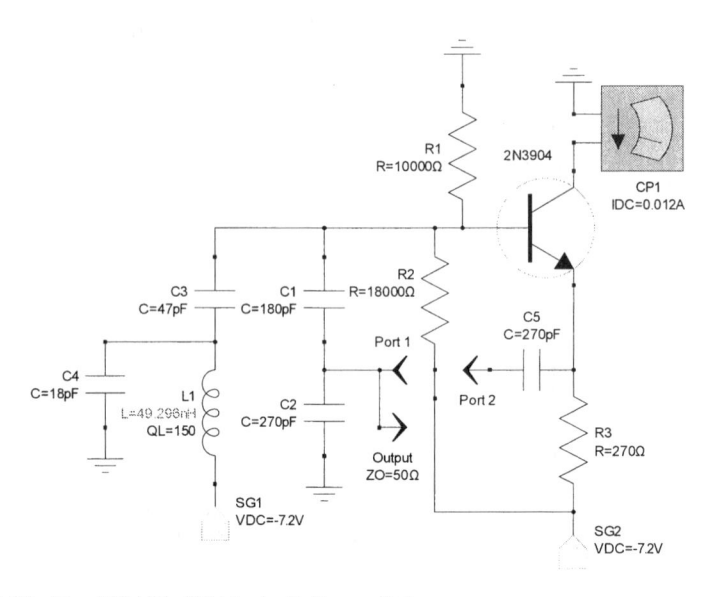

Figure 5.31 The 100-MHz CC bipolar Seiler oscillator.

In the past when leaded components were the norm, inductors were more prevalent and Hartley oscillators were more common. Modern manufacturing techniques significantly reduce the cost of chip capacitors but inductors become less competitive. The tapped inductor used with the Hartley is even more problematic. The Hartley is less common today. When tapped or multiple inductors are acceptable, the Hartley offers some advantages, particularly with JFET devices.

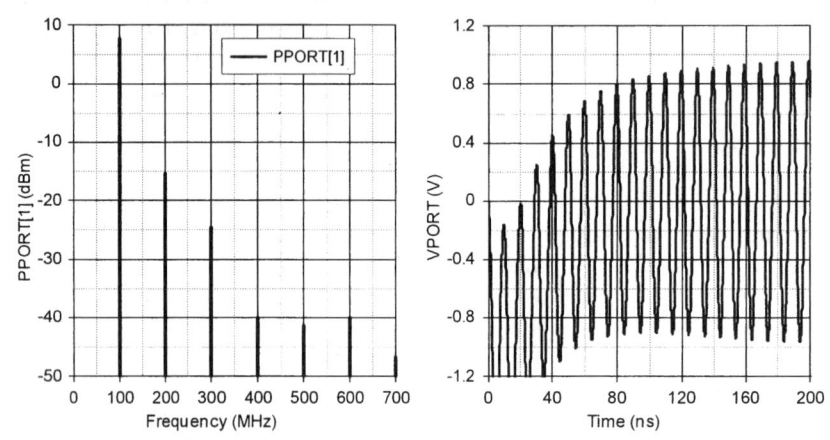

Figure 5.32 Output spectrum (left) and starting waveform (right) for the 100-MHz CC Seiler oscillator.

5.3.4.1 Hartley CB Bipolar

Fig. 5.33 shows a Hartley 100-MHz CB bipolar oscillator. The Hartley is normally depicted with a tapped inductor. It does not matter whether the inductor is tapped resulting in two mutually coupled inductors or whether two uncoupled inductors are used. This example oscillator uses uncoupled inductors. Notice that the value of L_2 is extremely small. This is typical of the bipolar Hartley. Coupled inductors require an even smaller value for this inductor. Lower frequency operation relaxes this problem.

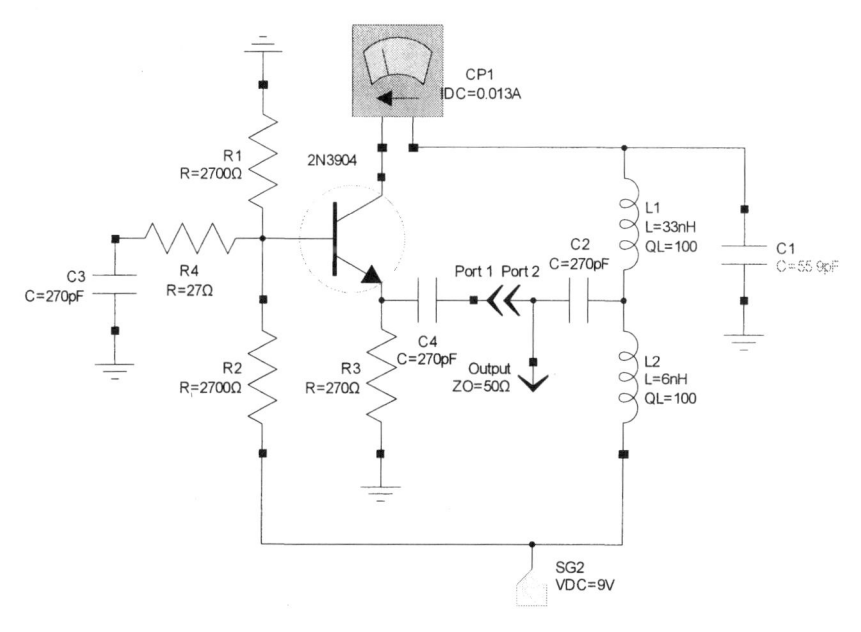

Figure 5.33 Hartley 100-MHz CB bipolar oscillator.

R_1 is a stabilizing resistor. As with the Colpitts, if spurious oscillations are observed in any of these Hartley oscillators, series base resistance is suggested. Capacitors C_2 and C_4 are required in the final oscillator to avoid bias disturbance by the load. The loaded Q of this oscillator is approximately 20. The bipolar CB Hartley has good output power but bias decoupling requirements raise the component count. It is not a particularly compelling topology. The output spectrum is given on the left in Fig. 5.34 and the starting waveform is on the right.

5.3.4.2 Hartley CD FET

Fig. 5.35 shows a 100-MHz Hartley CD FET oscillator. R_2 moderates and stabilizes the drain current. The resonator inductors provide a DC return for the gate. To avoid these inductors shorting R_2 at DC, capacitor C_2

is required when the loop is closed. The loaded Q of this oscillator is approximately 18. The higher impedance of the FET allows somewhat larger inductor values but they remain rather small at high frequency.

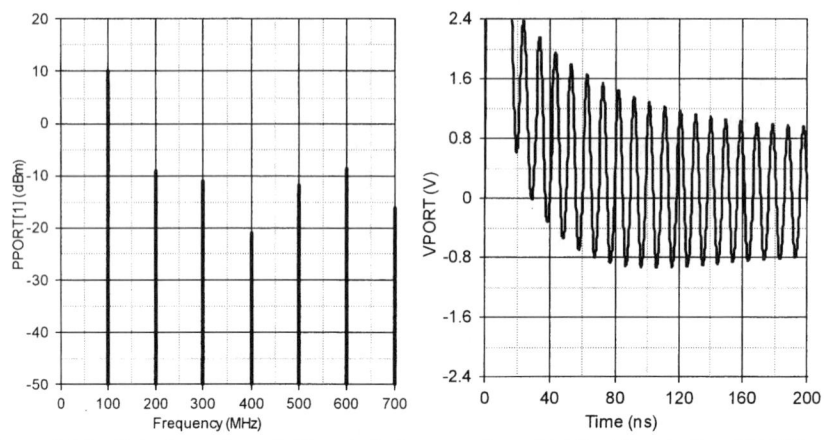

Figure 5.34 Output spectrum (left) and starting waveform (right) of the Hartley 100-MHz CB oscillator.

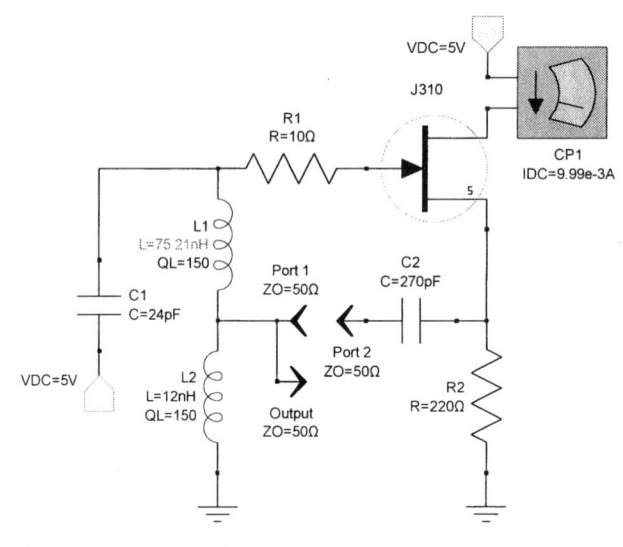

Figure 5.35 The 100-MHz Hartley CD FET oscillator.

The output spectrum is given on the left in Fig. 5.36 and the starting waveform, with triggering, is given on the right.

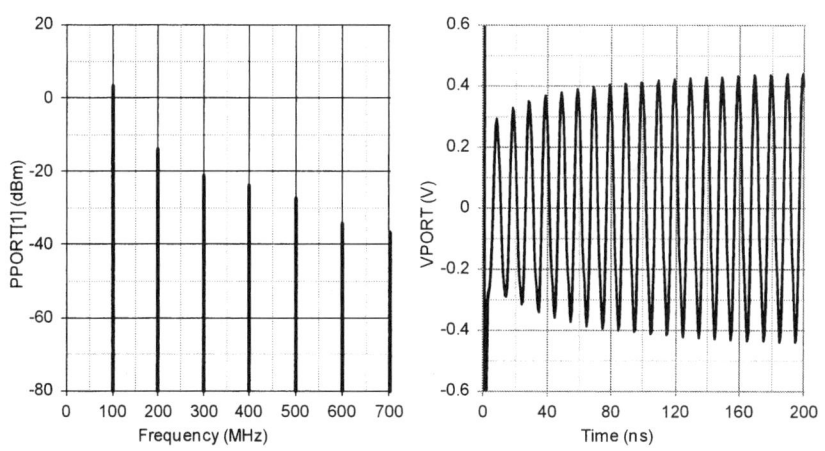

Figure 5.36 Output spectrum (left) and starting waveform (right) for the Hartley CD FET oscillator.

5.3.4.3 Harley CD FET Simple

The oscillator in Fig. 5.37 represents the ultimate in simplicity. Even so, at lower frequencies where the smallest practical inductor value has low reactance, the loaded Q is high and performance of this oscillator is excellent. The drain current is unstabilized and the drain current equals I_{dss}, which typically varies as much as a factor of 3.

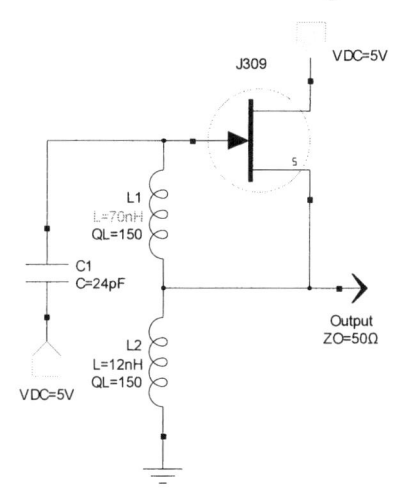

Figure 5.37 Simplified 100-MHz Hartley CD FET oscillator.

The starting waveform for this oscillator is shown in Fig. 5.38. If starting is not critical, the resonator capacitor is returned to ground.

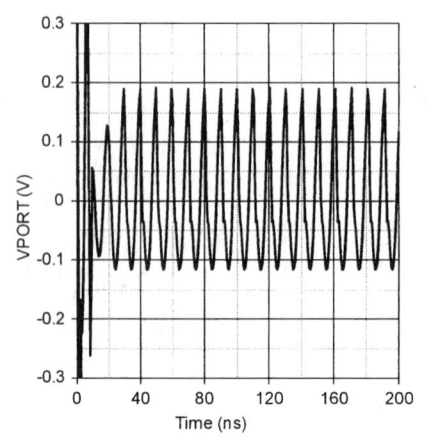

Figure 5.38 Output waveform of the simplified Hartley CD FET oscillator.

5.3.4.4 Hartley CG FET

Fig. 5.39 shows a CG 100-MHz Hartley oscillator. Capacitors C_2 and C_3 are required when the loop is closed. The loaded Q of this oscillator is approximately 21. The CG Hartley offers higher output power for a given supply power than the CG topology. Again, inductor L_2 is rather small.

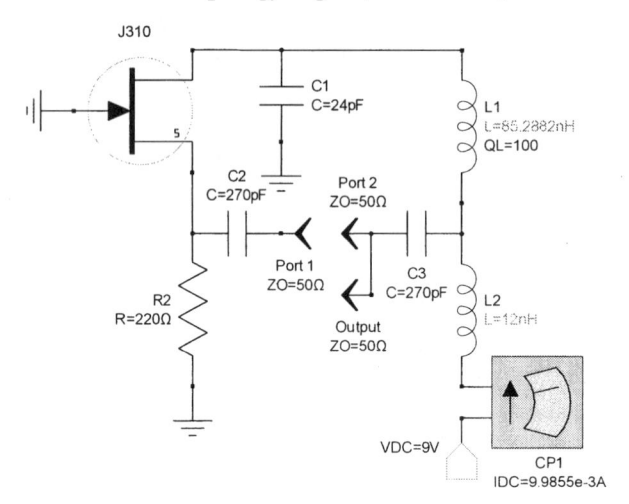

Figure 5.39 The 100-MHz Hartley CG FET oscillator.

The output spectrum and starting waveform are given in Fig. 5.40.

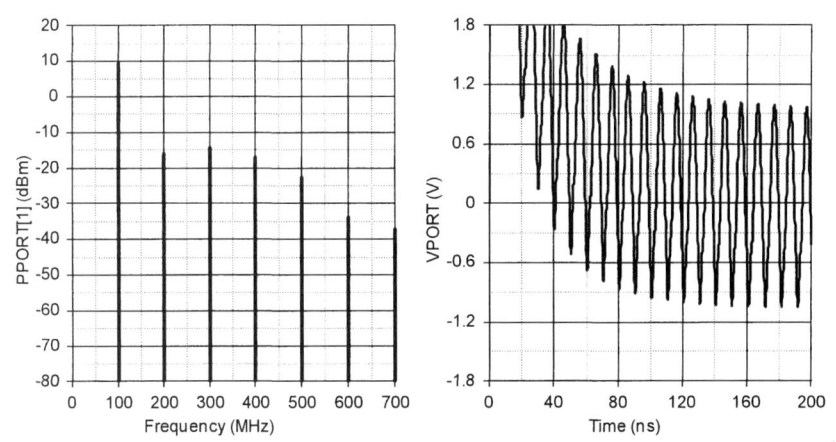

Figure 5.40 Output spectrum (left) and starting waveform (right) for the 100-MHz Hartley CG FET oscillator.

5.3.5 Pierce

The oscillator that Pierce developed in 1923 utilized a bulk quartz crystal resonator and a single vacuum tube. Solid-state derivatives of the Pierce crystal oscillator are described in Chapter 8. When the quartz resonator in the Pierce oscillator is replaced with an inductor, the resulting oscillator is often referred to as Pierce. The resonator form is a shunt capacitor, a series inductor, and a shunt capacitor. All transmission zeros occur at infinite frequency so the resonator is lowpass. At high Q, the response is pseudobandpass. The transmission phase shift in the limiting case of infinite Q is 180°, and the resonator input and output impedance are similar, so CE and CS active devices are utilized.

5.3.5.1 Pierce CE Bipolar

An example 100-MHz Pierce CE oscillator is shown in Fig. 5.41. The sustaining stage includes the resistor R_2 for biasing and shunt feedback. Approximately 3 mA flows in resistor R_2, so the quiescent collector current in this circuit is approximately 6.6 mA. The loaded Q of this oscillator is 16. With approximately 50-ohm source and load impedances from the amplifier, the capacitor and inductor values in the resonator are extreme for frequencies above about 20 MHz. The inductor value is too small to realize good unloaded Q and the shunt capacitors are so large that parasitic inductance is an issue. This issue is mitigated by using the higher impedance FET and the issue is completely resolved in Section 5.3.6.

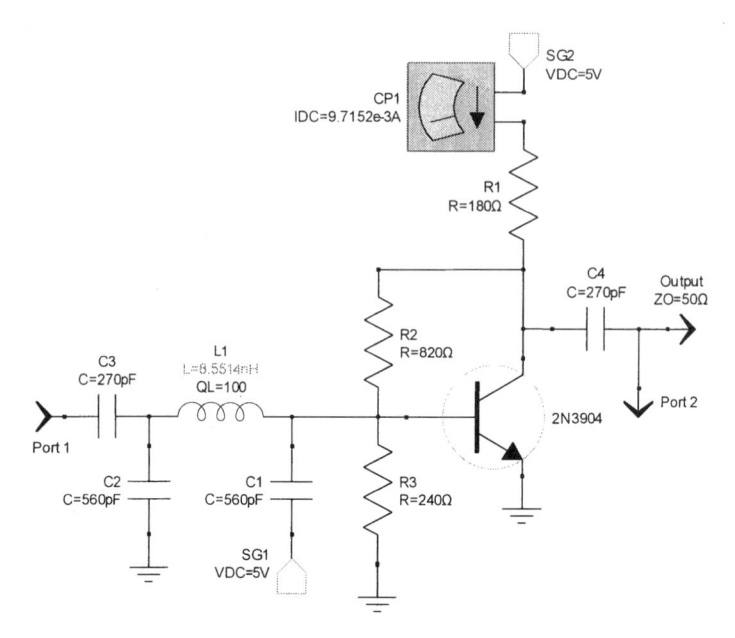

Figure 5.41 The 100-MHz Pierce CE oscillator.

The output spectrum and stating waveform are given in Fig. 5.42. Starting is delayed by the time required to reach bias by charging the capacitors through the resistors.

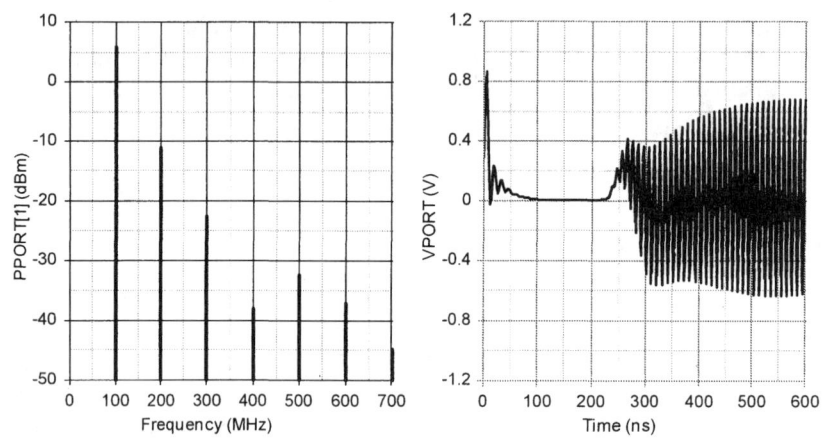

Figure 5.42 Output spectrum (left) and starting waveform (right) for the 100-MHz Pierce CE oscillator.

5.3.5.2 Pierce CS FET

Fig. 5.43 shows an example 100-MHz Pierce CS FET oscillator. The load and port reference impedances are 500 ohms. At this higher impedance the resonator values are more practical. The loaded Q of this oscillator is 18.

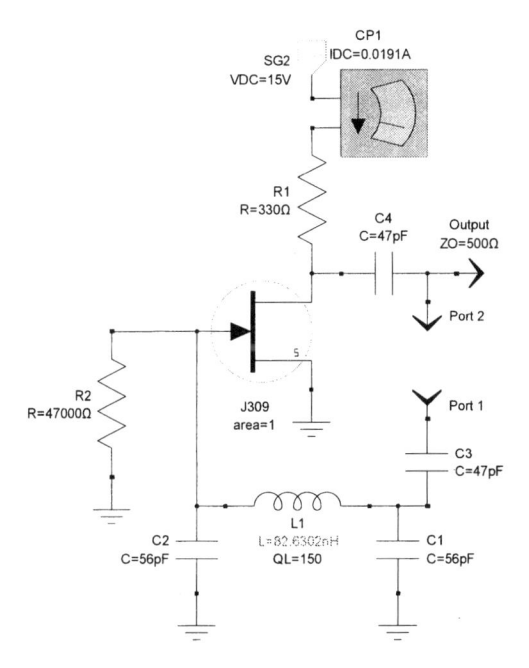

Figure 5.43 The 100-MHz Pierce CS FET oscillator.

The output spectrum and starting waveform are given in Fig. 5.44. Starting is faster than the CE Pierce because the higher impedance allows smaller-valued coupling capacitors that offer faster bias settling.

5.3.6 Coupled Series Resonator

The oscillator shown in Fig. 5.45 is similar to the Pierce CE bipolar topology except capacitor C_3 is placed in series with the inductor. The structure is a series resonator with shunt coupling capacitors at the input and output. It is bandpass with one transmission zero at DC and three at infinite frequency. It is inverting in the limiting case of infinite loaded Q. Since motional branch of a quartz crystal is a series resonator, the L-C circuit in Fig. 4.45 is more like the original Pierce topology than the L-C lowpass topology often referred to as Pierce.

A significant advantage of this modified Pierce is improved resonator inductor and capacitor values. This circuit also has a loaded Q of 18 but the inductor value is increased from 8.55 to 107 nH and the shunt capacitors are reduced by a factor of nearly 4. The additional capacitor provides a degree of freedom not available with the simpler Pierce L-C oscillator. Reducing C_3 further increases the inductor and decreases the capacitor values. The output spectrum and starting waveform are given in Fig. 5.46. The starting of this circuit benefits from triggering C_2 rather than C_1.

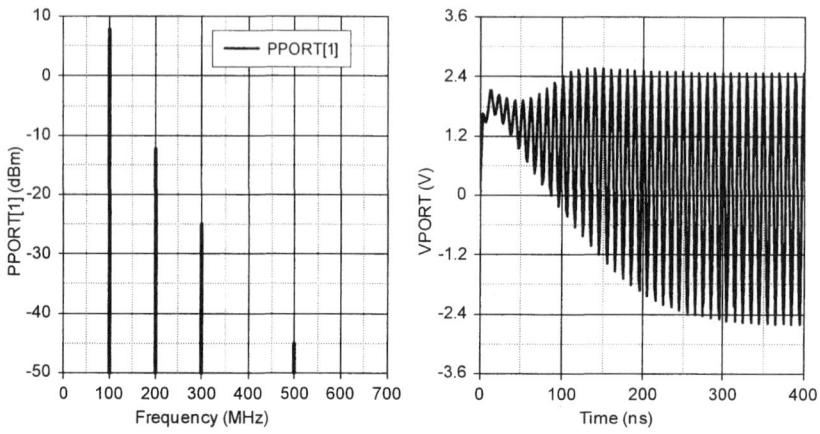

Figure 5.44 Output spectrum (left) and staring waveform (right) for the 100-MHz Pierce CS FET oscillator.

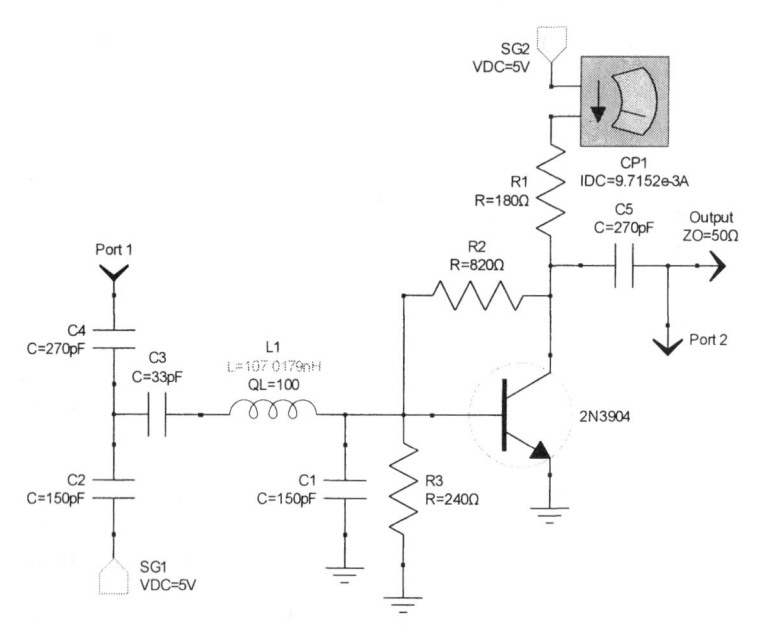

Figure 5.45 The 100-MHz coupled series resonator CE oscillator.

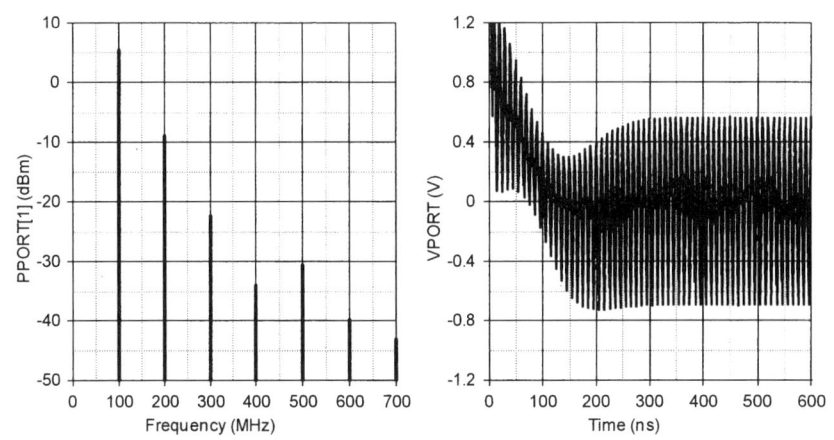

Figure 5.46 Output spectrum (left) and starting waveform (right) for the coupled series resonator oscillator.

5.3.7 Rhea

Fig. 5.47 shows an oscillator that to my knowledge was first described in the *Microwave Journal* [3] and then *Microwaves Magazine* [4]. A photograph of the test oscillator, simulated and measured output spectrums, and starting waveforms are given in Section 3.5. The sustaining stage is a CE bipolar with series and shunt feedback to stabilize the gain and improve the match. Defining this oscillator class is a highpass resonator consisting of series capacitor C_1, shunt inductor L_1 and series capacitor C_2. Recall that with the Pierce oscillator, the resonator is lowpass but at high loaded Q the transfer function is pseudobandpass. With this oscillator, the resonator is highpass but at high loaded Q the transfer function is also pseudobandpass. The phase shift of the resonator in the limiting case of high loaded Q is 180°. The loaded Q of this example is approximately 5.6.

Capacitor C_3 contributes little to the resonator phase slope. It steps down the load resistance presented to the inductor to reduce load coupling. Capacitor C_4 is not required when the loop is closed since the resonator capacitors block the collector bias voltage from the base. The elimination of a required blocking capacitor was the initial motivation for using a highpass resonator. With high loaded Q, the value of the resonator inductor is rather large. Therefore, inductor parasitic capacitance limits the useful upper frequency range of this oscillator.

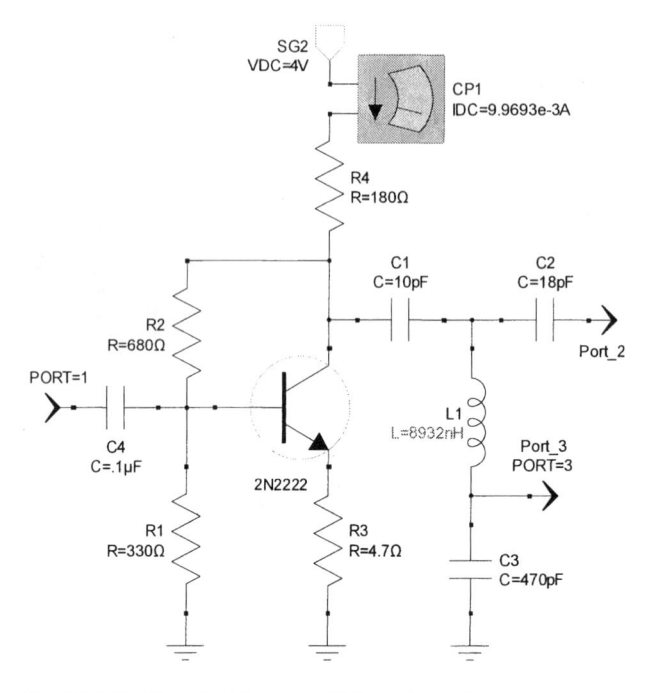

Figure 5.47 The 10-MHz Rhea CE bipolar oscillator schematic.

5.3.8 Coupled Parallel Resonator

Fig. 5.48 shows a 700-MHz CE bipolar oscillator using a parallel resonator coupled with series capacitors. The phase shift of this resonator in the limiting case of infinite loaded Q is 180°. It is typically used with CE sustaining stages that have similar input and output impedance. However, the resonator accommodates dissimilar impedances by adjusting the ratio of C_1 and C_2. A degree of freedom is available to choose a particular value of inductor L_1 or capacitor C_3. At high frequencies the values of the coupling capacitors or the resonator inductor become small, as is the case with this example. Inductor L_2 assists in aligning the phase zero crossing at the maximum phase slope. It is unnecessary and is replaced with a resistor at low frequencies where the phase shift of the sustaining stage is ideal and therefore inverting. This 700-MHz example uses a bipolar transistor with an F_t of approximately 8 GHz to assist in realizing the required inverting phase. The loaded Q of this oscillator is approximately 18.

The spectrum of the 700-MHz oscillator is given on the left in Fig. 5.49 and the starting waveform is given on the right. The high output power and efficiency is facilitated by coupling power at the collector and the collector inductor L_2 that avoids the signal dissipation of a collector resistor.

General-Purpose Oscillators 255

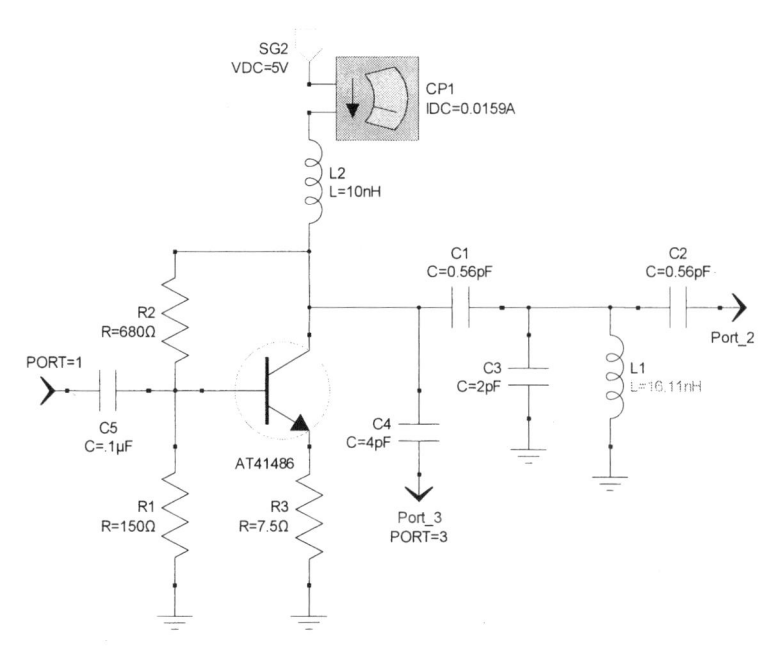

Figure 5.48 The 700-MHz CE bipolar oscillator with parallel resonator and series coupling capacitors.

The problem with difficult resonator values at high frequency is resolved by replacing the parallel inductor and capacitor with a low-impedance ceramic transmission line. This is covered in Chapter 6.

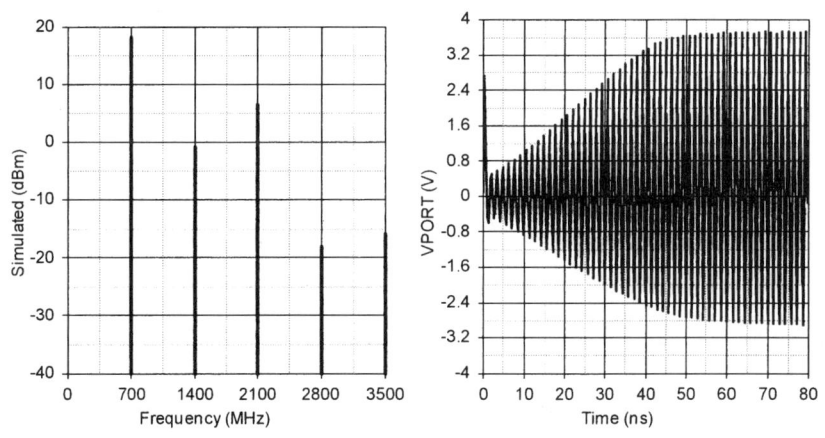

Figure 5.49 Output spectrum (left) and starting waveform (right) for the 700-MHz coupled parallel resonator oscillator.

5.3.9 Gumm

Fig. 5.50 shows an oscillator attributed to Linley Gumm and described by Hayward [5]. It was originally designed as a very low-noise mechanically tuned oscillator for 10 MHz. The loaded Q is extremely high due to loose coupling of L_1 to the 17 turn resonator winding L_2.

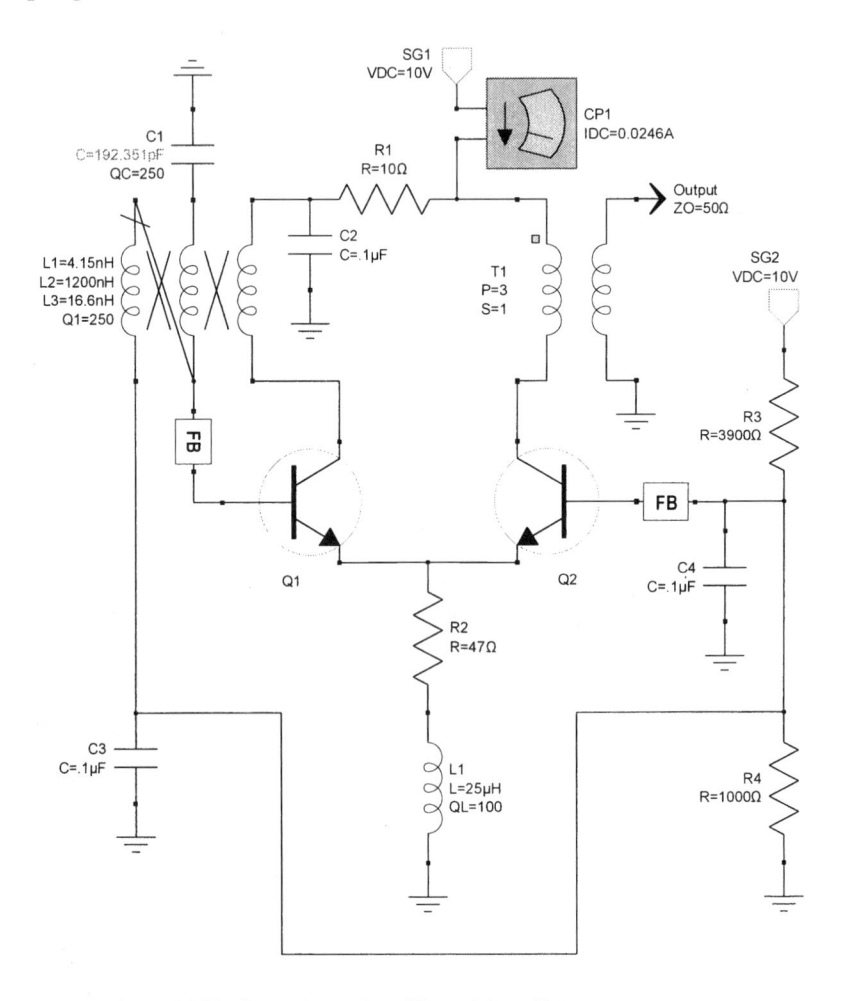

Figure 5.50 The 10-MHz Gumm low-noise differential oscillator.

The open-loop transmission amplitude, phase, and loaded Q are given in Fig. 5.51. The high gain margin of 21 db would be excessive except that differential amplifiers have benign limiting characteristics. The loaded Q of this circuit is approximately 80. This requires good unloaded Q in the resonator inductor. The primary inductor is 17 turns on a T68-6 toroid core with an initial permeability of 8. The outside diameter is 0.68 inches, the inside diameter is 0.37 inches, and the thickness is 0.19 inches. The unloaded Q is approximately 250. The collector winding is two turns and the base tap is one turn from the grounded end. The ferrite beads are modeled as 51 ohms in parallel with 215 nH.

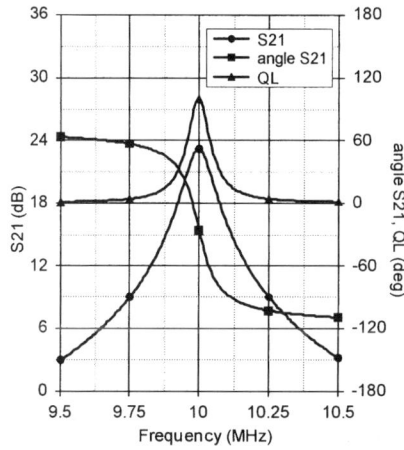

Figure 5.51 Gumm oscillator open-loop transmission gain (circular symbols), phase (square symbols) and loaded Q (triangular symbols).

The simulated output spectrum is given on the left in Fig. 5.52. The output power at 10 MHz is predicted to be 19 dBm. The measured output power is 17 dBm. The simulated starting waveform is given on the right. Notice the time scale is microseconds rather than nanoseconds as used in the previous examples. Starting is slow not because of the high loaded Q, but primarily because capacitors C_3 and C_4 are 0.1 uF and they are charged through the Thevenin equivalent resistance of R_3 and R_4, 796 ohms. The resulting time constant for the bias is 160 uS. While long, it is too short to significantly filter power supply noise below 1 kHz. Starting is not critical for the intended application, a VFO for a communications receiver. If starting is critical, smaller values for C_3 and C_4 are indicated.

5.3.10 Simplified Gumm

The Gumm oscillator inspired the design in Fig. 5.53. Tapped or multiple winding inductors are expensive. Furthermore, the Gumm oscillator requires very loose coupling of the base winding to the resonator winding and in the previous example the inductance of L_1 is 4 nH even though the frequency is only 10 MHz. The oscillator shown in Fig. 5.53 utilizes a modified topology that replaces the coupling inductor loop with a small-valued capacitor, C_3, to the base of Q_2. The resonator is parallel mode comprising L_1 and C_1. To simplify the bias networks, JFETs rather than bipolars are used. U310 JFETs with a high I_{dss} are used for good output power, approximately 15 dBm, to reduce the phase noise. The resistor R_2 serves the dual purpose of an approximate current source and stabilizing I_d below I_{dss}. The output is coupled from a 2:1 turns ratio transformer at the drain of Q_2.

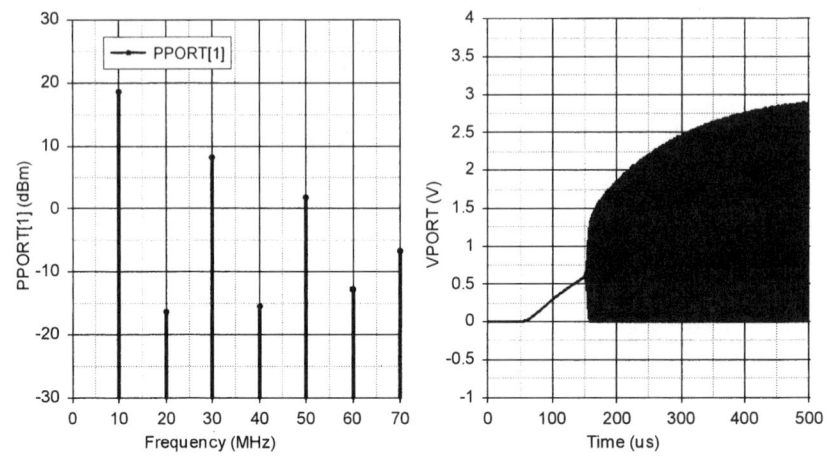

Figure 5.52 Simulated output spectrum (left) and starting waveform (right) of the Gumm.

The simulated output spectrum (left) and starting waveform (right) of the 10 MHz simplified Gumm oscillator are given in Fig. 5.54. The JFETs reach cutoff easily, resulting in a square positive-going waveform. The negative-going waveform is more sinusoidal. The resulting waveform is high in second harmonic content. Harmonic distortion is reduced at the expense of output power by using a load impedance higher than 50 ohms. Starting is extremely fast with steady-state operation reached within the first cycle.

5.4 Oscillator Topology Selection

There is an abundance of oscillator topologies. The example 27 general-purpose oscillator schematics presented in this chapter are only representative. The next three chapters include many topologies and variations on these general forms. Choosing a topology can be daunting. On first principle, it makes little difference whether a Colpitts or Hartly, a bipolar or FET, or a CB or CE topology is used. For example, Leeson's equation has no term requiring such a choice. What is important is how these choices influence parameters that do affect performance.

General-Purpose Oscillators

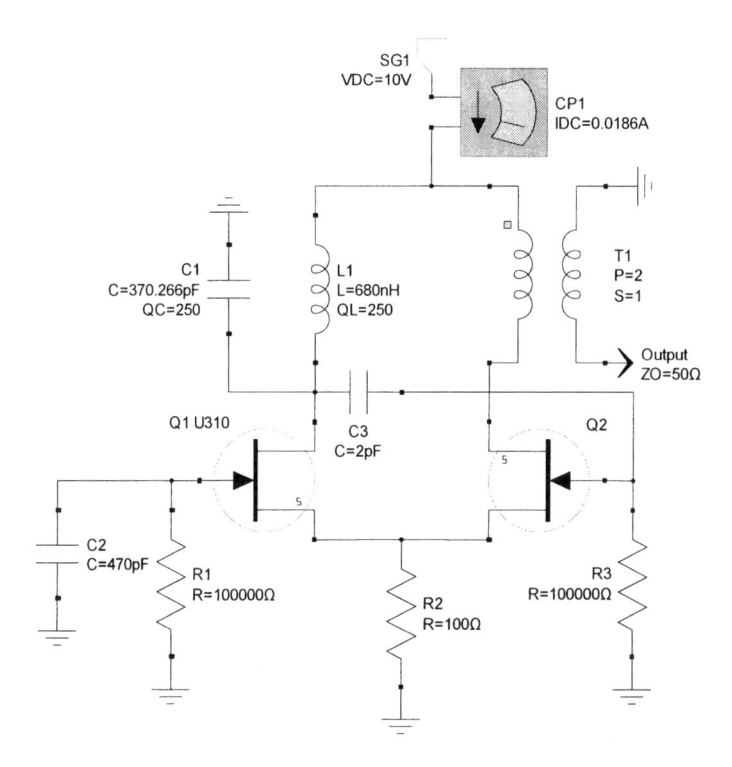

Figure 5.53 Schematic of the 10-MHz simplified Gumm.

For example, the Colpitts resonator is directly loaded by either the base or gate of the device, or the collector or drain. Because FETs generally have higher resistive impedance, they provide a higher loaded Q. However, this is compensated by decreasing the Colpitts resonator inductor. Therefore, it is the design details that are important rather than a selection based on generalizations. Practical issues rather than fundamental ones may dictate an optimum solution. For example, is the lower inductor value mentioned above a practical value?

In a prior era when machine wound, tapped inductors were common, the Hartley was more popular than it is today. If dual supplies are available, biasing the CB bipolar circuit with directly grounded base is economic. If only a negative supply is available, the CC circuit is an economic choice. If MMICs are used throughout the design, using one in the oscillator may be a practical solution. These and other numerous choices may dictate the optimum topology.

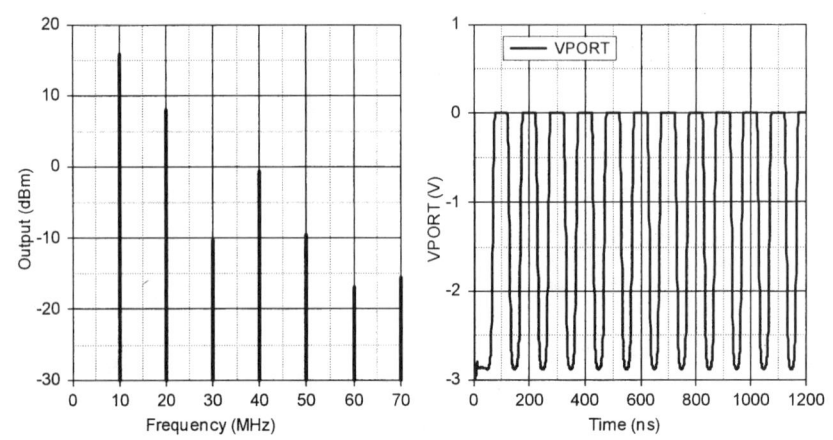

Figure 5.54 Simulated output spectrum (left) and starting waveform (right) of the 10-MHz simplified Gumm.

What is important, regardless of how these choices are managed, is achieving a design with high loaded Q. Increasing the loaded Q improves the phase-noise, improves long-term and environmental stability, reduces supply pushing, and reduces load pulling. If the nature of a design is such that improved loaded Q is achieved with practical element values, then that topology is the best choice.

The preceding 27 example schematics all involved design choices, for example, active devices in my parts bin were selected. Different design choices result in different performance outcomes. Nevertheless, for the topologies illustrated, here are a few generalizations about selecting a topology.

R-C oscillators are only selected at low frequency or when inductors are prohibitive. The resulting loaded Q is below 1.57 and the performance is limited. Of the R-C oscillator types discussed, the multivibrator is perhaps the most compelling. It starts fast, the duty cycle is easily controlled, and the waveform is nearly square, which is desirable for some applications. Above 10 MHz, latching may be a concern. Another candidate topology is the Wien bridge differential oscillator. It offers good harmonic performance, fast starting, tuning by either two resistors or two capacitors, and it is easily linearized with an incandescent-filament bulb.

When inductors are prohibitive at higher frequency, the ring oscillator is suggested. Again, the loaded Q is limited so short and long-term stability are compromised. Nevertheless, ring oscillators are used extensively through the microwave frequency range.

In terms of L-C oscillators, perhaps the most popular is the bipolar Colpitts. In reality, it is a fair choice only if dual supplies are available. With a single supply, the required base blocking capacitor can be replaced with a higher reactance capacitor at no additional cost, resulting in a Clapp with a more desirable inductor value. If varactor tuning is required, the Seiler is a good choice. If the designer is fixated on the Colpitts, a FET device is probably a better choice than bipolar as the blocking capacitor is not required and the parallel resonator loading is higher impedance. In many cases, the CB or CG offers higher DC to RF conversion efficiency and faster starting. Spurious modes may be more problematic for CB and CG topologies, but this is manageable via base or gate resistance or ferrite beads. The Pierce and shunt-coupled series-resonator topologies generally start slow and conversion efficiency is only fair. The parasitic inductance of the shunt coupling capacitors must be managed.

The top-C coupled parallel resonator is a good choice for frequencies in the VHF and higher frequency range. An extra degree of freedom in the resonator inductor value is useful in matching available varactor capacitance values. The shunt inductor naturally returns the tuning voltage to ground of back-to-back varactors. Starting is fast and conversion efficiency is high. The Rhea is a subset of the top-C coupled form without a parallel capacitor. It is only compelling at frequencies below 30 MHz when fast starting, low harmonics and maximum economy are required.

The Gumm and simplified Gumm are excellent choices as mechanically tuned, high-performance oscillators for communication and surveillance radios.

At frequencies above the VHF frequency range, Chapter 6 offers additional choices. If tuning bandwidth above 30% is required, Chapter 7 should be consulted. Finally, if long- and short-term stability drive the design, Chapter 8 should be consulted.

References

[1] J.K. Clapp, "An Inductance-Capacitor Oscillator of Unusual Frequency Stability," *Proc IRE*, Vol. 36, 1948, pp. 356-358.

[2] E.O. Seiler, "A Low-C Electron Coupled Oscillator," *QST*, American Radio Relay League, November 1941, p. 26.

[3] R. Rhea and B. Clausen, "Recent Trends in Oscillator Design," *Microwave Journal*, January 2004, pp. 22-24, 26, 28, 30, 32, 34.

[4] R. Rhea, "A New Class of Oscillators," *Microwave Magazine*, June 2004, pp. 72-83.

[5] W. Hayward and D. DeMaw, *Solid State Design for the Radio Amateur*, American Radio Relay League, Newington, CT, 1977, p. 126-127.

6 Distributed Oscillators

The maximum available unloaded Q of wire wound inductors is approximately 250 and typical values are 100 and less. This significantly limits achievable oscillator performance. Oscillators that use distributed transmission line resonators can achieve much higher unloaded Q. Oscillators with distributed resonators, and hybrid resonators such as helical, are considered in this chapter.

6.1 Resonator Technologies

Oscillators with L-C resonators are covered in the previous chapter. As pointed out, the maximum solenoid size is limited by parasitic capacitance and radiation. Therefore, the maximum available unloaded Q of the unshielded solenoid is approximately 200 to 250.

Modern photolithographic techniques allow for extremely small physical size. However, for devices that store all or a part of their energy in magnetic fields, unloaded Q increases with size. Distributed resonators offer higher unloaded Q because capacitance is integral to resonance and the resonator can be physically larger without invoking parasitic issues. The larger resonator stores more energy and provides higher unloaded Q. The higher unloaded Q of distributed resonators is not the result of wizardry but simply larger size.

Plotted versus frequency in Fig. 6.1 is the unloaded Q per square millimeter of footprint (figure of merit, FOM) for various technologies suitable for use as oscillator resonators [1]. The footprint area is used rather than device volume because PWB area is a critical factor in many designs and because normalization is difficult using volume with some technologies.

The solenoid FOM is the dashed line rising from 10 MHz to the upper right of the graph. Dashed traces represent unshielded or unpackaged devices. The FOM is potentially optimistic for these technologies because spacing may be required between devices. For example, an inductor may be immediately adjacent to resistors and capacitors but must be spaced from other inductors to avoid coupling. Solid traces represent packaged technologies that require no additional spacing.

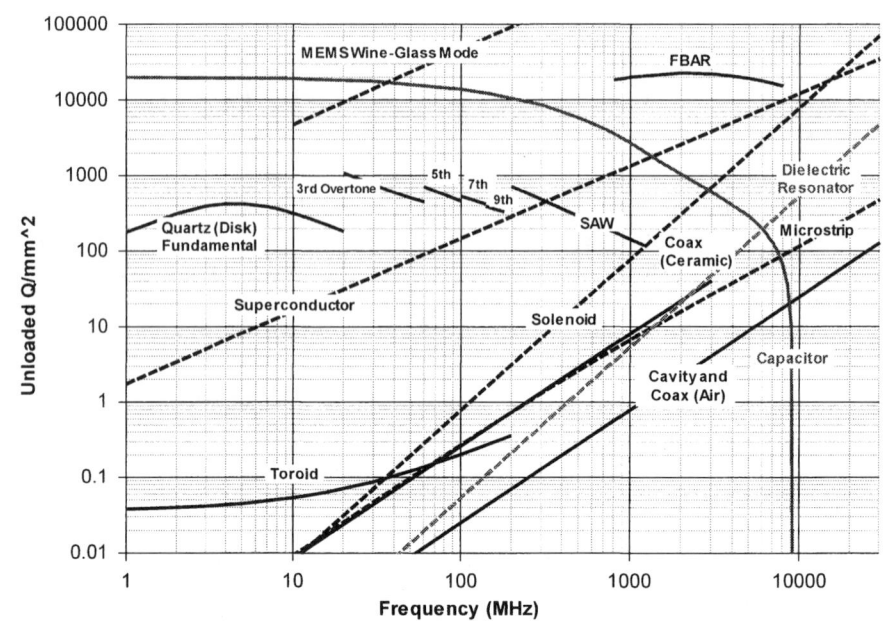

Figure 6.1 Unloaded Q per square millimeter of footprint versus frequency for various resonator technologies (reprinted with permission from the *Microwave Journal*).

It is interesting to note that the FOM for the solenoid is superior to distributed devices. Again, the advantage of distributed resonators is that with additional size the unloaded Q is higher than with the solenoid. The unloaded Q of a distributed resonator of length equal to a multiple of a quarter wavelength is

$$Q_R = \frac{\beta}{2\alpha_t} = \frac{\omega\sqrt{\varepsilon_r}}{2\alpha_t \upsilon_0} = \frac{\pi\sqrt{\varepsilon_r}}{\alpha_t \lambda_0} \qquad (156)$$

where β is the phase constant, α_t is the total line attenuation in nepers/unit length, ε_r is the line relative dielectric constant and λ_0 is the wavelength in free space. At 1 GHz, distributed resonator unloaded Q ranges from about 75 for microstrip on a 0.63-mm thick alumina substrate, to 300 for microstrip on a 1.57-mm thick PTFE substrate to 3000 for a 25-mm diameter coaxial cavity.

6.2 Lumped and Distributed Equivalents

Current flowing longitudinally in a TEM mode transmission line develops an encircling magnetic field that impedes the current flow. A

Distributed Oscillators

transmission line thus possesses series inductance. The conductor of a transmission line is spaced from the ground plane by a dielectric. The conductor thus possesses capacitance to ground. Fig. 6.2 depicts equivalent relationship between a series transmission line of characteristic impedance Z_o and electrical length θ, and an equivalent lumped inductor capacitor network [2, 3].

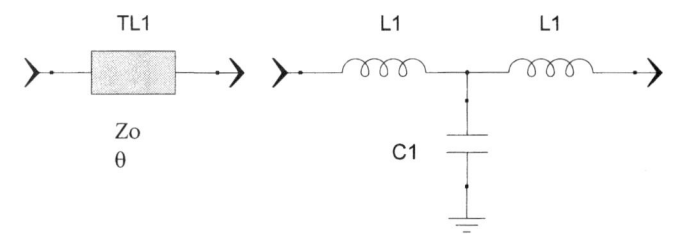

Figure 6.2 Transmission line with lumped element equivalent.

The value of the inductors and the capacitor are

$$L_1 = \frac{Z_0 \tan \frac{\theta}{2}}{2\pi f} \quad (157)$$

$$C_1 = \frac{\sin \theta}{2\pi f Z_0} \quad (158)$$

where f is the frequency at which the equivalence is required. If the transmission line impedance is high, the line is thin and the capacitance is small. In fact if the line is shorter than 30° and the impedance is high, the capacitance may be ignored and the total inductance is

$$L_1 \cong \frac{Z_0 \tan \theta}{2\pi f} \quad (159)$$

Fig. 6.3 shows the equivalent lumped networks for three other short transmission line configurations. All types are most effective with as short a length as is feasible. This results in high impedance for types 1 and 2 and low impedance for types 3 and 4. The equivalent lumped element values are

$$L_2 = \frac{Z_0 \tan \theta}{2\pi f} \quad (160)$$

$$C_3 = \frac{\tan \theta}{2\pi f Z_0} \quad (161)$$

$$C_4 = \frac{\sin \theta}{2\pi f Z_0} \qquad (162)$$

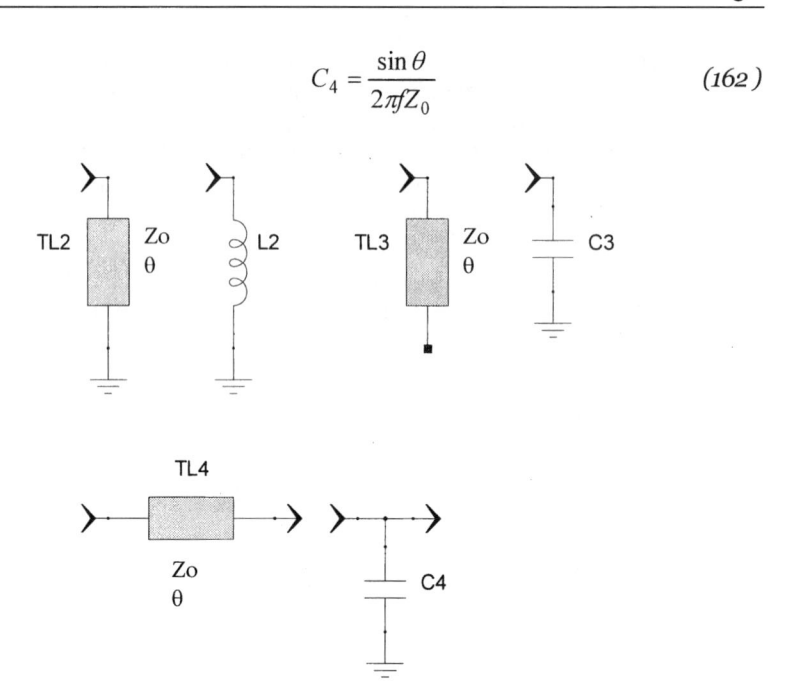

Figure 6.3 Additional short transmission lines and their lumped element equivalents.

Equivalences for resonant transmission lines and lumped elements are given in Fig. 6.4. The values of the lumped elements, equivalent over a limited bandwidth, are

$$L_5 = \frac{2Z_0}{\pi^2 f} \qquad (163)$$

$$L_6 = \frac{Z_0}{8f} \qquad (164)$$

$$L_7 = \frac{Z_0}{4f} \qquad (165)$$

$$C_x = \frac{1}{(2\pi f)^2 L_x} \qquad (166)$$

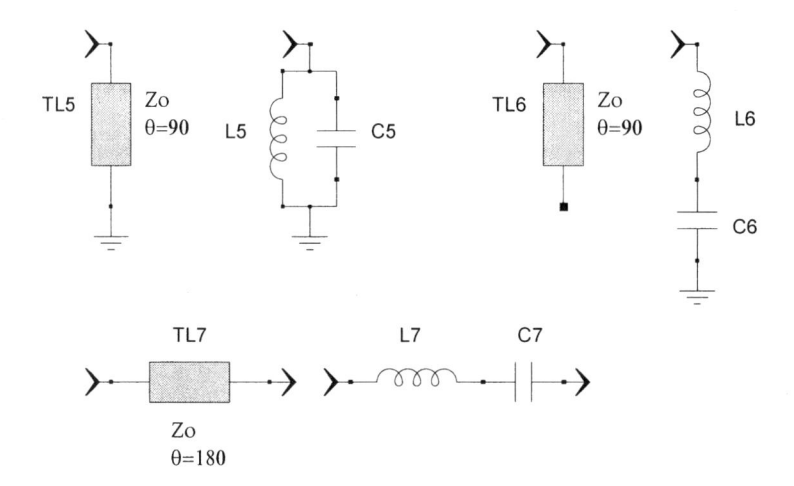

Figure 6.4 Resonant transmission lines and their lumped element equivalents.

All seven of these equivalences apply over a limited frequency range. Because oscillator resonators are narrowband, the equivalences are useful. Subject to the limitation of practical values, any of the lumped element networks depicted in Figs. 6.3 and 6.4 may be replaced with the distributed form. The fact that values may be extreme can be an advantage. For example, consider the popular ceramic-loaded coaxial resonator. A Trans-Tech standard profile coaxial resonator with a relative dielectric constant of 39 has a characteristic impedance of 9.5 ohms [4]. The equivalent value of inductor L_5 for a shorted quarter wavelength resonator at 900 MHz is 2.14 nH. This small value is difficult to realize as a solenoid. Good loaded Q for parallel resonators requires a small value of inductance so the coaxial resonator is an excellent replacement for the L-C parallel resonator. Furthermore, the unloaded Q of this coaxial resonator exceeds 450.

As illustrated in Figs. 6.3 and 6.4, series resonators, parallel resonators, series inductors, shunt inductors, and shunt capacitors may be replaced with distributed equivalents. Oscillator design may begin with an L-C topology converted to distributed whenever that form is more practical or offers a higher unloaded Q. However, conspicuously absent from the equivalences is the distributed form of the series capacitor. Because the capacitor is often less problematic than the inductor, the inductor in an L-C resonator may be converted to distributed while the capacitors remain lumped. A resonator that combines lumped and distributed elements is called a hybrid resonator.

6.3 Quarter-Wavelength Resonators

The type 5 lumped/distributed equivalence depicted in Fig. 6.4 suggests an oscillator topology similar to that shown in Fig. 5.48 with the parallel resonator replaced with a shorted quarter-wavelength transmission-line stub.

6.3.1 The Quarter-Wavelength Resonator

The loaded Q of a lossless quarter-wavelength resonator is

$$Q_L = \frac{\pi R_{par}}{4Z_0 \sin^2 \theta_{tap}} \qquad (167)$$

where R_{par} is the parallel resistance loading the resonator from both terminations, Z_0 is the characteristic impedance of the resonator transmission line, and θ_{tap} is the electrical length of the resonator tap point from the grounded end. For example, with input and output loads of 500 ohms connected 45° from the grounded end of a 50-ohm resonator, the loaded Q is 7.85. Eq. 167 is valid when the resonator physical length is shortened by the addition of capacitive loading at the open end of the resonator, provided of course that θ_{tap} is less than the length of the loaded resonator.

Converting Eq. 156 to decibel units, the unloaded Q of a distributed resonator that is a multiple of a quarter wavelength is given by

$$Q_R = \frac{8.686\pi\sqrt{\varepsilon_r}}{\alpha_{t(dB)}\lambda_0} \qquad (168)$$

There are two forms of loss in purely distributed resonators. Dielectric loss dominates near open ends where the electric field is highest and the conducted current is lowest. Conductor loss dominates near grounded ends where the electric field is lowest and the current is highest.

6.3.2 Ceramic-Loaded Coaxial Resonators

Resonators with air or low dielectric constant materials are lengthy below UHF frequencies. A quarter-wavelength resonator in air at 1 GHz is 8.32-cm long, hardly compatible with modern, small-sized applications. The use of high dielectric constant, nonpiezoelectric materials, significantly reduces the physical length required for resonance. Fig. 6.5 is a photograph of a variety of coaxial resonators popular for use in the VHF through UHF frequency range [5]. A block of ceramic material with a square perimeter and circular center is silver-plated on the sides and one end. The opposite end is open with a metal tab contacting the silver in the center cylinder.

Distributed Oscillators 269

These devices are modeled as shorted transmission lines with series tab inductance connecting to the open end. Circular outer surface resonators are also available but the square form is well suited for mounting on a PWB.

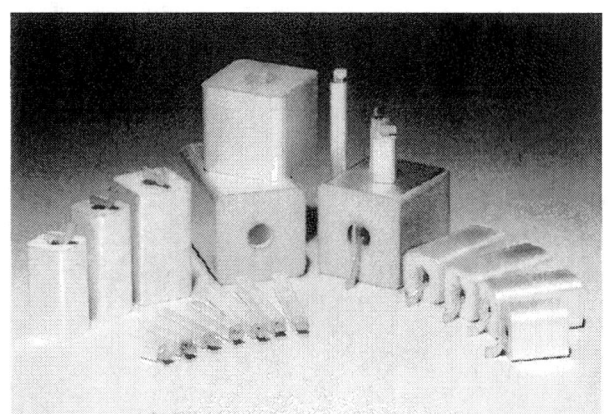

Figure 6.5 Photograph of various ceramic-loaded quarter-wavelength coaxial resonators with a square outer conductor and circular inner conductor (courtesy of Trans-Tech, Inc.).

The length and dielectric constant determine the resonant frequency. The side width determines the unloaded Q. Large sizes are used for high unloaded Q and small sizes are used when space is premium.

Ceramic-loaded coaxial resonators became practical with the development of low-loss, temperature-stable, high dielectric constant materials. Given in Table 6.1 are characteristics of four ceramic materials used in standard Trans-Tech coaxial resonators. These parameters, a surface roughness of 0.002 mm and a resistivity relative to copper of 0.92 are used to model the coaxial resonator in circuit simulators. These parameters are only approximate and vary with frequency and the profile of the resonator. More accurate unloaded Q values are obtained from Q-curve graphs published by ceramic resonator manufacturers.

Table 6.1 Characteristics of four ceramic materials used in Trans-Tech coaxial resonators

Material	ε_r	ε_r tol	Loss δ	$\Delta f/f$ (ppm $^\circ$C)	Suggested frequency range (MHz)
1000	10.5	±0.5	0.0002	0 ±10	1500-5700
2000	20.6	±1.0	0.0002	0 ±10	700-5800
8800	39	±1.5	0.0002	4 ±2	700-5200
9000	90	±3.0	0.0006	0 ±10	300-3400

Trans-Tech coaxial resonators are available in standard dimensions listed in Table 6.2. The tab inductance is given in column 6. Representative unloaded Qs are given in columns 7 through 9. Coaxial resonators have low impedance because of the high material dielectric constant. Coaxial resonators offer unloaded Q about 5 times higher than L-C resonators and they offer improved temperature stability, although well-designed L-C resonators with temperature compensating capacitors offer good temperature stability as well.

Table 6.2 Standard dimensions for Trans-Tech coaxial resonators with the characteristic impedance and representative unloaded Q

Profile	Side (mm)	Inn. dia (mm)	ε_r	Zo (ohms)	Tab L (nH)	Qu @ 900 MHz	Qu @ 2.4 GHz	Qu @ 5.6 GHz
SM	2	0.813	10.5	18.4	0.6		160	220
			20.6	13.1		190	230	310
MP	3	0.813	10.5	25.7	0.6		230	330
			20.6	18.4		260	380	430
			39	13.3		230	370	
			90	8.8		220		
LP	4	0.965	10.5	27.4	1.0		280	
			20.6	19.6		320	480	
			39	14.2		310	480	
			90	9.4		270		
SP	6	2.41	10.5	18.3	1.0		420	
			20.6	13.1		470	660	
			39	9.5		460		
			90	6.3		370		
EP	8	2.57	10.5	22.5	1.0		560	
			20.6	16.1		650		
			39	11.7		600		
HP	12	3.33	20.6	18.1	1.8	900		
			39	13.1		880		

Lower impedance reduces the RF voltage across a tuning varactor but also increases the required varactor capacitance. High varactor capacitance generally reduces the varactor unloaded Q. Generally, a lower characteristic impedance is used if the varactor RF voltage is an issue and higher characteristic impedance is used when wider tuning is required.

Notice in Table 6.2 that the specified dielectric constant tolerance is ±3% for high dielectric constant materials to ±5% for the lower dielectric constants. The resonant frequency is inversely proportional to $\sqrt{\varepsilon}$ so the standard frequency tolerance ranges from ±1.5 to 2.5%. The frequency down shift from capacitive loading by the landing pad for the center conductor tab is often more significant than the resonator tolerance. This error is compensated during prototyping.

The final center frequency may be tuned by removing metalization with a rotary abrasive device. Removing metalization at the open end of the resonator deceases the distributed capacitive and increases the resonant frequency. A frequency shift of as much as 20% has little impact on the unloaded Q. Removing metalization on the shorted end increases the distributed inductance and decreases the resonant frequency. Unfortunately, this increases the conductor resistance at the shorted end and degrades the unloaded Q.

Distributed Oscillators

The temperature coefficient of Trans-Tech coaxial resonators is given in Table 6.2. The worst-case specified temperature coefficient is ±10 ppm per °C. For a 900-MHz resonator, this is ±0.9 MHz for the temperature range of 0 to 100°C. Coaxial-resonator VCOs are easily tuned more than this. These oscillators are well suited for phased-locked applications.

Because the impedance is low, to achieve the best available unloaded Q, silver metalization is used to minimize resistance. Silver has a resistivity relative to copper of approximately 0.92. However, silver has a strong tendency to migrate into tin-lead solder. To avoid the silver plating being drawn from the ceramic block, a silver bearing solder such as SN62 (62Sn-36Pb-2Ag) is used. Prewarming with a brief 230°C regime is recommended.

6.3.3 Capacitor-Loaded Quarter-Wavelength Resonator

If the resonator is physically shortened by capacitive loading at an open end, current loss is unaffected but a portion of the electric field energy storage and loss transfers to the loading capacitance. The loading capacitance required to resonate a physically shortened quarter wavelength transmission line is

$$C_{loading} = \frac{1}{2\pi f_0 Z_0 \tan \theta} \qquad (169)$$

To determine the unloaded Q of a capacitively loaded resonator, the contributions to the unloaded Q by the loading capacitor, the dielectric, and the conductor must be quantified.

$$Q_R = \frac{1}{1/Q_{cap} + 1/Q_{diel,shortened} + 1/Q_{cond}} \qquad (170)$$

For high-quality capacitors the majority of the loss is plate and terminal series conduction loss. The unloaded capacitor Q is then given by

$$Q_{cap} = \frac{1}{ESR \omega C} \qquad (171)$$

Replacing a portion of the open-end transmission line with a lumped loading capacitor replaces that dielectric loss with *ESR* loss in the capacitor. With less loss in the dielectric, the effective unloaded Q of the dielectric is increased. The effective unloaded Q of the dielectric in a physically shortened resonator is determined heuristically to be approximately

$$Q_{diel,shortened} \cong \frac{1}{loss\, \delta} \left(\frac{90}{\theta} \right)^2 \qquad (172)$$

where *loss* δ is the loss tangent of the dielectric material and θ is the electrical length of the loaded line in degrees. $1/loss\ \delta$ is the unloaded Q of the dielectric material. For microstrip, because a portion of the electric field is in the air above the substrate, the "effective" unloaded Q is slightly higher than the bulk material dielectric unloaded Q. For example, with a bulk material *loss* δ of FR4 of 0.008, the bulk unloaded Q is 125. However, the unloaded Q of a resonant 50-ohm microstrip line with ideal conductors is about 137.7 as determined by the GENESYS T/LINE simulator and 139.2 as determined by the GENESYS EMPOWER simulator [6]. Microstrip unloaded Q is higher than the bulk material unloaded Q because a portion of microstrip electric fields are in air.

Replacing a length of the resonator line near the open end with a loading capacitor has little effect on the conductor loss. Therefore, the resonator unloaded Q apportioned to the conductor is given by

$$Q_{cond} = \frac{8.686\pi}{\alpha_{cond}\lambda_{g0}} \quad (173)$$

where α_{cond} is that portion of the loss in decibels per unit length of the transmission line attributable to the conductor. Combining Eqs. 169-173,

$$Q_R = \frac{1}{\dfrac{ESR}{Z_0 \tan\theta} + loss\ \delta\left(\dfrac{\theta}{90}\right)^2 + \dfrac{\alpha_{cond}\lambda_{g0}}{8.686\pi}} \quad (174)$$

The unloaded Q from Eq. 174 is plotted in Fig. 6.6 at 1 GHz versus the physical line length with a 6 mm outer diameter, 2 mm inner diameter copper coaxial resonator with a relative dielectric constant of 36 and a loss tangent of 0.0004. The characteristic impedance is 10.98 ohms. The trace with diamond symbols is with a lossless loading capacitor. At 90° physical line length, the loss is due to a combination of both dielectric and conductor loss. The loss is primarily conductor loss and the unloaded is 341, typical of a ceramic-loaded coaxial resonator at this frequency. At shorter physical lengths, the dielectric loss vanishes and the ultimate unloaded Q is 395, limited only by conductor loss. Plotted with triangular symbols is the unloaded Q versus physical length when the loading capacitor ESR is 0.02 ohms. This low value of ESR is only achieved with several high-quality capacitors connected in parallel. Nevertheless, the unloaded Q is severely degraded with loading capacitance.

The situation is different on FR4 substrate where the majority of the loss is in the dielectric material rather than the conductor. The microstrip plots in Fig. 6.8 are with a material loss tangent of 0.008, a relative dielectric constant of 4.69, and 8-mm wide strips. The wide strip lowers the conductor loss and results in a characteristic impedance of 24.4 ohms. An effective microstrip *loss* δ of 0.00754 is used. The round traces are again

Distributed Oscillators 273

with a loading capacitance *ESR* of 0.02 ohms. Because the predominant source of loss on FR4 is the dielectric, loading the resonator with a high-quality capacitor actually increases the unloaded Q by replacing a portion of the dielectric with the capacitor, thus improving the quality of a resonator on FR4. In this case, the maximum unloaded Q occurs at a physical length of 30°. This is an important technique for improving the quality of resonators on FR4 but it requires high-quality capacitors.

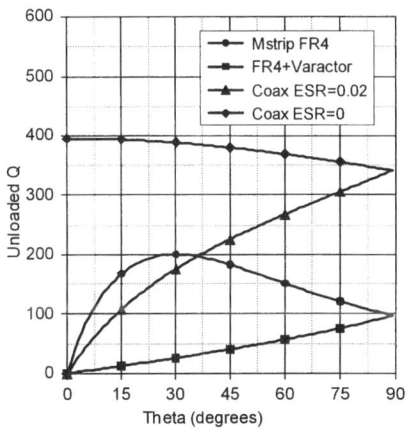

Figure 6.6 Unloaded Q for a resonator versus the physical length when loaded to quarter-wave resonance with a capacitor in coax with a lossless capacitor (diamond symbols), in coax with a capacitor ESR of 0.02 ohms (triangular symbols), in microstrip on FR4 with a capacitor ESR of 0.02 ohms (round symbols), and with a varactor ESR of 0.5 ohms (square symbols).

The square symbols in Fig. 6.8 illustrate the impact on resonator unloaded Q when varactors are used to tune the resonator. In this case a varactor *ESR* of 0.5 ohms is assumed. Even a modest amount of loading destroys the resonator Q.

6.4 Distributed Oscillator Examples

The following sections include several distributed oscillator examples.

6.4.1 Negative-Resistance Hybrid Oscillator

Fig. 1.48 depicts the basic negative-resistance oscillator with a series tuned L-C resonator at the base. Fig. 6.7 shows a basic hybrid-resonator negative-resistance oscillator created by replacing the resonator inductor with a transmission line. The line length is typically 45° to 60° at the highest operating frequency and the characteristic impedance of the line is typically higher than 50 ohms. A port on the left side of the line is used for analysis and this point is grounded for oscillation.

Power is typically coupled at the emitter through a capacitor as shown in Fig. 6.7 or at the resonator by tapping the transmission line. Fifty ohms coupled directly at the emitter loads the circuit heavily and may absorb the negative resistance, thus preventing oscillation. If so, an emitter coupling-capacitive reactance somewhat higher than 50 ohms is used. If resonator coupling is used the tap is near ground. Shifting the tap point toward the base reflects a high series resistance into the circuit and absorbs the negative resistance.

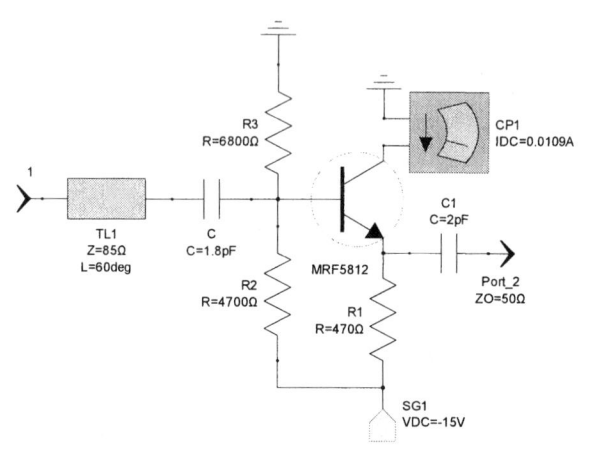

Figure 6.7 Basic hybrid-resonator negative-resistance oscillator.

The loaded Q of this circuit is poor unless an extremely small tuning capacitance and a high characteristic impedance line is used. The poor loaded Q is not an issue with wideband varactor tuning because varactor noise dominates. The poor temperature, device, and bias dependency of this circuit is often mitigated by using a PLL. The primary application for this circuit is UHF and microwave VCO's. Therefore, this circuit is considered in more detail in Chapter 7.

6.4.2 Negative-Resistance High-Power 1 GHz Oscillator

The collector of bipolar transistors is connected directly to the case for TO-39 and certain other package styles. This provides for efficient heat-sinking for power amplifiers and oscillators. In the 1-GHz negative-resistance oscillator shown in Fig. 6.8, the case of an NPN 2N5109 transistor is soldered directly to the copper ground plane of the PWB. A simple bias scheme uses a negative power supply. With a -24 volt supply, the quiescent bias current is just under 50 mA without oscillation. Due to the small resistance in the collector-emitter path, the supply current increases to approximately 100 mA during oscillation. This circuit delivers over 1/2 watt at 1 GHz.

Distributed Oscillators

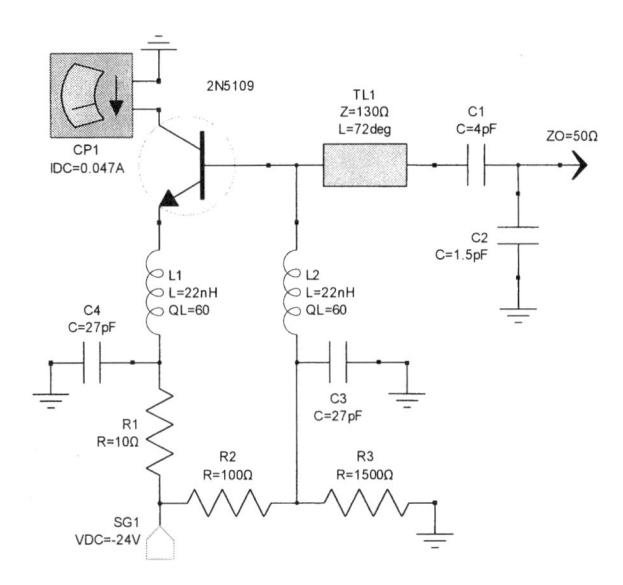

Figure 6.8 Negative-resistance bipolar high-power 1-GHz oscillator.

This circuit is modified from a 1.68-GHz radiosonde transmitter [4] that uses a 2N5108 bipolar transistor and delivers approximately 300 mW with a -24 volt supply. This circuit is useful at frequencies from 1 to 2 GHz with only small adjustments in the length of the transmission line in the resonator. At lower frequency, shifting the capacitor C_2 from the output to the common node between the line and C_1 improves performance. Below 1 GHz, added capacitance from the emitter to ground is helpful.

Table 6.3 lists the output frequency, output power, supply current, and DC to RF conversion efficiency versus supply voltage for the 1-GHz circuit shown in Fig. 6.8.

Table 6.3 Measured frequency, output power, and DC to RF conversion efficiency versus supply voltage for the negative-resistance, high-power, 1-GHz oscillator

Vs (volts)	Frequency (MHz)	Output (dBm)	Output (watts)	Is (mA)	η (%)
13	988	8.0	0.006	18	2.7
15	995	20.3	0.107	39	18.3
17	997	24.0	0.251	55	26.8
19	997	25.5	0.355	71	26.3
21	998	26.6	0.457	84	26.0
23	999	27.3	0.537	97	24.1
25	1001	27.6	0.575	112	20.5

6.4.3 Quarter-Wavelength Hybrid Oscillator

Consider the lumped-element 700-MHz capacitor-coupled parallel-resonator oscillator shown in Fig. 5.48. The 2-pF capacitor C_3 and 16.11-nH inductor L_1 resonate at 886.7 MHz. The coupling capacitors further load the resonant frequency to near 700 MHz. The lumped elements may be replaced with a stub one quarter-wavelength long at the unloaded resonant frequency. From Eq. 160 the characteristic impedance of the shorted quarter-wavelength stub is 70.5 ohms. With a resonant frequency of 886.7 MHz the stub is 1.5 mm wide by 47.8 mm long. Both the line open-end capacitance and the via hole inductance lower the resonant frequency so that the required physical length is approximately 45 mm. The coupling capacitors remain lumped, thus the resonator is a hybrid of lumped and distributed elements.

With parallel resonators, low inductor values with high capacitor values are desirable. However, achieving good unloaded Q is difficult with small-valued inductors. This limited the inductor in the 700 MHz oscillator to a minimum of about 16 nH. The relatively high impedance level of the resonator also resulted in rather small values of coupling capacitor, but still the loaded Q is only about 19. A transmission line with lower characteristic impedance provides a higher loaded Q with more reasonable coupling capacitors.

Fig. 6.9 shows the schematic of a 700-MHz oscillator open-loop cascade using a tapped quarter-wavelength resonator with capacitor coupling. The quarter-wavelength resonator uses 8-mm wide microstrip on a 1.57-mm thick FR4 substrate. The wide strips result in a characteristic impedance of 24.6 ohms.

The open-loop cascade gain, transmission phase and loaded Q are shown on the left in Fig. 6.10. The lower impedance and tapping the resonator result in an open-loop cascade loaded Q of 36, almost twice the loaded Q of the oscillator in Fig. 5.48, while using a larger value of coupling capacitor. $TL4$ and $TL5$ are merely connecting lines for the lumped coupling capacitors. The collector inductor choke is replaced with a thin microstrip line $TL3$, thus eliminating all of the inductors in the original design.

The harmonic-balance simulated output spectrum for the 700-MHz hybrid quarter-wavelength resonator oscillator is shown on the right in Fig. 6.10. The output power is high but the harmonic performance is poor. By taking the output from the resonator with a 50-ohm load tap point a few millimeters from the ground end of the resonator, this circuit offers approximately 10-dBm output with all harmonics at least 25 dBc. A microstrip quarter-wavelength oscillator similar to this example but with varactor frequency tuning is described in Section 7.4.7.

Distributed Oscillators 277

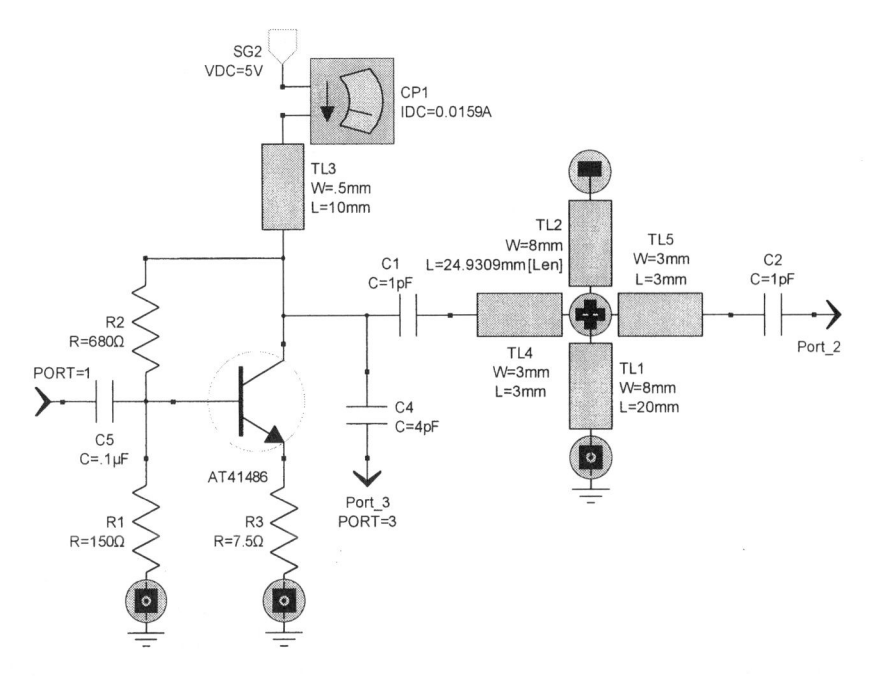

Figure 6.9 A 700-MHz oscillator open-loop cascade using a hybrid distributed-lumped resonator.

6.4.4 Simple Hybrid Coaxial Resonator MMIC

Fig. 6.11 shows the schematic of a simple oscillator using an MMIC sustaining stage and a hybrid resonator with a quarter-wavelength ceramic-loaded coaxial line and lumped coupling capacitors. Despite its simplicity, this oscillator offers good loaded Q, output power, and harmonic performance.

The sustaining stage is a Mini-Circuits MAR 3 gain block that uses Darlington-pair bipolar transistors. The 5-volt supply is connected to the MAR 3 through a 4-nH inductive choke that also serves as a phase-lead network to adjust the phase-zero crossing at maximum phase slope. The temperature and bias stability is improved by adding bypassed series resistance in the supply line and increasing the supply voltage to keep the quiescent supply current in the 20 to 35 mA range. The loop cascade match is excellent. The gain margin is low, approximately 3 dB. However, once the prototype verifies the design, the repeatability is good because the MMIC gain is tightly controlled by its resistive feedback.

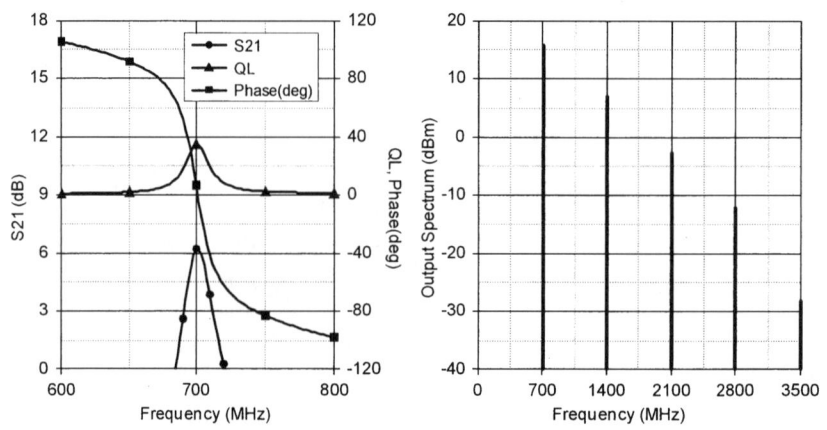

Figure 6.10 Open-loop cascade gain (circular symbols), transmission phase (square symbols), and loaded Q (triangular symbols) are shown on the left and the simulated output spectrum is shown on the right.

The loaded Q is approximately 180, resulting in good phase noise performance. This circuit is capable of producing SSB phase-noise levels below -90, -120, and -145 dBc/Hz at 1, 10, and 100 kHz, respectively.

The simulated output spectrum is shown in Fig. 6.12. The output power is approximately 9 dBm and the second harmonic is -19 dBc. The good harmonic performance is aided by a low gain margin that moderates the degree of gain compression.

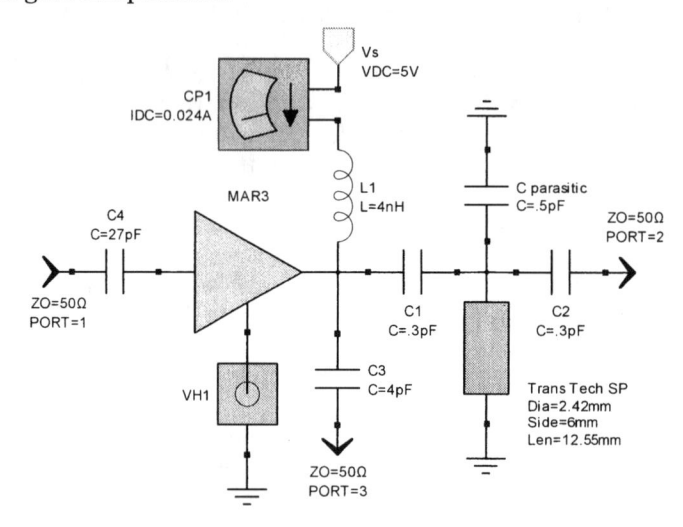

Figure 6.11 Simple hybrid coaxial-resonator MMIC oscillator.

Distributed Oscillators 279

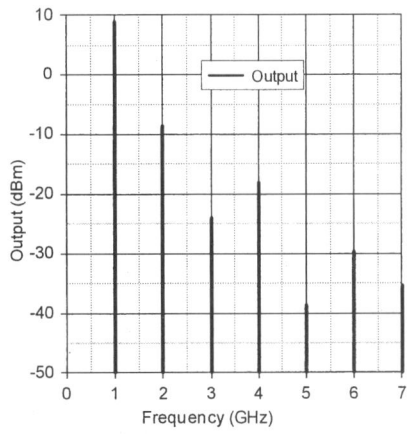

Figure 6.12 Simulated output spectrum of the hybrid coaxial-resonator MMIC oscillator.

6.4.5 Probe-Coupled Coaxial Resonator Bipolar

Fig. 6.13 is a probe-coupled coaxial resonator with a discrete resistive-feedback Darlington bipolar transistor amplifier. A system approach is used for the construction of this prototype 700-MHz oscillator.

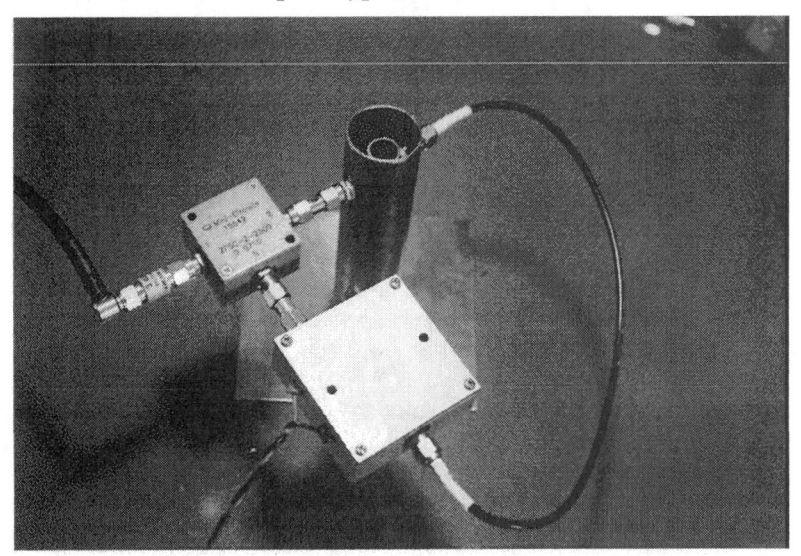

Figure 6.13 Probe-coupled coaxial resonator oscillator.

The amplifier in the aluminum anodized housing labeled WMC is a discrete resistive-feedback Darlington bipolar amplifier using two MRF901 transistors. The schematic is shown in Fig. 6.14. The 115° length of transmission line at 700 MHz is added to simulate the physical size of the amplifier, the PWB, and the connectors. The total phase shift of the amplifier is -40° at 700 MHz. The amplifier gain is approximately 18 dB and flat to 400 MHz. At 700 MHz the gain is 16.4 dB.

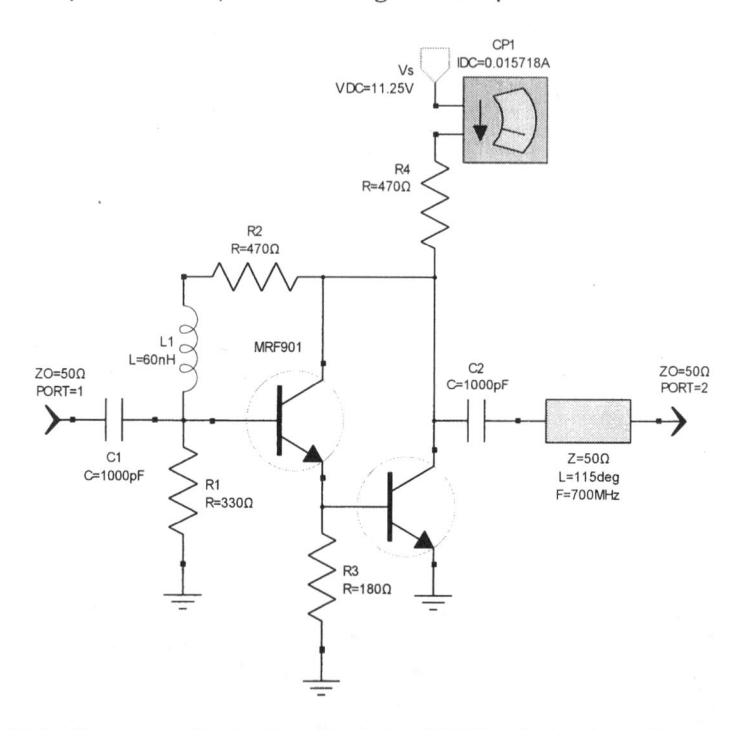

Figure 6.14 Discrete resistive-feedback Darlington MRF901 wideband amplifier.

The 3-port Minicircuits device is a ZFSC-2-2500 splitter with approximately 3.9-dB insertion loss and an electrical length of -56° at 700 MHz. The output of the amplifier drives this splitter. One output port drives the resonator and the other connects to an HP8568A spectrum analyzer through a 3-dB pad.

The outer conductor of the coaxial resonator is a hard-drawn copper tube with an inside radius of 13 mm. The center conductor is a soft-drawn copper tube 100 mm in length with an outside radius of 6.35 mm. The characteristic impedance of this resonator is approximately 42.8 ohms.

The coaxial cavity is coupled near the open end via the center pins of SMA connectors that act as capacitive probes. The estimated probe capacitance to the center conductor is 0.13 pF. The end effect capacitance, derived from the measured resonant frequency, is assumed to be approximately 0.28 pF. The measured loaded Q is 484. The measured insertion loss of the resonator is -4 dB at resonance and the electrical length is +150°. The amplifier, splitter, and resonator cascade is closed using a 306° length of coaxial cable shown on the right in Fig. 6.13. The resulting net transmission phase of the cascade is 360° at resonance.

The measured output power without the 3 dB pad is 1.9 dBm and the second harmonic is -11 dBc. The harmonic-balance simulated output power is 2.1 db and the second harmonic is -13.1 dBc.

Shown on the left in Fig. 6.15 is the measured output spectrum of the 700-MHz coaxial cavity oscillator under battery operation and mounted directly on the measurement bench. The scan width is 1 kHz per division and the residual FM noise modulation is roughly 1 kHz. This residual FM is caused by bearing noise in the cooling fans of the HP8568A spectrum analyzer vibrating the bench.

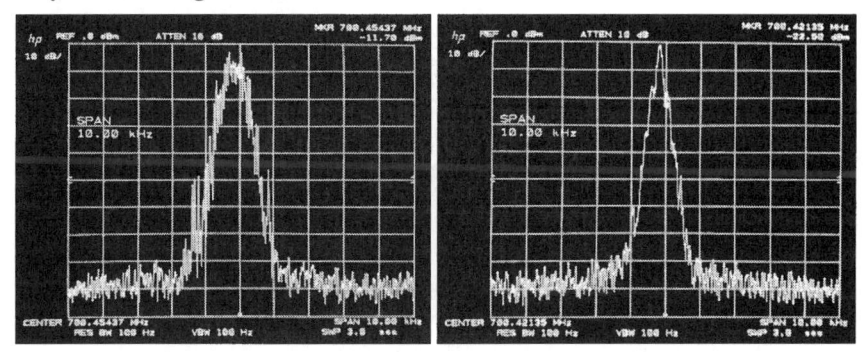

Figure 6.15 Measured output spectrum of the 700-MHz cavity oscillator mounted directly on the measurement bench (left) and shock-mounted on the bench (right).

On the right in Fig. 6.15 the oscillator is mounted on four small damped-foam standoffs. The resulting spectrum is significantly cleaner. However, the remaining residual FM is also caused by bearing noise. Manually lifting the oscillator off the bench eliminates all visible residual FM at this scan width. Although the center-conductor copper tube is 12.7 mm in diameter and only 100-mm long, the vibration induced by the bench vibration is sufficient to introduce significant residual FM.

6.4.6 End-Coupled Hybrid Half-Wavelength Bipolar

Fig. 6.16 shows the schematic of a bipolar end-coupled half-wavelength 2400-MHz oscillator. The substrate material is 0.635-mm thick Rogers 3006 ceramic-loaded PTFE with a relative dielectric constant of 6.15 and a loss tangent of 0.0025. At 2400 MHz, most of the gain of the AT41486 transistor is required so the shunt-feedback resistor R_2 provides only minimal RF feedback and serves primarily for biasing. A via hole is included in the emitter ground path because at this frequency inductance is critical. At 2400 MHz, the simulated gain of the amplifier with 50 ohms terminations is 9 dB and phase shift is +57 degrees.

The resonator uses lumped 0.4 pF capacitors to decouple the transmission line from the approximate 50-ohm terminations of the amplifier output and input, thus increasing the loaded Q. This value of capacitance is difficult to achieve by gap or even interdigital fingers. At 2400 MHz, the insertion loss of the resonator is -2.7 dB in a 50-ohm system and the phase shift is -47 degrees. The coupling capacitors load the resonant frequency of the resonator. The unloaded resonant frequency of the transmission line is just over 2900 MHz. The cascade phase shift is near 0° at the resonator amplitude peak, thus satisfying the oscillation criteria. The cascade gain margin is approximately 4 dB. The loaded Q is approximately 50. The unloaded resonator Q is approximately 100. The 0.635-mm thick substrate is too thin to provide a high unloaded Q.

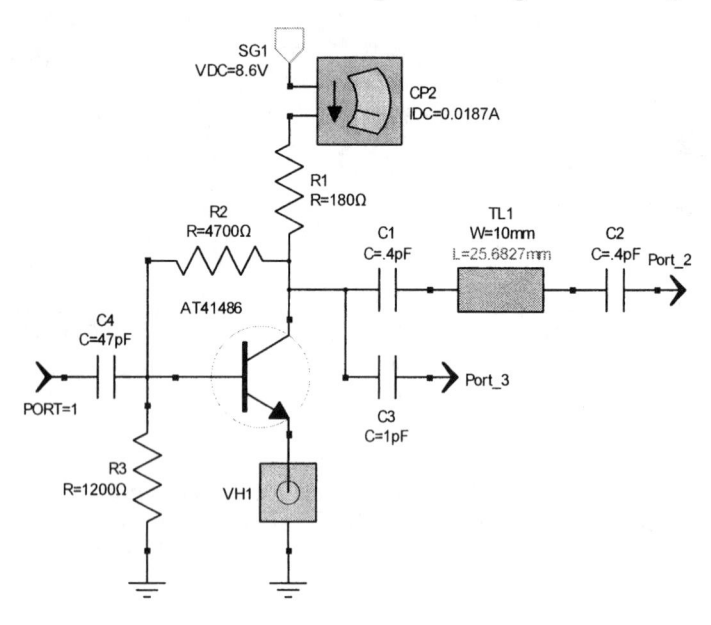

Figure 6.16 End-coupled half-wavelength 2400-MHz bipolar oscillator.

The simulated output spectrum is shown in Fig. 6.17. Although power is extracted at the collector with a coupling capacitor, fair harmonic performance is achieved because the gain margin is low and the device F_t is only approximately 4X the oscillation frequency.

6.4.7 Helical Transmission Line Resonator Bipolar

Transmission line resonators are often long. Loading with modern ceramic materials shortens resonators and is used down to about 300 MHz. Helical resonators offer operation down to the HF range with unloaded Q

Distributed Oscillators 283

that is 2 to 3 times that of solenoids. Quarter-wavelength helical resonators are typically used with a grounded and an open end.

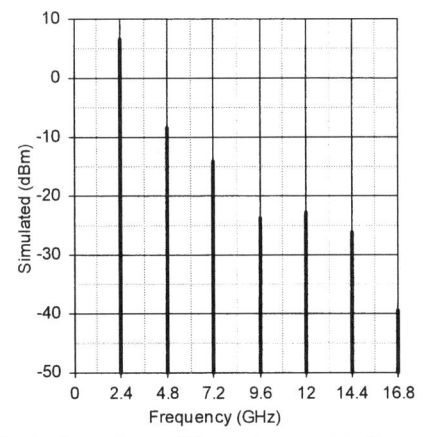

Figure 6.17 Simulated output spectrum of the end-coupled half-wavelength oscillator.

This example illustrates a 581-MHz oscillator using a half wave-length resonator ground at both ends. The structure is shown in Fig. 6.18. The side is 12.7 mm and the 6-mm diameter solenoid is wound with 0.64-mm diameter solid copper wire. The winding is 20.5 turns plus the end supporting wire with a mean length of 23 mm. The enclosure is brass. The resonator is coupled with two capacitive probes consisting of 2-mm² thin plates soldered to the ends of the SMA pins. The probes are located near the grounded end and are placed approximately 1 mm from the solenoid. The estimated capacitance to the solenoid is 0.3 pF.

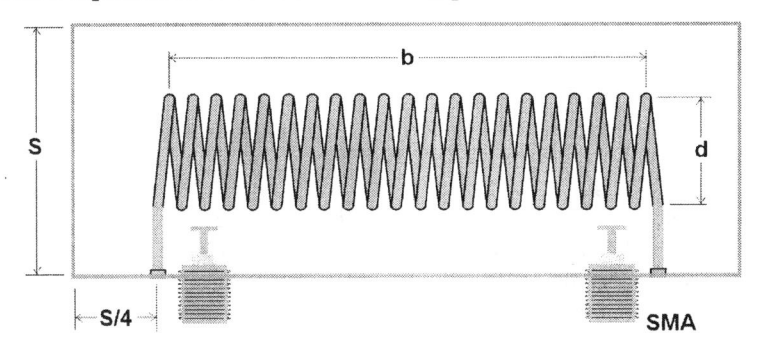

Figure 6.18 Half-wavelength helical resonator with capacitive coupling probes.

The measured resonant frequency is 581 MHz with a loaded Q of 184 and 4.1-db insertion loss. This infers an unloaded Q of 490. An oscillator was constructed using the discrete Darlington broadband amplifier shown in Fig. 6.14 and the Mini-Circuits splitter shown in Fig. 6.13. The transmission phase shifts of the amplifier, splitter, and resonator are 11°, -44°, and +166°, respectively, for a total cascade transmission phase of 133°. The oscillator is formed by closing the cascade with a 227° length of 50-ohm coaxial cable. The measured output power is 3.5 dBm and the second harmonic is -10 dBc. The simulated harmonic-balance output power is 3.0 dBm and the second harmonic is -8 dBc.

This example illustrates the effectiveness of helical resonators for reducing the size of distributed resonators. This resonator operates 17% lower in frequency than the 700-MHz coaxial resonator oscillator, yet this resonator occupies only 9% of the volume. The unloaded Q is significantly less than the coaxial resonator, but it still achieves an unloaded Q that is nearly 3 times that available with a conventional solenoid.

Unsupported helical resonators are extremely susceptible to acoustic induced phase noise and mounting form is typically required for oscillators. Low loss form material is required to avoid degrading the unloaded Q.

6.5 DRO Oscillators

With solenoids, microstrip transmission lines, and coaxial resonators, the unloaded Q is limited primarily by conductor loss. Dielectric resonators avoid conductor loss by storing fields in unmetallized modern ceramics that are high permittivity, low-loss, and temperature stabile. By avoiding conductor loss, unloaded Q as high as 10,000 is available. Oscillators constructed using unmetallized dielectric resonators are referred to as dielectric resonator oscillators (DROs).

6.5.1 Dielectric Resonator Basic Properties

Fig. 6.19 is a photograph of six dielectric resonators [5]. The resonant frequency is approximately inversely proportional to the volume of the resonator. The first dominant mode is $TE_{10\delta}$, which is typically used for oscillators. Additional modes exist. A center hole reduces these "spurious" modes. Improved spurious response is more critical for filter applications than for oscillators. Provided the center hole diameter is less than 25% of the resonator diameter, the impact of the center hole on the resonant frequency of the dominant mode is less than 2%. The $TE_{10\delta}$ mode requires that $h/D<1.015$. The height of commercial resonators is typically 35% to 50% of the diameter.

Distributed Oscillators

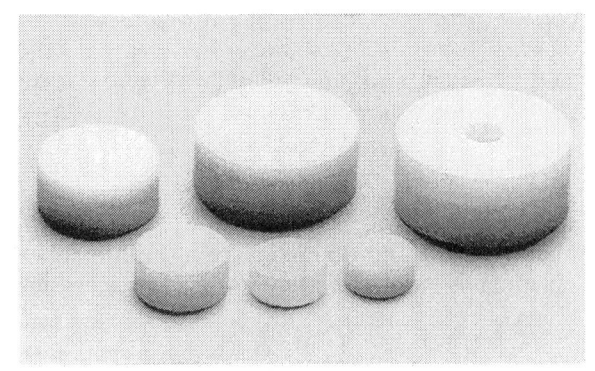

Figure 6.19 Dielectric resonators (courtesy of Trans-Tech, Inc).

Table 6.4 shows properties of various families of dielectric resonators from three manufacturers. Not all dielectric families of each manufacturer are included. There are many other manufacturers of ceramic resonators. Tighter specifications, frequencies outside the listed range, and custom designs are available by contacting the manufacturer. As listed in the table, within each family are often slightly different formulations with specific relative dielectric constants and temperature coefficients of the resonant frequency.

Table 6.4 Dielectric resonator ceramic material properties

Material	Available ε_r	Q×f (GHz)	Available Stability in f_o (ppm/°C)	Freq Range (GHz)
Trans-Tech				
2900	29.0±1 to 30.7±1	110,000	−2 to +4, ±2	1.5 to 6
3500	33.5±1 to 35.5±1	70,000	−3 to +6, ±2	1.5 to 25
8300	35.0±1 to 36.5±1	41,000	−3 to +9, ±2	0.7 to 25
4300	43±0.75	41,000	−6 to +6, ±2	0.7 to 13
4500	44.7±1.5 to 46.2±1.5	41,000	−6 to +6, ±2	0.7 to 3
Murata				
F	23.8±0.5 to 24.2±0.4	350,000	0 to 6, ±0.5, ±1, ±2	10 to 25.2+
B	27.9±0.5	150,000	0 to 6, ±0.5, ±1, ±2	4.8 to 25.9
R	29.7±0.8 to 31.5±0.85	120,000	0 to 6, ±0.5, ±1, ±2	4.6 to 24.2
V	33.5±0.5 to 35.1±0.5	100,000	0 to 8, ±0.5, ±1, ±2	2.9 to 13.2
U	36.6±0.5 to 38.9±0.5	42,000	−4 to 10, ±0.5, ±1, ±2	1.5 to 12.5
M	38.5±1 to 39.2±1	49,000	0 to 6, ±0.5, ±1, ±2	1.5 to 12.5
V	33.5±0.5 to 35.1±0.5	100,000	0 to 8, ±0.5, ±1, ±2	2.9 to 12.5
First Tech.				
Er21	20.5±1.5	60,000	+2.5±1.5	1.8 to 40
HQ29	28.75±0.5 to 29.05±0.5	120,000	−1.2±0.5 to 2.0±0.5	1.8 to 20
HQ34	34±0.3 to 34.4±0.3	87,500	0.0±0.5 to 2.0±0.5	1.8 to 20
Er45	44.5±0.5 to 45.5±0.5	45,000	−3.5±0.5 to 2.0±0.5	0.6 to 5
Er80	80.5±1 to 81±1	10,000	−2.5±1 to 1.5±1	0.6 to 3

6.5.2 Dielectric Resonator Resonant Frequency

The resonant frequency of an isolated dielectric resonator disk is related to the volume approximately by

$$f_0 \approx \frac{233}{\sqrt{\varepsilon_r}V^{1/3}} = \frac{252.5}{\sqrt{\varepsilon_r}D^{2/3}h^{1/3}} \; (GHz) \qquad (175)$$

where the resonator height and diameter are in millimeters. The accuracy of the above formula is approximately 10%. An expression accurate to about 2% is [7]

$$f_0 \approx \frac{34}{D\sqrt{\varepsilon_r}}\left(\frac{D}{h}+6.9\right) \qquad (176)$$

The frequency is tuned by approaching the ceramic resonator with a tuning slug. A metallic slug reduces the resonant frequency and a dielectric slug increases the resonant frequency. A tuning range up to 20% is possible but this significantly impacts the unloaded Q and the temperature stability. Shifting the frequency 10% can reduce the unloaded Q by 50%. Most applications require tuning of only a few percent, which is sufficient to correct the initial frequency accuracy.

6.5.3 Dielectric Resonator Unloaded Q

The frequency × unloaded Q product can exceed 100,000 GHz for high-quality ceramics, so the unloaded Q at 10 GHz can exceed 10,000. Fields are not totally constrained to the dielectric, therefore a surrounding housing is required to avoid radiation loss. The oscillator housing may serve this purpose. However, this housing affects the resonant frequency, unloaded Q, and temperature stability. The temperature characteristic of the dielectric resonator may be chosen to frequency compensate the temperature characteristics of the housing.

Increased housing-wall spacing reduces wall currents, thus reducing wall influence. Mounting on a dielectric pedestal, as shown in Fig. 6.20(a), also reduces frequency shift and degradation of the unloaded Q.

In addition to reducing spurious responses, a center hole in the dielectric resonator may be used to mount the resonator above the substrate using a dielectric screw. With the resonator well removed from the housing bottom and cover, the unloaded Q of a dielectric resonator mounted in a cylindrical cavity is given approximately by

$$Q_U = Q_0\left(1-\gamma e^{-\frac{\gamma}{2}\frac{D_{cavity}}{D}}\right) \qquad (177)$$

where Q_o is the unperturbed unloaded Q of the resonator and

$$\gamma = \sqrt[3]{\varepsilon_r} - 1 \qquad (178)$$

A housing with inside dimensions exceeding 3 times the diameter of the dielectric resonator has minimal impact on the unloaded Q.

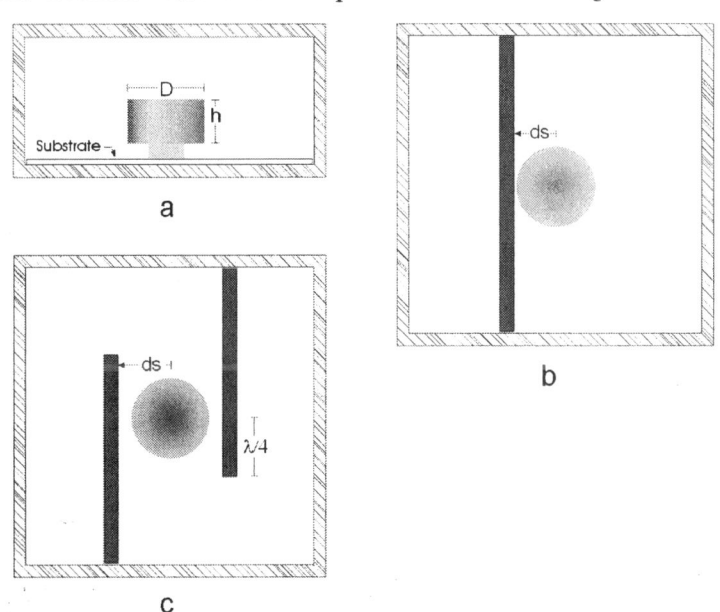

Figure 6.20 Dielectric resonator mounted over the substrate using a pedestal (a), microstrip bandstop coupling (b), and microstrip bandpass coupling (c).

6.5.4 Dielectric Resonator Coupling

The dielectric resonator is coupled to a circuit by orienting the magnetic field lines of the resonator with those of a microstrip line. One such configuration is depicted in Fig. 6.20(b). Maximum coupling occurs with d_s somewhat less than the radius of the dielectric resonator. In the case of oscillators, looser coupling with larger d_s is generally utilized to increase the loaded Q. The coupling depicted in Fig. 6.16(b) results in a bandstop response for the transmission path of the microstrip line. A circuit model is shown in Fig. 6.21(a).

The coupling coefficient between the resonator and the microstrip line is

$$k = \frac{R}{R_{ext}} = \frac{R}{2Z_0} = \frac{S_{11}}{S_{21}} \qquad (179)$$

$$Q_U = Q_L(1+k) = kQ_{ext} \qquad (180)$$

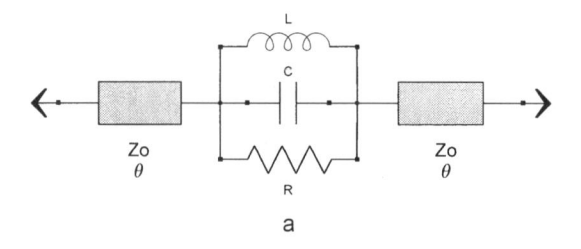

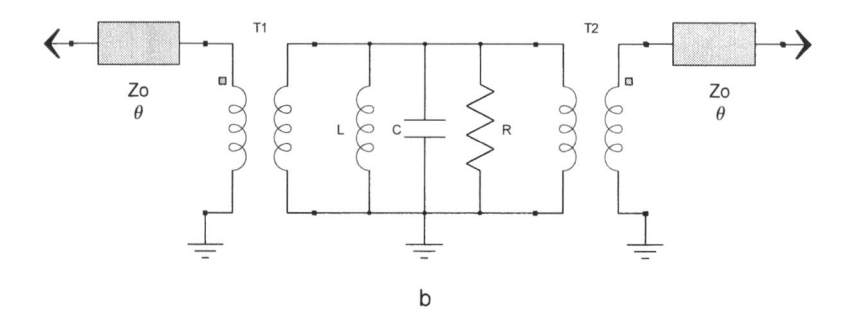

Figure 6.21 Circuit model for a dielectric resonator coupled to a single microstrip line (a) and coupled to two opposite microstrip lines (b).

Model parameters may be determined from the magnitude of the forward scattering parameter of the configuration depicted in Fig. 6.20(b). Fig. 6.22 shows the magnitude of S_{21} versus frequency with θ equal to 90 degrees at resonance. The coupling coefficient is

$$k = 10^{-S_{21}(dB)/20} - 1 \qquad (181)$$

where $S_{21}(dB)$ is the value of the forward scattering parameter at the resonant null. For example, $S_{21}(dB)$ is -28.3 in Fig. 6.22 and k is 25. The bandwidth of the response divided by the resonant frequency yields the loaded and unloaded Q of the dielectric resonator.

$$x = 3 - 10\log\left(1 + 10^{S_{21}(dB)/10}\right) \qquad (182)$$

With values of S_{21} less than -18 dB, x is approximately 3 dB.

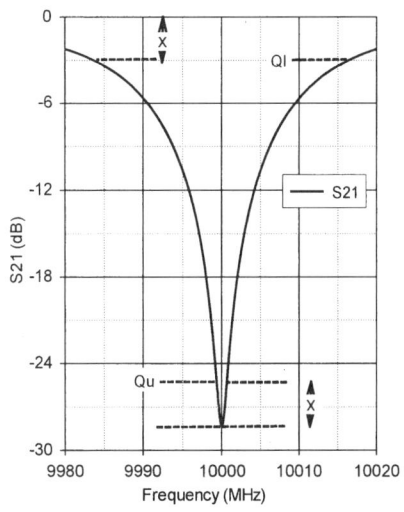

Figure 6.22 S_{21} of the dielectric resonator model shown in Fig. 6.17(a). S_{21} may be used to find the loaded and unloaded Q of the resonator.

The unloaded Q is the resonant frequency divided by the bandwidth at x decibels above the null, in this case 10000/1.25, or 8000. The loaded Q is the resonant frequency divided by the bandwidth at x decibels below 0 dB, in this case 10,000/32.6, or 307. Model parameters are then

$$R = 2Z_0 k \tag{183}$$

$$L = \frac{R}{2\pi f_0 Q_U} \tag{184}$$

$$C = \frac{1}{(2\pi f_0)^2 L} \tag{185}$$

The response of Fig. 6.22 S_{21} at the null is -28.3 dB. Therefore, k is 25, x is essentially 3 dB and R is 2500 ohms. The bandwidth 3 dB above the null is 1.25 MHz, so Q_U is 8000, L is 0.00497 nH, and C is 50.93 pF.

6.5.5 DRO Examples

A popular method for coupling the resonator to an active device to form an oscillator is shown in Fig. 6.23, the schematic of a bipolar 5600-MHz DRO. The quarter-wavelength transmission line TL_2 inverts the parallel-mode resonance of the dielectric puck coupled to the microstrip line. An open-end transmission line TL_1 provides a low impedance at the puck center-line reference plane so that the magnetic field of the resonator can induce current in the microstrip line. Alternatively, if the device develops sufficient negative resistance, TL_1 may be terminated with 50 ohms rather than an open end. With an open termination of TL_1, capacitor C_3 is not required.

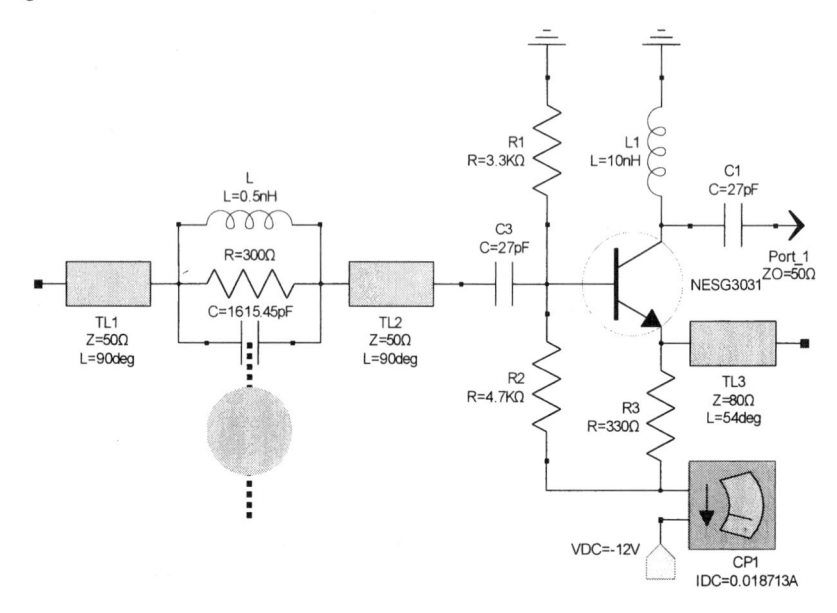

Figure 6.23 DRO formed by a quarter-wavelength transmission line TL_2 inverting the parallel-mode resonator into a series resonator in a CC negative-resistance oscillator.

Open terminated transmission line TL_3 provides capacitance at the emitter to enhance the negative resistance at the base. Power is extracted at the collector of the NEC/CEL bipolar NESG3031 by tight coupling using capacitor C_1.

Nonlinear simulation of this topology is provided using the schematics in Fig. 6.24. On the left, the device is biased and a port is added at the base to simulate the negative resistance. The simulator linearizes the device at the quiescent bias point.

Shown on the left in Fig. 6.25 are the device input resistance and reactance versus frequency. The negative resistance at 5600 MHz for this circuit is -48.4 ohms. The reactance is approximately -25.25 ohms, an effective series capacitance of 1.13 pF. This reactance increases the oscillation frequency approximately 0.8 MHz or 0.014%. As with other negative-resistance oscillators, to mitigate oscillation frequency dependence on the active device and the bias point, it is desirable to have a large effective series input capacitance and negative resistance that is flat with frequency.

Distributed Oscillators 291

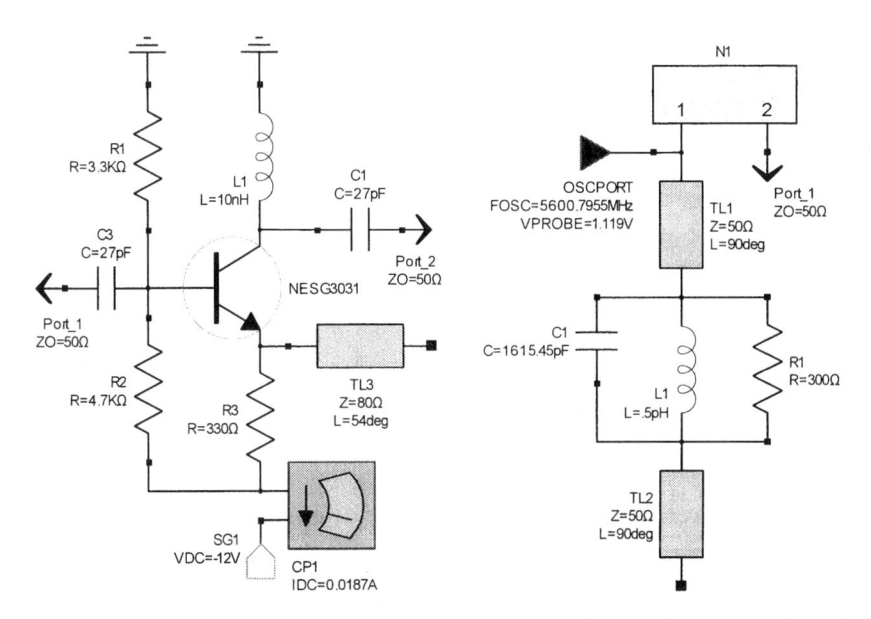

Figure 6.24 Device schematic for simulating negative resistance (left) and with added resonator and oscillator simulation port (right).

On the right in Fig. 6.24, the harmonic-balance oscillator simulation port is added to N_1, a copy of the device schematic on the left. Also added at the base is the model for the dielectric resonator coupled to a microstrip line. On the right in Fig. 6.25 is the simulated output spectrum for the 5600-MHz DRO.

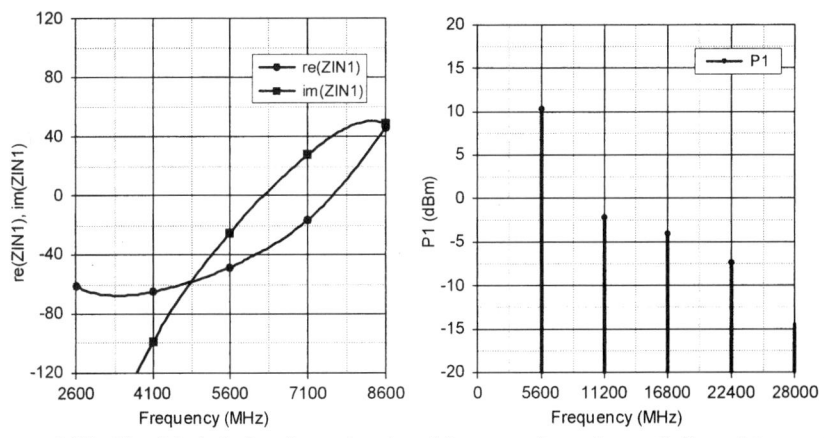

Figure 6.25 Simulated device linear input resistance and reactance (left) and harmonic-balanced output power (right).

Fig. 6.20(c) depicts a second method for coupling a dielectric resonator to microstrip. The response of this configuration is bandpass. A circuit model is given in Fig. 6.21(b). Fig. 6.26 shows a basic bandpass-type DRO using an MMIC gain block. Slightly improved amplifier NF and therefore oscillator SSB phase noise is achieved using a discrete active device. TL_1 and TL_3 are electrically a quarter wavelength. The open termination provides a low impedance at the resonator reference plane to support coupling the magnetic field of the resonator to the microstrip lines. The phase shift at resonance at the resonator reference plane is 0°. The gain of the MMIC overcomes the insertion loss of the bandpass dielectric resonator structure. The total phase shift of the MMIC and all the transmission lines is a multiple of 360°. Because of the high loaded Q of the dielectric resonator, the additional phase slope associated with the lines closing the loop has little impact on the cascade loaded Q. The temperature coefficient of the phase length of these lines can contribute to the temperature stability of the oscillator. If the layout allows, both of the microstrip coupling lines may approach the resonator from one side. In this event the phase shift of the resonant system is 180° at resonance and the total phase length of the MMIC and connecting lines are set to an odd multiple of 180°. This simple circuit is capable of excellent long- and short-term stability.

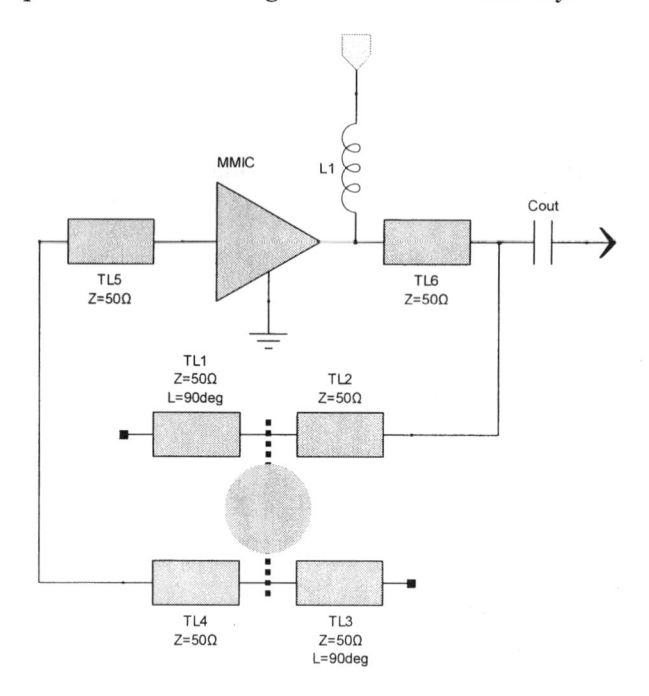

Figure 6.26 Basic MMIC bandpass DRO.

6.5.6 Coupling Test by Modulation

An excellent method for optimizing the spacing of a dielectric resonator to the microstrip coupling line was described in Trans-Tech, Inc. Application Note 1030 [8]. Before the dielectric resonator is cemented to the substrate it is easily moved in relation to a microstrip coupling line. As the resonator is moved toward the coupling line, the loaded Q is reduced. Fig. 6.27 shows a harmonic balance simulation of the oscillator in Fig. 6.23 with different loaded Q representing four different resonator spacings from the coupling line.

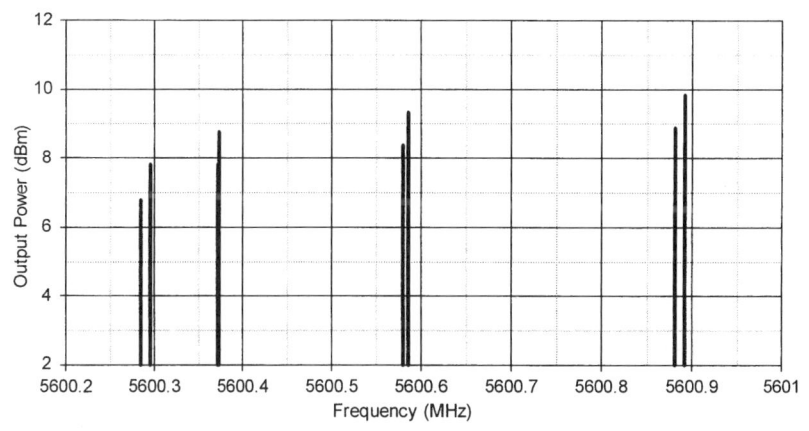

Figure 6.27 Simulated output of a DR oscillator with -11 and -12 volt supplies with decreasing loaded Q to the right.

The oscillator power supply is modulated at a low frequency. In this case the supply voltage is switched between -11 and -12 volts DC. The spectrum pair on the far left is for the maximum useful spacing of the resonator and coupling line and the highest loaded Q. Notice the output power is lower than the remaining cases. This is because the loaded Q is approaching the unloaded Q of the resonator and the resulting loss reduces the sustaining stage compression and therefore the output power. The spacing is increasingly reduced with cases to the right and the output power continues to increase.

The frequency shift between -11 and -12 volts for each case occur because sustaining stage reactance changes with biasing (pushing) and this impacts the oscillation frequency. With decreasing loaded Q to the right, pushing degrades and the frequency shift increases. The "sweet spot" with respect to pushing is the second case from the left. Even though the far left represents the highest loaded Q, the losses result in marginal oscillation conditions and the stability is degraded. The optimum spacing for phase noise is likely to be near this sweet spot, although the most rigorous approach is to measure the phase noise versus spacing.

The exact resonant frequency of the dielectric resonator is 5600 MHz. Increasing loaded Q to the left reduces the influence of the sustaining stage on the operating frequency, which therefore approaches 5600 MHz.

This empirical approach to optimizing dielectric resonator spacing is particularly helpful in designing DR oscillators because analytical expressions for the spacing are approximate and electromagnetic simulation is computationally expensive. However, Trans Tech's clever technique is also useful with any oscillator type where resonator coupling is adjustable.

References

[1] R. Rhea, "Technology Enables New Components," *Microwave Journal*, November 2006, pp. 20-30.

[2] G. Matthaei, L. Young, and E.M.T. Jones, *Microwave filters, Impedance-Matching Networks and Coupling Structures*, Artech House, Dedham, MA, 1980, p. 360.

[3] R. Rhea, *HF Filter Design and Computer Simulation*, Scitech Publishing (Noble Publishing), Raleigh, NC, 1994, pp. 88-89.

[4] RCA Silicon Power Circuits, 1967, pp. 542-550.

[5] www.trans-techinc.com.

[6] GENESYS Software, Agilent Technologies, www.agilent.com.

[7] D. Kajfez, et. al., *Dielectric Resonators*, Scitech Publishing (Noble Publishing), Raleigh, NC, 1998, second edition.

[8] Trans-Tech, Inc. (Skyworks, Inc.), *Optimize DRO's for Low Phase Noise*, AN 1030, 2006.

7 Tuned Oscillators

Any oscillator is frequency tuned by replacing a fix-valued reactor in the resonator with a variable reactor. With mechanically tuned high-frequency oscillators, parallel-plate variable capacitors or slug-tuned inductors are commonly used. High-quality broad-bandwidth electrically tuned oscillators often use YIG resonators tuned by a current generated magnetic field. However, the workhorse element of electrically tuned oscillators is the varactor, a nonconducting reverse-biased PN junction whose capacitance varies with the applied voltage. Varactor modeling is covered in Appendix A.

Oscillators that use piezoelectric resonators have limited tunability. Their tuning is discussed in Chapter 9. Most lumped and distributed oscillators are tunable up to about 30% bandwidth by using a varactor for the resonator capacitor. Broader tuning is aided by the use of certain oscillator topologies that are well suited for tuning. This chapter covers problems specific to broader frequency tuning of both lumped and distributed oscillators.

7.1 Resonator Tuning Bandwidth

Fig. 7.1 shows three topologies of varactor-tuned L-C resonators. The capacitor-coupled parallel resonator is popular because the coupling capacitors offer a degree of freedom so the inductor or varactor value is selectable, only one inductor is required, the coupling capacitors serve as bias decoupling for the active device, and ground returns for the varactor tuning voltage is provided. The series form also has a degree of freedom and requires only one inductor. However, bias decoupling and ground returns for the varactors require additional components.

There are two difficulties with the capacitor-coupled parallel resonator. When the resonator is tuned to higher frequency the reactance of the coupling capacitors decreases and therefore the loaded Q decreases. An additional difficulty is that the coupling capacitors add capacitance in parallel with the tuning varactors and therefore decrease the tuning range.

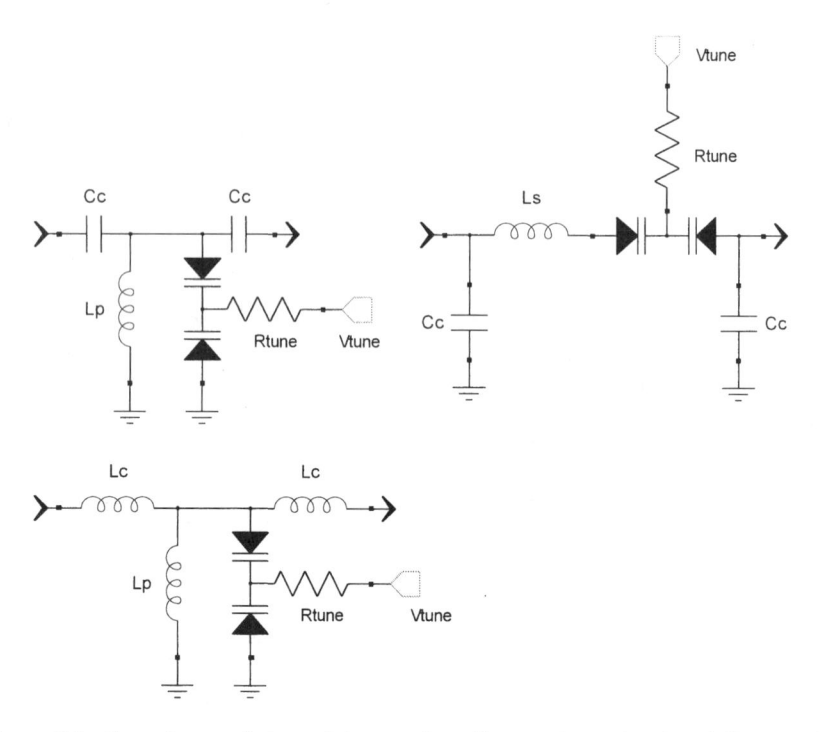

Figure 7.1 Capacitor-coupled parallel resonator with varactor tuning (top left), capacitor-coupled series resonator (top right), and inductor-coupled parallel resonator (bottom left).

Eqs. 48 and 49 in Chapter 1 calculate the reactance of each coupling capacitor that is reflected into the parallel resonator. For high loaded Q where the coupling capacitor reactance is much greater than the termination resistance, the effective parallel capacitance is approximately equal to the coupling capacitance. In this case, the fractional tuning range is given by

$$\frac{f_{high}}{f_{low}} \cong \sqrt{\frac{C_{max} + 2C_c}{C_{min} + 2C_c}} \qquad (186)$$

The tuning bandwidth limitation of the top-C and shunt-C coupled resonators is avoided by using inductive coupling as shown at the bottom left of Fig. 7.1. The reactance of the coupling inductors increases the required value of the resonator inductor. Additional capacitors are generally required to avoid shorting the bias circuitry of the active stage by the shunt inductor. However, the tuning bandwidth is determined by the full tuning capacitor ratio.

7.2 Resonator Voltage

In Section 3.9.2, the voltage across a resonator is given as a function of resonator resistance and drive voltage. Here, resonator voltage is given as a function of Q and drive power. The loaded Q of the resonator at the top left of Fig. 7.1 due solely to the external terminations is Q_{ext}. It is

$$Q_{ext} = \frac{X_{Cc}^2 + R_0^2}{2R_0 X_L} \quad (187)$$

where X_{Cc} is the coupling capacitor reactance, X_L is the inductor reactance, and R_o is the resistance of each termination. Q_{ext} versus the coupling capacitor reactance is plotted with cross symbols in Fig. 7.2, assuming 50-ohm terminations and an inductive reactance of 100 ohms. The loaded Q is reduced by the finite unloaded Q of the resonator.

$$Q_L = \left(\frac{1}{Q_{ext}} + \frac{1}{Q_R} \right)^{-1} \quad (188)$$

The loaded Q with a resonator unloaded Q of 200 is plotted with square symbols in Fig. 7.2. The insertion loss is computed using Eq. 38 and is plotted with triangular symbols. The peak voltage across the resonator is

$$V_p = \sqrt{P_i Q_{ext} X_L} \left(\frac{Q_R}{Q_R + Q_{ext}} \right) \quad (189)$$

where P_i is the source power level. Plotted with circular symbols in Fig. 7.2 is the resonator peak voltage with a source power of 13 dBm.

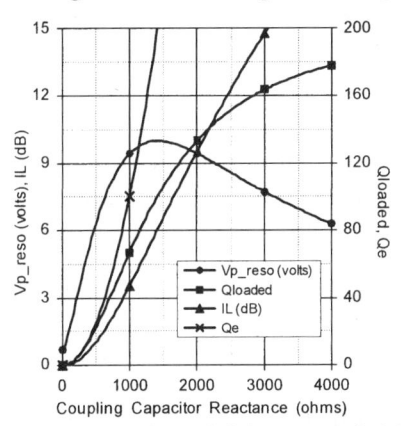

Figure 7.2 External Q (cross symbols), loaded Q (square symbols), insertion loss (triangular symbols), and resonator voltage (circular symbols) versus the coupling capacitor reactance for the resonator at the top left in Fig. 7.1, with 13-dBm drive, 100-ohms inductive reactance, a resonator unloaded Q of 200, and 50-ohm terminations.

With these parameters, at a coupling capacitor reactance of 1413 ohms, the insertion loss is 6.02 dB and the loaded Q equals 100, 1/2 of the unloaded Q. With a drive level of 13 dBm the resonator peak voltage is 10 volts. This would drive a varactor into heavy forward conduction and destroy oscillator performance. To manage this problem the following steps are taken.

1) Back-to-back varactors are used. This helps by more than a factor of two because even when one varactor is forward-biased, the other is reverse-biased and has a high resistance.
2) The reactance of the resonator inductor is reduced. Although this requires increased varactor values, it is often necessary.
3) The drive level is reduced by decreasing the active device quiescent bias current. If necessary, the output power is recovered using a buffer. However, the lower oscillator power level degrades the phase-noise performance.
4) Q_{ext} is reduced. However, this degrades the phase-noise performance even more than reducing the drive level.
5) A small-value capacitor is placed in series with the varactor. However, this reduces the tuning range.

It is imperative that the varactor is not driven into conduction. This causes reduced output power, erratic frequency jumps with tuning, spurious modes, and severely degrades phase-noise performance. Overdriving of the varactor must be avoided.

7.3 Permeability Tuning

The use of mechanically tuned rotating-plate variable capacitors eliminates the RF voltage problem of varactor-tuned oscillators. However, mechanically tuned variable capacitors are large and relatively expensive. In addition, the capacitance range is achieved in 180° of shaft rotation and fine tuning requires a reduction drive that is an additional expense.

An alternative is permeability tuned oscillators that utilize a threaded rod penetrating a solenoid inductor. A powdered iron or ferrite rod increases the permeability of the core and increases the inductance. Eddy currents in nonferrous metal such as copper, brass, or aluminum effectively short the turns of the solenoid and reduce the inductance. Long cores with high permeability are capable of very wide tuning bandwidth. Unfortunately, the temperature coefficient of powdered-iron and ferrite materials is relatively poor and temperature stability is an issue. Brass and aluminum have better temperature stability but the tuning range is less.

Table 7.1 shows the maximum and minimum inductance of various single-layer solenoid and metal core parameters. The solenoid is wound with enamel-coated 20-gauge copper wire. Brass material is standard

American UTS thread size and pitch per inch while the aluminum is solid rod. In these tests, for a given diameter, aluminum provided a wider tuning range than brass and had less unloaded Q degradation. Temperature performance was not measured. The unloaded Q of the 17 turn solenoids was not measured.

Table 7.1 Measured maximum (no penetration) and minimum (full penetration) inductance and unloaded Q with various solenoid and core parameters

Solenoid turns, ID X length	Core material, thread X length	Lmax (nH)	Lmin (nH)	Qmax	Qmin
23t, 5.3 X 20 mm	Brass,12-32 X 25 mm	650	260	63	<10
16t, 9.7 X 16 mm	Brass,12-32 X 25 mm	1160	980	102	55
16t, 9.7 X 16 mm	Brass,6-32 X 25 mm	1160	1075	102	71
16t, 9.7 X 16 mm	Alum rod, 5 X 25 mm	1160	916	102	64
17t, 7.6 X 16 mm	Brass,12-32 X 25 mm	850	645		
17t, 7.6 X 16 mm	Brass, 8-32 X 25 mm	850	700		
17t, 7.6 X 16 mm	Brass, 6-32 X 25 mm	850	761		
17t, 7.6 X 16 mm	Alum rod, 5 X 25 mm	850	584		

The first entry in Table 7.1 is a solenoid with a brass core only slightly smaller in diameter than the solenoid. The inductance range is 2.5:1. However, the unloaded Q is very poor at minimum inductance. The tuning range is reduced by using less penetration or smaller diameter core. Customizing the core diameter and pitch allows the designer to select the linear travel distance, number of core rotations, and the inductance shift.

7.4 Tunable Oscillator Examples

Next, a number of tuned oscillator examples are provided.

7.4.1 Permeability Tuned Colpitts JFET

Fig. 7.3 shows a 5.0- to 5.8-MHz Colpitts JFET oscillator tuned by inductor L_1. Capacitor C_4 is used for simulation and is not required in the final oscillator. Despite simplicity, this circuit offers a loaded Q of approximately 25, harmonics below -40 dBc and power supply pushing <200 ppm. Since power is extracted directly from the resonator, a buffer amplifier is suggested to avoid load pulling.

The solenoid inductor is 17 turns of 20-gauge enamel-coated copper wire with a 7.6 mm solenoid inside diameter. The solenoid length is 16 mm. The oscillator is tuned by a 5 X 25 mm solid aluminum rod centered axially within the solenoid. Core penetration is measured in millimeters from the end entry point of the solenoid. The measured oscillation frequency versus core penetration depth is plotted as the solid curve in Fig. 7.4. The frequency curve is relatively linear over the frequency range of 5.0 to 5.8 MHz with a penetration depth of 3 to 12 mm. With a UTS thread pitch of 32 per inch, this represents an average of 70.6 kHz per turn of the shaft.

The measured output power falls from 2.3 dBm at 5.0 MHz to -1.0 dBm at 5.8 MHz.

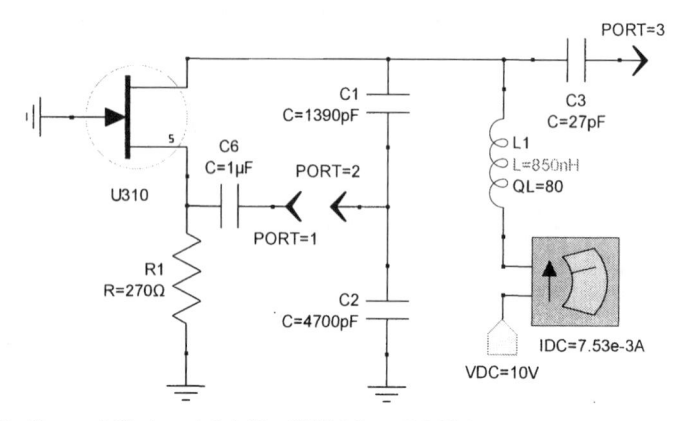

Figure 7.3 Permeability tuned Colpitts JFET 5.0- to 5.8-MHz oscillator.

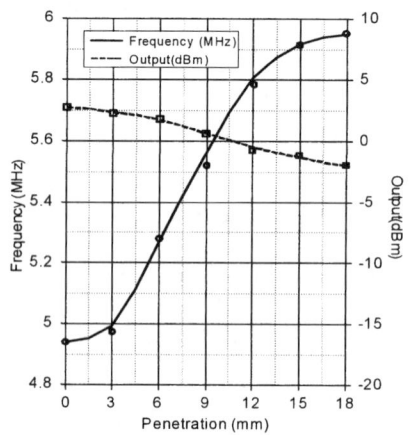

Figure 7.4 Output frequency (solid) and output power (dashed) versus core penetration in millimeters for the 5.0 to 5.8 MHz permeability tuned Colpitts JFET oscillator.

7.4.2 Vackar JFET VCO

The Vacker oscillator topology published in 1949 [1] is rooted in the development in Czechoslovakia of a stable, wide bandwidth vacuum-tube oscillator with mechanical tuning. The topology is legendary among HF communication radio designers. It offers improved long-term stability and constant amplitude output over a wide tuning range.

An example 45- to 75-MHz JFET oscillator is shown in Fig. 7.5. Capacitor C_2 tunes the frequency. Capacitors C_3 and C_4 form a voltage

Tuned Oscillators

divider that reduce the open-loop gain. Therefore, capacitor C_3 may be used to control the gain.

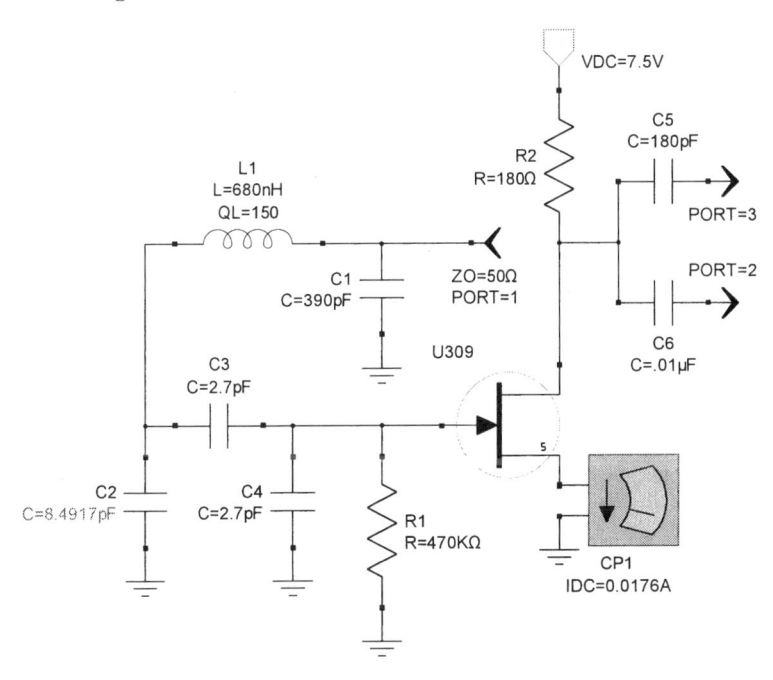

Figure 7.5 A 45- to 75-MHz JFET Vacker.

The loaded Q is also relatively flat with tuning, tending to rise somewhat with increasing frequency. However, the RF voltage across the tuning capacitor generally peaks midband. The RF voltage across capacitor C_3 at 60 MHz is 13 volts peak although the output power is only 0 dBm. This is partially due to the high loaded Q, approximately 90. The Vacker oscillator is appropriate for mechanically tuned capacitors where the RF voltage is less of an issue.

The simulated open-loop transmission gain and phase are shown on the left in Fig. 7.6 with tuning capacitor values of 17, 9, and 5 pF. The output spectrum is given on the right.

7.4.3 Hybrid Negative Resistance VCO

Fig. 7.7 shows a negative-resistance oscillator similar to the oscillator introduced in Section 1.7.1. The resonator inductor is replaced with a 32-mm long microstrip line realized on 1.57-mm thick FR4 with a relative dielectric constant of 4.6. The 3-mm wide line characteristic impedance is approximately 50 ohms and the electrical length is approximately 55°. This 710 to 825 MHz is phase-locked to a crystal reference oscillator and is frequency multiplied by 4 for use as the first LO in a commercial C-band satellite receiver that tunes 3720 to 4180 MHz.

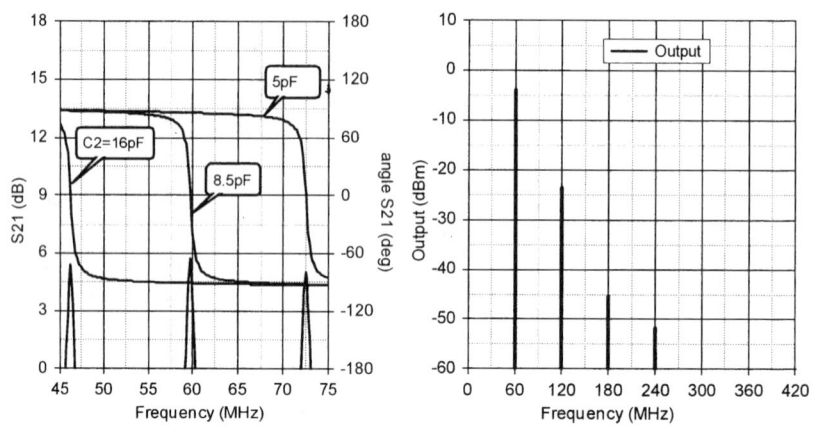

Figure 7.6 Simulated open-loop transmission gain and phase of the Vacker oscillator with tuning capacitor values of 17, 9, and 5 pF (left) and the output spectrum at 60 MHz (right).

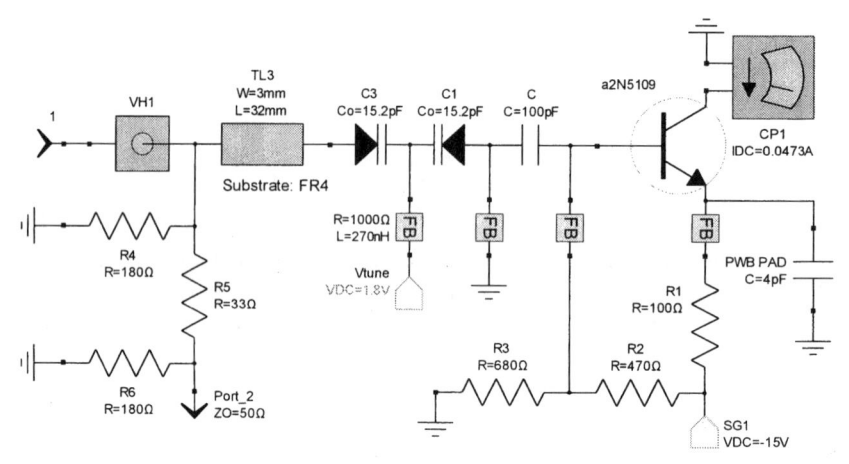

Figure 7.7 710- to 825-MHz negative-resistance VCO with a distributed resonator inductor on FR4.

A UHF transistor is used with the collector directly grounded by installing the metal case against a copper ground plane on the top of the PWB, thus providing excellent heat sinking. At approximately 10 volts V_{ce} and I_c equal to 50 mA, the device dissipation is 500 mW. The dissipation in R_1 is approximately 250 mW. Because of the relatively low value of R_1, a multiturn ferrite bead keeps the emitter load resistance high. The capacitance at the emitter of a negative-resistance oscillator is critical and a small PWB pad provides about 4 pF of capacitance to ground. The load of R_2 and R_3 is choked from the base with another multiturn ferrite bead. Tuning is derived by two back-to-back varactors modeled with a capacitance at 0 volts of 15.2 pF, a gamma of 0.5, series package inductance of 4 nH, and parallel package capacitance of 0.5 pF. A third ferrite bead returns the right side of C1 to DC ground and a fourth ferrite bead supplies the tuning voltage.

The distributed inductor is grounded by a via hole to the bottom ground plane. For simulation analysis, the ground is lifted and port 1 is used to observe the resistance and reactance. The linear simulation resistance and reactance are presented in Fig. 7.8 with the tuning voltage at 1.8 volts (solid traces) and 12.5 volts (dashed traces). The negative resistance ranges from -8.5 to -3 ohms, rather low but functional. A transistor somewhat higher F_t provides a more negative resistance; however, this VCO functioned reliably in thousands of units.

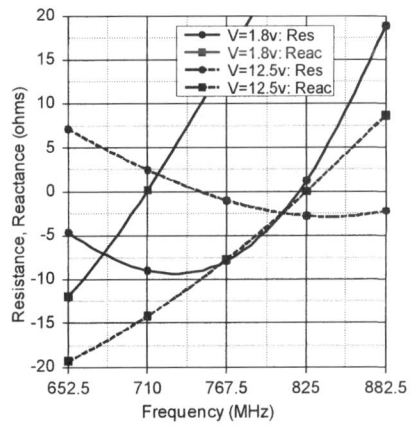

Figure 7.8 Simulated resistance and reactance of the 710- to 825-MHz VCO with the tuning voltage at 1.8 volts (solid traces) and 12.5 volts (dashed traces).

Power is coupled from the resonator by tapping the distributed inductor. In this circuit, the tap location is close to the ground point. In this case, the inductance of the via hole provides inductance to ground for coupling output power.

The high quiescent bias and the efficiency of the negative-resistance oscillator provide relatively high output power. This allows the use of a passive pad rather than an active buffer, resulting in improved phase noise and simplicity. The power at the via hole is typically 13 dBm and the 5-dB pad provides an output power of 8 dBm.

Fig. 7.9 shows output spectrums taken with an HP8568A spectrum analyzer. The spectrum on the left is with a PLL uA741 operational amplifier (opamp). The spectrum on the right is with a low-noise HA911-5 loop opamp. The loop bandwidth for each case is approximately 4.2 kHz, well inside the resolution bandwidth of the measurement. The phase noise of the HA911-5 circuit is near the noise floor of the spectrum analyzer so the actual phase noise is uncertain. Opamps in PLL circuits typically limit phase-noise performance, thus validating the trend to passive charge pumps.

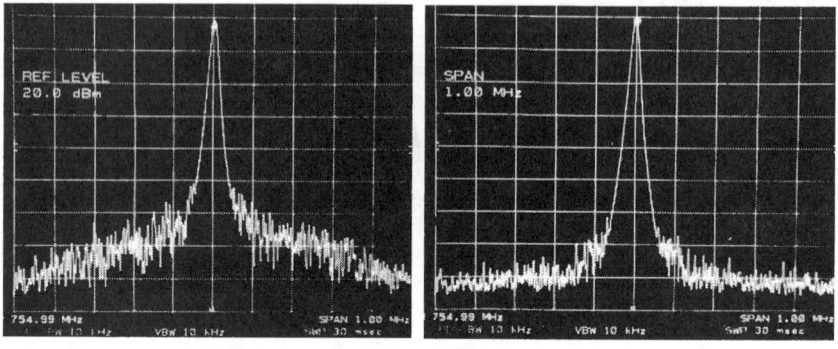

Figure 7.9 Output spectrum of the VCO at 755 MHz with a uA741 phase-locked loop operational amplifier (left) and an HA911-5 low-noise opamp (right).

7.4.4 Capacitor-Transformed Negative-Resistance VCO

Fig. 7.10 shows a modified version of the 710 to 825 MHz VCO. This simplified version tunes a similar frequency range but uses a single varactor and two fewer ferrite beads.

Although the effective loaded Q of the negative resistance oscillator is not easily determined, the phase-noise performance is improved by transforming the negative-resistance lower. This is accomplished in this oscillator by using a shunt capacitor at the input of the transistor. This also transforms the device reactance down, thus effectively increasing the capacitance in series with the negative resistance. This lowers the resonant frequency for a given value of tuning capacitance and reduces the dependence of the oscillation frequency on the device characteristics and the temperature. The capacitance of a single varactor is twice that of the back-to-back varactors. These shifts are compensated by shortening the transmission line to 24.5 mm.

The ferrite bead between the transistor base and the common node of the base bias resistors is eliminated by increasing the value of the resistors so they do not absorb significant negative resistance at the base.

7.4.5 Negative-Resistance VCO with Transformer

Fig. 7.11 shows a 650- to 870-MHz VCO with a 4:1 impedance transformer used to transform down the base impedance. This significantly improves the negative-resistance oscillator by increasing the tuning range,

Tuned Oscillators

improving the phase-noise, and reducing device dependence. The increased tuning range is achieved using the same varactors and tuning voltage as the previous examples.

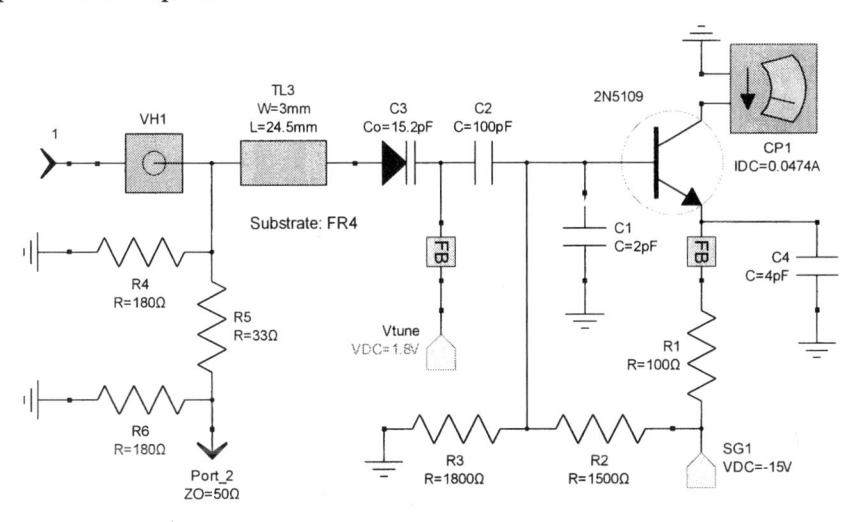

Figure 7.10 Capacitor-transformed 710- to 825-MHz VCO with a single varactor.

The transformer provides a natural DC return for the tuning voltage so back-to-back varactors are used. This circuit offers superior performance to the typical negative-resistance VCO. Unfortunately, the transformer adds cost and is difficult to realize at higher frequency. This configuration was the basis for a 400- to 700-MHz VCO designed by the author for use in a broadband CATV modem.

7.4.6 Negative-Conductance VCO

Fig. 7.12 shows a high-efficiency, negative-conductance CB MRF559 bipolar VCO. The simple bias scheme uses two resistors. The voltage drop in resistor R_1 provides a moderate degree of bias stabilization. A more complex bias network is used if greater temperature or device-to-device stability is required. The low total resistance from the supply path through the collector to ground path results in high DC to RF conversion efficiency. A value of 10 ohms for R_1 was empirically found to be optimum. A lower value of resistance provides even greater efficiency but causes spurious instabilities at certain supply voltages. The inductive choke coupling the supply to the oscillator couples the load directly to the collector, which improves efficiency but increases load pulling. The base inductor L_3 optimizes the negative conductance and increases the effective shunt inductance at the emitter. Since this is a negative-conductance oscillator, a parallel resonator is used at the emitter.

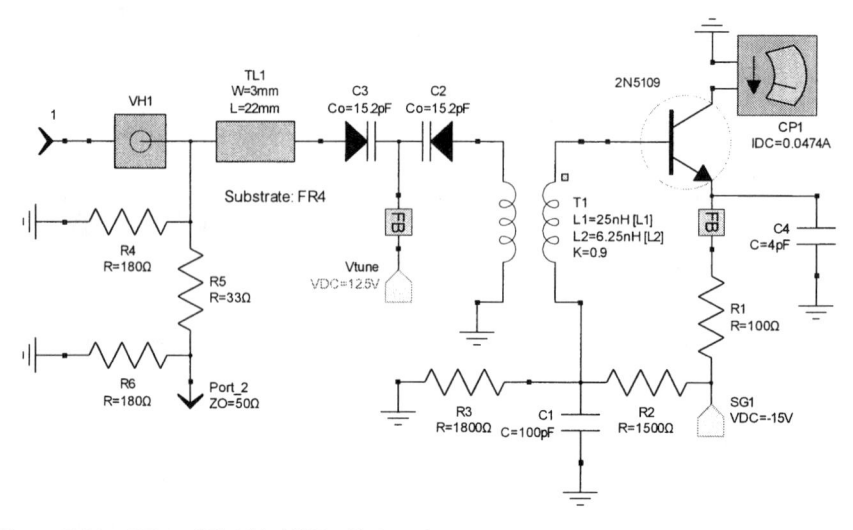

Figure 7.11 650- to 870-MHz VCO with transformer.

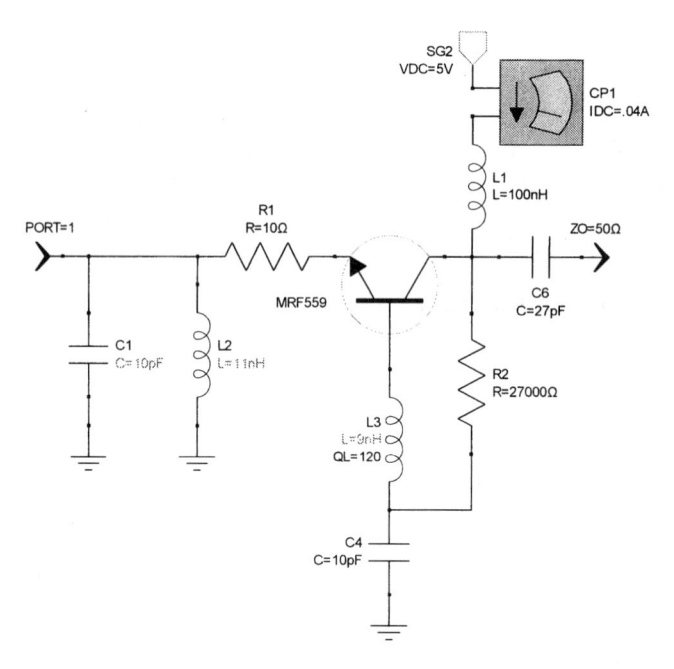

Figure 7.12 High-efficiency, negative-conductance CB bipolar VCO.

Tuned Oscillators

Fig. 7.13 shows the port 1 input conductance and susceptance. This circuit tunes from about 700 to 1200 MHz with a 10:1 capacitance change centered at approximately 10 pF. The CB negative-conductance oscillator provides a wider tuning range than the negative-resistance oscillator because the parasitic reactance in parallel with the negative conductance is inductive. The parasitic reactance of the negative-resistance oscillator is capacitance in series with the tuning varactor, which reduces the effective capacitance ratio. Negative-conductance oscillators provide tuning bandwidths up to an octave when the device is chosen properly and the base inductance is optimized.

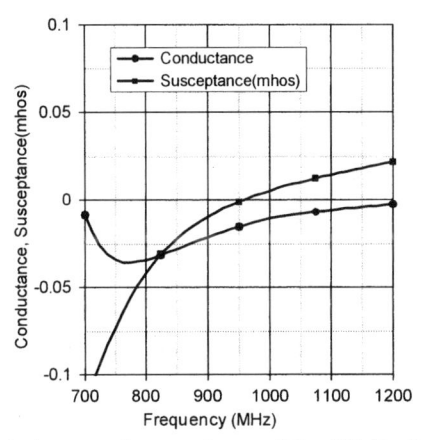

Figure 7.13 Input conductance and susceptance of the CB bipolar negative-conductance oscillator.

Table 7.2 shows the measured performance of the oscillator in Fig. 7.12 versus the supply voltage and with a fixed 10-pF capacitor C_1. With a 5-volt supply the gain margin and quiescent bias level is low and the output power and efficiency are low. At 7.5 volts the gain margin is improved and the output power and efficiency increase significantly. At 15 volts the oscillator is operating well, the output power is just over 200 mW, and the conversion efficiency is 32%. The high conversion efficiency is partly the result of low resistance in the collector path to ground. At higher supply voltage the efficiency increases somewhat but device dissipation significantly elevates the operating junction temperature.

As the supply voltage is increased from the initial 5 to 7.5 volts, the operating frequency drops 7 MHz. However, thereafter the pushing performance is good. At a nominal operating voltage of 15 volts pushing is approximately 0.5 MHz/volt or 0.6 parts per thousand. Harmonic performance is good for a high-efficiency oscillator.

Table 7.2 Measured performance of the high-efficiency, negative-conductance oscillator versus the supply voltage

Vs (volts)	Is (mA)	Po (dBm)	η (%)	Fo (MHz)	2nd Harm. (dBc)	φ-noise@10 kHz (dBc/Hz)
5.0	18	5.0	4	840.5	-24.5	-86
7.5	21	15.0	20	833.3	-21.5	-97
10.0	30	18.9	26	834.1	-17.0	-101
12.5	38	21.3	28	835.2	-16.8	-105
15.0	44	23.3	32	836.7	-17.0	-107
17.5	50	24.9	35	837.5	-17.3	-102
20.0	60	26.2	35	837.8	-18.0	-95

Notice that the SSB phase-noise performance improves with increasing supply voltage up to 15 volts. Up to 15 volts there is tight correlation between improved phase-noise and increased output power. The specified device noise figure is 2.7, 3, and 4 dB with collector currents of 10, 30, and 75 mA, respectively, so only a small portion of the degraded phase-noise above 15 volts supply is attributable to the noise figure. A significant portion of the degradation is likely attributable to a much higher junction temperature and nonoptimum operating current.

7.4.7 Hybrid Coaxial Resonator MMIC

Fig. 7.14 shows a 915-MHz coaxial-resonator oscillator with lumped element coupling capacitors C_1 and C_2 and an MAR3 MMIC amplifier. Capacitors are AVX 0805 chip capacitors. The coupling capacitors and the loading capacitance consisting of C_3 in series with the Metelics MSV34075 tuning varactor shorten the required electrical length of the quarter-wavelength coaxial resonator.

The coaxial resonator is a Trans-Tech standard-profile TEM-mode coaxial resonator with a square outer conductor and a circular center conductor. The material is 8800 with a relative dielectric constant of 38.6. The unloaded quarter-wavelength resonant frequency is 961 MHz.

The MSV34075 varactor SPICE model is provided by Metelics [6]. Parameters are given in Table 7.3. C_p in Fig. 7.14 models the E28 package 0.4 pF parallel capacitance. L_p models the 0.8-nH package series inductance plus the inductance of the path to ground from the top of the resonator to ground.

The unloaded Q of the varactor is specified by the manufacturer as 3600 at 50 MHz. This translates to 196 at 915 MHz, ignoring the package inductance. The 1-pF capacitor, C_4, significantly reduces the tuning range and the voltage across the tuning varactor. It also increases the effective unloaded Q of the tuning capacitance.

Tuned Oscillators 309

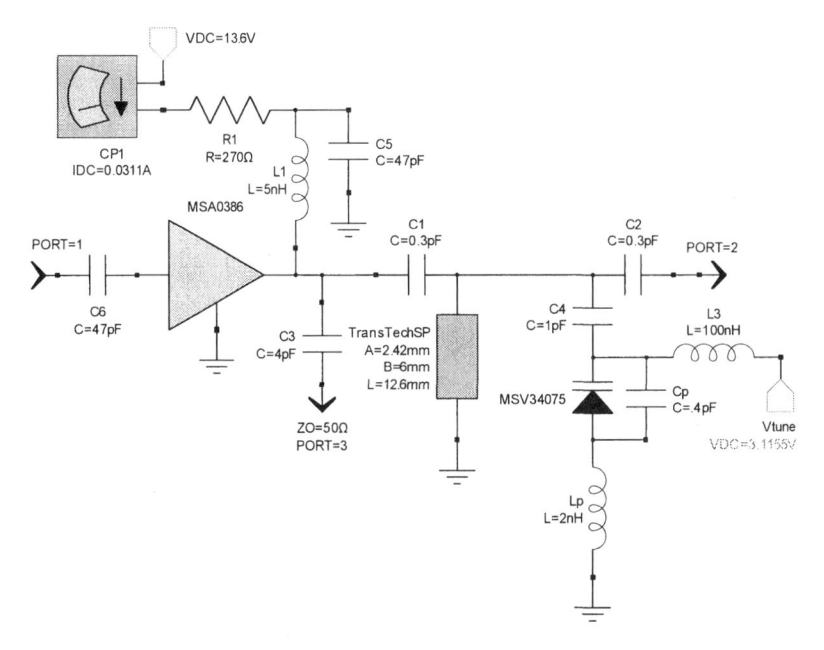

Figure 7.14 Coaxial resonator hybrid MMIC 915-MHz oscillator.

Fig. 7.15 shows the simulated open-loop transmission amplitude, phase, and loaded Q with tuning voltages of 0 and 12 volts. The cascade 50-ohm input and output matches are 10 dB or better. The gain margin is rather low at 3 dB. However, the MAR3 MMIC uses resistive feedback that produces repeatable gain and the characteristics of the resonator are stabile. The gain margin may be increased at the expense of loaded Q by increasing the values of the coupling capacitors C_1 and C_2. Alternatively, a higher gain MMIC may be used. The loaded Q is nearly 200 across the tuning range. An aggressive approach was taken for the loaded Q to improve the phase noise.

Table 7.3 SPICE model parameters for the MSV34075 varactor (courtesy of Metelics)

Description	Name	Value
Saturation current	Is	1E-12 amps
Ohmic resistance	Rs	1.7 ohms
Emission coefficient	N	1.65
Transit time	Tt	10E-9 nS
Junction capacitance	Cjo	3.78 pF
Junction potential	Vj	0.6 volts
Grading coefficient	M	0.45
Activation energy -eV	Eg	1.12
Reverse breakdown voltage	Bv	30 volts
Current at Bv	Ibv	10E-6 amps

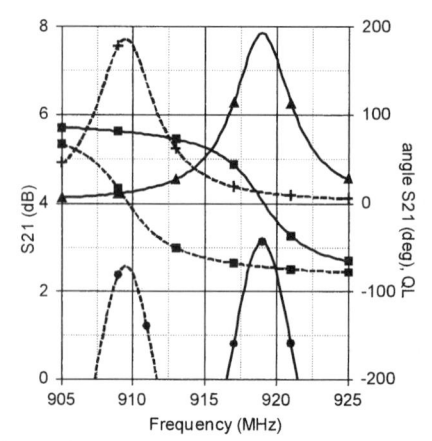

Figure 7.15 Simulated open-loop transmission amplitude (circular symbols), phase (square symbols), and loaded Q (triangular symbols) with a tuning voltage of 0 volts (dashed) and 12 volts (solid).

Table 7.4 compares key simulated and measured performance characteristics versus a tuning voltage of 0 to 12 volts. The oscillator tunes from approximately 909 to 919 MHz. The measured oscillation frequency is 0.6 to 0.9 MHz higher than the simulated oscillation frequency. While 0.9 MHz seems to be a rather large difference, it is only 0.1% and well within the errors of the lumped component tolerances and the coaxial resonator dielectric and dimensional tolerances. The output power agreement is good. The simulated second harmonic level is somewhat higher than measured.

Table 7.4 Simulated and measured performance characteristics of the 915-MHz coaxial resonator MMIC oscillator

Vt (volts)	Freq (MHz)	Power (dBm)	2nd Har (dBc)	Freq (MHz)	Power (dBm)	2nd Har (dBc)
	Simulated			Measured		
0	909.3	4.9	-12.1	909.9	4.8	-14
2	913.8	5.2	-11.4	913.8	5.1	-14
4	915.7	5.3	-11.1	915.8	5.3	-14
6	917.0	5.4	-10.9	917.3	5.4	-14
8	917.9	5.4	-10.8	918.5	5.4	-14
10	918.6	5.5	-10.8	919.4	5.4	-14
12	919.3	5.5	-10.7	920.2	5.6	-14
14	919.7	5.5	-10.6			

Shown on the left in Fig. 7.16 is the simulated output spectrum. Given that power is taken at the MMIC output, the harmonic performance is better than might be expected. This is likely attributed to low gain margin and therefore only moderate compression in the amplifier.

Tuned Oscillators

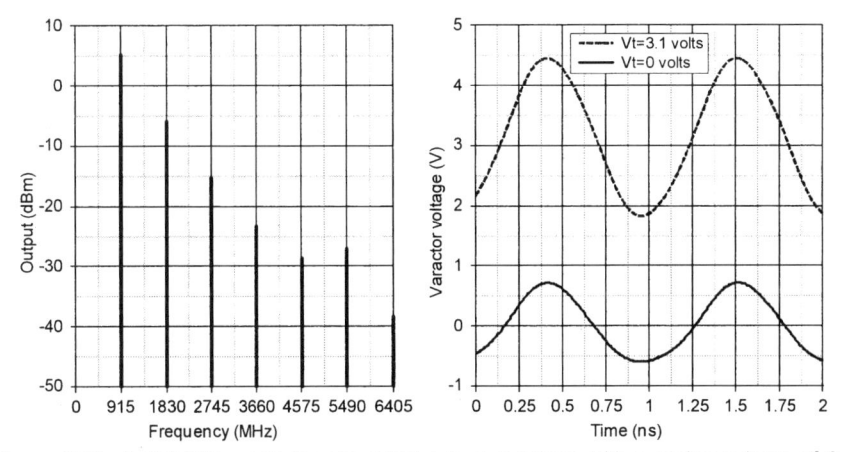

Figure 7.16 A 915-MHz oscillator simulated output spectrum with a tuning voltage of 3.1 volts (left) and the varactor RF voltage (right) with a tuning voltage of 3.1 volts (dashed) and 0 volts (solid).

Shown on the right in Fig. 7.16 is the RF voltage at the top of the tuning varactor.

Table 7.5 shows the phase noise predicted by Leeson's equation and the measured phase noise. Parameters assumed for Leeson's equation were the simulated power, loaded Q, and tuning sensitivity, an MAR3 noise figure of 6 dB and flicker corner of 30 kHz, and a varactor effective noise resistance of 3300 ohms. The predominate source of noise at 1-kHz offset is Leeson noise and agreement is very good. The predominate source of noise at 10-kHz offset is varactor modulation noise. At the higher offset, since the measured noise is worse than the simulated noise, it is possible that the effective noise resistance is higher than the assumed 3300 ohms.

Table 7.5 Simulated and measured SSB phase-noise performance of the 915-MHz coaxial resonator MMIC oscillator at tuning voltages of 0 and 13.6 volts

Vt (volts)	Offset (kHz)	Phase noise (dBc/Hz) Simulated	Phase noise (dBc/Hz) Measured
0	1	-92	-90
0	10	-116	-115
13.6	1	-94	-90
13.6	10	-123	-114

Coaxial resonator oscillators are an excellent choice for low phase-noise oscillators if the required tuning range is small. For a larger tuning range, the high parallel capacitance requires high-valued capacitance varactors, which have low unloaded Q. Furthermore, with wide tuning oscillators the phase noise is dominated by varactor modulation noise and there is little justification for the high unloaded Q offered by the coaxial resonator. An example of a wider tuning oscillator using a coaxial resonator is the Seiler coaxial resonator CC bipolar oscillator described in Section 7.4.9.

7.4.8 Loaded Quarter-Wavelength MMIC

Fig. 7.17 shows the schematic of an 840- to 880-MHz loaded quarter-wavelength MMIC oscillator constructed on FR4 [2]. This oscillator is similar to the example in the previous section except the coaxial resonator is replaced with a microstrip line printed on FR4. The microstrip resonator is shorter than a quarter-wavelength due to loading by fixed capacitor C_4 and the tuning circuit consisting of fixed capacitor C_5 and the tuning varactor C_6. Models for the via holes to ground are included to improve simulation accuracy. TL_1 is a short length of a high-impedance line that serves as a phase-lead network to adjust the phase-zero crossing at maximum phase slope.

The substrate parameters assumed for this design were ε_r = 4.6, *loss* δ = 0.02, and a thickness of 1.57 mm. The characteristic impedance of a 7.9 mm wide strip on this substrate is approximately 24 ohms. With a *loss* δ = 0.02 the unloaded Q of a quarter-wavelength resonator would be 50. Copper loss predominates in microwave laminates that have a reduced *loss* δ. With FR4 the poor *loss* δ dominates the unloaded Q and copper loss only slightly decreases the unloaded Q. However, the capacitive loading in this circuit reduces the required physical length from 44.6 to 12.7 mm. As described in Section 6.3.3, with *ESR* of 0.1 ohms, the capacitor loading also increases the effective unloaded Q from 50 to 80.1. Capacitive loading essentially replaces a portion of the line at the open-end where substrate loss has the greatest effect. This improves both the unloaded Q and the temperature stability of the substrate. However, capacitive loading reduces the tuning range.

Resistor R_2 reduces the unloaded Q about 10%, with increased degradation at higher frequency. Increasing the value of R_2 reduces this effect but introduces additional varactor-modulation noise.

A photograph of a prototype of the oscillator is shown in Fig. 7.18. The SMA output connector is at the upper left of the unit. The power supply feedthrough is on the left and the tuning feedthrough is on the right of the lower left wall.

Table 7.6 shows the measured output level and the measured and simulated operating frequency versus the tuning voltage for this VCO. The frequency versus tuning voltage is highly nonlinear, primarily due to the rapid change of capacitance with voltage at low tuning voltage.

Tuned Oscillators 313

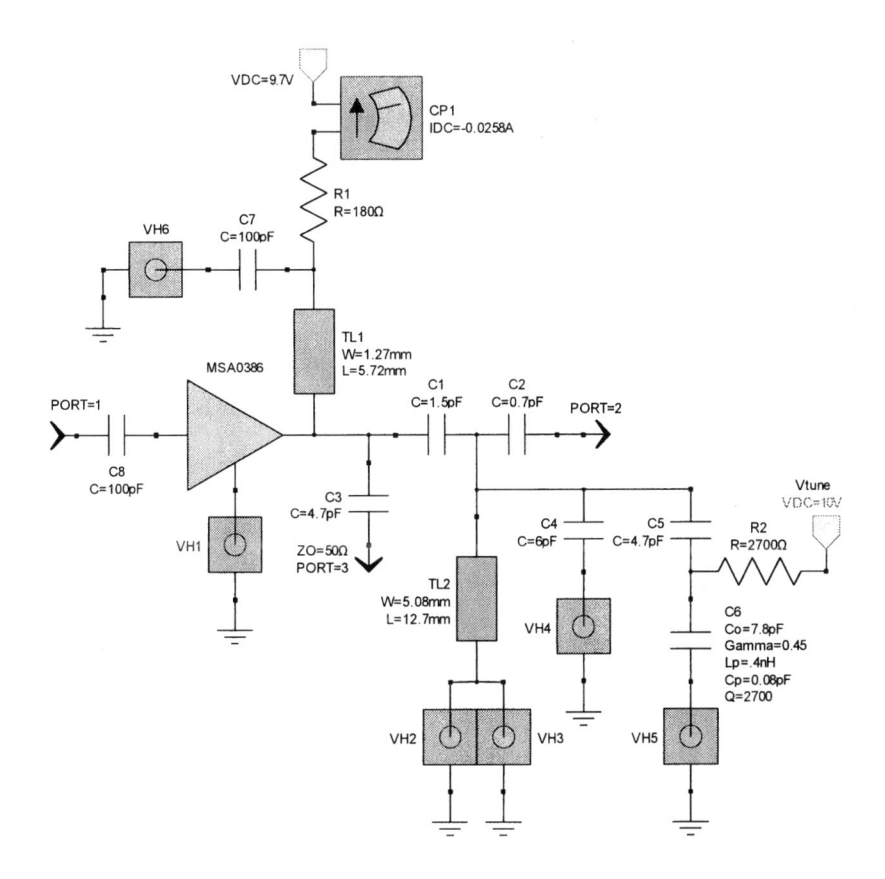

Figure 7.17 840- to 880-MHz oscillator with a capacitor loaded microstrip resonator.

Shown as the solid trace in Fig. 7.19 is the predicted SSB phase-noise of the loaded quarter-wavelength 840- to 880-MHz VCO assuming a loaded Q of 28, a flicker corner frequency of 100 kHz, and an MMIC noise figure of 6 dB. The dashed curve is noise predicted by Leeson and the dotted curve is the predicted varactor-modulation noise with a tuning sensitivity of 5 MHz/volt and an effective noise resistance of 3300 ohms. The circular symbols are measured noise at 10, 30, and 100 kHz offset. The dash-dot trace is the measurement system noise floor.

Figure 7.18 Photograph of the prototype 900-MHz loaded quarter-wavelength oscillator.

Table 7.6 Simulated and measured operating frequency of the loaded quarter-wavelength 840- to 880-MHz VCO versus the tuning voltage

Vt (volts)	Frequency (MHz) Simulated	Frequency (MHz) Measured	Output Level (dBm) Measured
0	837	836	4.0
2	862	855	6.0
4	873	864	6.4
6	880	873	6.0
8	885	882	6.8
10	888	888	7.0

7.4.9 Seiler Coaxial-Resonator CC VCO

Fig. 7.20 shows a Seiler coaxial-resonator 865- to 965-MHz oscillator using an AT41486 bipolar transistor. This oscillator has excellent tuning characteristics for a coaxial-resonator oscillator. These desirable characteristics are attributable to capacitor C_3 that converts this topology from a Colpitts to a Seiler. In reality, this does not decrease the economy of the CC Colpitts because a bypass capacitor is required anyway to avoid shorting the bias.

The coaxial resonator is a Trans-Tech, Inc. standard profile-size coaxial resonator using 8800 material with a relative dielectric constant of 38.6 [3]. The coaxial resonator provides a convenient DC return for back-to-back varactors. The varactor values are reasonable with 7.7- to 1.4-pF tuning this oscillator from 865 to 965 MHz. Despite the wide tuning range for a ceramic coaxial resonator, a loaded Q of near 150 is maintained throughout the tuning range. Because the impedance level is somewhat high at the

Tuned Oscillators 315

common node of the varactors, inductor L_1 must be high quality. Resistor R_2 stabilizes the oscillator by avoiding a negative-resistance mode with capacitors C_1 and C_2 and the inductance of the path from the base through these capacitors.

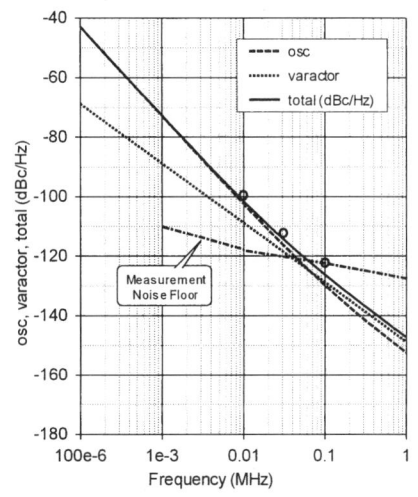

Figure 7.19 Predicted phase noise (solid), measured phase noise (circular symbols), and measurement noise floor (dash-dot) of the loaded quarter-wavelength 840- to 880-MHz VCO.

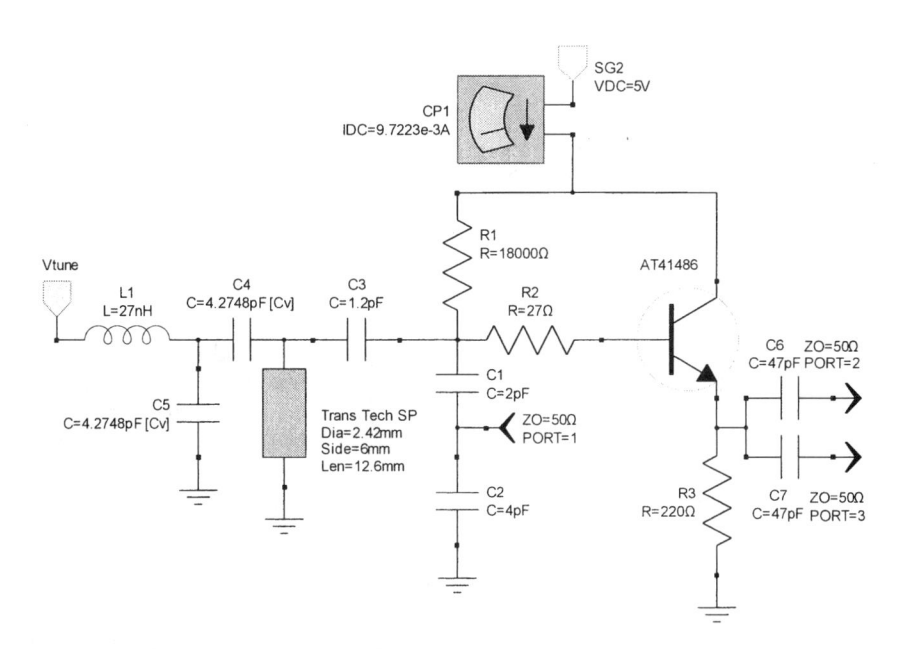

Figure 7.20 Seiler coaxial resonator CC bipolar oscillator.

The quiescent bias is a modest 9.7 mA with a 5-volt supply to keep the RF voltage across the varactors below 2 volts peak. The resulting output power is approximately 2.5 dBm. If the varactor tuning voltage is kept at a minimum of 2 volts to avoid varactor conduction, the output power of this oscillator can be increased by increasing the bias current to 20 mA. The simulated output spectrum at 9.7 mA bias and tuned to 915 MHz is shown in Fig. 7.21.

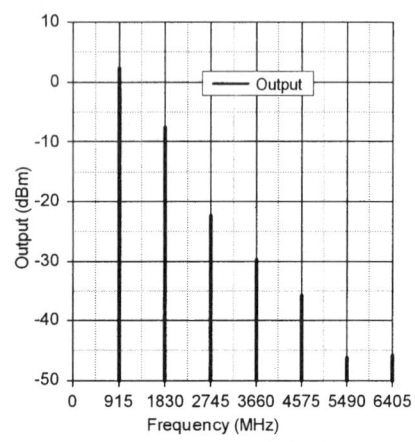

Figure 7.21 Simulated output spectrum of the Seiler coaxial-resonator Colpitts oscillator tuned to 915 MHz.

7.5 YIG Oscillators

Oscillators built using yttrium-iron-garnet (YIG) sphere resonators offer multioctave tuning and excellent phase-noise performance. YIG films can be used, but spheres are currently more common. YIG oscillators are normally acquired as assemblies from corporations specializing in their manufacture. A brief description of YIG oscillators is provided next for comparison with other oscillator technologies and to assist in their specification.

YIG spheres are manufactured by growing synthetic crystalline structures in platinum crucibles. The crystal is then diced, tumbled, and polished into spheres 1 mm and less in diameter. The sphere is typically mounted at the end of a ceramic rod. When immersed in a magnetic field the sphere resonates at a microwave frequency that is linearly proportional to the magnetic field intensity. For pure YIG resonators the resonant frequency is given by

$$f_0 = yH_0 \qquad (190)$$

where y is the gyromagnetic ratio equal to 2.8 MHz/Oe and H_o is the magnetic field intensity in oersteds. In this case, a magnetic field of 3570

Tuned Oscillators

oersted is required to resonate the sphere at 10 GHz. This biasing magnetic field is produced by placing the YIG sphere between the poles of a permanent magnet or an electromagnet with a DC current. Because the winding resistance of an electromagnet is a function of temperature, the winding is driven with a current source. The YIG crystal may be doped, for example with Gallium, to reduce the required magnetic field to achieve resonance.

Fig. 7.22 shows a photograph of a ceramic rod mounted YIG with a loop band to couple the RF signal to the YIG sphere [4]. The coupling loop is oriented to produce a magnetic field that is perpendicular to the static biasing field.

Figure 7.22 YIG sphere mounted at the end of a ceramic rod and coupled to the RF circuit via a loop band (courtesy VHF Communications magazine).

Oscillator open-loop loaded Q in excess of 1000 is achieved with YIG resonators. This results in excellent phase-noise performance. Furthermore, since the resonant frequency is linearly proportional to the biasing magnetic field, the frequency range is limited only by the ability to generate a strong magnetic field and the bandwidth of the sustaining stage. Multioctave YIG tuned oscillator (YTO) bandwidths are achievable. The combination of excellent high loaded Q and wide tuning bandwidth result in performance unparalleled by other oscillator structures. YIG oscillators are often found in high-performance instrumentation.

Table 7.7 shows performance data for an electromagnet MLMB-0208 YIG oscillator from Micro Lambda Wireless [5]. This relatively low-cost YIG oscillator is designed for PWB mounting. It utilizes a bipolar transistor coupled to a YIG sphere in a thin-film oscillator circuit. YIG heaters are used to maintain temperature stability. The main coil develops the primary bias magnetic field and the FM coil is used for modulation and high-speed tuning. The YIG heater is used to temperature stabilize the YIG. The overall size of this model is 1" x 1" x .5" and the weight is 1 ounce. Given in Table

7.8 are key specifications for a small, representative sample of YIG oscillators that are available.

Table 7.7 Performance parameters of a typical, low-cost, 2- to 8-GHz electromagnet YIG oscillator for PWB mounting from Micro Lambda Wireless

Specification	Value
Frequency range, minimum	2-8 GHz
Power output, minimum	+13 dBm
Power output variation, maximum	+/-2 dB
Frequency drift over 0 to 65°C, maximum	15 MHz
Pulling (12dB RL), typical	1 MHz
Pushing +12 Vdc supply, typical	0.1 MHz/V
Pushing -5 Vdc supply, typical	1 MHz/V
Magnetic susceptibility @ 60 Hz, typical	110 kHz/gauss
Second harmonic, minimum	-12 dBc
3rd harmonic, minimum	-20 dBc
Spurious output, minimum	-70 dBc
Phase noise @ 10 kHz offset, minimum	-98 dBc/Hz
Phase noise @ 100 kHz offset, minimum	-120 dBc/Hz
Main coil	
Sensitivity, typical	20 MHz/mA
3 dB bandwidth, typical	5 kHz
Linearity, typical	+/-0.1%
Hysteresis, typical	5 MHz
Input impedance @ 1 kHz, typical	17 ohms + 17 mH
FM coil	
Sensitivity, typical	310 kHz/mA
3 dB Bandwidth, typical	400 kHz
Deviation @ 400 kHz rate, minimum	+/-50 MHz
Input impedance @ 1 MHz, typical	1 ohm + 2 uH
DC circuit power +12 Vdc +/-5%, maximum	100 mA
DC circuit power -5Vdc +/-5%, maximum	50 mA
YIG heater power	
Input voltage range	24 +/-4 Vdc
Current, maximum surge/steady-state	250 mA/25 mA
Case style	81-049

Table 7.8 Key performance parameters of a few representative YIG-tuned oscillators from Micro Lambda Wireless

Model	Type	Freq (GHz)	φ-Noise (dBc/Hz) 10 kHz	100 kHz	Notes
MLMB-0204	EM	2-4	-103	-125	PWB mount
MLMB-0208	EM	2-8	-98	-120	PWB mount
MLOS-0306P	EM	3-6	-108	-130	Low noise
MLOS-0613P	EM	6-13	-105	-125	Low noise
MLXS-0220	EM	2-20		-120	Broadband, PN<12 GHz
MLXS-0220	EM	2-20		-112	PN<18 GHz
MLXS-0220	EM	2-20		-107	PN<20 GHz
MLPB-0204	PM	2-4	-103	-125	PWB mount, low cost
MLPB-0608	PM	6-8	-98	-120	PWB mount, low cost
MLLP-1200	PM	12	-95	-120	Fixed freq, 2-12 GHz range
MLPF-2000F	PM	20	-80	-105	Fixed freq, 4-20 GHz range

Tuned Oscillators 319

YIG oscillators are available from a variety of manufacturers, including, but not limited to, Micro Lambda Wireless, MicroSource, Omniyig, and Teledyne.

References

[1] J. Vackar, "LC Oscillators and their Frequency Stability," *Tesla Technical Reports*, December 1949, Praha, Czechoslovakia, pp. 1-9.

[2] R. Rhea, "Designing a Low-Noise VCO on FR4," *RF Design*, September 1999, pp. 72, 74, 76-77.

[3] Trans-Tech Inc., www.trans-techinc.com.

[4] B. Kaa, "A Simple Approach to YIG Oscillators," *VHF Communications*, April, 2004, pp. 217-224.

[5] Micro Lambda Wireless, www.microlamdawireless.com.

[6] Aeroflex Metelics, www.aeroflex.com, Aeroflex Microelectronic Solutions

8 Piezoelectric Oscillators

Certain crystals and ceramics, when subjected to mechanical stress, develop an electric potential between two electrodes contacting the material surface. This effect, piezoelectricity, was first observed by Antone Cesar Becquerel in 1820 [1]. The converse piezoelectric effect is the production of a deflection or strain when a potential field is applied. Useful resonators are produced from piezoelectric materials that are environmentally stable.

8.1 Bulk Quartz Resonators

Not all piezoelectric materials are suitable for electronic oscillators. However, quartz, the crystalline form of silicon dioxide, is highly stable with excellent temperature characteristics. It is mechanically rigid, which provides for exceptional unloaded Q with properly designed resonators. The U.S. Army funded significant research on the use of natural-quartz crystals in resonators to stabilize field radios during WW II. Today quartz crystals are cultured and quartz-resonator oscillators are a cornerstone of virtually every electronic device, with over a billion crystal units shipped annually.

Quartz blanks with plated electrodes can operate in a number of modes. When the rigid quartz material vibrates at high frequency in response to excitation, tremendous energy is stored resulting in high unloaded Q. Bulk-mode resonance is associated with vibration of a volume of quartz. Bulk-mode quartz resonators are often referred to simply as "crystals." Resonators associated with vibrational surface modes on a piezoelectric substrate are referred to surface acoustic wave (SAW) resonators. They are covered in a later section of this chapter.

Matthys [2], Frerking [3], and Parzen [4] cover crystal oscillators extensively. Other excellent tutorials are Vig [5] and Agilent [6]. The Digital Archive 31 CD set of the Ultrasonics Ferroelectrics Frequency Control Society of the IEEE is an invaluable resource for designers working with accurate frequency and time applications [7]. This chapter reviews some concepts presented in those references, adds additional material, and uses terminology consistent with this book.

8.1.1 Quartz Blank Cuts

Thin wafers of quartz cut from a crystal are referred to as blanks. The orientation angle of the cut with respect to the quartz crystal axis affects the gross properties of the resonator. A diagram of a quartz crystal [8] depicting important cuts is shown on the left in Fig. 8.1. The AT-cut is popular for frequencies of 1 MHz and higher.

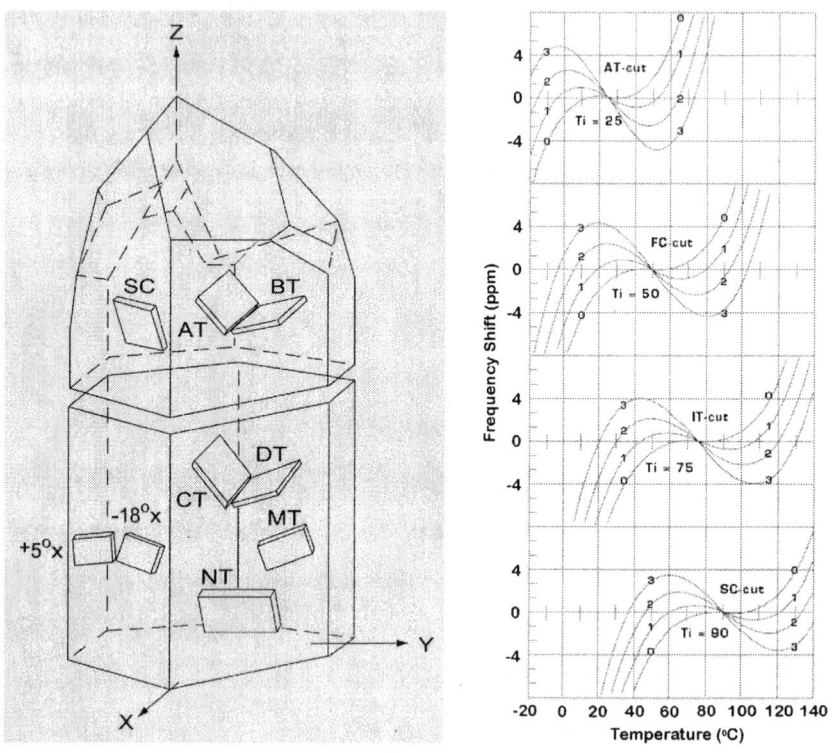

Figure 8.1 Typical crystal cuts (left) and frequency shift versus temperature for four types of cut (right) (Courtesy of Croven Crystals, Wenzel International).

The precise angle of the cut controls the temperature characteristics of resonators manufactured from the blank. The graph on the right in Fig. 8.1 shows the temperature characteristics of various cuts with minute variation in the cut angle. The center inflection points in these curves are referred to as the turnover temperatures. It is 25°C for the AT-cut. For resonators operated near 25°C the lowest frequency variation with temperature is provided by a cut adjusted for characteristic "0." For resonators operated over a wider frequency range, a smaller overall variation in frequency is achieved using a cut that achieves the higher numbered traces. Higher turnover temperature resonators embedded in small temperature-controlled ovens provide oscillators with excellent temperature stability.

The AT-cut blank vibrates in a face-to-face sheer mode as depicted on the lower left in Fig. 8.2. This figure also illustrates the positioning of electrodes plated onto the blank (upper left), the mounting of the plated blank on a base with terminals (left in photograph), and with a weld-sealed cover (right in photograph).

FUNDAMENTAL   OVERTONE

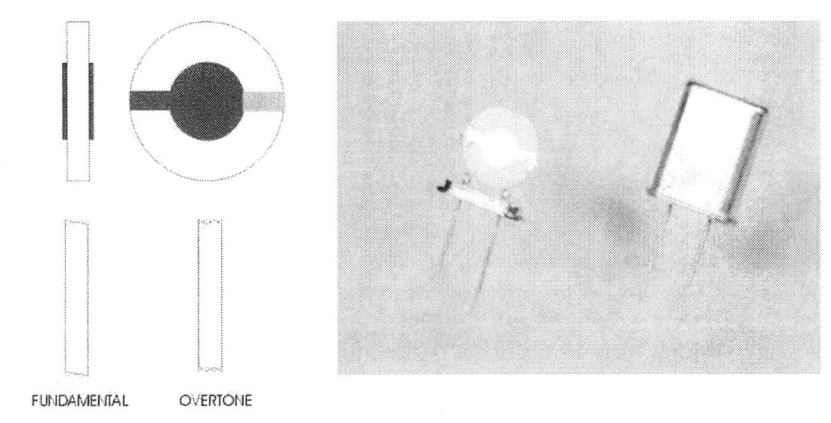

Figure 8.2 Electrode spots on crystal blank (upper left), fundamental and overtone shear vibration modes (lower left), and mounted bulk-quartz resonator (right).

The thickness of the blank controls the resonant frequency. Above 20 MHz, the blank becomes excessively thin and fragile. Crystal resonator oscillators may be constructed for higher frequencies by operating the crystal on an overtone vibration mode as depicted in Fig. 8.2 (line art, lower right). Odd overtones are practical through about the eleventh overtone thus providing resonators to over 200 MHz. Resonator overtone modes should not be confused with oscillator harmonics. An overtone oscillator has no signal components at the fundamental crystal resonator frequency. Overtone oscillators require special techniques to ensure operation on the correct overtone. This is considered using example oscillators later in this chapter.

In theory, the motional resistance increases with the square of the overtone. In practice, the motional resistance penalty for overtone operation is less severe. In addition, in theory, a crystal can be operated on any overtone. However, at higher overtones, the response becomes cluttered with a plethora of spurious modes (see Section 8.1.10 for additional remarks on spurious modes). Techniques are available for reducing the number of spurious modes at a particular overtone. In addition, the frequency of overtone resonance is not a precise multiple of the fundamental frequency. Therefore, the intended overtone mode should be specified to the manufacturer.

8.1.2 Crystal Resonator Model

The relationship between mechanical vibration and the electrical properties of the resonator is modeled effectively using lumped elements. This supports all of the design and simulation techniques described in this book. An electrical model for a crystal resonator is shown in Fig. 8.3.

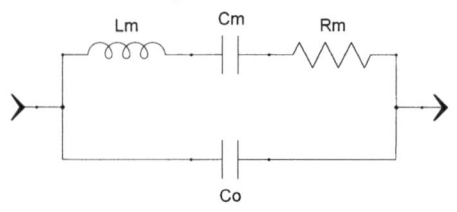

Figure 8.3 Electrical model for a bulk-mode piezoelectric resonator such as a quartz crystal.

The capacitance C_o models the static capacitance of the electrodes separated by the quartz material, plus stray capacitance. L_m, C_m, and R_m model the "series-resonant" piezoelectric effect and are referred to as motional parameters. The motional resistance is not particularly low. The value of L_m is exceptionally large and the value of C_m is exceptionally small. The effective reactance typically exceeds 1 million ohms at 10 MHz. The exceptional unloaded Q available with crystal resonators is due to high-energy storage, not low-loss resistance.

Table 8.1 shows some typical parameter values for high-frequency AT and SC-cut crystals [4]. The values for R_m are maximum specifications. Typical values for R_m are about 1/2 to 1/3 the maximum values. The unloaded Q values are computed from C_m and the typical values for R_m. Numerous tradeoffs are available to the crystal manufacturer including the blank diameter, the electrode spot size, blank face curvature, the precision of manufacture, the type of packaging, and whether the surrounding atmosphere is air or vacuum. Model parameters vary significantly. Unloaded Q exceeding 1 million is achievable with appropriate manufacturing techniques. The crystal manufacturer should be contacted for more accurate model parameters.

The amplitude transmission response of a fundamental mode 9.6-MHz crystal resonator in a metal HC-49/U holder is shown in Fig. 8.4. The horizontal scale is 10 kHz/div and the vertical scale is 10 dB/div. The first peak in transmission at 9.600 MHz on the left occurs at the series resonant frequency of the motional branch. Above series resonance, the motional branch becomes inductive. This inductive reactance parallel resonates with the static capacitance C_o to form a null in the transmission response at 9.620 MHz. This parallel resonance is alternatively referred to as antiresonance. Spurious modes for this crystal exist at 9.64, 9.689, and 9.709 MHz. Spurious modes are further discussed in Section 8.1.10. The crystal parameters may be calculated from the response.

Piezoelectric Oscillators

Table 8.1 Typical parameters for bulk-quartz resonators in adequately sized holders

Freq (MHz)	Cut	Mode	Rm,max (ohms)	Co (pF)	Cm (fF)	Q
1.0-1.5	AT	Fund	525	4	10	73,000
1.5-2.0	AT	Fund	250	4	10	109,000
2.0-3.0	AT	Fund	150	4	11	116,000
3.0-5.0	AT	Fund	80	4	12	124,000
5.0-7.0	AT	Fund	45	4	13	136,000
7.0-10	AT	Fund	35	5	14	115,000
10-15	AT	Fund	30	5	16	80,000
15-30	AT	Fund	27	6	18	56,000
15-60	AT	3rd	40	5	1.6	187,000
45-100	AT	5th	60	5	0.6	166,000
100-140	AT	7th	120	5	0.3	111,000
140-180	AT	9th	180	5	0.2	83,000
5.0-7.0	SC	Fund	45	6	4	442,000
7.0-10	SC	Fund	35	6	5	321,000
10-15	SC	Fund	30	6	5	255,000
15-30	SC	Fund	27	6	6	182,000
15-60	SC	3rd	40	6	0.5	597,000
45-100	SC	5th	60	6	0.2	497,000

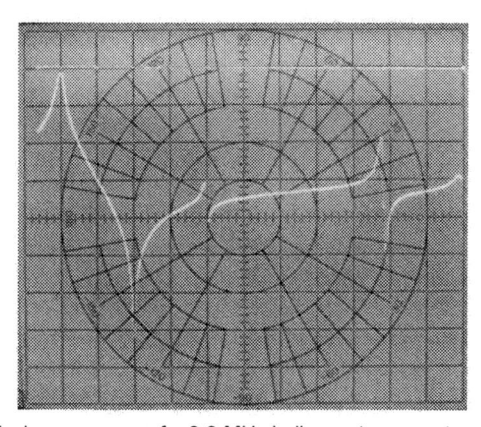

Figure 8.4 Transmission response of a 9.6-MHz bulk-quartz resonator.

8.1.3 Calculating Crystal Resonator Parameters

First, the static capacitance is measured using a low-frequency capacitance meter. Then the motional capacitance is given by

$$C_m = C_0 \left[\left(\frac{f_p}{f_s} \right)^2 - 1 \right] \quad (191)$$

where f_p is the frequency of the parallel resonant null and f_s is the frequency of the series resonant transmission peak. Then the motional inductance is calculated from C_m.

$$L_m = \frac{1}{(2\pi f_s)^2 C_m} \qquad (192)$$

Assuming the reactance of C_o is much greater than R_m, then

$$R_m = 2Z_0 \left(10^{IL/20} - 1\right) \qquad (193)$$

Here, the static capacitance is 5.45 pF and the insertion loss in a 50-ohm system is 0.95 dB. From this measured data, the calculated motional resistance is 11.6 ohms, the motional capacitance is 22.73 fF, and the motional inductance is 12.0920062 mH.

8.1.4 Crystal Resonator Frequency Pulling

The oscillator operation frequency is generally not precisely either the crystal series or parallel-mode frequency. The frequency of an oscillator with a crystal resonator is the frequency at which the oscillation criteria are satisfied. Crystal coupling reactors and the transmission-phase characteristics of the sustaining stage affect the frequency. An oscillator design objective is often minimizing these effects on the operating frequency so that the highly stable crystal is the primary controlling element. However, in certain applications it is desired to intentionally pull the oscillation frequency, to either set the frequency or to frequency modulate the oscillator.

Fig. 8.5 illustrates three pulling schematics. A common technique depicted as circuit A utilizes a capacitor in series with the crystal. This simple technique reduces the effective total capacitance in series in the motional branch and increases the overall series-resonant frequency.

Fig. 8.6 shows the transmission amplitude (circular symbols) and phase responses (square symbols) of the pulling circuit of Fig. 8.5(A) with a large value of pulling capacitor. The series and parallel resonant modes are evident. The motional inductance of the crystal is 16.886864 mH, the motional capacitance is 15 fF, the motional resistance is 10 ohm, and the static capacitance is 4 pF. The dashed trace (triangular symbols) is the transmission amplitude response with the pulling capacitor set to 2.3 pF.

The 2.3-pF pulling capacitor reduces the net series-branch capacitance and raises the resonant frequency by 12 kHz above the crystal series resonance. However, as the frequency is pulled toward the crystal parallel resonance, the insertion loss at the transmission peak becomes infinite. In a typical quartz-crystal resonator, the parallel resonant frequency is only a few hundredths of a percent higher than the series resonant frequency. From Eq. 191, the absolute upper pulling limit is

$$\frac{\Delta f}{f_s} = \sqrt{1 + \frac{C_m}{C_0}} - 1 \qquad (194)$$

Piezoelectric Oscillators

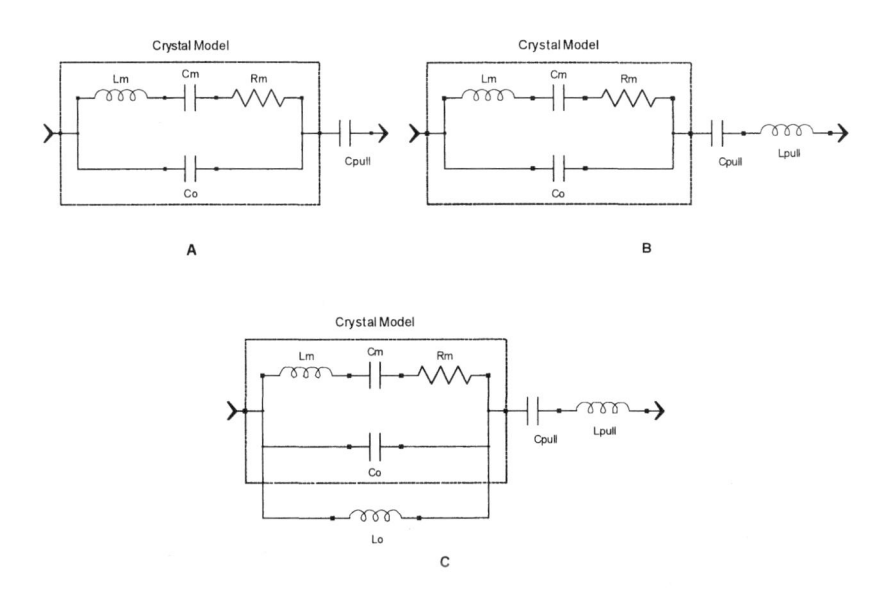

Figure 8.5 Circuits for pulling the operating frequency of a crystal oscillator.

Since C_o/C_m ranges from 180 to 450 for fundamental AT-cut crystals above 1 MHz, the absolute maximum pullability of the circuit of Fig. 8.5(A) is 0.11 to 0.33%. Because of the increased insertion loss, the practical pulling fraction is significantly less.

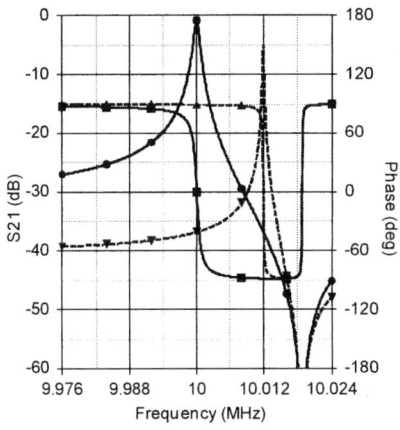

Figure 8.6 Transmission amplitude (circular symbols) and phase (square symbols) responses of the circuit in Fig. 8.5(A) with a large value of pulling capacitor and the transmission amplitude and phase (dashed, triangular symbols) with a 2.3-pF pulling capacitor.

The frequency may be pulled both below and above the original series resonance by utilizing circuit in Fig. 8.5(B). L_{pull} is generally a fixed inductor and C_{pull} is tunable. With a large value of C_{pull} the pulling circuit is net inductive and the overall series resonance is below the crystal series resonance. With a small value of C_{pull} the pulling circuit is net capacitive and the overall series resonance is above the natural crystal resonance. The lowest frequency is limited by the largest practical inductor value and the highest frequency is again limited by the parallel resonant frequency.

The limitation imposed by parallel resonance is nullified by resonating C_o with a parallel inductor as illustrated by circuit C in Fig. 8.5. These techniques allow pulling of several times that limited by Eq. 175. Nevertheless, the effectiveness of circuits B and C is limited by practical issues. The purpose of using a quartz crystal is its excellent stability. To the degree that the frequency is pulled, stability is compromised by the less stable lumped elements. In addition, the motional capacitance is very small and the motional inductance is very large for HF components, thus the degree of pulling is limited by practical sizes for these reactors.

Other factors being equal, the value of motional capacitance is inversely proportional to the square of the overtone mode. For example, a third overtone crystal has 1/9 the motional capacitance of a fundamental mode crystal. Since the static capacitances of overtone and fundamental mode crystals are similar, then

$$\frac{\Delta f}{f_0} \propto \frac{1}{overtone^2} \qquad (195)$$

Higher overtone-mode oscillators are progressively more difficult to pull.

8.1.5 Inverted-Mesa Crystal Resonators

A clever method of extending the practical frequency range of a fundamental-mode resonator, the inverted mesa blank geometry, was commercially developed by Innovative Frequency Control Products, later acquired by Valpey Fisher Corporation. Inverted mesa crystals are now available from a number of manufacturers. The geometry is depicted in Fig. 8.7 [9]. A thick outer ring provides structural strength while a very thin, chemically milled inner ring supports a high fundamental-mode resonance. Useful devices to a few hundred MHz are available. The pullability of a fundamental-mode resonator is achieved at much higher frequencies.

Piezoelectric Oscillators

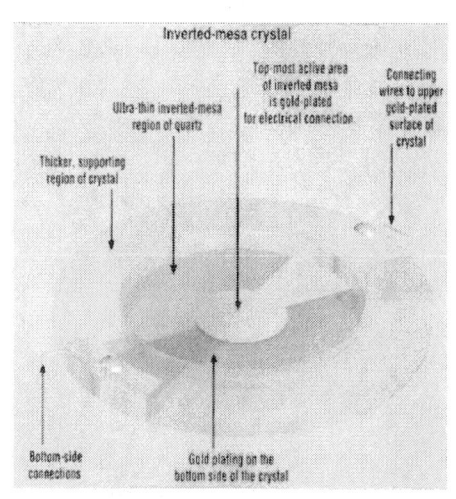

Figure 8.7 Inverted mesa crystal for increasing the frequency of fundamental mode resonators (courtesy Electronic Design).

Micro Crystal Switzerland produces 30- to 250-MHz inverted mesa crystals in a 3.7 by 8 mm SMD ceramic package with a maximum series resistance at 50 MHz of 50 ohms and a typical value of 15 ohms [10]. For 50- to 155-MHz crystals, the typical motional capacitance is 5.6 fF. C_o is 2.0 pF at 50 MHz and 2.9 pF at 155 MHz. The electrodes are typically small, resulting in a low C_o despite the thin dielectric.

8.1.6 Crystal Oscillator Operating Mode

Fig. 8.8 shows a bipolar 10-MHz Pierce crystal oscillator. The crystal is in series with the cascade with a peak of transmission near crystal resonance. However, the phase-shift of the crystal at series resonance is 0° and the CE amplifier is inverting; therefore the transmission phase of the cascade would not be correct. Capacitor C_1 driven by the output impedance of the transistor and resistor R_1 produces additional phase shift, as does capacitor C_2 driven by the motional resistance of the crystal. This additional phase shift causes the total cascade transmission phase to be 0° near the transmission peak. Furthermore, without capacitors C_1 and C_2, the total resistance in series with the crystal is much higher than the motional resistance, thus degrading the cascade, loaded Q. The values of C_1 and C_2 control the loaded Q and resistor R_1 adjusts the transmission phase.

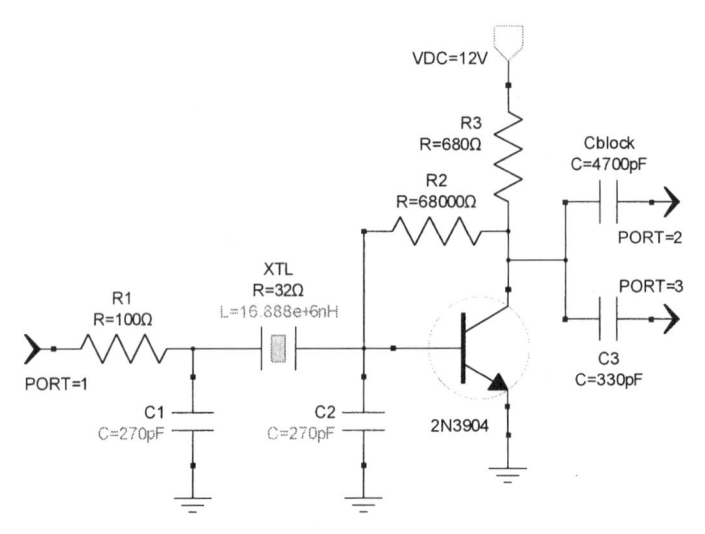

Figure 8.8 A 10-MHz Pierce oscillator with crystal operating near series resonance.

The responses of the open-loop cascade are shown in Fig. 8.9. Resistor R_1 is adjusted so that the maximum slope of the transmission phase occurs near ϕ_o. The motional capacitance of the crystal is 15 fF, the static capacitance is 4 pF, and the motional resistance is 32 ohms. The motional inductance is tuned so that ϕ_o occurs at 10 MHz. The series resonance of 15 fF and 16.8883 mH is 9,999.575 kHz or 425 Hz lower than ϕ_o. When the amplifier is truly inverting, the oscillation frequency approaches the crystal series resonance as the values of C_1 and C_2 are increased. The Pierce may be made to oscillate at the crystal exact series resonance by placing in series with the crystal an inductor with a reactance of magnitude equal to the reactance of C_1 in series with C_2. An approach that eliminates this inductor is working in concert with the crystal manufacturer to specify a crystal series resonance below the desired oscillation frequency. Except for critical applications, oscillators designed with high, loaded Q are suitably close to series resonance.

Also plotted in Fig. 8.9 is the loaded Q of the Pierce oscillator divided by 1000. The cascade loaded Q peaks at approximately 20,000. The unloaded Q of this crystal resonator as defined by the reactance of the motional inductance or capacitance divided by the motional resistance is 33,200. Therefore, the loaded Q is 60% of the unloaded Q.

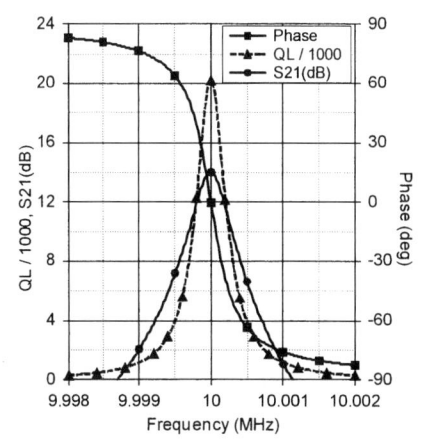

Figure 8.9 Transmission amplitude (circular symbols), phase (square symbols), and loaded Q (triangular symbols, dashed trace) of the series-mode Pierce oscillator.

Fig. 8.10 shows a bipolar 10-MHz Colpitts parallel-mode crystal oscillator. When the loop is closed by connecting ports 1 and 2, C_1, C_2, and the crystal form a parallel resonator cascaded with a CC amplifier. The crystal is operating inductive above series resonance and below the natural parallel resonance of the crystal. The crystal parameters are identical to the crystal parameters used in the previous Pierce example.

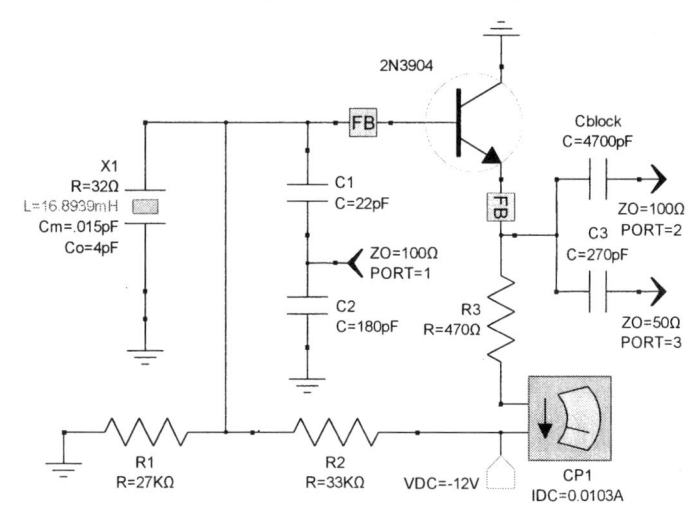

Figure 8.10 A 10-MHz bipolar Colpitts with crystal operating near parallel resonance.

The open-loop input and output S-parameter magnitudes of this cascade in a 100-ohm system are only -2.8 and -6.0 dB, respectively, so the

Randall-Hoch corrections are used. The open-loop amplitude and angle of G are plotted in Fig. 8.11. The corrected open-loop gain is 13.7 dB at ϕ_o. The motional inductance is adjusted to set ϕ_o at 10 MHz. With this motional inductance, the crystal natural parallel resonance is 10.0166 MHz and the natural series resonance is 9.9979 MHz. The oscillator is operating 16.6 kHz below the crystal's natural parallel resonance. The loaded Q at ϕ_o of this circuit is approximately 10,000, which is only 50% of the Pierce oscillator loaded Q. The crystal Colpitts oscillator requires a large impedance in parallel with the resonator. Later in this chapter it is shown that an FET is superior to a bipolar transistor for this application.

The large shift from the natural parallel resonance is caused by the capacitance of C_1 and C_2 in parallel with the crystal C_o. The capacitance in parallel with the crystal is referred to as load capacitance. For the Colpitts oscillator, the load capacitance is approximately the series combination of C_1 and C_2, in this case 19.6 pF. Because the shift is rather large, load capacitance is typically specified when ordering crystals for parallel-mode oscillators. Two commonly specified values of load capacitance are 20 and 32 pF.

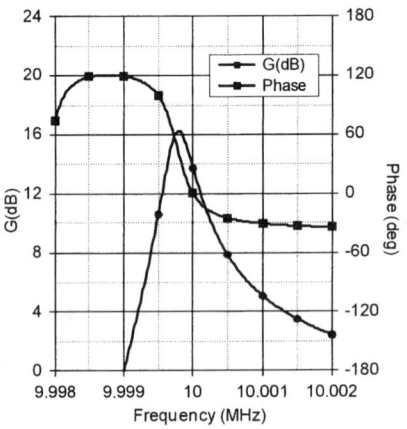

Figure 8.11 Randall-Hoch open-loop G amplitude (circular symbols) and angle (square symbols) for the 10-MHz Colpitts.

8.1.7 Crystal Oscillator Frequency Accuracy

The frequency accuracy of a crystal oscillator is defined as compliance to a specific desired frequency. Precision is defined as repeatability around any oscillation frequency. An oscillator design may have good accuracy (average frequency is the desired frequency) but low precision (wide frequency spread), or a design may have poor accuracy (off the frequency target) but little spread (high precision). An example of a design that has poor accuracy but high precision is one with accurately constructed crystals

Piezoelectric Oscillators 333

with a consistent resonant frequency but a sustaining stage with a load capacitance other than expected by the crystal manufacturer.

The variation of the oscillation frequency with time is referred to as instability. Specifically, over a short time period, typically 1 second or less, instability is referred to as noise. Over a time period of days or longer, instability is referred to as aging. Aging is considered in Section 8.1.11.

Intermediate time-period variations are often associated with oscillator turnon and turnoff, ambient temperature changes, and acceleration or changes in circuit orientation. The influence of temperature on crystal oscillators is considered in the next section.

Accuracy, precision, and intermediate to long-term variation is typically expressed in fractional terms in parts per million (ppm). For example, the accuracy of an oscillator that is within ±100 Hz at 10 MHz is accurate within ±10 ppm.

One manufacturing specification is the accuracy of the resonant frequency of the crystal. Inexpensive crystals offer accuracy of ±100 to ±50 ppm while moderately priced crystals offer accuracy of ±10 ppm at a given temperature. This performance significantly exceeds that available using 1% tolerance lumped-element reactors that offer essentially ±10,000 ppm. However, achieving ±10 ppm accuracy in a crystal oscillator also requires tight control of the loading effect of the sustaining stage. More accurate frequency control than ±10 ppm requires good quality control and close communication with the crystal manufacturer.

Shown on the left in Fig. 8.12 is a 25-sample Monte Carlo simulation of the open-loop cascade transmission angle of the 10-MHz bipolar Colpitts crystal oscillator in Fig. 8.10 with capacitors C_1 and C_2 at ±5% tolerance with uniform distributions. The ϕ_0 ranges ±65 Hz, or ±6.5 ppm.

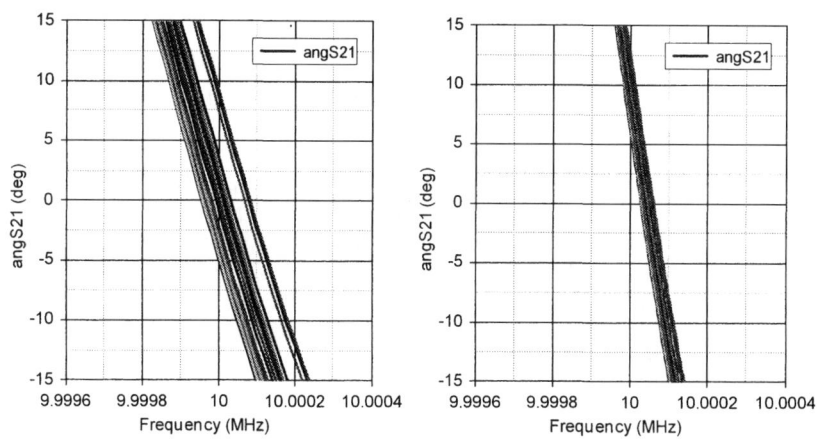

Figure 8.12 Monte Carlo analysis of the open-loop cascade transmission phase of the Colpitts crystal oscillator with ±5% variation in the loading capacitor C_1 (left) and for the Pierce crystal oscillator with ±5% in the shunt coupling capacitors C_1 and C_2.

Shown on the right in Fig. 8.12 is a 25-sample Monte Carlo simulation of the open-loop cascade transmission angle of the 10-MHz bipolar Pierce crystal oscillator in Fig. 8.8 with capacitors C_1 and C_2 at 5% tolerance, uniform distribution. The influence of the noncrystal components on the oscillation frequency is clearly superior for this Pierce oscillator. Nevertheless, the potential influence of the noncrystal elements on oscillator stability is clear.

8.1.8 Temperature Effects on Crystal Oscillators

Because of the initial accuracy and the high long-term stability of quartz, the major influence on the frequency stability of a quartz crystal oscillator is the operating temperature. The stabilization of crystal oscillators with temperature is a refined and mature art, and the subject is only briefly reviewed here.

As illustrated on the right in Fig. 8.1, the turnover temperature is controlled by the type of cut and the specific temperature curve is controlled by slightly different angles of cut. The AT-cut is popular partly because the turnover temperature is 25°C ambient. The tightest tolerance over a wide temperature range is achieved with a characteristic such as that labeled 3. If the crystal is operated in a tightly controlled environment near 25°C, the highest accuracy is achieved with the characteristic labeled 0. To realize a specific characteristic requires control of the cut angle within minutes of arc. Tight specification of the cut characteristic increases crystal cost.

Significant improvement in stability over a wide temperature range is achieved by temperature compensation. The environmental temperature is sensed and an open-loop compensation network corrects for the expected frequency characteristic curve of the crystal cut. This is referred to as a temperature compensated crystal oscillator (TCXO). State-of-the-art TCXOs today achieve temperature stability to ±0.2 ppm over a -20 to +70°C range in a 3 x 5 x 7 mm package operating at 5 volts and drawing little current [11].

Further improvement in temperature stability is achieved by placing the crystal in an oven. The temperature of the oven is controlled by closed loop. This is referred to as an oven controlled crystal oscillator (OCXO). An oven temperature equal to the crystal turnover temperature and higher than the highest encountered ambient temperature is normally employed.

8.1.9 Crystal Resonator Drive Level

As the excitation of a crystal increases, the frequency shifts upward for AT-cut and downward for SC-cut crystals. Excessive excitation can exceed the quartz elastic limit and fracture the crystal. The dissipation level in quartz-crystal resonators should generally be limited to 2 mW or less for frequencies above 100 kHz. Lower frequency resonators and applications with strict aging requirements are limited to significantly lower dissipation levels.

On the left in Fig. 8.13 is the motional resistance of a crystal versus the drive level. A well-behaved resonator, as depicted by the solid trace, displays a flat curve until drive level effects come into play, in this case at approximately 1-mW drive. The increase in motional resistance with higher drive level is normal and is referred to as a drive level effect. A blank with surface contamination may display excess resistance, sometimes rising abruptly, with decreasing drive. This leads to starting problems. This is not desirable and is avoided by quality control and specification for critical applications.

On the right in Fig. 8.13 is the crystal displacement amplitude versus frequency with the drive level as a running parameter. Another drive level effect is the skewing of an AT-cut crystal resonance higher in frequency with increasing drive. The amplitude curve becomes skewed. With higher drive, the curve may even become multivalued. This is avoided by limiting the drive level.

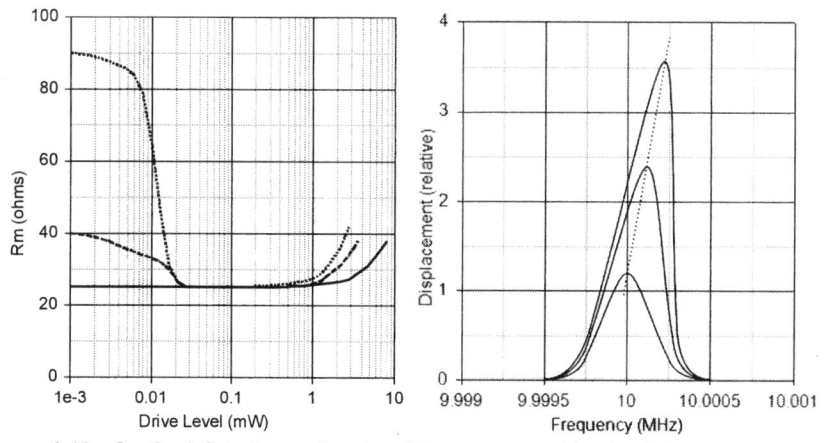

Figure 8.13 On the left is the motional resistance versus drive level for a well behaved crystal (solid traces) and for defective crystals (dashed traces). On the right is the oscillation displacement versus frequency with three drive-level amplitudes.

The frequency shift with drive level is given approximately [4] by

$$\frac{\Delta f}{f} = \alpha I^2 \qquad (196)$$

where α is 0.20 to $0.25/A^2$ for 5- to 100-MHz AT-cut crystals with typical electrode size. SC-cut crystals designed for low-drive dependence are about an order of magnitude better.

Crystal dissipation is best calculated using I^2R_m by finding the voltage across an external series resistance much less than R_m and computing the current. R_m is calculated using Eq. 193 or is otherwise known.

Shown on the left in Fig. 8.14 is a model of the quartz crystal used in the 10-MHz Colpitts oscillator of Fig. 8.10 with current probes to measure the current in the motional branch (CP2) and the total crystal current (CP3). The supply voltage is reduced to -8.6 volts and R_l is increased to 47 kohms to reduce the emitter current. The simulated emitter current is 4.74 mA and the oscillator output power is 3.4 dBm.

Only the motional current produces dissipation in the crystal. The total current includes current in the static capacitor. For oscillators operated near series resonance, little current flows in the static capacitor, and the total current is an accurate measure of the motional current. For oscillators with crystal operation near parallel resonance, current flows in the static capacitor, and the total current is not an accurate measure of the motional current. The currents are easily separated by simulation, however, for measured circuits, only the total current is accessible. The error is generally small and the measurement overestimates the motional current, thus introducing a safety margin.

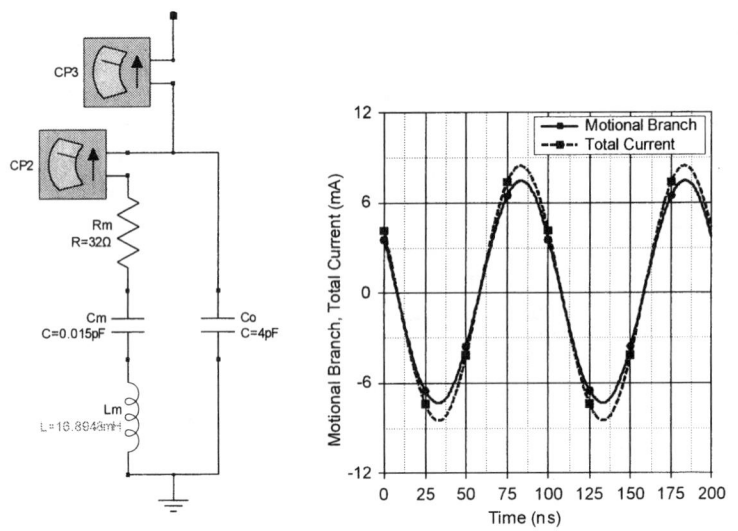

Figure 8.14 Crystal model with current probes measuring the motional branch current and the total crystal current (left) and harmonic-balance simulation waveforms (right).

Shown on the right in Fig. 8.14 are the currents in CP2 and CP3 as simulated by a steady-state harmonic-balance of the oscillator. The peak motional current is 7.2 mA, and since the waveform is sinusoidal, the rms current is 5.09 mA. The crystal dissipation with 21.2 ohms of motional resistance is 0.549 mW, within the normal operating range for an oscillator designed for best phase noise rather than best aging.

A lower crystal drive level reduces the available oscillator power level, thus limiting the phase-noise performance. However, very low drive level is required for the best aging performance. The combination of very high loaded Q but low oscillator power level results in excellent close-in phase noise but poor phase noise at higher offset. As a result, the phase-noise characteristic of typical crystal oscillators is flatter with offset frequency than are higher power, lumped and distributed-resonator oscillators.

8.1.10 Crystal Resonator Spurious Modes

Fig. 8.15 shows X-ray topography and the response versus frequency of a 3200-kHz fundamental-mode resonator excited at certain spurious mode frequencies [12]. Dark areas in the topography represent maximum displacement of the crystal. The topography provides insight into the nature of the desired and the spurious modes. The fundamental-mode topography illustrates how the thicker outer ring of an inverted-mesa crystal only marginally disturbs the resonance process, since little energy resides at the edge of the blank. The ratio of the spot diameter and the blank diameter may be used to help control spurious modes.

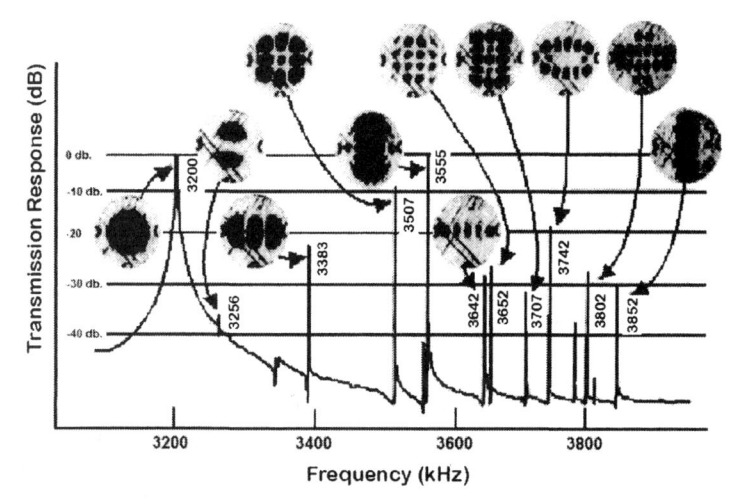

Figure 8.15 X-ray topography of a quartz AT-cut resonator excited at various spurious mode resonances (redrawn from [12]).

The level of the worst-case spurious mode is generally specified by requesting that the series resistance at any spurious mode frequency be higher than the fundamental mode series resistance. Alternatively, the response insertion loss at any spurious mode frequency may be specified.

In general, for oscillator applications, spurious mode responses 10 dB or more below the desired response are acceptable. Spurious mode control is more critical for filter crystals.

8.1.11 Crystal Resonator Aging

Aging refers to the long-term stability of a crystal oscillator. Aging is generally expressed as ppm per day, month, or year. Aging is typically cumulative at an exponential rate, with the rate of change decreasing with time. A major initial source of aging is foreign matter on the blank and electrodes being redistributed by vibration to other locations within the housing. The acceleration on the crystal surface caused by movement at megahertz rates is phenomenal. Redistributed material reduces the blank mass and therefore increases the oscillation frequency. Initial aging is often a positive frequency change. Other sources of aging include:

1) enclosure leaks;
2) relieved mounting and electrode stresses;
3) moisture absorption;
4) electrode corrosion;
5) outgassing of materials within the enclosure.

The best long-term aging performance is achieved with low drive level, as low as 10 µW. Unfortunately, this degrades short-term stability (phase-noise). Accelerated preaging is achieved by elevated temperature and cycled burn-in procedures. High-volume production can achieve 0.01 ppm/day. Careful design, quality-control procedures, and device selection can achieve better than 0.1 parts per billion per year. The type of seal used can affect aging. Cold weld parts are generally superior to resistance weld parts. Solder sealing is no longer used for high-performance applications because of quality control issues and poor aging characteristics.

8.1.12 Crystal Resonator *1/f* Noise

The Leeson noise model assumes the source of noise is the active device. To be sure, the resonator plays a significant role and the designer's strongest lever for improving noise performance is via higher resonator loaded Q. However, the resonator is only transforming noise initially generated in the active device. Leeson's assumption is valid for L-C and distributed resonator-based oscillators. However, this assumption is not necessarily valid with piezoelectric quartz bulk or surface wave resonators [13]. The resonance of quartz resonators exhibits flicker ($1/f$) and random walk ($1/f^2$) noise. The random walk frequency noise is associated with temperature fluctuations but quartz-resonator frequency flicker noise is not well understood [14].

Temperature changes strongly influence the random walk noise that masks the flicker noise unless the crystal resonator is ovenized. The influence of the resonator flicker noise on the overall oscillator noise performance is greatest close to the carrier. At greater offset, the active device noise dominates and the classic Leeson model is valid. The crossover-offset frequency depends on crystal parameters, but it is generally 100 Hz or lower.

Crystal resonator flicker noise is a strong inverse function of unloaded Q. The flicker noise of high-quality crystals is proportional to $1/Q^4$, but individual crystals vary at least an order of magnitude. Data in reference [14] was taken with 2.5-, 5-, and 10-MHz resonators with unloaded Q ranging from 3×10^5 to 4×10^6. Crystal resonator flicker noise is also drive-level dependant. The cited reference indicates that increasing the drive from 10 µW to 2.5 mW increased the flicker noise in a 2.5-MHz overtone resonator by an order of magnitude. Techniques for measuring quartz resonator frequency instability are given in [15].

8.1.13 Crystal Resonator Acceleration Effects

Acceleration affects the natural resonant frequency of a piezoelectric resonator. The magnitude of the effect is linear with acceleration up to at least 50g [5]. The magnitude of the effect is also a function of direction of the acceleration with respect to the crystal axis.

The acceleration sensitivity of high-quality SC-cut resonators manufactured for low sensitivity is a few parts in 10^{10}. Tuning fork 32-kHz watch crystal sensitivities are a few parts in 10^7. Otherwise, for most AT, BT, IT, and SC-cut, fundamental and overtone crystals from 1 to 500 MHz, the maximum sensitivity is within a factor of three of $1 \times 10^{-9}/g$ [5].

Because of Earth's gravitational acceleration, even the orientation of the crystal influences the natural resonant frequency. This is referred to as tipover. Therefore, when the crystal is oriented with the direction of maximum effect, for a typical resonator, tipover is 2×10^{-9}. This deviation exceeds that caused by aging for high-performance oscillators.

Vibration (dynamic acceleration) introduces sideband noise in the oscillator output spectrum. Real environments are not vibration-free. The spectral density of vibration sources can be sinusoidal or noisy. For sinusoidal vibration, the peak phase deviation is given by

$$\theta_d = \frac{\left(\overline{\Gamma} \bullet \overline{A}\right) f_0}{f_m} \quad (197)$$

where $\overline{\Gamma}$ is the crystal sensitivity vector and $\overline{A}$ is the acceleration vector. For noisy vibration

$$\theta_d = \frac{(\overline{\Gamma} \bullet \overline{A})f_0}{2f_m} \qquad (198)$$

where

$$|\overline{A}| = \sqrt{2 \times PSD} \qquad (199)$$

where the power spectral density, PSD, is measured in g/Hz. For small deviation, the SSB component caused by the vibration is

$$\mathscr{L}(f_m) = 20\log\frac{\theta_d}{2} \qquad (200)$$

The SSB phase noise of a typical crystal oscillator in an aircraft with a PSD of 0.04 g/Hz from 5 to 1000 Hz is easily degraded by 40 dB over that baseband frequency range. If natural resonance modes exist in the resonator, packaging, or enclosure, vibration effects are amplified.

Table 8.2 shows typical acceleration levels in representative environments [5]. Even a seemingly benign environment such as a building has vibration induced by occupants, elevators, HVAC equipment, and outside vehicle traffic. The author has experienced degraded noise sidebands from bearings in cooling fans on bench-mounted instruments.

Table 8.2 Acceleration level of representative environments

Environment	Acceleration (typical, g)
Buildings	0.02 rms
Tractor-trailer rig	0.2 peak
Armored personnel carrier	0.5 to 3 rms
Ship, calm seas	0.02 to 0.1 peak
Ship, heavy seas	0.8 peak
Propeller aircraft	0.3 to 5 rms
Helicopter	0.1 to 7 rms
Jet aircraft	0.02 to 2 rms
Missile, boost phase	15 peak
Railroads	0.1 to 1 peak

In all but the most benign environments, vibration-induced sideband noise dominates Leeson predicted phase noise in the lower audio frequency range. Vibration isolation mounts and damped foam are marginally effective in reducing vibration-induced sidebands. Active compensation using acceleration detecting sensors is more effective.

Shock introduces a large, one-time frequency excursion and a smaller permanent frequency offset probably caused by stress relief or contaminant removal from the surface. Improved shock survival performance is achieved by chemical etching to produce a scratch-free surface. Crystals processed this way can survive being fired from large guns.

8.1.14 Crystal Resonator Standard Holders

Table 8.3 is a list of popular quartz-crystal resonator holders. HC is an abbreviation for holder crystal. The once-popular solder-sealed HC-6/U through HC-33/U holders are now replaced with metal cold weld, resistance weld, and smaller ceramic and plastic holders for surface mount applications. The W and D dimensions of the metal can holders are for the base that is slightly larger than the top portion of the holder. The column labeled *Freq* is the fundamental-mode frequency range commonly used for that holder.

Table 8.3 Standard wire through-hole crystal holders

Type	Seal	Terminal (dim in mm) Type	Dia	Spacing	W (mm)	D (mm)	H (mm)	Freq (MHz)
HC-6/U	Solder	Pin	1.27	12.34	8.9	19.3	19.69	1-35
HC-13/U (tall)	Solder	Pin	1.27	12.34	8.9	19.3	38.76	0.09-1
HC-18/U	Solder	Wire	0.43	4.88	4.65	11.05	13.46	2.9-35
HC-25/U	Solder	Pin	0.99	4.88	4.65	11.05	13.46	2.9-35
HC-33/U	Solder	Wire	0.76	12.34	8.9	19.3	19.69	1-35
HC-36/U	Cold	Pin	1.27	12.34	8.9	19.3	19.69	1-35
HC-42/U	Cold	Pin	0.99	4.88	4.65	11.05	13.46	2.9-35
HC-43/U	Cold	Wire	0.43	4.88	4.65	11.05	13.46	2.9-35
HC-45/U	Cold	Wire	0.43	3.75	3.3	8.15	8.69	9-175
HC-47/U	Cold	Wire	0.43	12.34	8.9	19.3	19.69	1-35
HC-48/U	Resist	Pin	1.27	12.34	8.9	19.3	19.69	1-35
HC-49/U	Resist	Wire	0.43	4.88	4.65	11.05	13.46	2.9-35
HC-50/U	Resist	Pin	0.99	4.88	4.65	11.05	13.46	2.9-35
HC-51/U	Resist	Wire	0.43	12.34	8.9	19.3	19.69	1-35
HC-52/U	Resist	Wire	0.43	3.75	3.3	8.15	8.69	9-175
HC-35/U(TO-5)	Cold	Wire			9.75	Circular	5.59	7-175
HC-37/U(TO-7)	Cold	Wire			15.75	Circular	5.95	2.5-25
HC-40/U	Cold	Wire			21.35	Circular	9.01	1-10

8.2 Fundamental Mode Crystal Oscillators

The first oscillator stabilized by a quartz crystal resonator is attributed to Cady around 1921. His career included many important contributions to the piezoelectric art. The IEEE-UFFC Society annually bestows the W. G. Cady Award to an individual "to recognize outstanding contributions related to the fields of piezoelectric or other classical frequency control, selection and measurement; and resonant sensor devices." The references in this chapter include several recipients of the award.

The temperature properties of the crystal resonator are controlled by resonator manufacture. A goal of the oscillator designer is the development of a coupling circuit and sustaining stage that minimizes the influence of the device, the supply, and circuit temperature change on overall stability.

Several crystal oscillator circuit examples follow. In each of the fundamental mode oscillators a 9.6 MHz AT-cut HC-49/U crystal is used. The transmission amplitude response of this crystal is shown in Fig. 8.4. The series resonant frequency is 9.600 MHz, the parallel resonant frequency is 9.620 MHz, the static capacitance is 5.45 pF, and the insertion loss in a 50 ohm system is 0.95 dB. From this measured data, the calculated motional resistance is 11.6 ohms, the motional capacitance is 22.73 fF, and the motional inductance is 12.0920062 mH. This is an unloaded Q of 62,900.

With each of these examples, a schematic is given with the loop opened to illustrate useful break points. For consistency, the design operating voltage for each is 8.6 volts. The device currents are adjusted to provide 3 ±0.25 dBm simulated output at the fundamental frequency. For each circuit, the current measured by probe CP_1 is the open-loop DC quiescent current before the onset of oscillation. CP_2 is used to measure crystal current. For lower-power operation, devices should be selected with good performance at low V_{ce} or V_{ds} and low current. Some of the bias configurations given are better for low-voltage operation. To avoid the open-loop terminations from disturbing the DC bias, in certain cases blocking capacitors are used. If the bias is not disturbed when the loop is closed, these blocking capacitors may be eliminated when the oscillator is formed by closing the loop. To facilitate confirmation of the open-loop response with standard test equipment, open-loop termination impedances of 50 ohms are used where feasible. In some cases, other termination impedances are required to improve simulation accuracy. For cases where a poor match results for any reference impedance, the Randall/Hock correction is used.

For bipolar circuits, the device is a common 2N3904 NPN bipolar transistor with an F_t specification of 300 MHz and β of 150. A Phillips QMMBT3904 model is used for simulation. For phase-noise critical applications, a device with lower noise figure is suggested. However, the use of a high-gain device with F_t greater than 100 times the operating frequency potentially causes stability problems. For discrete FET circuits, an N-channel J309 JFET is used with an ON Semiconductor company device model.

In general the open-loop cascade is adjusted for 5 to 8 dB of gain margin. It is noted when the natural gain margin is not within this range. External lumped elements are adjusted to achieve as high a loaded Q as possible. This aggressive approach to maximizing loaded Q is appropriate when the crystal motional resistance is well known. When the motional resistance is not tightly controlled, the loaded Q should be reduced to guarantee a positive gain margin for higher resistance. In general, the reactor values are scaled linearly to accommodate other oscillation frequencies. However, an open-loop simulation of the cascade gain, phase, match, and loaded Q is advised for all oscillators.

Piezoelectric Oscillators

For each oscillator, the loop is closed and a harmonic balance simulation is used to predict the output spectrum and steady-state waveform. Coupling to a 50-ohm load is generally tight so a buffer circuit is advised for critical applications to avoid load pulling. Where a 50-ohm termination overloads the circuit, a coupling capacitor is used to raise the effective load impedance.

8.2.1 Miller JFET Crystal

The schematic of the 9.6-MHz Miller JFET crystal oscillator is given in Fig. 8.16. The CS JFET provides approximately 180° phase shift. The resonant frequency of L_1 and C_1 is below the operating frequency so this tank appears inductive. This inductance driven by the device output resistance provides approximately 90° phase shift. C_2 with the crystal operating above series resonance provides an additional 90° phase shift. As it operates between the crystal series and parallel resonant frequency, the Miller is called a parallel mode oscillator. This circuit operates 2869 Hz above series resonance.

The effective feedback capacitance is C_2 in parallel with the Miller capacitance. Since the Miller capacitance is device-dependant, the stability of this circuit is poor. The circuit is included here for a historical perspective. It is not recommended for new designs.

The selected cascade reference impedance is 300 ohms. Even so, the open-loop cascade matches are poor and Randall/Hock corrections are required for analysis. The open-loop gain margin is only about 2.5 dB, well below the target gain margin of 5 to 8 dB. The loaded Q of the Miller oscillator is very poor, in this case just 2130, only 3.4% of the crystal unloaded Q.

Output is taken at the drain through a small coupling capacitance to raise the effective load resistance. Capacitor C_3 raises the effective load impedance to 7.59 kohms. The output spectrum and waveform are given in Fig. 8.17. This Miller circuit offers a clean output waveform with low harmonic content. It also offers high DC to RF power conversion efficiency, in this case 21.2%, the highest of all the example oscillators. This is because L_1, C_1, and C_3 operate much like a class-E power amplifier. Above a few megahertz, a bipolar Miller circuit is possible. However, the lower input impedance of bipolar amplifiers makes an already marginal Miller circuit even more questionable.

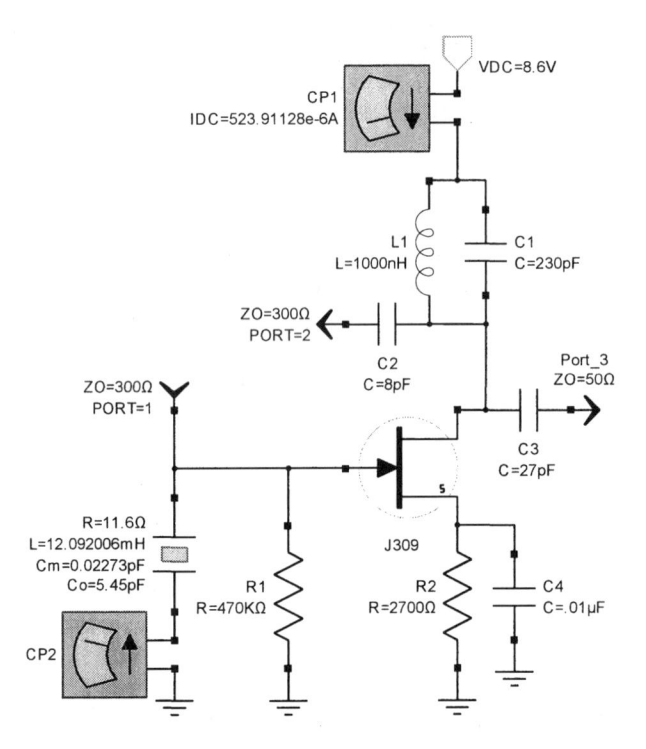

Figure 8.16 Miller JFET 9.6-MHz crystal oscillator.

8.2.2 Colpitts Bipolar Crystal

The Colpitts bipolar is a common crystal oscillator topology. The example design is shown in Fig. 8.18. Above a few megahertz, the Colpitts bipolar is relatively easy to design and it offers adequate performance for many noncritical applications. Unfortunately, the cascade match and the loaded Q are poor and aligning the maximum phase slope with ϕ_o is difficult. The stability of the CC amplifier is poor so a ferrite bead is often used in series with the base to prevent spurious oscillation. A bead in series with the emitter improves the cascade match. The beads are modeled as a 100-ohm resistor in parallel with a 400-nH inductor.

Piezoelectric Oscillators 345

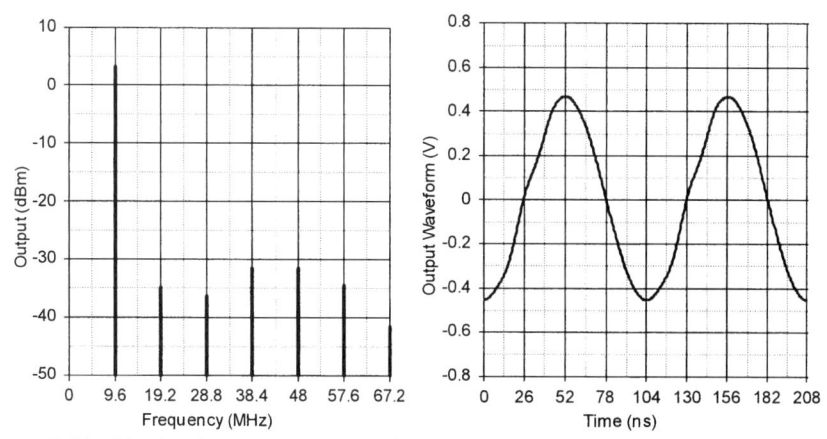

Figure 8.17 Simulated output spectrum (left) and waveform (right) of the Miller JFET crystal oscillator.

Biasing is provided by an alternate form of the Bias 11 network described in Section B.1.7. Capacitors C_1 and C_2 transform the low output impedance of the CC amplifier up to the high base impedance. The crystal operates above series resonance and is inductive, parallel resonating with C_1 and C_2. This circuit operates 3282 Hz above crystal series resonance. The crystal load capacitance is approximately equal to C_1 in series with C_2, in this case 20 pF.

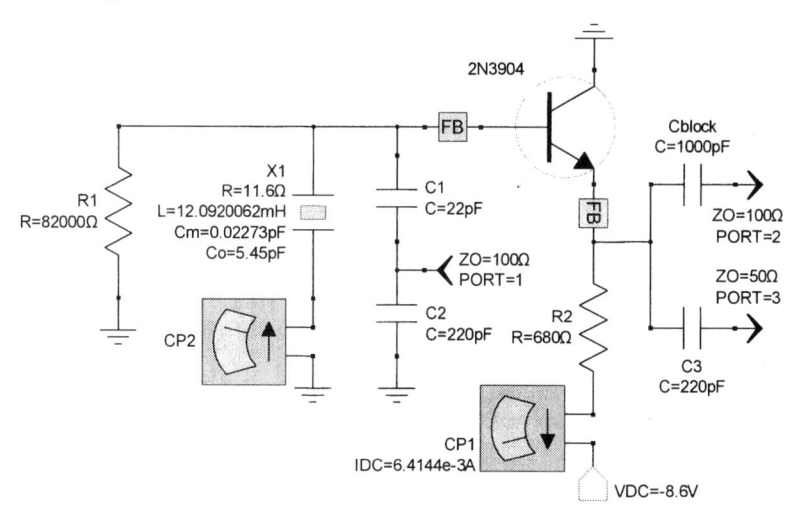

Figure 8.18 Colpitts bipolar 9.6-MHz crystal oscillator.

The selected cascade reference impedance is 100 ohms. Nevertheless, the open-loop cascade match is poor and the Randall/Hock correction is advised. The open-loop gain margin is much higher than the 5- to 8-dB target, suggesting further stabilization with beads is warranted. The loaded Q of this Colpitts is 14,200, 23% of the crystal unloaded Q.

Output power is taken at the emitter through a 220-pF coupling capacitor, C_3, that raises the effective load impedance to 163 ohms. The output waveform is shown on the left in Fig. 8.19 and the output spectrum is given on the right. The second harmonic is 3.7 dB below the fundamental. The harmonic-rich spectrum results in a rather poor waveform. The output power and DC to RF power conversion efficiency are more than doubled by taking output power directly into a 50-ohm load that grounds the collector. Unfortunately, the harmonic performance is even worse and the output waveform is nearly an impulse. The high harmonic content is supported by a transistor with an F_t over 30 times the oscillation frequency. Although the 2N3904 might be considered a marginal transistor for RF applications, for crystal oscillators in the HF frequency range, the 2N3904 is more than adequate.

Crystal dissipation for this example is 775 μW and is higher than most of the other example oscillators.

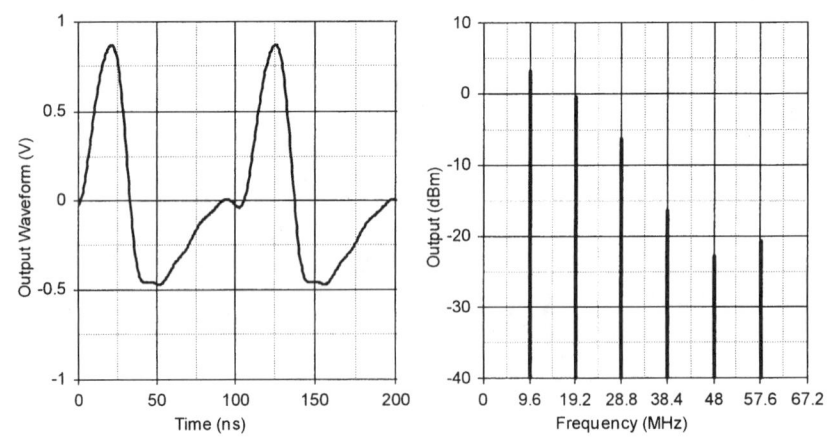

Figure 8.19 Simulated output waveform (left) and spectrum (right) of the Colpitts bipolar 9.6-MHz oscillator.

8.2.3 Colpitts JFET Crystal

For optimum performance, the Colpitts oscillator requires high impedance across the crystal. The performance of the Colpitts oscillator is significantly improved by replacing the bipolar transistor with a JFET, as illustrated in the schematic of Fig. 8.20. The JFET substitution may be necessary at frequencies lower than a few megahertz because of the higher

Piezoelectric Oscillators

series resistance of lower-frequency crystals. The basic operation is similar to the bipolar form.

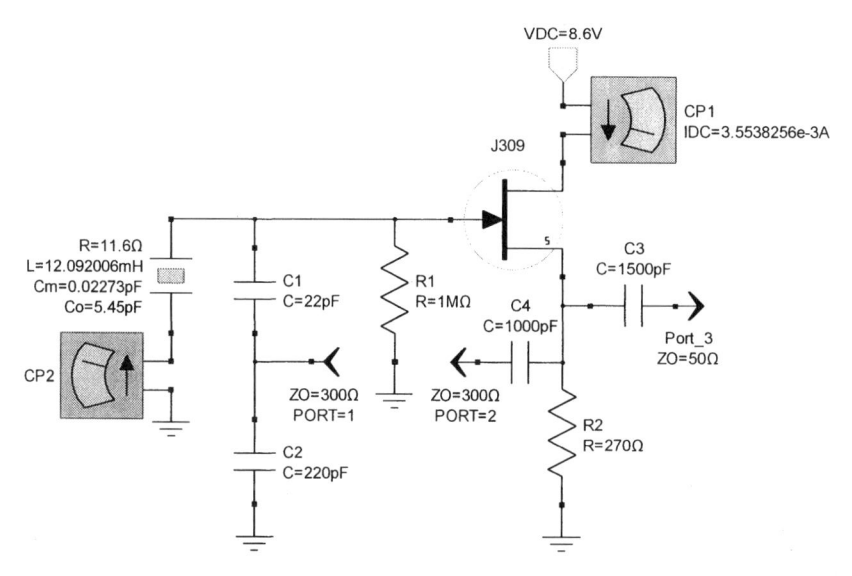

Figure 8.20 Colpitts JFET 9.6-MHz crystal oscillator.

The optimum cascade reference impedance is higher than the bipolar form, in this case 300 ohms. The open-loop cascade match for this example is better than the bipolar form, but analysis accuracy still benefits from the Randall/Hock correction. The open-loop gain margin is on the high end of the 5- to 8-dB target. If spurious modes are an issue, gate and source beads are used. The loaded Q of this JFET Colpitts is 42,000, a respectable 67% of the crystal unloaded Q and a significant improvement over the bipolar form.

Output power is taken at the emitter through a 1500-pF coupling capacitor, C_3. This is essentially full coupling into the 50-ohm load. The output waveform is shown on the left in Fig. 8.21 and the output spectrum is given on the right. The second harmonic is 6.3 dB below the fundamental. The harmonic-rich spectrum results in a nonsinusoidal waveform. The DC to RF power conversion efficiency is somewhat better than the bipolar Colpitts. Crystal dissipation for this example is 123 µW, which is among the lowest dissipations of the examples.

8.2.4 Pierce Bipolar Crystal

The Pierce bipolar crystal oscillator is shown in Fig. 8.22. The Pierce uses an inverting amplifier and a crystal resonator with shunt coupling capacitors C_1 and C_2. The output impedance of the amplifier driving C_1 and the crystal motional resistance driving C_2 add the additional required 180° phase shift. A series resistor, R_1, may be used to increase the phase shift in C_1 to better align the maximum phase slope with the zero crossing. C_1 and C_2 also reduce the effective input and output impedance of the amplifier and therefore increase the loaded Q.

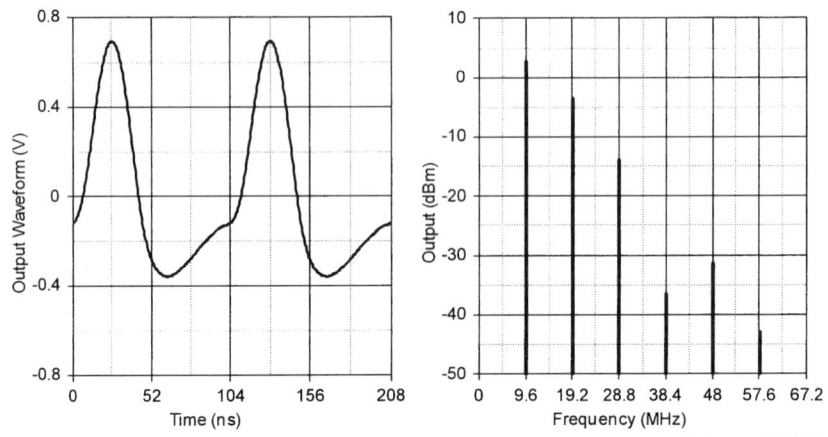

Figure 8.21 Simulated output waveform (left) and spectrum (right) of the Colpitts JFET 9.6-MHz crystal oscillator.

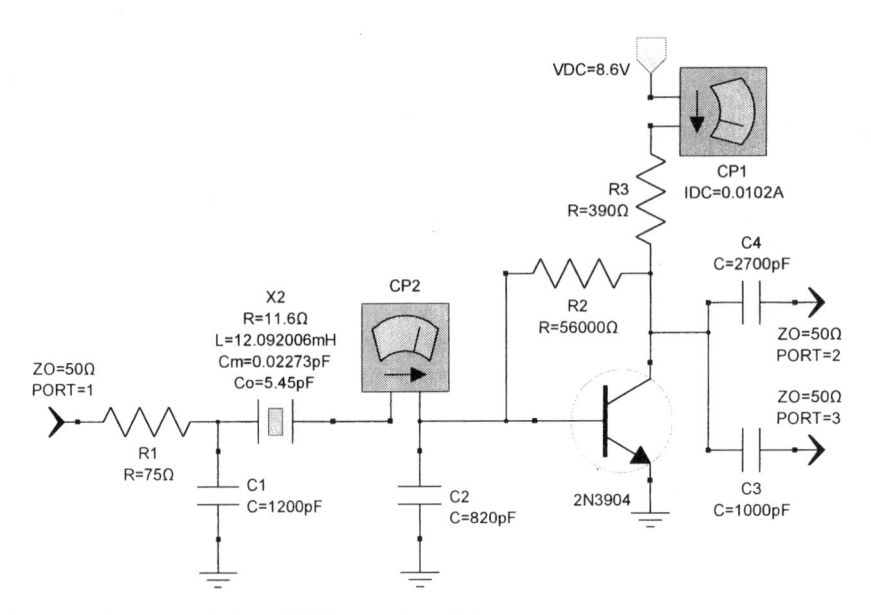

Figure 8.22 Pierce bipolar 9.6-MHz crystal oscillator.

The cascade input and output S-parameter magnitudes are better than -10 dB referenced to 50 ohms and the Randall/Hock correction is unnecessary. The open-loop gain margin is just over 5 dB. The loaded Q of this bipolar Colpitts is 52,100, 83% of the crystal unloaded Q and the highest of the example oscillators. The gain margin is reduced by the aggressive pursuit of loaded Q. Increasing C_1 and C_2 increases the loaded Q to unloaded Q ratio and increases the resonator insertion loss.

C_1 and C_2 are sometimes referred to as loading capacitors, with the effective loading capacitance equal to their series combination, in this case 487 pF. However, the goal of Pierce oscillator design is to maximize the values of C_1 and C_2 consistent with adequate gain margin while the goal with parallel-mode crystal oscillators is typically to adjust the loading capacitance to a smaller standard value, such as 20 or 32 pF. As C_1 and C_2 are increased, the oscillation frequency approaches the crystal series resonant frequency. This example oscillates 209 Hz above the crystal series resonant frequency.

Output power is taken at the collector through a 1000 pF capacitor. This is nearly full coupling to a 50-ohm load. The simulated output waveform is shown on the left in Fig. 8.23. The second harmonic is 6.4 dB below the fundamental and the third is 17.8 dB down. A measured waveform for this oscillator is shown on the right in Fig. 8.23. The measured waveform is slightly more sinusoidal, resulting in a slightly lower negative peak voltage.

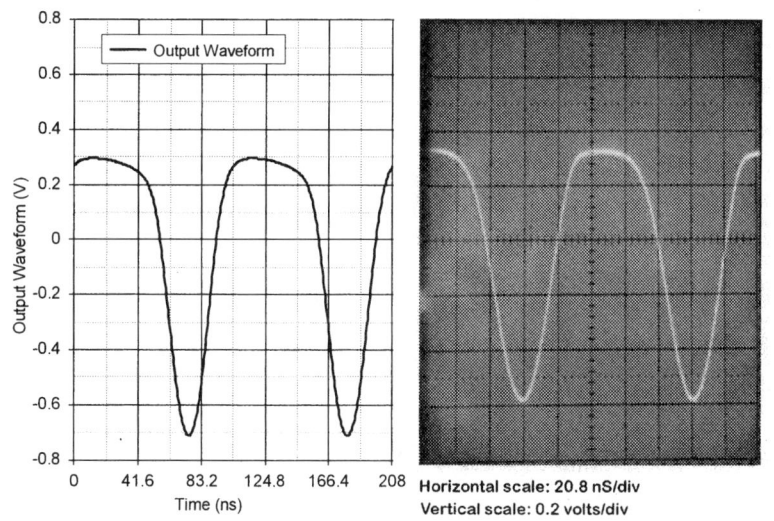

Figure 8.23 Simulated (left) and measured (right) output waveform of the Pierce bipolar fundamental-mode crystal 9.6-MHz oscillator.

8.2.5 Pierce MMIC Crystal

The Pierce in Fig. 8.24 is similar to the Pierce bipolar except that a silicon monolithic microwave integrated circuit (MMIC) is used for the active device. The Mini-Circuits MAR6 and Avago Technologies MSA06 are Darlington pair 10-GHz F_t transistors with shunt and series resistive feedback and internal biasing resistors. The open-loop gain is 20-dB flat to 200 MHz and 12 dB of gain at 2 GHz. The advantage of MMIC gain blocks is temperature stabilized gain, repeatability, improved device stability, and better than -12 dB input and output S-parameter magnitudes up to 1 GHz at 50 ohms. The resistive feedback degrades the noise figure to about 3 dB, and excellent match is only marginally beneficial for an oscillator, so this circuit offers little advantage over the discrete bipolar form. The loaded Q is less and the crystal dissipation is higher than the bipolar form. Nevertheless, this is a useful and easy to design alternative when the use of an MMIC is desired.

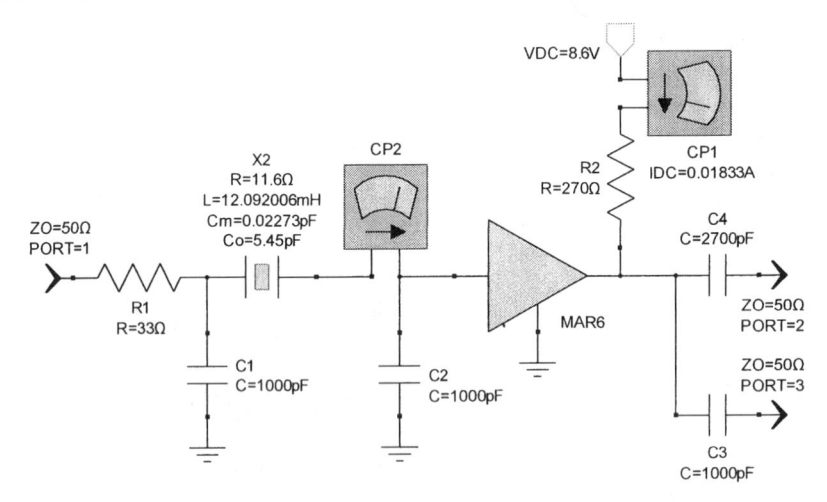

Figure 8.24 Pierce MMIC 9.6-MHz crystal oscillator.

The simulated output waveform is shown on the left in Fig. 8.25 and the spectrum is shown on the right.

8.2.6 Pierce Inverter TTL Crystal

Fig. 8.26 depicts a Pierce 9.6-MHz crystal oscillator using a TTL inverter. The simplified model for the inverter was introduced in Section 5.2.4. Resistor R_5 biases the TTL gate in a linear operating mode. It is adjusted by observing that the input or output voltage is near the state transition voltage.

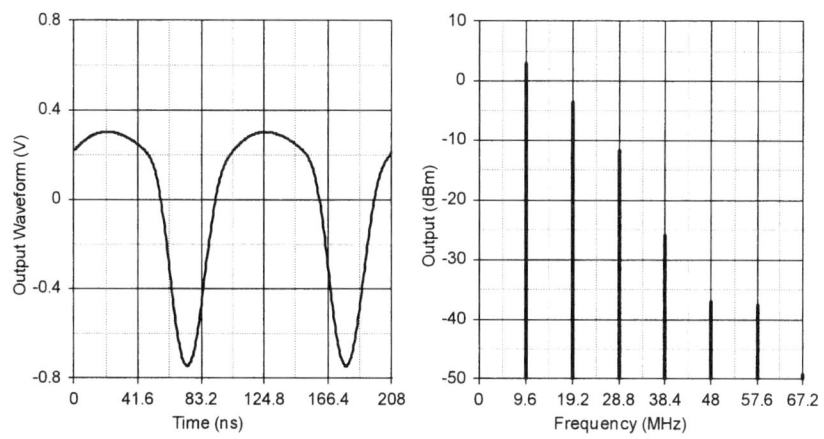

Figure 8.25 Simulated output waveform (left) and spectrum (right) of the Pierce MMIC 9.6-MHz crystal oscillator.

Operation of this oscillator is similar to the bipolar and MMIC Pierce oscillators. The TTL gate is inverting and the additional 180° phase shift is provided by the gate output resistance and R_6 driving C_2 and the crystal motional resistance driving C_1. As with the previous Pierce circuits, increasing the value of C_1 and C_2 improves the loaded Q but reduces the gain margin. The simulated responses are shown in Fig. 8.27.

The simulated output waveform and spectrum for the TTL inverter Pierce crystal oscillator are shown in Fig. 8.28. The waveform is more sinusoidal and the second and third harmonics are over 15 dB below the fundamental. The operating frequency is only 120 Hz above the crystal series resonant frequency. The cascade loaded Q is 36,500, 58% of the crystal unloaded Q. The crystal dissipation is a moderate 284 µW.

8.2.7 Pierce Inverter CMOS Crystal

An example CMOS Inverter Pierce 9.6-MHz crystal oscillator is shown in Fig. 8.29 using a simplified CMOS gate model. An additional gate is used to buffer the output to facilitate delivering 3-dBm output power to a 400-ohm load. A 1-Mohm feedback resistor biases the oscillator gate in the active region. Its value is not critical. A 1000-hm open-loop cascade reference impedance matches the output impedance at port 2. The input impedance is significantly higher so the Randall-Hoch correction is used to predict the open-loop gain margin of 6 dB.

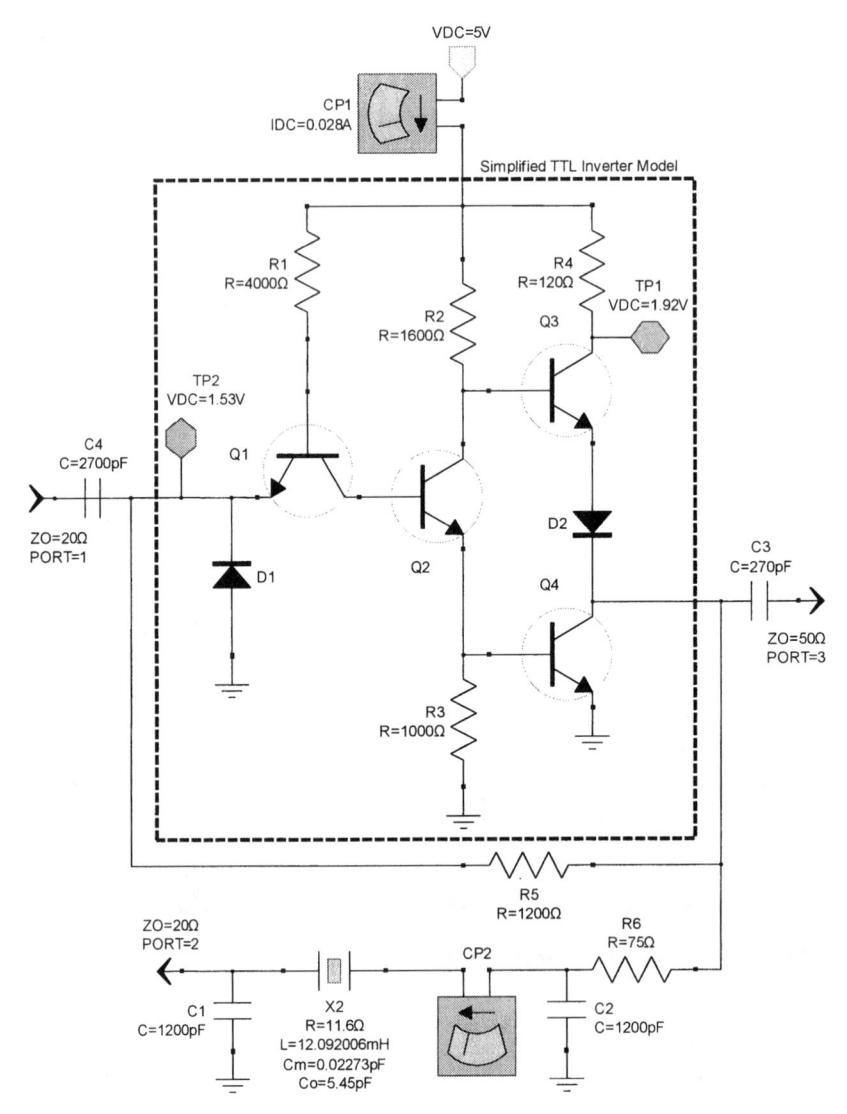

Figure 8.26 Pierce TTL Inverter 9.6-MHz crystal oscillator.

Because of the high impedance level of this circuit, much smaller values of shunt coupling capacitors are required to realize good loaded Q. The loaded Q of this example as determined by the Randall/Hock corrected transmission phase is 54,600, an impressive 87% of the crystal unloaded Q. If the crystal motional resistance is not tightly controlled, smaller and less aggressive values of C_1 and C_2 are used. The smaller values of C_1 and C_2

Piezoelectric Oscillators 353

result in a higher operating frequency, in this case 1457 Hz above the crystal series resonant frequency.

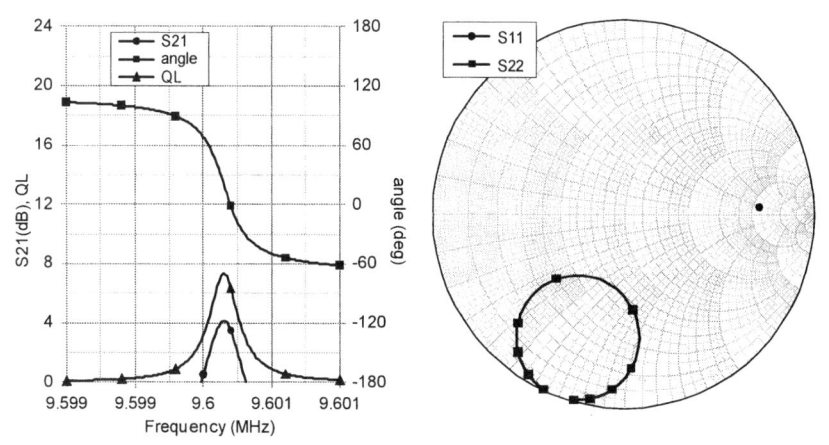

Figure 8.27 Simulated open-loop responses (left) and match (right) for the TTL inverter gate Pierce 9.6-MHz crystal oscillator.

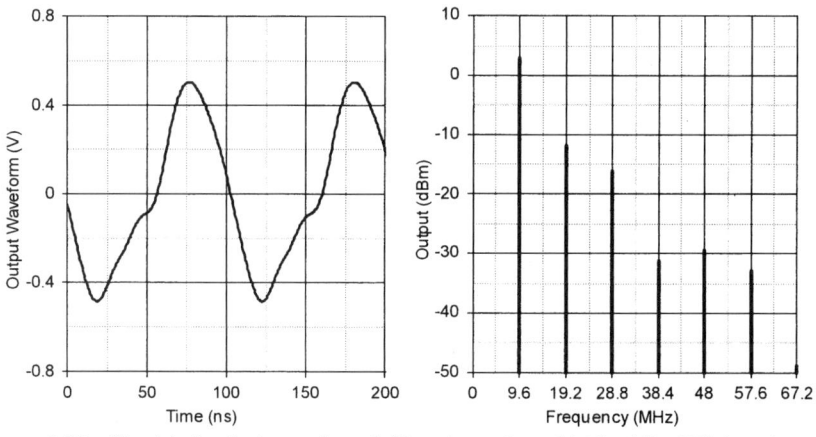

Figure 8.28 Simulated output waveform (left) and spectrum (right) of the TTL inverter gate Pierce 9.6-MHz crystal oscillator.

The high offset above the crystal series resonant frequency is a result of the small values of C_1 and C_2. The Pierce is often considered a series-mode oscillator because of operation close to the crystal series resonant frequency. However, the loading capacitance in this case is effectively 75 pF, only a factor of 2.3 greater than the standard value of 32 pF often used with what are considered parallel-mode oscillators. The small values of C_1 and C_2 result in a higher sensitivity to their values, much as with a Colpitts oscillator, rather than the typical low sensitivity of a Pierce oscillator.

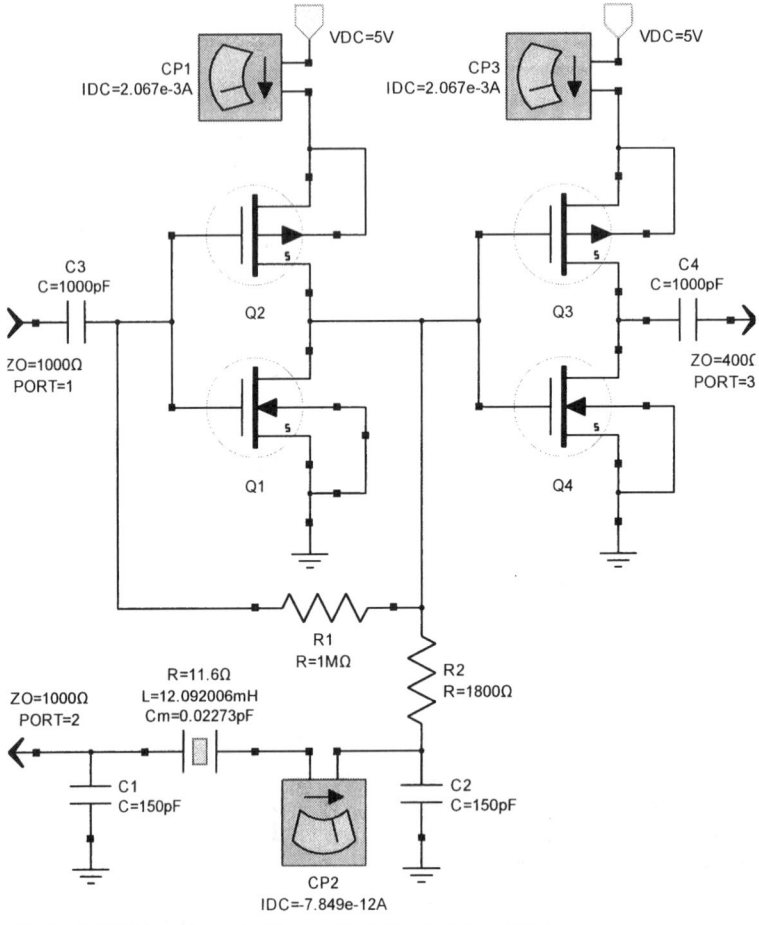

Figure 8.29 CMOS inverter gate Pierce 9.6-MHz crystal oscillator.

The output waveform and spectrum of this example are shown in Fig. 8.30. The higher impedance results in a higher output voltage for a given output power so the voltage scale for this example is increased. The output waveform is nearly symmetrically clipped, resulting in a spectrum with primarily odd harmonics. The ripple in the waveform in the clipped regions is a result of truncation caused by using only seven harmonics in the harmonic balance simulation. Increasing the number of harmonics balanced in the simulation reduces the level of these ripples. The number of balanced harmonics is left at seven for consistency with the other simulations.

Piezoelectric Oscillators 355

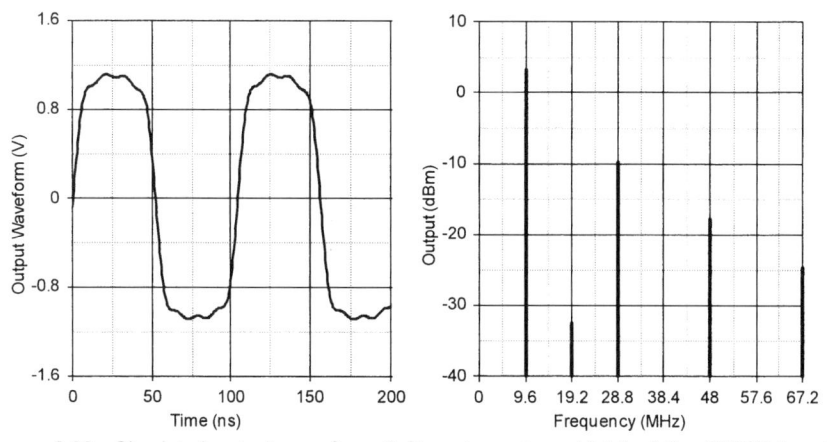

Figure 8.30 Simulated output waveform (left) and spectrum (right) of the CMOS inverter gate Pierce 9.6-MHz crystal oscillator.

8.2.8 Butler Dual Bipolar Crystal

A Butler dual bipolar transistor 9.6-MHz crystal oscillator is shown in Fig. 8.31. The left transistor operates as a noninverting CB amplifier. The right transistor operates as a noninverting CC amplifier. The loop is closed by the series resonance of the crystal resonator. Other crystal oscillators that operate with ideal active devices exactly at the series resonant frequency of the crystal are also referred to as Butler oscillators. An advantage of this circuit is that no reactive elements are required to establish the cascade phase response. Therefore:

1) initial calibration is not compromised by reactor tolerance;

2) stability is not compromised by reactor temperature characteristics;

3) a wide frequency range of crystals is accommodated without modifying the circuit.

These advantages make this circuit useful as a fundamental mode crystal test circuit.

Because harmonic currents flow in the crystal static capacitance, the total crystal current waveform is not particularly sinusoidal. Therefore, to accurately simulate crystal dissipation, C_o is placed external to the model so the true motional current is measured. The low impedance of the CB input stage and the CC output stage result in good loaded Q. The loaded Q is particularly good with lower frequency crystals, which naturally have higher motional resistance. For higher frequency crystals, such as this 9.6-MHz unit, the low motional resistance places more importance on amplifier impedance. The impedance is reduced by increasing the device currents but this increases crystal dissipation. The loaded Q of this example is 28,800, 46% of the resonator unloaded Q.

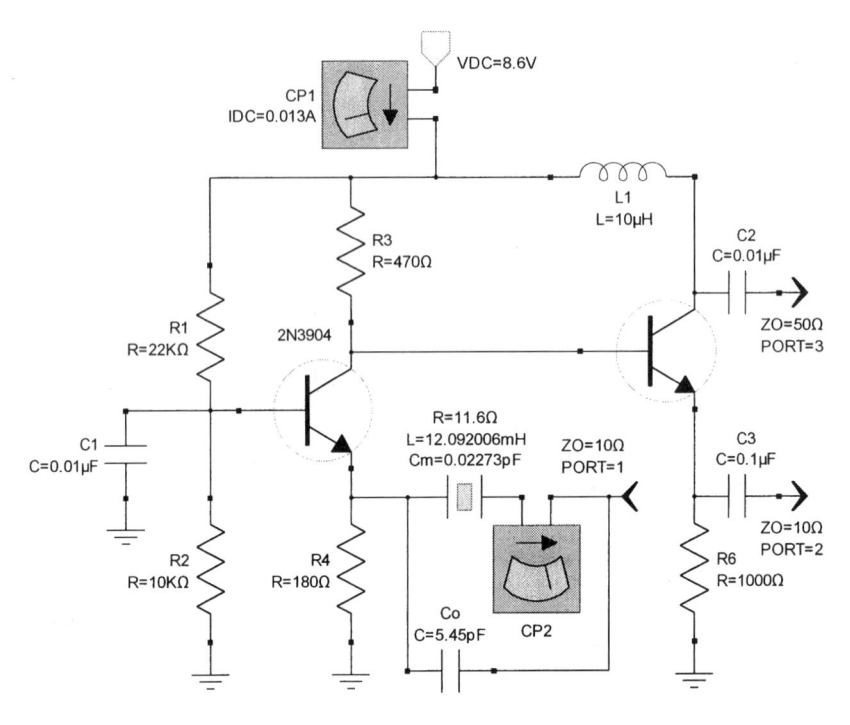

Figure 8.31 Butler dual bipolar 9.6-MHz crystal oscillator.

The simulated output waveform and spectrum are shown in Fig. 8.32. Significant clipping occurs on the positive waveform peak. The peak is rather square. The waveform is rippled at the voltage peak because seven harmonics are an insufficient quantity to balance. Nevertheless, the spectrum through the sixth harmonic is within 2 dB of the value predicted when balanced with a large quantity of harmonics. In general, simulation accuracy of the spectrum is limited by compliance of the device model to the specific device, provided the quantity of harmonics balanced exceeds the highest frequency of interest. The simulation was set to seven harmonics for consistency with the other simulations.

The operating frequency of this example is 170 Hz above the crystal series resonant frequency. The operating frequency is exactly the crystal series resonant frequency if the amplifiers present purely resistive terminations to the crystal. The crystal dissipation for this example is 543 µW.

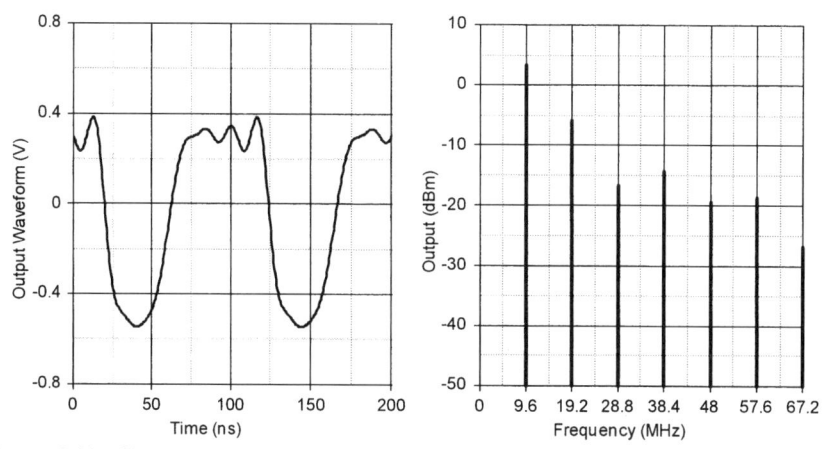

Figure 8.32 Simulated output waveform (left) and spectrum (right) of the Butler dual transistor 9.6-MHz crystal oscillator.

8.2.9 Driscoll Bipolar Crystal

The oscillator topology shown in Fig. 8.33 is due to Driscoll [16]. The left transistor serves as an inverting CE amplifier with high emitter degeneration except at the crystal series resonant frequency. At series resonance the emitter is grounded through the motional resistance. Resistor R_4 is a biasing element that is decoupled by inductor L_1. The right transistor serves as a noninverting CB amplifier stabilized by resistor R_8. The C_6, L_2, and C_2 network steps down the high collector impedance of the right transistor and provides a feedback path back to the base of the left transistor. This network also provides the additional required 180° phase shift. Capacitor C_1 is a blocking capacitor. The quiescent current in the right transistor is set well below the left transistor. The left transistor operates linearly as the right transistor limits, therefore providing a more constant impedance load for the crystal. Driscoll also described a second topology that uses a tapped inductor for the collector to base impedance transformation.

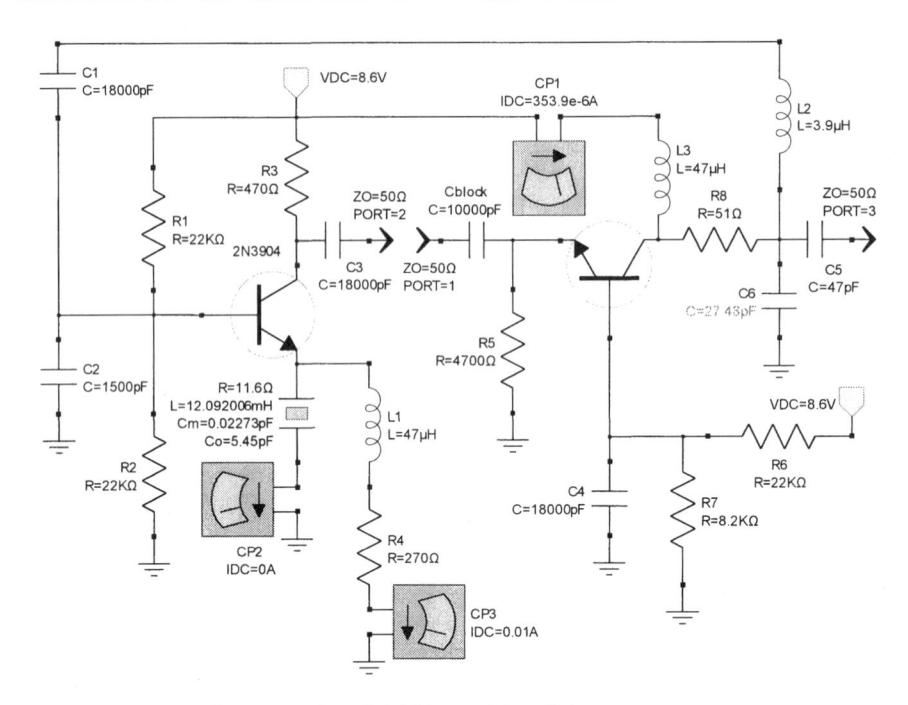

Figure 8.33 Driscoll dual transistor 9.6-MHz crystal oscillator.

Capacitor C_6 tunes the impedance-transforming network so that the operating frequency is the crystal series resonant frequency. The loaded Q of this oscillator is degraded by the left transistor emitter resistance in series with the crystal motional resistance. As with the previous Butler oscillator, lower frequency crystals with a naturally higher motional resistance are degraded less. Higher quiescent current in the left transistor improves the loaded Q. The loaded Q of this example is 48,200, 77% of the crystal unloaded Q.

Output power is taken through the coupling capacitor C_5 that increases the load impedance presented to the collector. The simulated output waveform and spectrum are shown in Fig. 8.34. The tuned-collector output results in a relatively sinusoidal output with low harmonics. The crystal dissipation is only 60 μW because the output power is provided by the right transistor and the signal level in the left transistor is low. It should be noted that the right transistor is essentially a buffer and the power level in the left transistor is used in Leeson's equation to predict the phase-noise performance.

Piezoelectric Oscillators

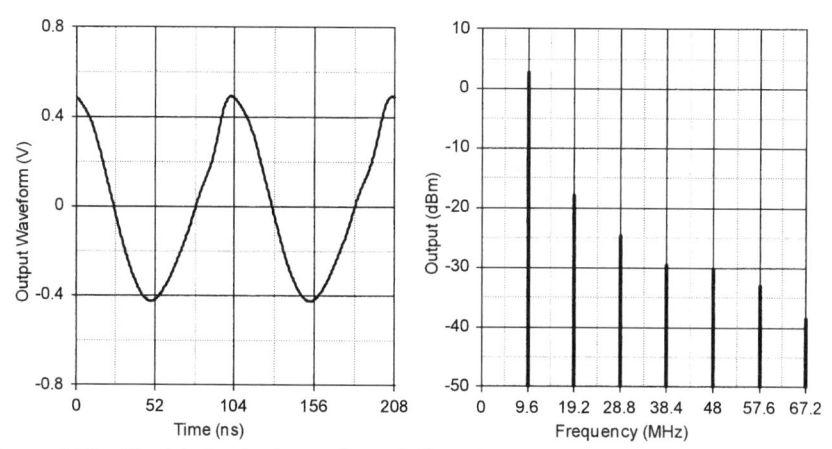

Figure 8.34 Simulated output waveform (left) and spectrum (right) of the Driscoll dual transistor 9.6-MHz crystal oscillator.

Fig. 8.35 shows the measured SSB phase-noise of a tapped-inductor Driscoll 5-MHz oscillator [16]. The crystal resonator is a third overtone BT-cut with a motional inductance of 6.8 henry and a motional resistance of 85 ohms. This is an unloaded Q of 2.51×10^6! The reported loaded Q is 80% of the unloaded Q, or 2×10^6. The output power is 4 dBm and the crystal dissipation is 85 µW. With this loaded Q and operating frequency, the $f_0/2Q$ frequency is only 2.5 Hz, lower than the lowest frequency measurement of the SSB phase noise! Therefore, the inflection in the measured phase-noise curve is the result of flicker noise. The curve would be expected to have a 10 dB/decade slope but it is closer to 13 dB/decade, suggesting an additional source of noise or a measurement issue. The SSB phase is reasonably modeled using Leeson's equation with a flicker corner of 10 kHz and a power level of -4 dBm suggesting the right transistor is in fact simply a buffer.

8.2.10 Inverted-Mesa Pierce Bipolar Crystal

The following example is a bipolar Pierce oscillator with parameters for a typical inverted-mesa 100-MHz crystal resonator. The motional resistance is 25 ohms, the motional inductance is 0.452326713 mH, the motional capacitance is 5.6 fF, and the static capacitance is 2.4 pF. The unloaded Q is therefore 11,400.

The oscillator is shown in Fig. 8.36. The inverted-mesa Pierce oscillator is essentially identical to the Pierce oscillator for conventional AT and other cuts except the higher frequency of operation requires smaller values of coupling capacitors and generally a higher F_t device. The device used here is an MRF901 device, small-signal bipolar transistor with an F_t of 4 GHz.

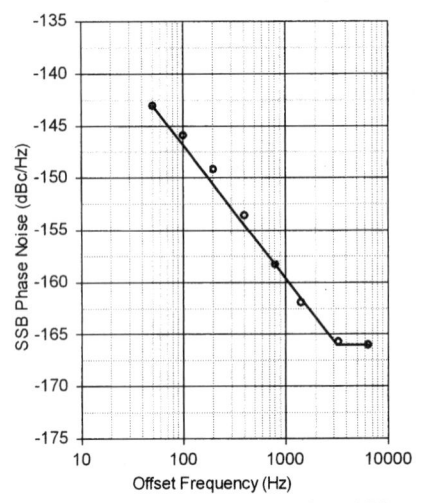

Figure 8.35 Measured SSB phase-noise performance of a 5-MHz tapped-inductor Driscoll.

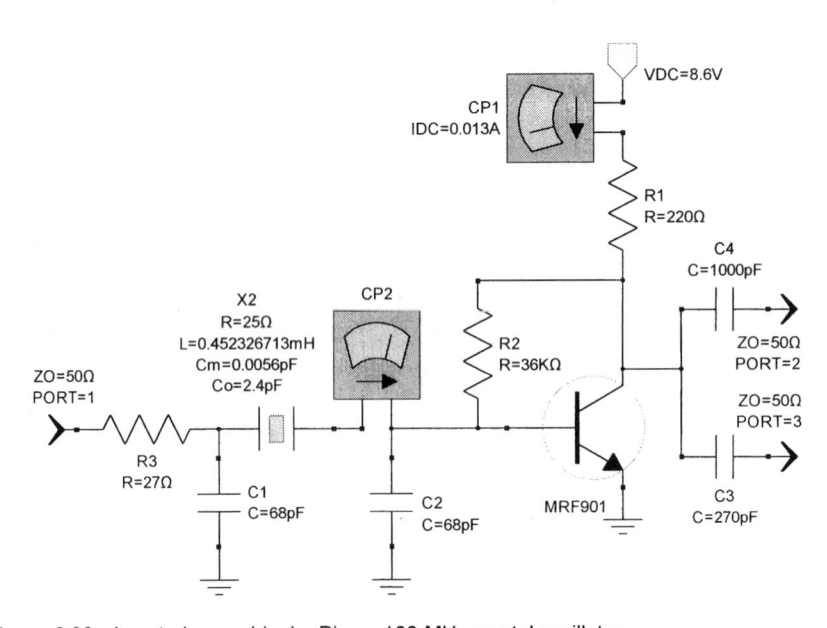

Figure 8.36 Inverted-mesa bipolar Pierce 100-MHz crystal oscillator.

The loaded Q of this example is 4250, 37% of the unloaded Q. The crystal dissipation is 531 µW. The simulated output waveform and spectrum are shown in Fig. 8.37. The output waveform is similar to the previous bipolar Pierce oscillators with unsymmetrical clipping on the positive peak.

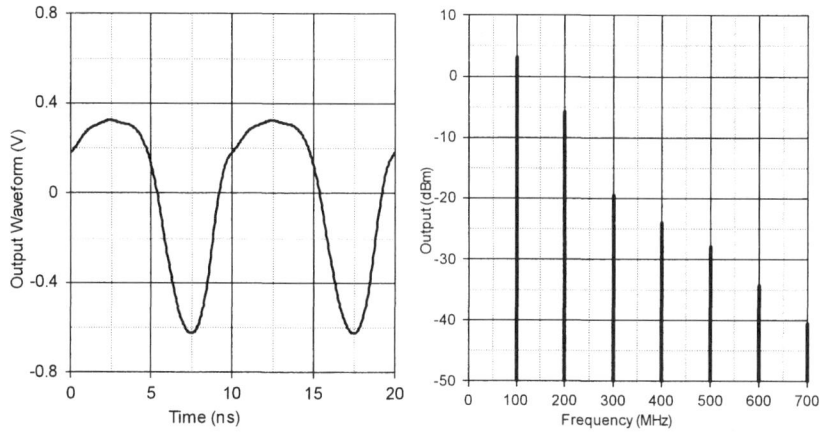

Figure 8.37 Simulated output waveform (left) and spectrum (right) for the Pierce inverted-mesa crystal oscillator.

8.3 Overtone Mode Crystal Oscillators

The following examples illustrate overtone mode crystal oscillators. They utilize an HC-49/U, fifth overtone, 100-MHz resonator. The measured static capacitance is 4.55 pF, the insertion loss at series resonance is 4.3 dB, the series-resonant frequency is 100.005 MHz, and the parallel resonant frequency is 100.011 MHz. From this data the calculated motional parameters are C_m=0.546 fF, L_m=4.6387844 mH, and R_m=64 ohms. The crystal unloaded Q is 45,500. The overtone mode oscillator must not only satisfy the oscillation criteria at the desired overtone frequency, it must also ensure those criteria are not satisfied at the fundamental and other overtone modes.

To achieve adequate gain at the higher frequency of overtone operation, a 4-GHz F_t MRF901, high-gain, low-noise bipolar transistor is used. Simulation uses a Motorola SPICE model.

8.3.1 Colpitts Overtone Bipolar Crystal

Fig. 8.38 shows a Colpitts bipolar fifth overtone 100-MHz oscillator. It operates much like a Colpitts fundamental mode crystal oscillator except that below the desired overtone the parallel inductor L_1 and capacitor C_2 look inductive and ensure the cascade phase is not 0° at other overtones and the fundamental frequency.

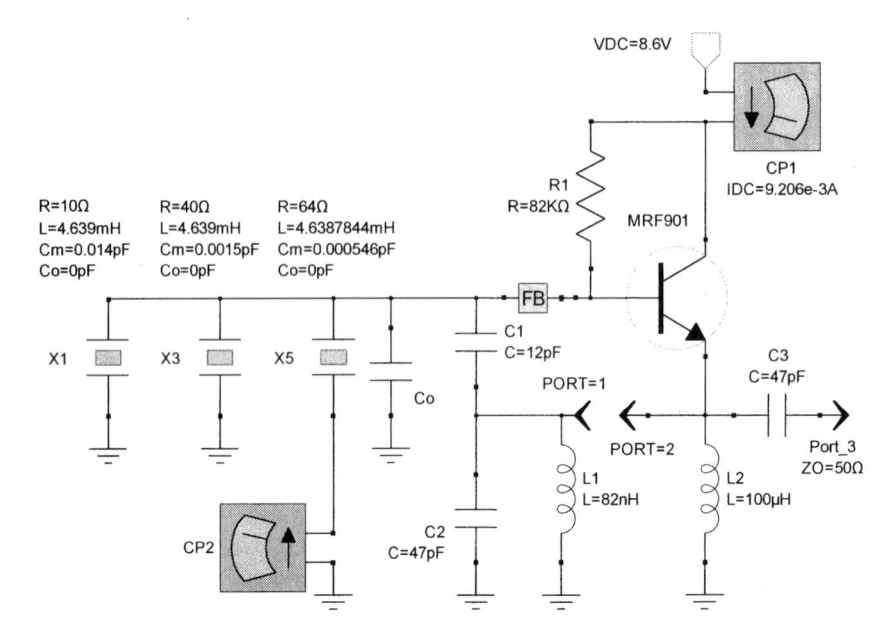

Figure 8.38 Colpitts fifth overtone bipolar 100-MHz oscillator.

Fig. 8.39 shows the fifth overtone mode responses of this oscillator. The ϕ_o occurs below the maximum phase slope, thus reducing the effective loaded Q and the gain margin. This would appear to be a potentially successful oscillator design.

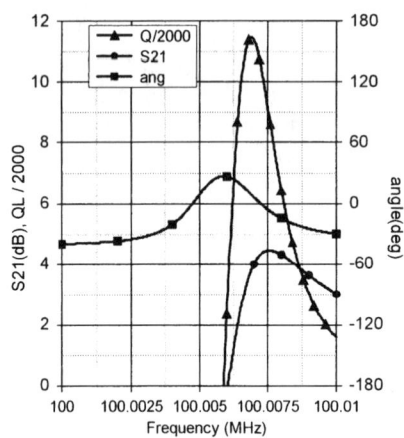

Figure 8.39 Simulated Colpitts fifth overtone open-loop cascade gain (round symbols), transmission phase (square symbols), and loaded Q (triangular symbols).

However, consider the wideband sweep of the open-loop cascade in Fig. 8.40. Notice in Fig. 8.38 that crystal models are included for the fundamental and third overtone modes. The static capacitance is included as a separate lumped element so that it is included only once for all crystal modes. The additional mode models are not required to simulate the fifth overtone oscillator, but they simulate other overtone modes. The fundamental at 20 MHz, the third overtone at 60 MHz, and the desired fifth overtone resonance at 100 MHz are clearly observed in the gain and phase responses. Expanding the frequency sweep around each resonance confirms that the oscillation criteria are satisfied only at the desired fifth overtone. However, notice that a broad phase-zero crossing not associated with a crystal resonance occurs at approximately 75 MHz and the gain margin is 1.7 dB at this frequency, representing a potential undesirable oscillation mode. Off resonance, the motional branch of the model is high impedance and the crystal appears as a simple capacitance equal to C_o. Then, C_1, L_1, and C_2 form a resonator producing this undesired response. This example illustrates how to model additional overtone modes and the importance of wideband sweeps to ensure an unambiguous design. Because of the low gain margin, the reduced loaded Q and this potential spurious oscillation mode, the Colpitts overtone mode oscillator is not a preferred topology. Nevertheless, the simulated output waveform and spectrum for this example are shown in Fig. 8.41.

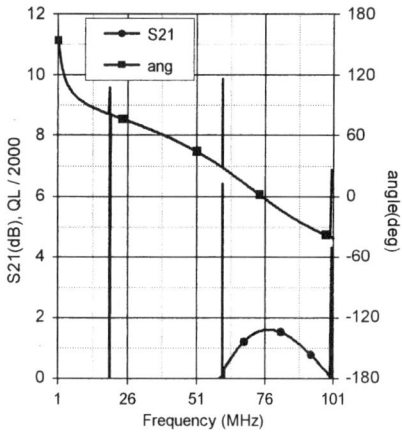

Figure 8.40 Simulated wideband frequency sweep of the Colpitts fifth overtone crystal cascade open-loop response.

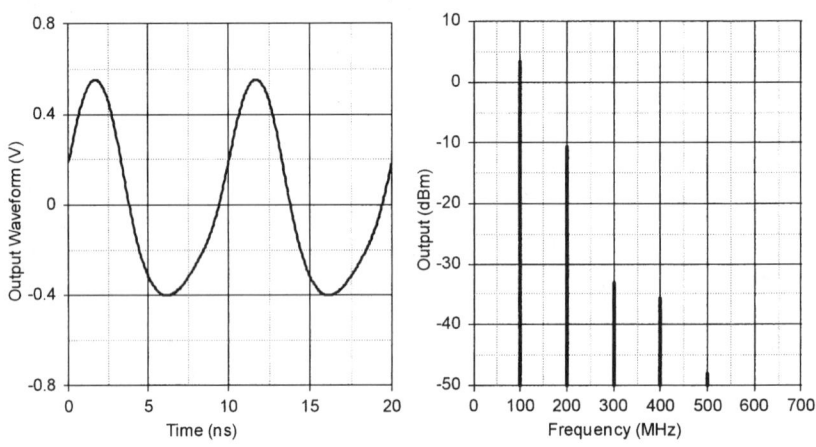

Figure 8.41 Simulated output waveform (left) and spectrum (right) of the Colpitts bipolar 100-MHz fifth overtone crystal oscillator.

8.3.2 CB Butler Overtone Bipolar Crystal

Fig. 8.42 shows a Butler common-base (CB) bipolar fifth overtone 100-MHz crystal oscillator. As with an L-C CB Colpitts, the amplifier is noninverting and the tank consisting of L_1, C_1, and C_2 transforms the collector impedance down to drive the low-impedance emitter. In this case, series resonance of the motional branch of the crystal closes the loop and provides the majority of the selectivity and phase slope. Inductor L_1 trims the phase shift so that oscillation occurs at exact crystal series resonance. The tank serves the additional function of preventing positive gain margin at undesired overtones. Inductor L_2 provides an important function: off series resonance the static capacitance of the crystal also closes the loop and oscillation can occur at a frequency determined by the tank rather than by the crystal. The value of L_2 is selected to parallel resonate with the static capacitance of the crystal. While the value of L_2 is not critical and it does not need to be tuned in production, the value of the static capacitance must be known to find the approximate value of L_2. Again, the static capacitance C_o is separated from the model so current probe CP_2 measures only the crystal current in the motional branch.

Because the emitter is capacitively loaded, this circuit is similar to a negative-resistance oscillator and beads or resistance in series with the base and collector are required to ensure stability at UHF frequencies. Capacitor C_6 is not required when the loop is closed to form the oscillator. The open-loop cascade gain and phase responses are clean, text-book examples of oscillator design with a convenient reference impedance of 50 ohms. The cascade match is good and the gain margin is approximately 4.6 dB. The

Piezoelectric Oscillators 365

loaded Q of this example is 22,000, 48% of the crystal unloaded Q. The crystal dissipation is 319 μW.

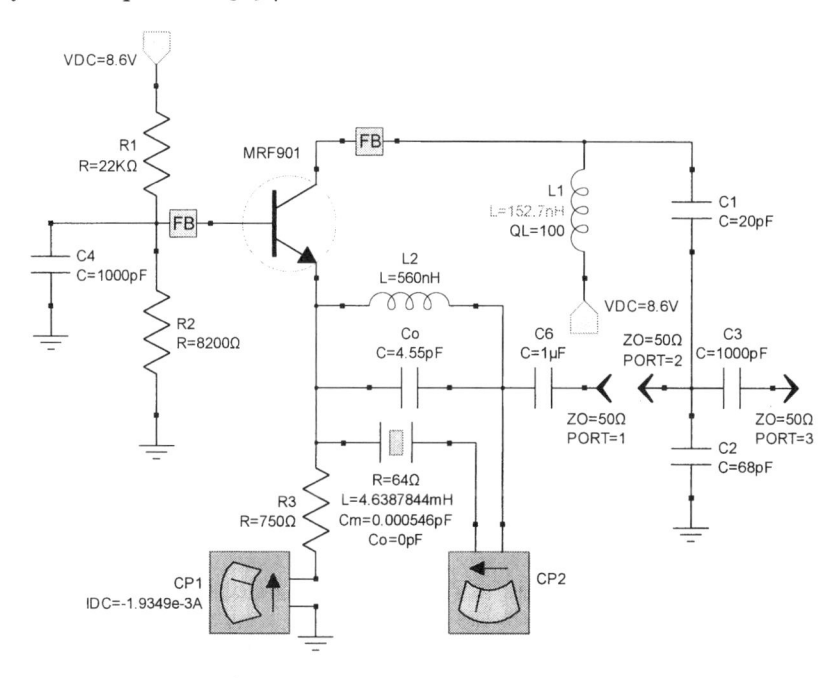

Figure 8.42 Butler CB bipolar fifth overtone crystal oscillator.

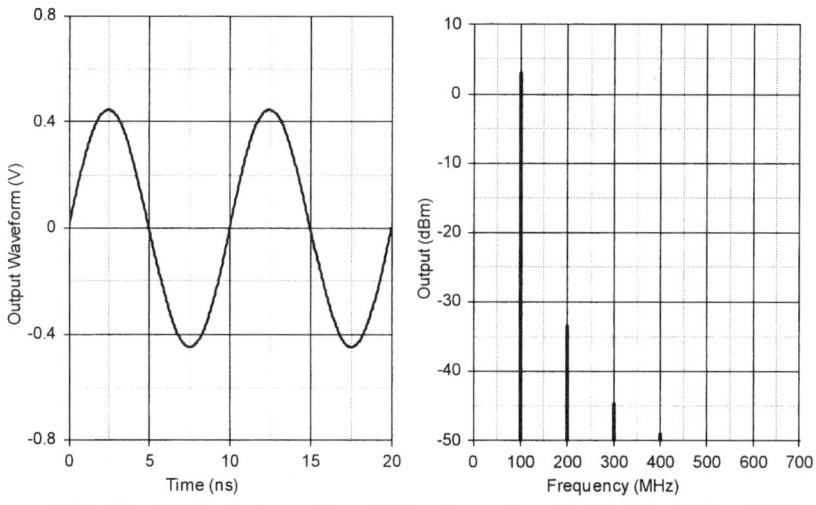

Figure 8.43 Simulated output waveform (left) and spectrum (right) of the Butler CB bipolar fifth overtone crystal oscillator.

Output is taken at the capacitive tap through DC blocking capacitor, C_3. The waveform is nearly sinusoidal and the harmonic performance is excellent as illustrated in Fig. 8.43.

8.3.3 CC Butler Overtone Bipolar Crystal

Fig. 8.44 shows a Butler common-collector (CC) bipolar fifth overtone 100-MHz crystal oscillator. Operation is similar to the Butler CB overtone oscillator except the tank steps the low-impedance emitter follower output up to the high impedance base. Capacitor C_5 is not required when the oscillator is formed by closing the loop. As with the CB Butler overtone oscillator, a ferrite bead or series resistor is required in the base to insure stability. If a high Q inductor is used for L_1, insuring stability may require a high value of resistance in parallel with the inductor.

As with the Butler CB overtone oscillator, the open-loop cascade gain, phase response, and match are good with a 50-ohm reference impedance. The gain margin is approximately 7 dB.

This example uses a negative supply that works well with the NPN CC topology. If a negative supply is not available, a PNP transistor or a different bias network is used. The loaded Q is 22,000, 48% of the crystal unloaded Q. The crystal dissipation is 200 µW.

Output is taken at the emitter through a DC blocking capacitor. The waveform is severely clipped on the negative peak, resulting in poor harmonic performance as illustrated in Fig. 8.45. The ripple in the simulated waveform is the result of an insufficient quantity of balanced harmonics. The harmonic performance of this oscillator is improved by taking power at the capacitive tap but this may result in excessive crystal dissipation.

8.4 Crystal Oscillator Examples Summary

Table 8.4 summarizes the simulated performance characteristics for the example quartz-crystal oscillators. The second column labeled *Offset(Hz)* is the operating frequency offset from the crystal series resonant frequency. The larger offset of the Miller and Colpitts fundamental-mode oscillators is expected as these types are typically classified as parallel-mode oscillators that operate between the crystal series and parallel-mode frequency. The value of the offset is a function of the lumped loading capacitance and the characteristics of the transistor. The Pierce type is often referred to as operating in series-resonant mode but they also have shunt impedance and phase transforming capacitors that function like high-valued loading capacitors. The Butler fundamental mode oscillator operates at the crystal series-resonant frequency if the devices are ideal. The Butler overtone mode oscillators operate at series resonance because the tank inductors are adjusted to make this happen.

Piezoelectric Oscillators

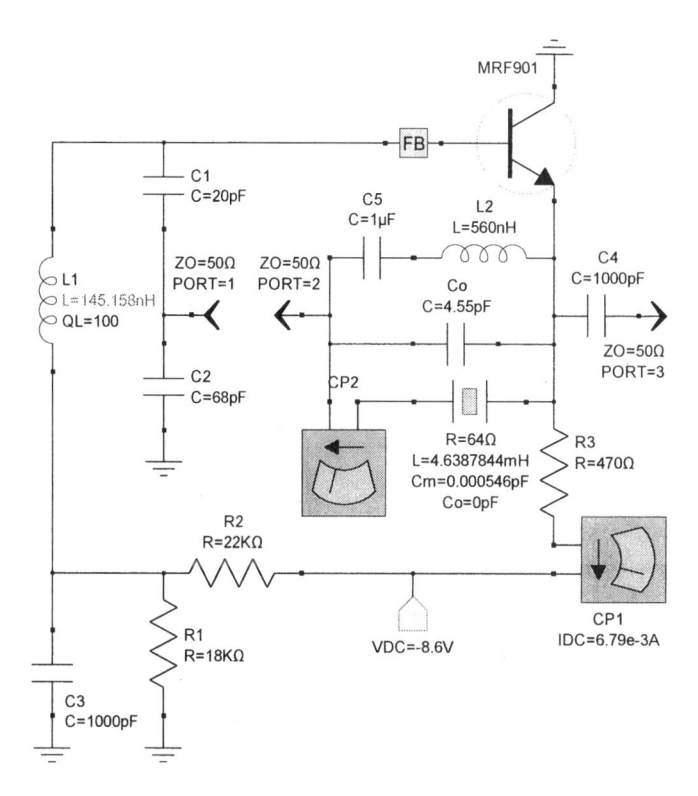

Figure 8.44 Butler CC bipolar fifth overtone crystal oscillator.

The column labeled *Pushing(Hz/v)* is the frequency shift of the oscillator when the operating voltage is increased from 8.6 to 9.6 volts. The pushing performance may be better or worse for a smaller change in the operating voltage. The frequency accuracy of the simulation is roughly 1 Hz, so a rather large voltage change was used to improve the resolution of the measurement. The pushing performance of all of the fundamental mode oscillators is good. The bipolar designs have better pushing performance, essentially better than the simulation resolution. The worst performer in this regard was the Miller FET oscillator, which is not a recommend design in any event.

The column labeled *Circuit(Hz)* is the frequency shift caused by increasing by 5% all loading reactor values. This frequency-shift specification is an important measure of good aging and temperature stability performance. A 5% detuning of loading reactors may lead to oscillation failure as with the fundamental Miller FET and the Butler CB overtone bipolar examples. This does not necessarily signify that the design is unsuitable for any application but it probably does indicate that production tuning is required. It should also be noted that in comparing designs, the absolute shift should be normalized by the operating frequency. The operating frequency of the overtone examples is greater than 10X that of the fundamental oscillators.

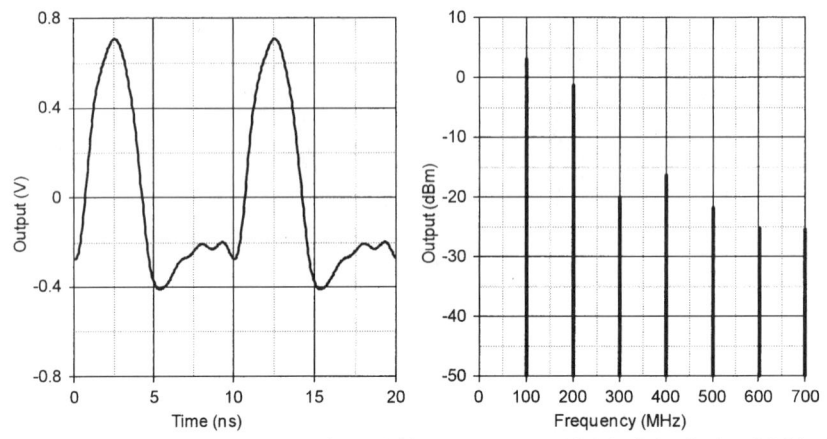

Figure 8.45 Simulated output waveform (left) and spectrum (right) of the Butler CC bipolar fifth overtone crystal oscillator.

Table 8.4 Simulated performance summary of the crystal oscillators

Type	Offset (Hz)	Pushing (Hz/v)	Circuit (Hz)	Q_{LOADED}	Drive (μW)	η (%)
Miller FET	+2869	+61	FAIL	2130	240	21.2
Colpitts Bip	+3282	0	-99	14200	775	2.4
Colpitts FET	+3597	+14	-111	42000	123	3.3
Pierce Bip	+209	0	+1	52100	175	2.3
Pierce MMIC	+249	0	-12	39100	866	1.2
Pierce Inverter TTL	+120	0	-8	36500	284	2.7
Pierce Inverter CMOS	+1457	0	-69	54600	194	12.6
Butler Bip (Dual)	+170	0	0	28800	543	1.2
Driscoll BIP	0	-1	+20	48200	60	2.5
Mesa Pierce Bip	+7450	-56	-373	4250	531	1.9
Colpitts OT Bip	+3010	+30	-340	~13000	326	2.8
Butler OT CB Bip	0	+66	FAIL	22000	319	9.9
Butler OT CC Bip	0	-107	-3000	22000	200	3.0

The column labeled Q_{LOADED} is one of the more important performance characteristics. It is critical for phase-noise performance. The fundamental Pierce and overtone Butler oscillators offer the best performance in this regard.

The column labeled $Drive(\mu W)$ is the crystal dissipation with the bias levels adjusted for roughly 3-dBm oscillator output. The fundamental Colpitts and Pierce MMIC oscillators have the highest crystal dissipation of these examples. Increasing the quiescent bias current, and low resistance in the collector-emitter or drain-source path, increases the output power.

Increasing the output power improves short-term stability (phase-noise) but it increases crystal dissipation, which worsens long-term stability. Therefore, one figure of merit for oscillators is the ratio of the output power to the crystal dissipation. Because each of these example oscillators was adjusted for 3-dBm output, the types with the lowest crystal dissipation are preferred. In this regard, the fundamental Colpitts FET, Pierce bipolar and Pierce CMOS inverter, and the Butler CC overtone bipolar are preferred. The fundamental Driscoll oscillator looks attractive but the power level of the CE oscillator stage is much lower than 3 dBm. The output limiting stage is acting as a buffer that amplifies both the intrinsic signal and the noise.

The column labeled $\eta(\%)$ is the DC power required by the supply divided by the output power delivered by the oscillator. This is an important characteristic for battery-operated devices. However, if efficiency is a critical characteristic, other topologies should be explored. Of these examples, the FET oscillators tend to have the best efficiency.

The performance of these oscillators is a result of the specific design choices made for each oscillator. For example, power could be taken at different nodes, different bias schemes could be used, or resistors in the collector-emitter or drain-source path could be replaced with inductors for higher output power. In addition, the frequency of operation also influences the best choice for a given application. Table 8.4 should not be interpreted literally but rather it is used as a guideline for selecting a general oscillator type. That having been said, good choices are the Colpitts FET for low part count and simplicity, the Butler bipolar for frequency independent applications, the Pierce bipolar and Pierce CMOS inverter for performance, and the Butler CC bipolar for overtone oscillators.

8.5 Oscillator with Frequency Multiplier

Fig. 8.46 shows a Butler oscillator with a built-in X5 frequency multiplier. The oscillator operates the crystal in fundamental mode at 20 MHz with the fifth harmonic extracted at the collector using a parallel resonator to filter undesired harmonics. The Butler overtone oscillator may be used with the crystal operated in either the fundamental or an overtone mode. In this case, for best pullability, the crystal is operated in fundamental mode and fifth harmonic is at 100 MHz. In this way, the pullability of a fundamental mode oscillator is realized with an output frequency of a fifth overtone, 100-MHz oscillator. A series resistor, R_4, rather than a ferrite bead is used to stabilize the device. Inductor L_2 cancels the static capacitance at the fundamental frequency. Capacitor C_6 is not required in the final oscillator.

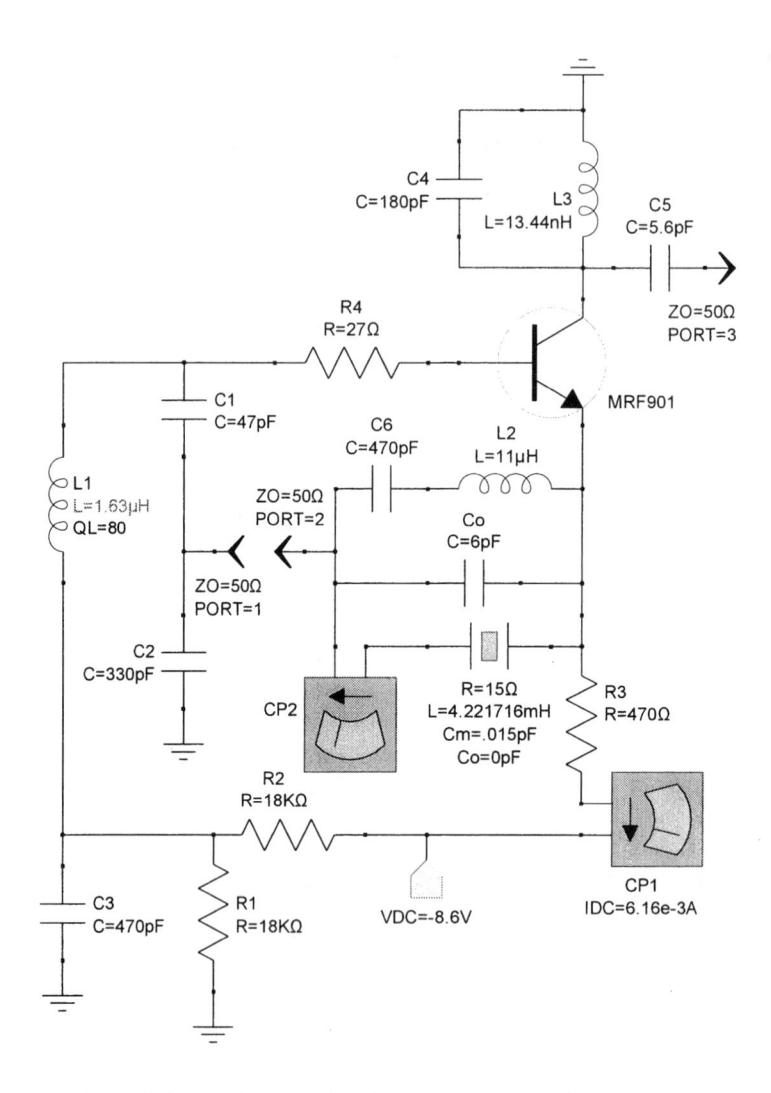

Figure 8.46 Butler CC oscillator with built-in X5 frequency multiplier.

CC oscillator circuits have an impulsive current in the collector, resulting in high harmonic content. A tuned circuit consisting of L_3 and C_4 resonate near 100 MHz and filter the remaining harmonics of 20 MHz. Capacitor C_5 raises the effective impedance of the 50-ohm load presented to the resonator and increases the resonator loaded Q and therefore its filtering effectiveness. A device with high F_t and that has good gain at the peak of the impulsive current improves the output power.

Shown on the right in Fig. 8.47 are the output waveform (solid trace) and the crystal current waveform (dashed trace). The output filter is reasonably effective and the waveform is nearly sinusoidal. The crystal current is a 20-MHz sinusoid with a peak value of 12.5 mA. The crystal dissipation is therefore 1.17 mW. The output spectrum on the right in Fig. 8.47 reveals that the output power is -6.4 dBm and the harmonics at 80 and 120 MHZ are -27 and -19 dBc, respectively. This oscillator circuit provides the pullability of a fundamental mode crystal at higher frequency. An alternative that avoids fundamental mode frequency components in the spectrum is the use of inverted-mesa crystals as described in Section 8.2.10.

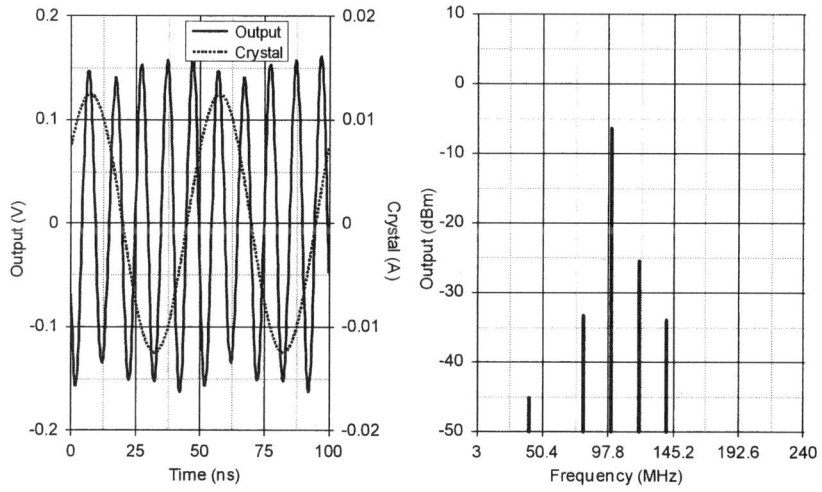

Figure 8.47 Simulated output waveform of the oscillator with multiplier (left, solid trace), crystal current waveform (left, dashed trace), and output spectrum (right).

8.6 Crystal Oscillator Starting

Crystal oscillators with high loaded Q start slowly. State-of-the-art crystal oscillators with loaded Qs over 100,000 may require a second or more to start. This requires special simulation techniques as discussed in Section 3.8. When temperature stabilization by oven is used, the final operating frequency is not reached for a substantially longer time.

8.7 Surface Acoustic Wave Resonators

The surface acoustic wave (SAW) resonator utilizes an acoustic wave traveling on the surface of an elastic substrate of piezoelectric material such as quartz, lithium nibate, lithium tantalite, lanthanum gallium silicate, and others. Quartz is often used for high-Q resonators and narrow bandwidth filters where temperature stability is important. Lithium nibate and lithium tantalite are often used for wide bandwidth filters because of their high electromechanical coupling coefficient. For oscillator resonators, Y-cut quartz is often used. The amplitude of the wave decays exponentially with depth. Photolithographic techniques are used to place conducting interdigital transducer (IDT) structures on the substrate surface. Various geometries implement SAW resonators and filters and they are available in 1-port and 2-port forms as illustrated in Fig. 8.48.

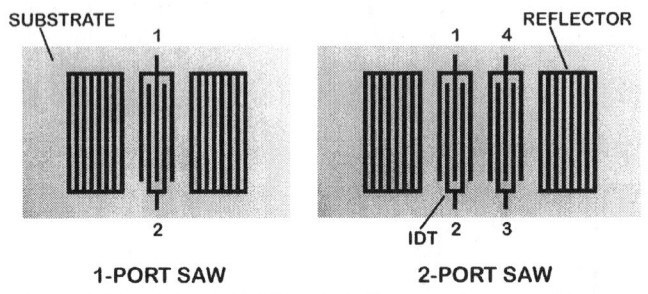

Figure 8.48 1-port (left) and 2-port (right) basic SAW resonator geometries.

A signal applied to the IDT produces strain on the substrate surface and acoustic waves propagate from the IDT. When the signal frequency is equal to V_s/p, where V_s is the surface propagation velocity and p is the interdigital period, the waves generated by each finger are in phase. This represents resonance. Reflectors help confine energy to the substrate.

The practical lower frequency range is defined by the maximum substrate size and the upper frequency limit is defined by the fine-pattern resolution. SAW resonators are useable in the frequency range of 10 to 2500 MHz with common application in the 300 to 1000 MHz range. SAW resonators have somewhat lower unloaded Q and looser stability specifications than bulk quartz crystals. SAW resonator loaded Q and stability are superior to L-C resonators and SAW resonators offer higher frequency operation than conventional bulk quartz crystals. SAW resonators safely dissipate higher power than bulk quartz crystals. Typical maximum dissipation level specifications for inexpensive SAW resonators are 0 or 10 dBm but resonators with higher maximum dissipation are practical.

8.7.1 SAW Resonator Models

Fig. 8.49 shows models for 1-port (top), 0° 2-port (middle), and 180° 2-port SAW resonators. The 1-port model is similar to the bulk quartz crystal resonator with added capacitors Cs for parasitic capacitance to ground for the IDT and terminal connections. 2-port SAW resonators are available in two forms with either 0° or 180° at resonance, depending on which IDT fingers are grounded and which are connected to the device ports.

Typical values for the model parameters for RF Monolithics SAW resonators [17] are given in Table 8.5. The motional resistance for these units is significantly less for the 1-port resonators but the motional inductance is correspondingly lower, therefore the unloaded Q are similar. For a given resonator design, the unloaded Q tends to decrease with increasing frequency. These resonators have an unloaded Q of 12,000 at 300 MHz dropping to 6,000 at 1000 MHz, with the exception RO2071-2 model that has an unloaded Q of 18,000 at 980 MHz.

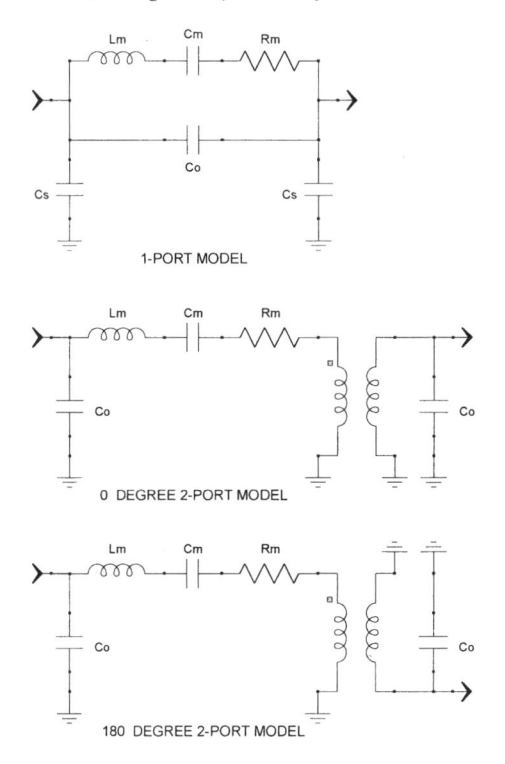

Figure 8.49 SAW resonator 1-port model (top), 0^0 2-port model (middle), and 180^0 2-port model (bottom).

8.7.2 SAW Resonator Frequency Stability

Quartz SAW resonator accuracy and temperature stability are approximately an order of magnitude worse that bulk quartz resonators. Nevertheless, SAW resonator accuracy and unloaded Q are 2 orders of magnitude better than L-C resonators. The initial frequency accuracy at 25°C for inexpensive quartz SAW resonators is typically in the range of

±200 ppm. Series reactance may be used to set the initial frequency. A typical specified aging rate is <10 ppm/yr. A remaining factor in frequency accuracy is operating temperature. Fig. 8.50 plots the frequency change versus the operating temperature relative to the turnover temperature T_o. The frequency versus temperature is given by

$$f = f_0 \left[1 - FTC(T_0 - T_c)^2 \right] \qquad (201)$$

where the frequency temperature coefficient, FTC, is 0.037 for the SAW resonators given in Table 8.5. However, there is tolerance in the turnover temperature. It can nominally be set ±10°C but is often specified as ±15°C to avoid testing and to reduce expense. A typical turnover temperature target is 25°C but SAWs are available for higher turnover temperatures for use with temperature-controlled ovens.

Table 8.5 Typical model parameters for RF Monolithics quartz SAW resonators

Model	Phase	Freq (MHz)	Rm (ohms)	Lm (μH)	Cm (fF)	Co	Cs (pF)
One-Port							
RO2073	0°	315.00	19	127.667	1.99943	2.6	0.5
RO2101	0°	433.92	18	86.0075	1.56417	2.0	0.5
RO3164	0°	868.35	11	11.0	3	1.9	0.5
RO2071-2	0°	980.00	12	35.0	0.76	2.5	0.5
Two-Port							
RP1239	180°	315.00	84	758.027	0.336771	2.2	
RP1308	180°	433.92	107	481.378	0.279470	1.7	
RP1104	180°	824.25	182	248.091	0.150284	1.5	
RP1094	180°	915.00	166	191.3434	0.158119	1.4	

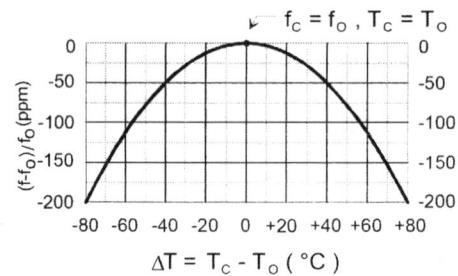

Figure 8.50 Typical frequency versus temperature curve for a SAW resonator (courtesy of RF Monolithics).

8.8 SAW Oscillators

As with the bulk quartz crystal oscillator examples, the following schematics are given with the loop opened to illustrate break points. The design operating voltage for each is 8.6 volts. The current measured by probe CP_1 is the open-loop DC quiescent current before the onset of oscillation. If the bias is not disturbed when the loop is closed, certain blocking capacitors may be eliminated when the oscillator is formed by closing the loop. Again, the chosen open-loop termination impedances are 50 ohms. For cases where a poor match results for any reference impedance the Randall/Hock correction is used.

Two 1-port SAW and one 2-port SAW oscillator examples are given. All use 315-MHz devices with model parameters from RF Monolithics datasheets [17]. The 1-port SAW model motional resistance is 20 ohms, the motional capacitance is 2 fF, and the static capacitance is 2.6 pF. The motional inductance is set to resonance at the desired series operating frequency. The unloaded Q is 12,600. The 2-port 180° SAW model motional resistance is 84 ohms, the motional capacitance is 0.34 fF, and the static shunt capacitance is 2.2 pF. The motional inductance is set to resonance at the desired series operating frequency. The unloaded Q is 17,700.

The parallel resonant frequency of these SAWs is only 121 kHz above the series resonant frequency. The proximity of the parallel resonant phase reversal to the desired series resonant frequency results in a reduction of the phase slope thus reducing the effective loaded Q. The use of an inductor L_o to cancel the static capacitance is often seen in bulk crystal oscillator circuits but it is less common in SAW oscillators. Therefore, L_o is not used in the following examples. However, including L_o would increase the loaded Q of these circuits.

A 4-GHz F_t MRF901, high-gain, low-noise type bipolar transistor is used. Simulation uses a Motorola SPICE model.

8.8.1 SAW 1-Port Colpitts Bipolar

Fig. 8.51 is a Colpitts bipolar 1-port SAW oscillator that uses a CB amplifier topology with low-input impedance and high-output impedance. Various forms of this oscillator topology are given in SAW manufacturer application notes and this simple circuit has become popular in keyless entry systems.

The capacitive tap consisting of C_1 and C_2 steps the high collector impedance down to match the low emitter impedance. At SAW resonance the base is grounded through the motional resistance. Off resonance the SAW impedance is high, thus reducing the loop gain. The ferrite bead stabilizes the CB amplifier configuration. The open-loop cascade match is rather poor and the Randall/Hock correction is advised. The loaded Q of this circuit is approximately 6,000 but it is heavily dependant on the tuning of the collector tank circuit.

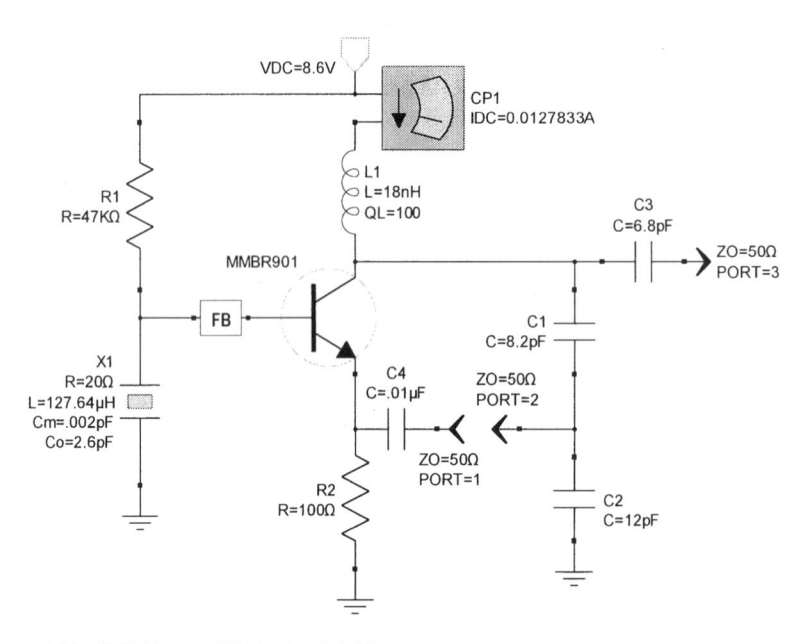

Figure 8.51 SAW 1-port CB bipolar Colpitts.

Power is extracted through a coupling capacitor at the collector. An antenna may be connected at this node, or inductor L_1 can be a loop antenna. Power may also be extracted at the common node of C_1 and C_2. The simulated output spectrum is shown in Fig. 8.52. The output power is approximately 7 dBm. Because power is extracted from the resonant collector tank the harmonic performance is good. The SAW dissipation is 930 μW.

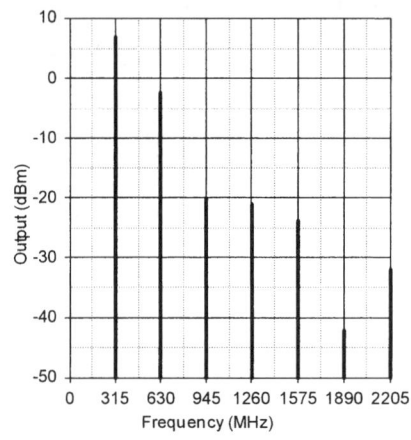

Figure 8.52 Simulated output spectrum of the 315-MHz SAW 1-port CB bipolar Colpitts.

Piezoelectric Oscillators

8.8.2 SAW 1-Port Butler Bipolar

Fig. 8.53 shows a SAW 1-port Butler CB bipolar oscillator. As with the previous SAW Colpitts oscillator, the tank steps the high impedance at the collector down to the low-impedance emitter input. The loop is closed via series resonance of the SAW. In this example the frequency is set to 315 MHz and inductor L_1 is optimized, along with the excitation source voltage, by harmonic-balance simulation to minimize the current flowing in the excitation source. Resistor R_3 rather than a ferrite bead stabilizes the amplifier.

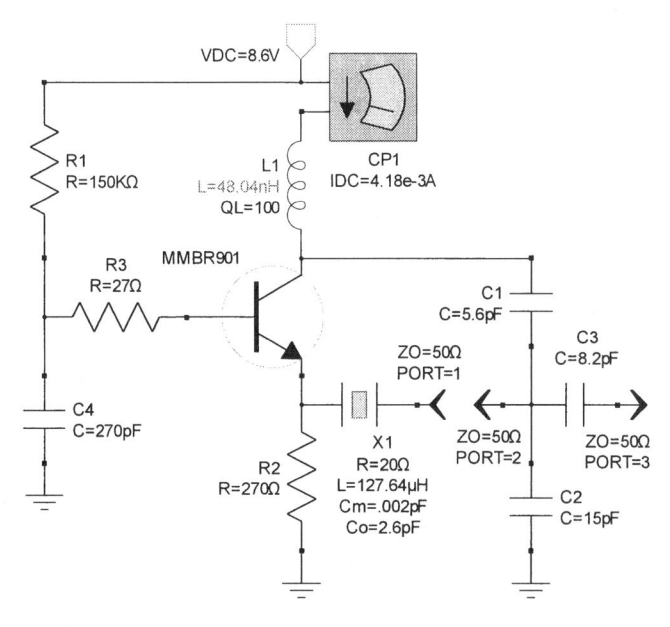

Figure 8.53 SAW 1-port CB bipolar Butler.

Output power is extracted from the common node of C_1 and C_2. The output spectrum is shown in Fig. 8.54. The output power is 10.4 dBm and SAW dissipation is 1.33 mW. The loaded Q of the output tank is higher than the previous example and the harmonic performance is excellent with the second harmonic nearly -30 dBc. However, the cascade loaded Q is only 2460, 19% of the SAW unloaded Q. With a SAW motional resistance of only 20 ohms, very low emitter and capacitive tap resistances are required to achieve good loaded Q. This circuit realizes a higher fraction of the SAW unloaded Q when the SAW motional resistance is higher.

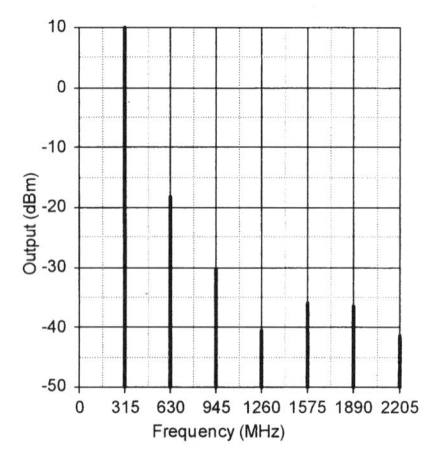

Figure 8.54 Simulated output spectrum of the SAW 1-port CB bipolar Butler.

8.8.3 SAW 2-Port Pierce MMIC

Fig. 8.55 shows a SAW 2-port Pierce MMIC oscillator. Capacitor C_4 is not simply a blocking capacitor but is used when the loop is closed. It has a reactance of 50 ohms at 315 MHz. L_1 and C_4 form a highpass phase-lead network to adjust the zero crossing at the maximum phase slope. The MMIC is an Avago Technologies MSA06 or a Mini-Circuits MAR6 Darlington-pair CE resistive feedback amplifier. An Avago Technologies model is used. Resistor R_1 sets the current in the MMIC at the manufacturer specified DC quiescent value of 16 mA.

The gain margin of this circuit is somewhat high, approximately 10 dB. Optional shunt coupling capacitance in parallel with C_1 and C_2 raises the loaded Q and reduces the gain margin by increasing resonator loss. The loaded Q of this circuit without optional capacitance is 8540, 48% of the SAW unloaded Q. The higher loaded Q of this circuit and 10-dBm output power should result in the best phase-noise performance of the three example SAW oscillators. The SAW dissipation is approximately 2 mW.

Fig. 8.56 shows the simulated output spectrum of the 2-port Pierce SAW oscillator. Output is taken directly at the collector of the MSA06 MMIC through capacitive reactance of approximately 50 ohms. Output power is just over 10 dBm but the harmonic performance is only -6 dBc.

Piezoelectric Oscillators

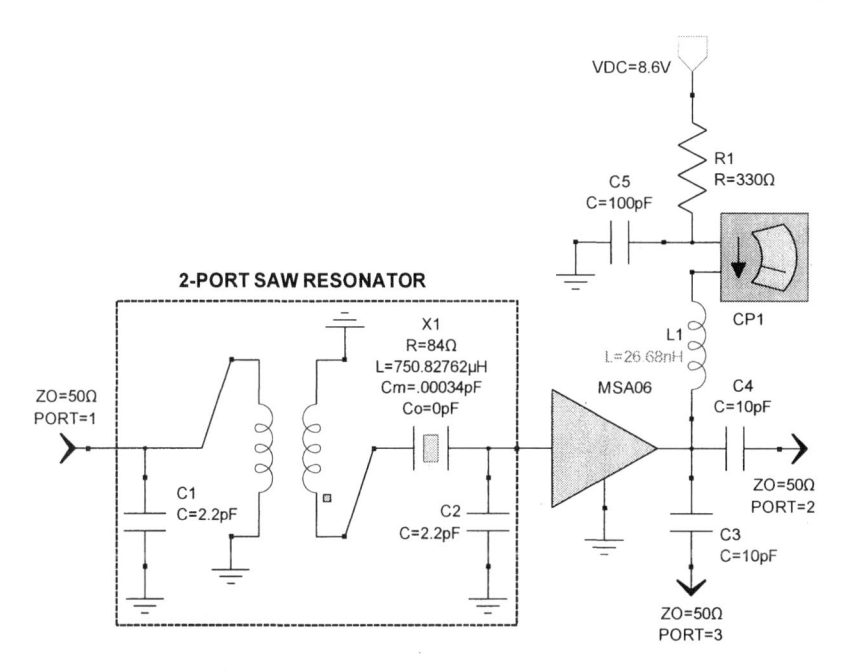

Figure 8.55 SAW 2-port Pierce MMIC oscillator.

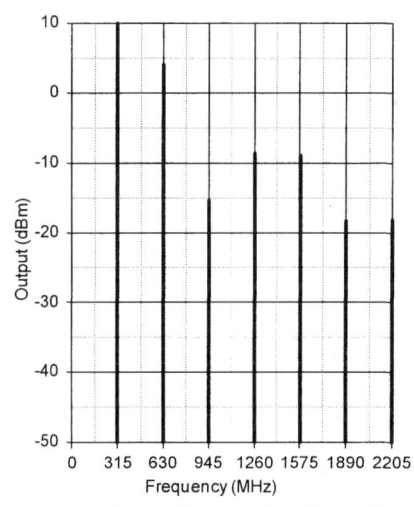

Figure 8.56 Simulated output spectrum of the 315-MHz 2-port Pierce SAW oscillator.

8.9 Piezoelectric Ceramic Resonators

Resonators using ceramic piezoelectric materials bridge the gap between relatively expensive bulk quartz resonators and inexpensive L-C resonators with relatively poor stability and low unloaded Q. Lead zirconium titanate is a common material used for ceramic resonators. Ceramic resonators are typically used in the 0.4- to 60-MHz frequency range. They should not be confused with coaxial transmission-line resonators loaded with nonpiezoelectric ceramic to shorten the physical length required for resonance. These latter devices are referred to as ceramic coaxial resonators or coaxial resonators.

Monolithic, multiple-section ceramic resonators were initially used as 455 kHz and 10.7 MHz IF filters and FM discriminators for consumer broadcast receivers. Single-pole ceramic resonators are now often used to replace quartz bulk crystal resonators in inverter gate and microprocessor oscillators when stability is not as critical. They are less expensive, are more tolerant of shock and vibration, and operate to higher frequency than fundamental mode quartz crystals. Ceramic resonators can be designed to suppress fundamental mode resonance and therefore higher frequency oscillators are feasible without the use of tank circuits. Faster starting is also claimed by ceramic resonator manufacturers but this is simply because the loaded Q is lower than with quartz crystal oscillators.

8.9.1 Ceramic Resonator Models

The equivalent-circuit model for ceramic resonators is identical to bulk quartz resonators (Fig. 8.3), but the parameter values are significantly different. Ceramic resonators are also available with built-in shunt coupling capacitors, which is convenient for Pierce oscillators. Table 8.6 gives model parameters computed using Eqs. 191-193 from measured data for ECS International ZTA series ceramic resonators [18].

Table 8.6 Model parameters for three ECS International ceramic resonators quartz SAW resonators computed from measured data

Model	Freq (MHz)	Rm (ohms)	Co (pF)	Cm (pF)	Lm (µH)	Qu
ZTA-4.00MG	4.00	8.4	25.2	4.58	376.367	1034
ZTA-8.00MT	8.00	5.9	44.5	6.81	63.9111	495
ZTA-12.00MT	12.00	4.1	28.6	4.18	44.9805	774

8.9.2 Ceramic Resonator Accuracy and Stability

Initial frequency accuracy, temperature stability, and aging are orders of magnitude worse than quartz crystals but ceramic resonators are better in those regards than L-C resonators. The normal initial frequency accuracy

Piezoelectric Oscillators 381

at 25°C for inexpensive ceramic resonators is ±0.5%. This is significantly better than the tolerance available for inductors and capacitors. With a cost of less than US $0.10 in volume, and unloaded Q better than L-C resonators, ceramic-resonator oscillators are attractive when the accuracy and stability of quartz resonators is unnecessary. Initial frequency accuracy of ±0.1% is available. Frequency stability with temperature is typically ±0.5% from -20 to 80°C with ±0.3% readily available. Aging specification is typically ±0.3% per 10 years.

Because ceramic resonators are often used in inverter-gate oscillators, such as shown in Fig. 8.57, the resonator manufacturers calibrate the frequency with specific values of shunt capacitors C_1 and C_2. Values typically range from 330 pF at 200 kHz, 30 pF at 4 MHz, and 5 pF at 50 MHz. Some manufacturers specify different loading capacitance for different ICs. Since the resonator motional capacitance is only an order of magnitude less than the loading capacitors, the loading capacitors result in a significant frequency shift. The manufacturer of the ceramic resonator should be consulted when calibration is critical.

8.9.3 Ceramic Resonator Oscillators

The following three example oscillators use an ECS International 4-MHz ceramic resonator with parameters given in Table 8.6. In these examples, because the oscillating frequency is well below the maximum operating frequency of the active devices, the gain margin is rather high, and since power is extracted at the device outputs, the harmonic content is relatively high. Therefore the number of harmonics balanced is increased to 12.

8.9.3.1 Ceramic Resonator Pierce CMOS Inverter

Fig. 8.57 shows a Pierce CMOS inverter oscillator that uses a ceramic resonator. An additional gate is used to buffer the output to a 400-ohm load. Additional information on the use of CMOS inverter gates for oscillators is given in Section 8.2.7. The discussion of crystal oscillator design in that section is applicable to ceramic resonator oscillators as well.

The open-loop port 2 impedance of this circuit is approximately 4000 ohms at resonance. However, the CMOS gate input impedance is much higher so the Randall/Hock correction is advised. Although the manufacturer specified values for loading capacitors C_1 and C_2 were 30 pF, the resulting operating frequency for this ceramic resonator is approximately 50 kHz too high. To shift the frequency down to 4 MHz, the loading capacitors are increased to 50 pF. This has the additional benefit of increasing the loaded Q. The loaded Q of this example is 249, or 24% of the unloaded Q. The loaded Q is increased by increasing the values of C_1 and C_2 but the operating frequency is then below 4 MHz.

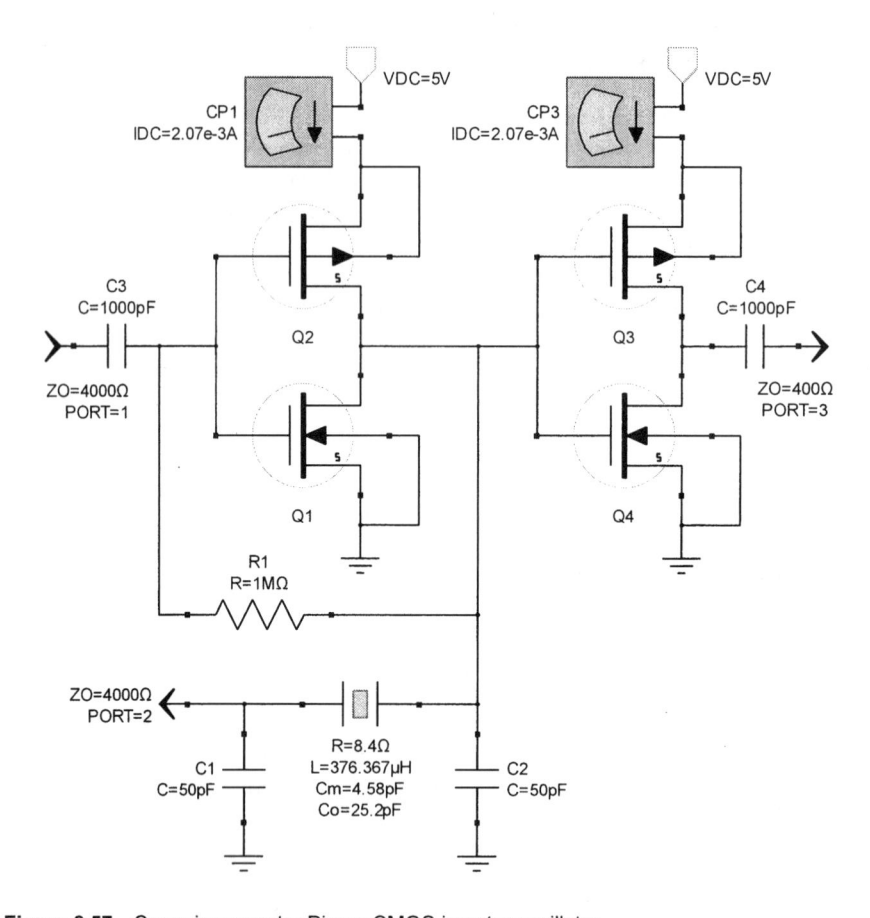

Figure 8.57 Ceramic resonator Pierce CMOS inverter oscillator.

The simulated output waveform and spectrum are shown in Fig. 8.58. The waveform is approximately square-wave, resulting in reduced even-order harmonics. The output power delivered to a 400-ohm load by the buffer is approximately 4 dBm. Ceramic resonators with C_1 and C_2 built-in are available.

The simplicity of the circuit, compatibility with common ICs, improved performance over L-C circuits and the low cost of ceramic resonators contribute to the popularity of this type of oscillator.

8.9.3.2 Ceramic Colpitts FET

Fig. 8.59 shows a ceramic resonator Colpitts FET oscillator. Resistor R_2 stabilizes the device operating current below I_{dss} of the U310 N-channel

JFET. The drain could be directly RF grounded but resistor R_3 supports extracting power at the drain for increased output power.

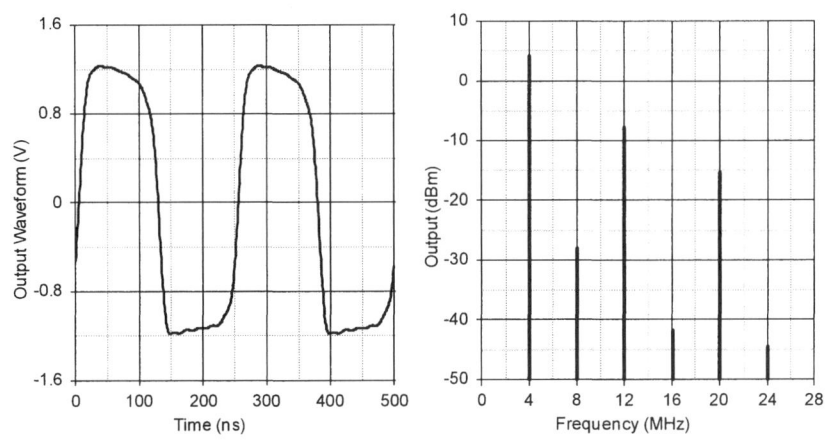

Figure 8.58 Simulated output waveform (left) and output spectrum (right) of the 4-MHz ceramic resonator Pierce CMOS inverter oscillator.

This circuit offers significantly improved loaded Q over the previous CMOS inverter gate oscillator. The high gate impedance and large tap ratio of capacitors C_1 and C_2 provide only light loading of the ceramic resonator. The loaded Q of this circuit is approximately 700, or 68% of the resonator unloaded Q. The larger value of loading capacitance reduces the operating frequency below 4 MHz, in this case approximately 3.878 MHz. The cascade input and output match to 50 ohms is good, facilitating confirmation of the design using a vector network analyzer.

Fig. 8.60 shows the simulated output waveform and spectrum. This circuit offers high output power but rather poor harmonic performance. The DC to RF conversion efficiency is better than many oscillators, in this case 6.6%.

8.9.3.3 Ceramic Butler Bipolar

Fig. 8.61 shows a 4-MHz bipolar Butler ceramic resonator oscillator. The left transistor is operated CB and the output transistor is CC. The impedance level is low and 20-ohm terminations provide an excellent match, and the Randall/Hock correction is not required. This oscillator does not use reactive-coupling elements so it is broadband, and therefore useful as a general-purpose ceramic resonator tester. Capacitor C_2 is not required when the loop is closed. The frequency of operation is 3.834 MHz, 166 kHz below the specified oscillation frequency because loading capacitance is not used.

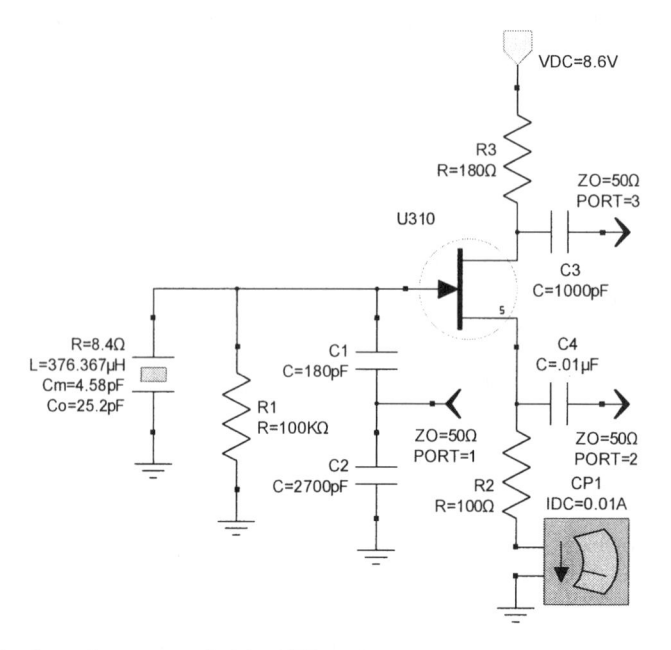

Figure 8.59 Ceramic resonator Colpitts FET oscillator.

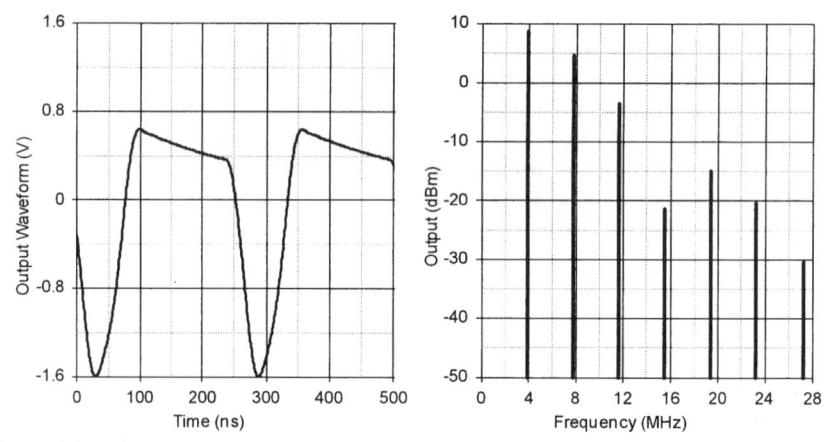

Figure 8.60 Output waveform (left) and spectrum (right) for the Colpitts ceramic resonator oscillator.

Fig. 8.62 shows the simulated (left) and measured (right) output waveforms for the ceramic Butler oscillator. The output level is -0.4 dBm into a 50-ohm load. The second harmonic is -12.4 dBc and the third harmonic is -20.4 dBc.

Piezoelectric Oscillators 385

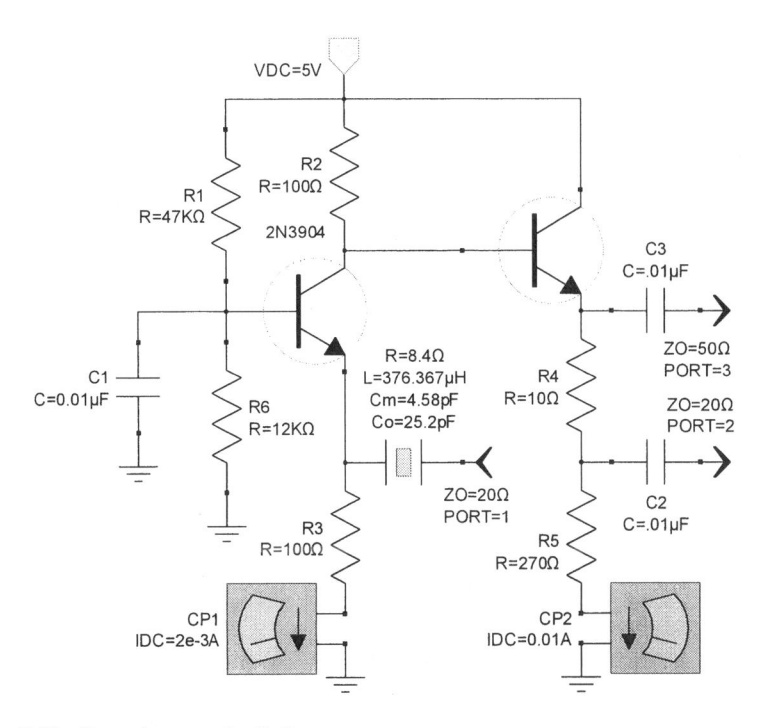

Figure 8.61 Ceramic resonator Butler.

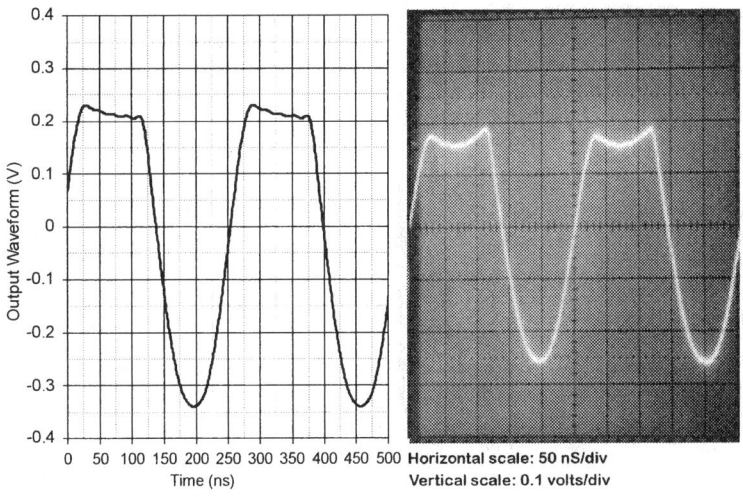

Figure 8.62 Simulated (left) and measured (right) output waveform of the ceramic resonator Butler.

8.10 MEMS and FBAR Resonators

Historically, the limited availability of on-chip IC inductors resulted in the use of off-chip inductors or ring oscillators with relatively poor phase-noise performance. Now, microelectromechanical systems (MEMS) technology offers a wide range of solutions.

In its simplest form, it refers to the micromachining of essentially lumped elements. Inductors, once confined to surface patterns in IC and MMIC, are constructed as 3-dimensional structures using MEMS with a resultant increase in unloaded Q. Better utilization of MEMS technology includes vibrational beams and disks, yielding unloaded Q greater than 10,000 in the VHF and UHF frequency range [19]. Figure 8.63 shows a MEMS wine-glass mode vibrational resonator.

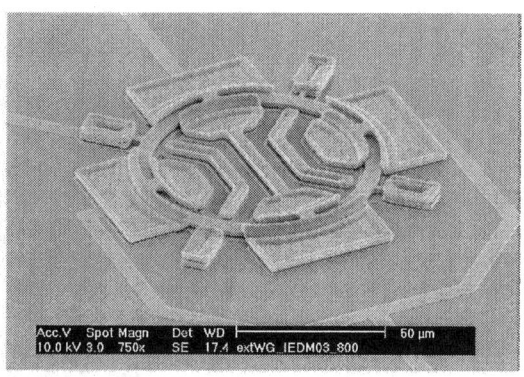

Figure 8.63 Wine-glass mode ring resonator (courtesy Discera).

Because advanced processing and micromachining are required, film bulk-wave acoustic resonators (FBARs) are classified as MEMS devices. FBARs are similar to SAWs in that electrodes and piezoelectric materials are used to generate acoustic waves, but energy propagates in FBARs as longitudinal waves in bulk material rather than surface waves. A single-crystal material such as aluminum nitride (AlN) is typically used. SAW devices predated FBAR devices because of simpler manufacturing processes. However, once the required processing techniques are perfected, FBAR technology is attractive because bulk-mode operation offers performance advantages [20]. Unloaded Q of several thousand at UHF frequencies is available in small packages. FBAR devices have been successfully implemented as both filters and resonators. Four FBAR filters in a wafer-level package are shown placed on a grain of rice in Fig. 8.64.

Piezoelectric Oscillators

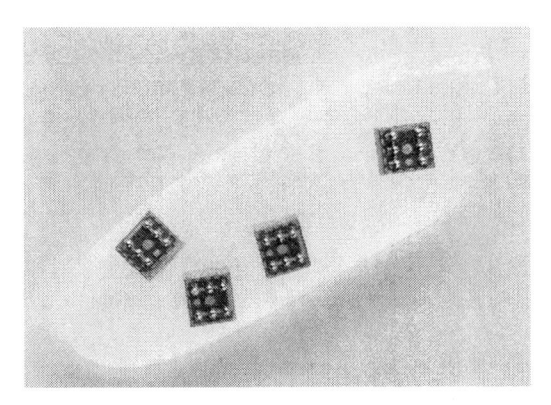

Figure 8.64 Four microcap FBAR filters with wafer level packaging (WLP) displayed on a grain of rice. The resonators are hermetically sealed in an all-silicon package (photo courtesy of Rich Ruby, Avago Technologies).

Fig. 8.65 shows an equivalent-circuit lumped-element model for a 2-GHz FBAR resonator [21].

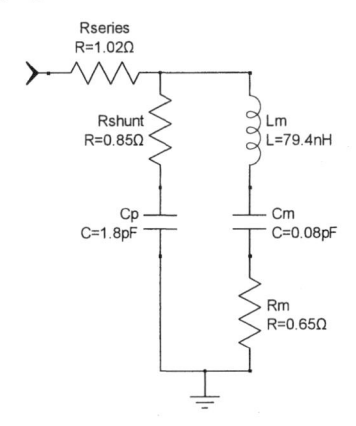

Figure 8.65 Equivalent-circuit model for a 2 GHz FBAR.

An excellent review of MEMS technology is given by De Los Santos [22].

References

[1] T. Sakar, et. al., *History of Wireless*, John Wiley & Sons, New Jersey, 2006, p. 57.

[2] R.J. Matthys, *Crystal Oscillator Circuits*, John Wiley & Sons, 1983.

[3] M. E. Frerking, *Crystal Oscillator Design and Temperature Compensation*, Van Nostrand Reinhold, 1978.

[4] B. Parzen and A. Ballato, *Design of Crystal and Other Harmonic Oscillators*, John Wiley & Sons, New York, 1983.

[5] J.R. Vig, "Quartz Crystal Resonators and Oscillators: For Frequency Control and Timing Applications, a Tutorial," www.ieee-uffc.org, January 2007.

[6] Fundamentals of Quartz Oscillators, Application Note 200-2, Agilent Technologies, www.agilent.com.

[7] IEEE-UFFC Society, www.ieee-uffc.org.

[8] Croven Crystals, Wenzel International, Inc., www.crovencrystals.com.

[9] M. Finklestein, "Inverted-Mesa Crystals Carry Oscillators into the Internet Age," *Electronic Design*, March 2000.

[10] Micro Crystal Switzerland, www.microcrystal.com.

[11] Greenray Industries, www.greenrayindustries.com.

[12] K. Haruta and W.J. Spencer, "X-ray Diffraction Study of Vibrational Modes," *20th Frequency Control Symposium Proceedings*, 1966, pp. 1-13.

[13] J.J. Gagnepain, "Fundamental Noise Studies of Quartz Crystal Resonators," *Proc. of the 30th Annual Frequency Control Sym.*, 1976, pp. 84-91.

[14] JJ. Gagnepain, M. Oliver and F.L. Walls, "Excess Noise in Quartz Crystal Resonators," *Proc. of the 37th Annual Frequency Control Sym.*, 1983, pp. 218-225.

[15] A.E. Wainwright, F.L. Walls and W.D. McCaa, "Direct Measurements of the Inherent Frequency Stability of Quartz Crystal Resonators," *Proc. of the 28th Annual Frequency Control Sym.*, 1974, pp. 177-180.

[16] M.M. Driscoll, "Two-Stage Self-Limiting Series Mode Type Quartz Crystal Oscillator Exhibiting Improved Short-Term Frequency Stability," *Proc. of the 26th Annual Frequency Control Sym.*, 1972, pp. 43-49.

[17] RF Monolithics, www.rfm.com.

[18] ECS International, www.ecsxtal.com.

[19] J. Wang, J. E. Butler, T. Feygelson, and C. T.-C. Nguyen, "1.51-GHz Polydiamond Micromechanical Disk Resonator with Impedance-Mismatched Isolating Support," *Proc. 17th Int. IEEE Micro Electro Mechanical Systems Conf.*, The Netherlands, 2004, pp. 641-644.

[20] R. Ruby and P. Merchant, "Micromachined Thin Film Bulk Acoustic Resonators," *Proc. IEEE 48th Symposium on Frequency Control*, 1994, pp. 135-138.

[21] R. Ruby, P. Bradley, Y. Oshmyansky, A. Chien and J. Larson, "Thin Film Bulk Wave Acoustic Resonators (FBAR) for Wireless Applications," *Proc. IEEE 55th Symposium on Frequency Control*, 2001, pp. 813-821.

[22] H.J. De Los Santos, *RF MEMS Circuit Design for Wireless Communications*, Artech House, Norwood, MA, 2002.

Appendix A: Modeling

The majority of the computer simulations in this book utilize models that are ideal or ideal with unloaded Q for lumped components, measured S-parameter data, or nonlinear models to represent a typical active device and analytical models for distributed components. Distributed models are sufficiently accurate for oscillator applications and nonlinear model accuracy often exceeds the repeatability of the devices. Active device and distributed models are beyond the scope of this appendix. Lumped element models are discussed in this appendix.

The basic analytical model of the capacitor includes only the negative reactance that decreases with frequency resulting from a frequency-independent value of capacitance. This is referred to in this appendix as a level 1 model. Including the effective series resistance (ESR) models the unloaded Q and is referred to as level 2 modeling. In this appendix, capacitor models to level 3 are considered, with each level providing better conformance to measured data. Analogous inductor models are also considered, as are a few other common lumped elements.

Level 1 and level 2 models are employed in examples in this book for a number of reasons. The first is to not obfuscate the theory of oscillator design with component details. The second is that the required level of component modeling is application-specific. Level 2 models are generally adequate to 1 GHz and higher for oscillators when small chip components and compact PWB layouts are used. On the other hand, narrow passband filters with critical stopband requirements need higher-level models. Discrepancies between simulated data and measured data in the many oscillator examples in this book are attributable to a combination low-level modeling, component tolerances, and the adequacy of the active device models. The many examples with measured data validate the methods and quantify the errors. If tighter conformance is required than is reported in these examples, then modeling higher than level 2 is suggested.

A.1 Capacitors

A.1.1 Capacitor: First-Level Model

A capacitor is formed by two or more plates separated by a dielectric. The capacitance of two plates is given by

$$C = \frac{\varepsilon_0 \varepsilon_r A}{t} \ (farads) \qquad (202)$$

where A is the area of each plate in meters and t is the plate separation or dielectric thickness in meters. With dimensions in millimeters and capacitance in picofarads

$$C = \frac{0.00885 \varepsilon_r A}{t} \ (pF) \qquad (203)$$

Modern chip capacitors are high-quality and level 1 models are sufficient for coupling and bypass applications. Level 2 or level 3 modeling is recommended for capacitors in the resonator. Level 3 models are suggested for emitter bypassing of CE amplifiers.

A.1.2 Capacitor: Second-Level Model

The second-level capacitance model considers parasitic series resistance and inductance. Dielectric materials used in high-quality capacitors for RF and microwave applications are low loss. The majority of loss, leading to a reduction in unloaded Q, is caused by terminal and plate conductor loss. It is effectively in series with the capacitive reactance. Capacitor models used in simulation of oscillators are improved by adding the ESR. For a given capacitor type and value, the ESR is relatively constant with typically a slight increase with increasing frequency. Fig. A.1 shows the ESR of representative values of AVX Corporation AQ13/14 series multilayer ceramic chip capacitors of value 1, 15, and 100 pF [1].

The capacitor unloaded Q is then

$$Q_{cap} = \frac{1}{\omega C \times ESR} \qquad (204)$$

From Eq. 204 it is evident that the unloaded Q falls at least linearly with increasing frequency. Therefore, modeling of capacitor unloaded Q should be with resistance in series with the capacitor rather than a specified constant Q.

The capacitor displacement current must flow in series with the terminals, plates, and any connecting path or leads. This current introduces series inductance. This inductance is approximately proportional to the length of the path and therefore the physical size of the capacitor. As a rough rule of thumb, the series inductance is approximately 0.8 nH per millimeter of path length. The inductance is somewhat less for paths with ground spacings less than 1.5 mm and somewhat higher for paths further removed from a ground plane. Wide conduction paths have less inductance than narrow paths.

Modeling

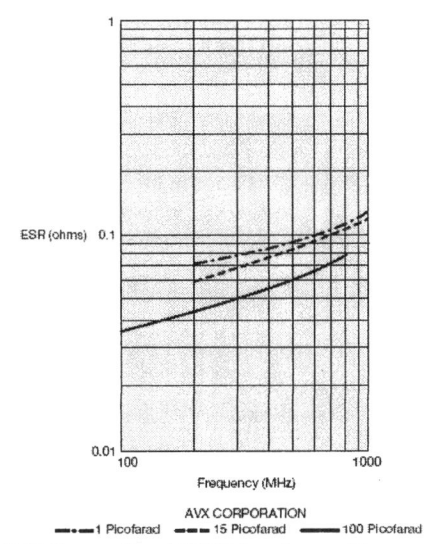

[Figure: ESR vs Frequency plot, ESR (ohms) from 0.01 to 1, Frequency (MHz) from 100 to 1000, showing three curves for 1 pF (dash-dot), 15 pF (dash), and 100 pF (solid). AVX CORPORATION.]

Figure A.1 ESR of a AVX Corporation AQ13/14 series multilayer ceramic chip capacitors of value 1 pF (dash-dot), 15 pF (dash), and 100 pF (solid) (courtesy of AVX Corporation).

Fig. A.2 shows the unloaded Q of representative values of AVX Corporation AQ13/14 series multilayer ceramic chip capacitors. The rapid decrease of unloaded Q with increasing frequency is evident in the figure. Smaller values of capacitance have a higher reactance and therefore higher unloaded Q. Notice the unloaded Q falls precipitously at higher frequencies, particularly with the higher-valued capacitors. This is because the reactance of the series inductance cancels a portion of the capacitive reactance and further degrades the unloaded Q.

Small-valued capacitors with higher unloaded Q are compatible with series resonators. Unfortunately, parallel resonators require large values of capacitance.

Above the series resonant frequency, the capacitor appears inductive. The use of level 3 models that include the series inductance is particularly important in parallel resonators and emitter and source bypassing. The series resonant frequency of the 2.79-mm long 47-pF AQ13 capacitor is approximately 700 MHz. The inferred series inductance is 1.3 nH.

A.1.3 Capacitor: Third-Level Model

Measured S-parameter data is available for a wide range of devices from a variety of manufacturers. This trend began with linear data for transistors. The successful use of this data in popular circuit simulators encourages the extension to passive devices such as capacitors and inductors. However, S-parameters were developed partly because the shorts and opens required in the earlier definition of transistor Y-, Z-, G-, and H-parameter sets are not possible at high, broadband frequencies. Active devices at high frequency have significant resistive components and S-parameters are well-suited for the measurement and characterization of these devices. However, S-parameters are ill-conditioned for accurate measurement of primarily reactive devices such as capacitors and inductors. As pointed out in Section 1.6.10, an S-parameter measurement error as small as a hundredth of a decibel introduces significant error in the unloaded Q of reactive devices.

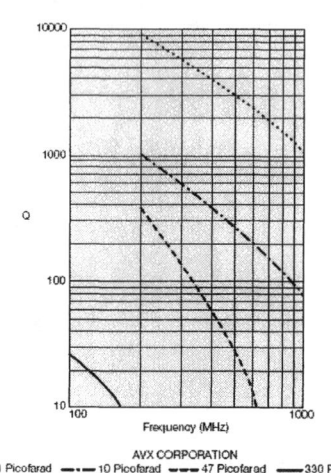

Figure A.2 Unloaded Q of a AVX Corporation AQ13/14 series multilayer ceramic chip capacitors of value 1 pF (dot), 10 pF (dash-dot), 47 pF (dash), and 330 pF (solid) (courtesy of AVX Corporation).

In addition, the use of S-parameter data to characterize the many passive devices in a typical circuit creates a cumbersome simulation environment, complicates tuning and optimization, and requires interpolation of data with frequency that introduces additional error.

Most importantly, Herculean efforts at de-embedding the device test fixture from the measured data ignore the fact that devices are embedded on a PWB in the circuit. Fig. A.3 plots the S_{11} magnitude of a 1.5 pF ATC-0402 capacitor as characterized by a Modelithics model [2] when mounted on a 4 mil Rogers 4350B substrate (square symbols), 15-mil TMM10 substrate (+ symbols), and 60-mil 4003 substrate (diamond symbols). The lines represent the Modelithics model and the symbols represent the measured data to a maximum upper frequency limit.

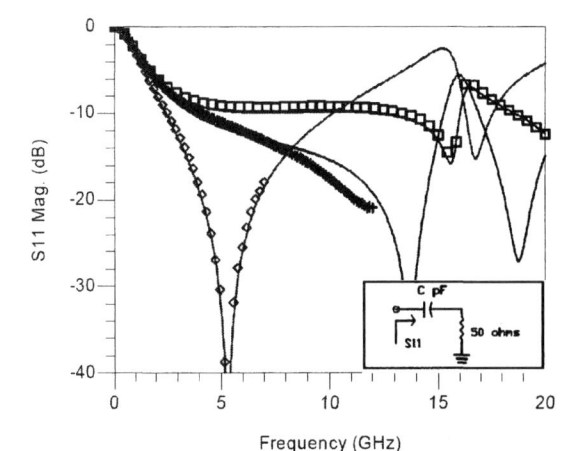

Figure A.3 S_{11} magnitude of the Modelithics model for a 1.5 pF ATC-0402 capacitor on a 4-mil Rogers 4350B substrate (square symbols), 15-mil TMM10 substrate (+ symbols), and 60-mil 4003 substrate (diamond symbols) (courtesy of Modelithics, Inc).

Fig. A.3 demonstrates that the response is a strong function of the substrate. In this case, the responses are similar to approximately 1.5 GHz. At these frequencies, the level 2 model is sufficient. At higher frequency, only substrate-dependent models are accurate. Measured S-parameter data on a specific substrate may be used to develop substrate-dependent models. Otherwise, S-parameter data does not play an important modeling role for reactive devices. Modelithics models for many circuit simulators are available for active and passive devices from a variety of manufacturers.

A.2 Varactors

The classic lumped-element equivalent linear model for a voltage-controlled capacitor (varactor) is shown in Fig. A.4. Increasing a reverse-bias voltage of a diode PN junction increases the depletion width that reduces the capacitance. L_s and C_p are the package parasitic inductance and capacitance, respectively. R_s is the effective series resistance that limits the unloaded Q of the varactor. This resistance is a slight function of frequency and the applied reverse-bias voltage, but the resistance is often assumed constant.

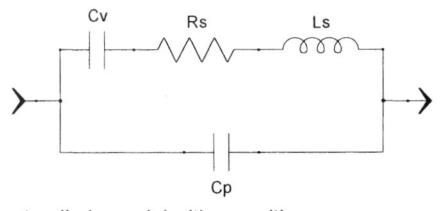

Figure A.4 Linear varactor diode model with parasitics.

The junction capacitance, C_V, is approximated by [3]

$$C_V = \frac{C_{JO}}{\left(1 + \frac{V_R}{V_J}\right)^M} + C_p \qquad (205)$$

where C_{JO} is the junction capacitance at 0 volts, V_R is the applied reverse-bias tuning voltage, V_J is the diode junction potential, and M is the doping-profile grading coefficient. The grading coefficient is also referred to as gamma. The grading coefficient varies from 1/3 for a linear gradient, to 1/2 for an abrubt gradient to 1 or more for a hyperabrubt gradient. Because of the manner in which the varactor is doped to achieve a hyperabrupt profile, the unloaded Q is reduced. Unless a wide tuning range is required, a nonhyperabrubt profile is recommended. The junction potential of silicon is approximately 0.7 volts.

Provided that L_s and C_p are small, the unloaded Q is given by

$$Q_U = \frac{1}{2\pi f C_V R_s} \qquad (206)$$

Alternatively, the unloaded Q is specified, normalized to a given measurement frequency, often 50 MHz. Again provided L_s and C_p are small, the reactance is a linear function of frequency, so the unloaded is inversely proportional to frequency. For example, if the unloaded Q measured at 4 volts reverse-bias and 50 MHz is 5000, the unloaded Q at 4 volts reverse-bias and 926 MHz is 270.

Given in Table A.1 are model parameters for selected Skyworks Solutions, Inc., varactor diodes mounted in SOD-323 or SOT-23 plastic packages [3]. The reference includes additional varactors. The parameter values are based on an empirical fit to measured capacitance curves. From the data it is apparent that some values do not represent physical properties; specifically, the nonphysical values of the junction potential. This is likely at least partially caused by the grading profile, M, being a function of the reverse-bias voltage. The effective series resistance is also often a function of the reverse-bias voltage. While the capacitance data is a good fit, nonlinear simulation using these models should be accepted with reservation. A nonlinear model for varactor diodes involves replacing C_V in the model with a SPICE diode model.

A.3 Inductors

All conducted current develops an encircling magnetic flux that attempts to reduce any change in that current. The ratio of flux to current is inductance. The opposing reactance increases with increasing frequency. Energy is also stored in the magnetic field. A simple wire conductor is inductive. Conductor configurations such as printed spirals and wire-wound coils link the flux among turns and increase inductance for a given length of conductor.

Table A.1 Model parameters for selected Skyworks Solutions, Inc. varactor diodes

Part	C_{JO} (pF)	V_J (V)	M	C_P (pF)	R_S (Ω)	L_S (nH)
SMV1127	23.9	2.2	1.0	0	0.5	1.7
SMV1129	27.5	2.8	1.1	0	0.4	1.7
SMV1139	8	1.2	0.65	0	0.6	1.7
SMV1142	13.38	2.2	1.0	0	0.7	1.7
SMV1145	41.8	2.5	1.1	0	0.6	1.7
SMV1148	104.7	2.25	1.1	0	0.5	1.7
SMV1232	4.2	1.7	0.9	0	1.5	1.7
SMV1234	8.75	2.3	1.1	1.2	0.8	1.7
SMV1236	21.63	8.0	4.2	3.2	0.5	1.7
SMV1237	66.16	10	5.3	9	0.13	1.7
SMV1245	6.9	3.5	1.7	0.47	2	1.7
SMV1248	21.54	13	10.5	0	1.8	1.7
SMV1251	60	17	14	0	1.3	1.7

A.3.1 Single-Layer Wire Solenoid

The single-layer, wire-wound solenoid on a cylindrical form is ubiquitous and well studied. It is the starting point for any investigation of high-frequency inductors.

A.3.1.1 Solenoid: First-Level Model

The classic formula by Wheeler [4] for the value of inductance of a single-layer, wire-wound solenoid on a circular, nonmagnetic form using dimensions in inches is

$$L = \frac{n^2 r^2}{9r + 10l} \text{ microhenry} \qquad (207)$$

where n is the number of turns, r is the mean radius of the solenoid, and l is the mean length of the solenoid. With dimensions converted to millimeters, Wheeler's formula is

$$L = \frac{n^2 r^2}{0.2286r + 0.254l} \text{ nanohenry} \qquad (208)$$

For small wire diameter, the accuracy is within 1% for radius to length ratios less than 1.5 and within 4% for radius to length ratios less than 2.5. As the wire diameter becomes appreciable with respect to the solenoid diameter, Wheeler's formula overestimates the inductance. Using the inside diameter of the solenoid partially corrects this error.

When the solenoid is wound on a cylindrical core of magnetic material the inductance increases proportionally to the effective permeability [5].

$$\mu_e = \frac{\mu_r}{1 + N_f(\mu_r - 1)} \quad (209)$$

N_f is the demagnetization factor given approximately by

$$N_f \approx \frac{0.28}{l/d} - 0.0158 - \frac{0.0915}{\mu_r} + \frac{0.00063 l/d}{\log(\mu_r)} \quad (210)$$

where l/d is the length to diameter ratio of the core, the winding occupies 85% to 95% of the core, there is no spacing between the winding and the core, $1 < l/d < 100$ and $1 < \mu_r < 1000$. For short cores, the effective permeability is a weak function of the material permeability, therefore reducing the temperature instability of the core. Below 200 MHz, a core can provide increased inductance and unloaded Q with smaller solenoids.

A.3.1.2 Solenoid: Second-Level Model

The second-level model for the single-layer solenoid includes the unloaded Q. The self-repelling nature of a current filament causes the current distribution within the wire of a solenoid to be complex, and so attempts at mathematically deriving from first principles the unloaded Q of the simplest solenoid have been unsuccessful. Instead, empirical curve fits to measured data are used. Medhurst published an expression for the unloaded Q of a single-layer, round-copper wire solenoid on a lossless form [5]. After conversion to millimeter units

$$Q_U = 0.015 r \Psi \sqrt{f} \quad (211)$$

where r is the mean radius of the solenoid in millimeters and f is the frequency in hertz. ψ is a function of the length to diameter and spacing ratios with tables and formula given in [5, 6]. ψ ranges from approximately 0.5 to 0.9 with length to diameter ratios from 0.5 to 8, assuming an optimum wire spacing so that copper occupies 55% to 95% of the solenoid length. Callender [7] offered a simple curve fit to the Medhurst data accurate to a few percent, again with units of millimeters

$$Q_U = \frac{\sqrt{f}}{\frac{69}{r} + \frac{54}{l}} \quad (212)$$

The formula for the unloaded Q assume the solenoid is unshielded, the radius is sufficiently small to ignore the effects of parasitic capacitance and radiation, and the frequency is sufficiently high that skin effect is significant.

A.3.1.3 Solenoid: Third-Level Model

Derivation attempts to compute inductor parasitic capacitance failed and again Medhurst used empirical data to formulate solenoid inductor capacitance. Tables and curve fits for Medhurst's parasitic capacitance are available [5, 6].

Historically, capacitance derivations failed because it was assumed parasitic capacitance is winding-to-winding. However, a multiturn solenoid operated below resonance has little voltage phase shift between adjacent turns and capacitive coupling is nonexistent. Inductor models that assume interwinding capacitance fail because the primary source of parasitic capacitance is from the winding to ground. Medhurst's capacitance formulation was valid for the specific case of one inductor terminal grounded. However, if neither inductor terminal is grounded, the popular model fails.

Consider the 12.9 turn solenoid mounted over a ground plane [8] shown in Fig. A.5. It is wound with 18-gauge copper wire with a solenoid mean radius of 6.81 mm, a mean length of 22.6 mm, and a height of 3.43 mm from the outside of the wire to the aluminum ground plane.

Fig. A.5 Wire-wound solenoid mounted over a ground plane.

The inductor is in series with the transmission path via two SMA connectors. The measured transmission amplitude and phase responses are shown on the left in Fig. A.6. The inductance from Wheeler is 916 nH and the parasitic capacitance from Medhurst is 0.69 pF. On the right in Fig. A.5 are the simulated transmission amplitude and phase of a series mounted 916-nH inductor with 0.69 pF of parallel capacitance.

Notice that the frequency of first resonant mode is near 216 MHz for both the measured data and the simulated model. However, the measured and model characteristic shape of both the amplitude and phase responses are uncorrelated. The popular solenoid model fails to predict the solenoid behavior because it assumes one end of the solenoid is grounded.

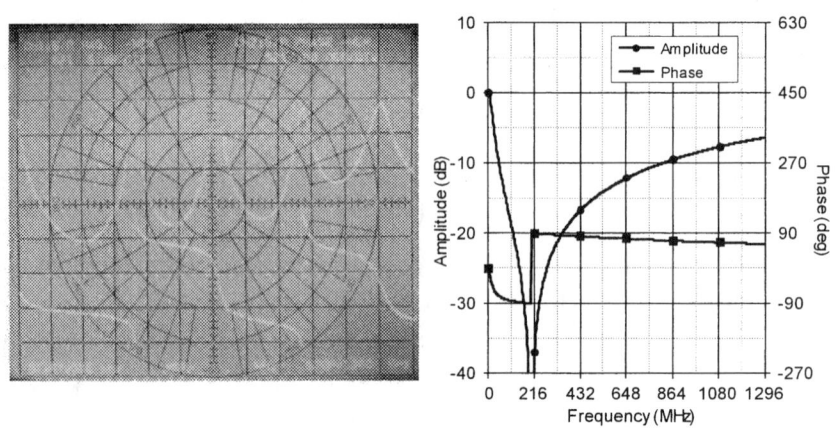

Fig. A.6 Measured transmission amplitude (upper) and phase (lower) responses of the 12.9 turn solenoid (left) and predicted by the classic parallel capacitance model (right). The sweep is 0.5 to 1300 MHz (108 MHz/div), the amplitude scale is 5 dB/div and the phase scale is 90°/div.

Examination of the measured response reveals periodic behavior suggestive of a transmission line. In fact, the solenoid above ground is essentially a helical transmission line. The classic formulae for distributed transmission lines are

$$Z_0 = \sqrt{\frac{L_0}{C_0}} \qquad (213)$$

$$\theta = 2\pi f_0 \sqrt{L_0 C_0} \text{ radians} \qquad (214)$$

$$\theta = 360 f_0 \sqrt{L_0 C_0} \text{ degrees} \qquad (215)$$

where L_o is the static inductance, C_o is the static capacitance, and f_o is the modeling frequency.

Fig. A.7 shows the simulated transmission amplitude and phase for a transmission line with Z_0 of 796 ohms adjusted to match the depth of the first resonance of the solenoid measured data and a θ of 600° at 1300 MHz to match the frequency of the first transmission peak of the solenoid. The solenoid clearly behaves similarly to a series transmission line. In fact, the inductor is accurately modeled using this technique with the static inductance and capacitance measured at low frequency! Dispersion causes nonperiodicity of the higher frequency resonances and unequal transmission amplitude valleys. These effects are absent in well-shielded solenoids, such as a helical inductor with a square or cylindrical shield.

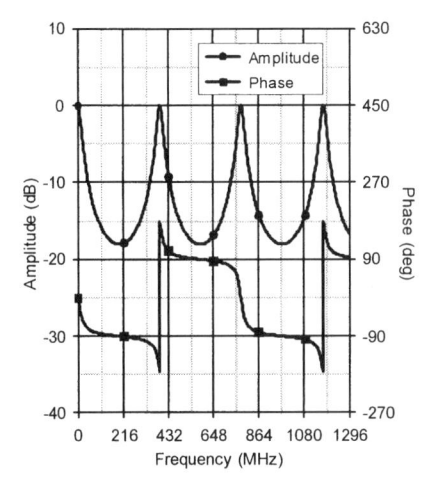

Figure A.7 Simulated response of a transmission line with a Z_0 of 796 ohms and θ of 600^0 at 1300 MHz.

Since the transmission-line model truly characterizes the physical solenoid, mathematical solutions based on a transmission-line model for the solenoid capacitance to ground are successful. Relatively accurate capacitance calculations are made by simply assuming the solenoid is a cylinder above a ground plane. For example, formula for the characteristic impedance of common transmission-line structures may be used, such as a solenoid above a ground plane with a dielectric (round microstrip), a solenoid surrounded by a cylindrical ground (coax), a helical inductor with a ground above and below (slabline), and others. The characteristic impedance of various physical structures is given in Section A.7. The capacitance is then found from the characteristic impedance.

$$C_0 = l \frac{\sqrt{\varepsilon_r}}{v_0 Z_0} \qquad (216)$$

where l is the length of the cylinder. For the solenoid shown in Fig. A.5, the calculated capacitance is 1.34 pF. The ends of the solenoid have additional capacitance to ground; a similar phenomena to antenna end effect. This is estimated using Snow's [10] formula for the capacitance of a thin disk with radius equal to the solenoid radius.

$$C_{disk} = 8\varepsilon_0 r = 70.83r \,(pF) \qquad (217)$$

where r is the radius of the disk in meters. Half of this capacitance is effectively present at each end of the solenoid. The total end-effect capacitance equals Snow's value. For the solenoid in Fig. A.5, the calculated end-effect capacitance is 0.52 pF. The total cylinder and end-effect capacitance is then 1.86 pF. The capacitance measured with a low-frequency capacitance meter is 1.73 pF.

A.3.1.4 General Inductor: Fourth-Level Model

These procedures could be extended to other popular inductor structures such as chip inductors. However, the internal physical structure of a chip inductor may be complex and unknown to the designer. Detailed issues such as the effect of the ceramic surrounding the winding may be difficult to access. Furthermore, analytical models for radiation effects on the unloaded Q are not published. As with chip capacitors, the most accurate models for chip and other commercially available inductors are substrate-dependant, such as those offered by Modelithics [2].

A.3.2 Toroid

The toroid is similar to a solenoid except the cylindrical form is curved until the opposite ends of the solenoid meet. For toroids, the form often has a rectangular rather than a circular cross-section. An advantage of the toroid is a certain degree of self-shielding.

A.3.2.1 Toroid: First-Level Analytical Model

The inductance of a single-layer wire-wound inductor on a toroidal core of rectangular section is given by [9]

$$L = \frac{n^2 t \mu_r}{2171} \log \frac{r_{outer}}{r_{inner}} \; microhenry \qquad (218)$$

where r_{outer} and r_{inner} are the radius and t is the thickness of the toroid in millimeters and μ_r is the core permeability relative to free space.

The above formula assumes the effective permeability equals the core permeability. This is a reasonable assumption provided the wire radius is small, many turns are evenly distributed around the core, and $\mu_r \gg 1$.

A.3.2.2 Toroid: First-Level Index Model

Toroidal inductance is also modeled by

$$L = A_L \left(\frac{n}{n_A} \right)^2 \qquad (219)$$

where A_L is an inductance index provided by the toroid core manufacturer. It assumes a small diameter wire distributed evenly around the toroid. The inductance of a toroid is tuned by compressing or spreading the winding. Toroids wound on low permeability cores tune over a wider range.

Modeling

A.3.2.3 Toroid: Higher-Level Models

The next-level toroid model is similar to the solenoid third-level model in that it incorporates capacitance to ground by treating the toroid as a helical transmission line above a ground plane. The mean diameter of the solenoid is the length of the rod over ground or rod over ground with a dielectric, and the techniques described in Section A.4.1.3 are used to model the toroid. For the adjacent winding ends, an additional capacitance parallel to the inductance is suggested. As with other component models, the use of measured S-parameter data is discouraged. S-parameter data is better used to develop a model for the toroid.

A.3.3 Ferrite Beads

Ferrite beads are lossy inductive elements useful for both filtering of high-frequency power supply noise and eliminating active device instability. The ferrite material has a high permeability with little concern for loss. The high loss avoids potential resonance and instability in active devices. A ferrite bead model is shown in Fig. A.8. Inductor L_1 controls the reactance at low frequency. At the resonance of L_1 and C_1, the impedance is resistive as controlled primarily by R_1.

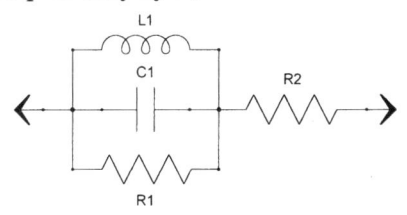

Figure A.8 Model of a ferrite bead.

Given in Table A.2 are model parameters for the TDK MMZ0603 series of multiplayer chip ferrite beads. TDK offers multiplayer ferrite beads in the MMZ series in sizes 0402 through 2012. The larger sizes offer higher impedance and lower DC resistance for higher current [11]. The TDK series is representative of ferrite beads available from a variety of manufacturers.

A.3.4 Mutually Coupled Inductors

When two inductors are coupled the total inductance is given by [7]

$$L_T = L_1 + L_2 \pm 2M \qquad (220)$$

where L_1 and L_2 are the uncoupled individual inductances and M is the mutual inductance between windings. If the magnetic fields aid, the sign of the mutual term is positive. In this case, the mutual inductance is given by

Table A.2 Model parameters TDK series chip ferrite beads

Model	R1 (ohm)	L1 (μH)	C1 (pF)	R2 (ohm)	Freq Range (MHz)
MMZ0603S100C	13	0.03	0.05	0.04	35-450
MMZ0603S800C	105	0.16	0.14	0.30	35-450
MMZ0603S121C	190	0.31	0.16	0.50	35-450
MMZ0603S241C	370	0.63	0.29	0.80	35-450
MMZ0603S471C	670	1.12	0.35	1.10	35-450
MMZ0603S601C	800	1.51	0.41	1.30	35-450
MMZ0603Y121C	190	0.24	0.20	0.31	85-450
MMZ0603Y241C	480	0.45	0.25	0.80	85-450
MMZ0603Y471C	780	0.95	0.33	1.27	85-450
MMZ0603D330C	420	0.06	0.24	0.69	300-1000
MMZ0603D560C	630	0.12	0.26	1.00	300-1000
MMZ0603D800C	800	0.15	0.30	1.18	300-1000
MMZ0603F100C	310	0.02	0.21	0.39	650-1800

$$M = \frac{L_T - L_1 - L_2}{2} \qquad (221)$$

The coupling coefficient, k, is given by

$$k = \frac{M}{\sqrt{L_1 L_2}} \qquad (222)$$

From Eq. 208 and Eqs. 220–222, the coupling coefficient of a tapped inductor with constant pitch is derived

$$k = \frac{\dfrac{(l_1+l_2)^2}{0.2286r+0.254(l_1+l_2)} - \dfrac{l_1^2}{0.2286r+0.254l_1} - \dfrac{l_2^2}{0.2286r+0.254l_2}}{2\sqrt{\dfrac{l_1^2 l_2^2}{(0.2286r+0.254l_1)(0.2286r+0.254l_2)}}} \qquad (223)$$

where units are in millimeters, r is the radius of the winding, and l_1 and l_2 are the lengths of the two windings.

Fig. A.9 shows the coupling coefficient of a tapped inductor versus the total length to diameter ratio for a center tap. The coupling increases only slightly as the tap point is moved toward the end of the winding, so Fig. A.9 is useful to estimate the coupling for any tap point. For a more accurate coupling estimate versus tap point, Eq. 223 is used. The coupling of typical solenoids whose length is similar to or longer than the diameter is quite low. Short coils offer better coupling but poorer unloaded Q.

A.4 Helical Transmission Lines

While a solenoid behaves much like a transmission line, when well shielded, it is indistinguishable from a transmission line. The development of a model for a helical transmission line then becomes mathematically straightforward. Consider the helical coaxial transmission line in Fig. A.10. The solenoid is 16 turns of 18-gauge wire with an inside radius of 1.83 mm and a mean length of 25.4 mm. The inside shield radius is 4.45 mm.

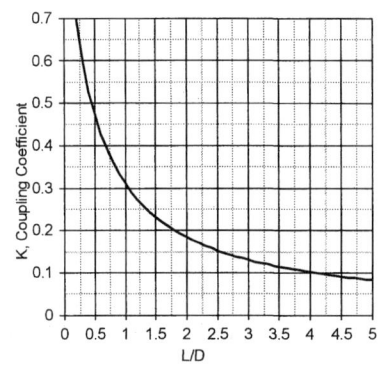

Figure A.9 Coupling coefficient of a tapped inductor versus the total length to diameter ratio for a center tap.

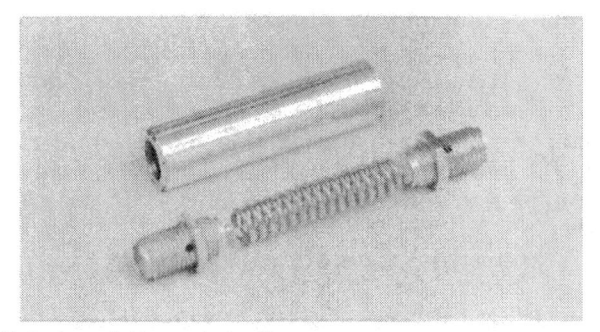

Figure A.10 Helical coaxial transmission line.

The static inductance for the helical transmission line model is computed using Wheeler's formula. The effect of a nonmagnetic, conducting cylindrical shield on the inductance is given by Bogle [12]. The inductance is decreased by the factor

$$LF = 1 - \frac{(r/r_s)^2}{1 + \frac{1.55(r_s - r)}{l}} \qquad (224)$$

where r is the solenoid radius and r_s is the shield radius. The capacitance is easily found using Eq. 216 and the characteristic impedance of a coaxial line assuming the inner conductor is a cylinder with a radius equal to the radius of the solenoid.

The static inductance for this example, from Wheeler's formula, is 125 nH, which is reduced to 107 nH by Bogle's shield factor. The coaxial model characteristic impedance for the solenoid is 26.77 ohms, resulting in a static capacitance of 3.17 pF. The resulting line model is Z_o equal to 198 ohms and an electrical length of 90° at 432 MHz. The measured first resonant frequency for this helical transmission line is 421 MHz. This example illustrates both the accuracy and intuitively satisfying nature of the transmission-line model for well-shielded inductors.

A.5 Signal Control Devices

Oscillators are generally limited bandwidth devices. Bandwidth is nil for crystal and other reference oscillators. Tuning bandwidths of 2:1 are not rare, but decade tuning ranges are limited to special applications. Nevertheless, wide bandwidth signal control devices such as splitters, couplers, and matching transformers are sometimes useful in oscillator design. The following sections review modeling for a few of the more important devices.

Conventional transformers transfer energy through flux linkage. Transmission line structures are also used to transfer energy. At high frequency, core loss, parasitic capacitance, and leakage inductance limit the performance of conventional transformers and transmission-line transformers are preferred. Transmission-line transformers are realized using twin lead, twisted pair and coaxial lines wound on rod, toroid, or shotgun cores. The line effectively cancels flux in the core while winding the line on a core suppresses undesired current modes.

A.5.1 Bifilar Transformer Operating Modes

The benchmark work on broadband transformers was by Guanella in 1944 [13]. Fig. A.11 depicts five different forms, each with a bifilar winding on a core, but with differing configurations of the windings and grounds.

The basic circuit is a balanced to unbalanced (balun) transformer shown at the upper left in Fig. A.11. When the load is floating, the currents in each wire of the bifilar transmission line are equal in amplitude and opposite in phase, thus eliminating flux in the core. At low frequency, the reactance of the winding is inadequate to prevent conventional transformer current. The design objective is to have a sufficient number of turns and core permeability to develop a reactance of a few times the load resistance (low reactance limits the low frequency performance) while keeping the winding length less than a quarter wavelength (line length limits the high-frequency performance).

Modeling 405

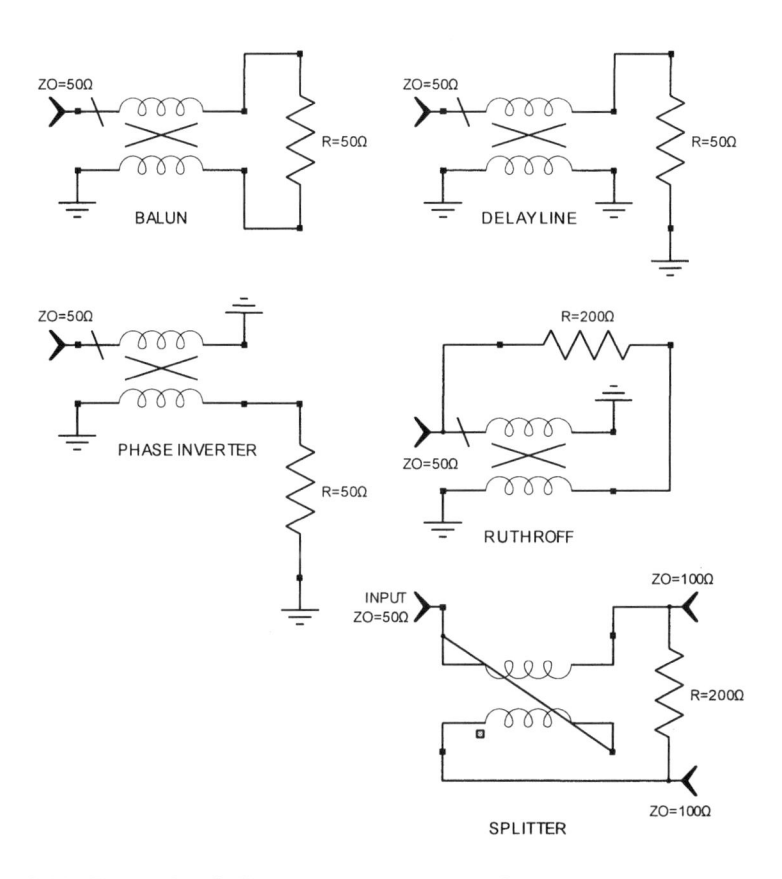

Figure A.11 Five modes of bifilar wound broadband transformer.

The splitter at the lower right both splits the input signal and raises the impedance level. Unfortunately, for all ports to be matched, this simple splitter requires that the two loads are twice the impedance of the input. In addition, some of the energy reflected back from an input port must be absorbed by the balance resistor.

A.5.2 Ruthroff Impedance Transformer

The bifilar transformer mode depicted in the middle on the right in Fig. A.11 was described by Ruthroff [13]. It serves as an unbalanced to unbalanced (unun) 1:4 impedance transformer. As with other bifilar transformers, the operating mode is transmission line rather than flux linkage. The low-frequency choking effect is realized by winding the line on a core with high permeability, consistent with material loss at the highest frequency of operation.

The schematic of Fig. A.12 supports the simulation of both the high-frequency transmission line and low-frequency choking behavior of the Ruthroff 1:4 transformer.

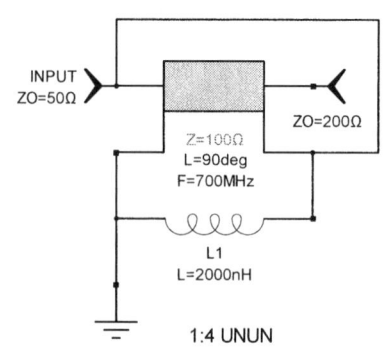

Figure A.12 Schematic for simulation of the low- and high-frequency response of the Ruthroff 1:4 unun impedance transformer.

The S_{21} and S_{11} magnitudes of the Ruthroff 1:4 impedance transformer are shown on the left in Fig. A.13. The high-frequency transmission roll-off occurs as the transmission line length becomes appreciable with respect to a wavelength. The solid traces are with the schematic values shown in Fig. A.12. The dashed traces are with the transmission line impedance reduced to 50 ohms and the choking inductance reduced to 1 µH. The high-frequency bandwidth reduction results from a nonoptimum characteristic impedance for the transmission line. The optimum line impedance for the Ruthroff 1:4 impedance transformer is $\sqrt{R_S R_L}$, or 100 ohms in this case.

The low-frequency bandwidth reduction results from decreasing the choking inductance from 2 to 1 µH. A longer line wound on the core increases the choking inductance and improves the low-frequency bandwidth but shifts the high-frequency corner downward. Higher permeability also increases the choking inductance; therefore, a material with the highest possible permeability provides the widest bandwidth. For low-power applications, manganese-zinc ferrites with permeability over 1000 work well. These ferrites have higher loss, and at higher power dissipation becomes a factor. In this case, nickel-zinc ferrites with permeability up to 250 are used. For most oscillator applications, decade bandwidths are not required and design of broadband transformers, splitters, and couplers is relaxed.

Modeling 407

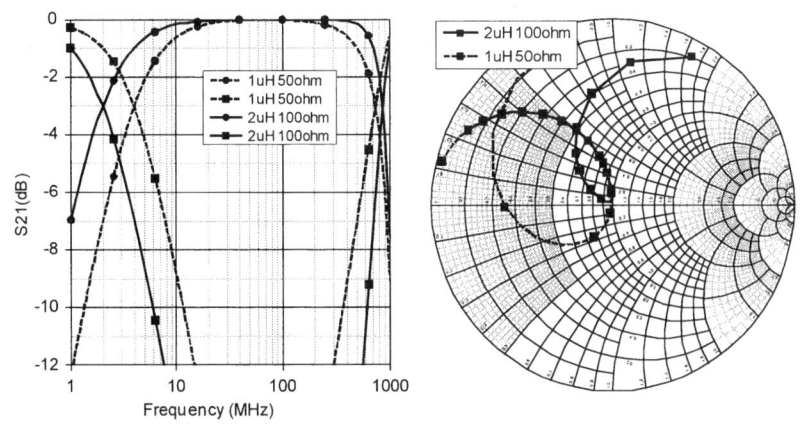

Figure A.13 On the left are S_{21} (circular symbols) and S_{11} (square symbols) of the Ruthroff 1:4 transformer. The solid traces are with a line impedance of 100 ohms and chocking inductance of 2μH. Dashed traces are with a line impedance of 50 ohms and chocking inductance of 1μH.

A.5.3 Wire-Wound Couplers

Up to approximately 1000 MHz, practical broadband directional couplers are realizable using wire-wound transformers on powdered iron or ferrite cores. Reasonable directivity is achievable for high coupling values. A schematic of one type is shown in Fig. A.14.

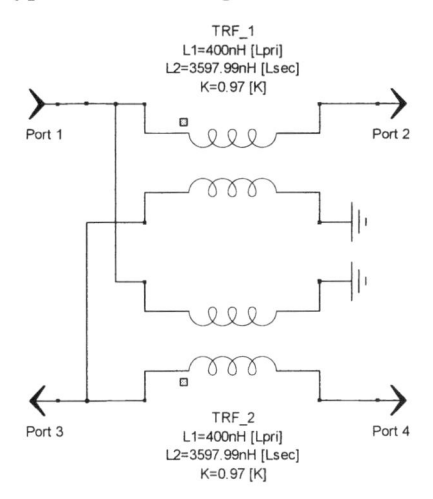

Figure A.14 Transformer based -9.54-dB directional coupler.

The input is port 1, the through signal is derived at port 2, the coupled port is port 3, and the isolated port is 4. The directivity improves with looser coupling. A 15-dB directivity is achievable midband at -5-dB coupling. Directivity is generally limited by parasitics for coupling values looser than -10 dB.

If the primary turns ratio, N_p, is 1, the secondary turns ratio is

$$N_p = \frac{1}{\sqrt{C}} \qquad (225)$$

where

$$C = 10^{C(dB)/10} \qquad (226)$$

For example, for a -9.54-dB coupler, C is 0.11, and N_p is 3 for an impedance ratio of the turns of 9. The primaries in Fig. A.14 are signified by the transformer polarity dots. The transmission phase of both the through port 2 and the coupled port 3 is 0° near midband.

Fig. A.15 shows the responses of the -9.54-dB coupler in Fig. A.14. The low frequency responses are limited by the reactance of the windings. In this example a reactance of 2 times the 50 ohm reference impedance was selected at a midband frequency of 39.8 MHz. The upper frequency responses are limited by the leakage inductance introduced by finite coupling between the transformer windings. The winding coupling coefficient used in this example is 0.97.

Ideally, the isolated port termination impedance is equal to the reference impedance. In practice, the coupling value is generally closer to the predicted value, particularly for tighter coupling values, if the isolated port is terminated in a resistance equal to

$$R_t = Z_0(1 - C) \qquad (227)$$

The design of these devices is an art involving the selection of the core material and the handling of the windings. Practical issues share much in common with broadband impedance matching transformers.

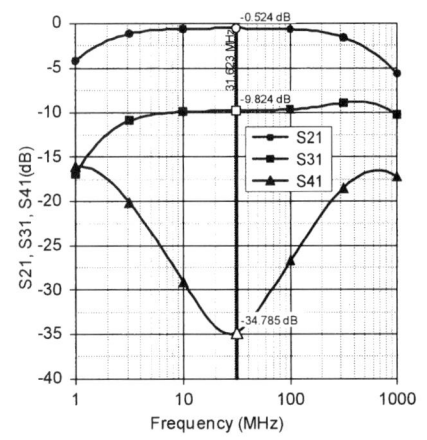

Figure A.15 Coupled response (port 1 to 3:circular symbols), through response (port 1 to 2:square symbols) and isolated port response (port 1 to 4:triangular symbols) of the -9.54 dB coupler in Fig. A.14.

A.6 Characteristic Impedance of Transmission Lines

In oscillator design, the characteristic impedance of transmission lines is needed to design resonators, to compute the static parasitic capacitance of inductors, and to design transmission-line transformers. This section gives formulas from Gunston [14] for the characteristic impedance of the physical structures shown in Fig. A.16. The characteristic impedance of free-space, η, is

$$\eta = \sqrt{\frac{\mu_0}{\varepsilon_0}} = 376.73 \ (ohms) \qquad (228)$$

Some authors round the characteristic impedance of free-space to 377 ohms. This number divided by π is almost exactly 120 and divided by 2π is 60. 376.73 divided by 2π is 59.96. Both 59.96 and 60 are found in transmission line literature. In the following formula the more accurate numbers consistent with the cited reference are used.

A.6.1 Coax

The characteristic impedance of conventional coax with a relative permeability of 1 is

$$Z_0 = \frac{\eta}{2\pi\sqrt{\varepsilon_r}} \ln\frac{D}{d} = \frac{59.96}{\sqrt{\varepsilon_r}} \ln\frac{D}{d} = \frac{138.06}{\sqrt{\varepsilon_r}} \log\frac{D}{d} \ (ohms) \qquad (229)$$

where ε_r is the relative permittivity of the dielectric. The cutoff frequency for higher order modes is

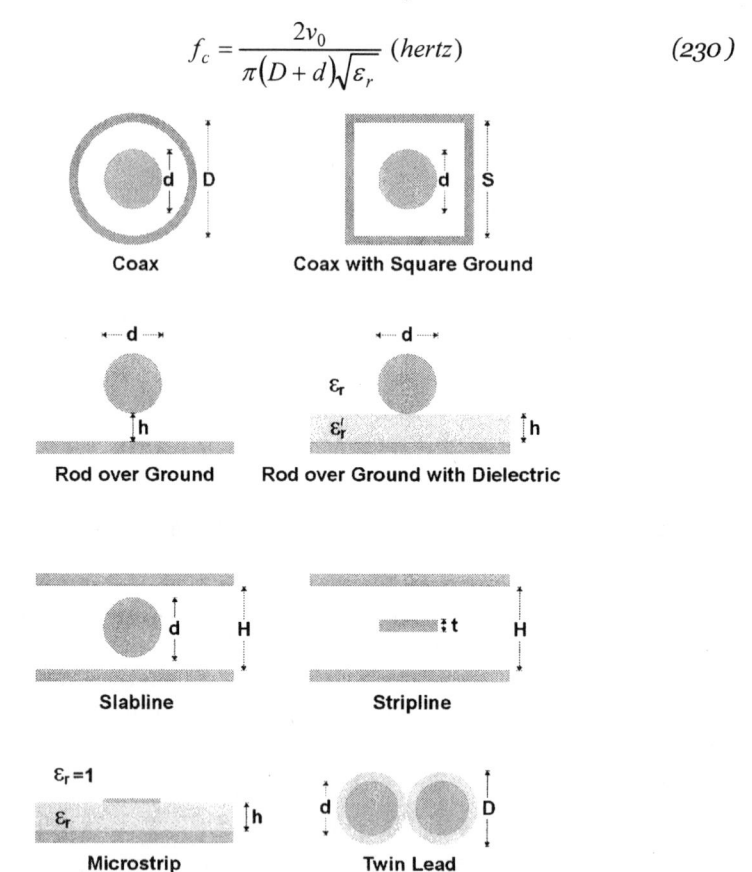

$$f_c = \frac{2v_0}{\pi(D+d)\sqrt{\varepsilon_r}} \ (hertz) \tag{230}$$

Figure A.16 Common transmission line structures and definitions used for the calculation of the characteristic impedance.

A.6.2 Coax with Square Ground

The characteristic impedance of coax with a circular center conductor and a square outer conductor is given approximately by

$$Z_0 \approx \frac{138.06}{\sqrt{\varepsilon_r}} \log \frac{1.0787S}{d} \ (ohms) \tag{231}$$

where S is the length of the side of the outer conductor. The 1.0787 factor is often approximated as 1.08. This factor is actually a function of the S/d ratio. For S/d of 1.1, corresponding to a characteristic impedance of 9.45 ohms in a vacuum, the factor is 1.064. For S/d of 1.3, corresponding to a characteristic impedance of 20.1 ohms in a vacuum, the factor is 1.076. For larger S/d, the factor approaches 1.0787.

A.6.3 Rod over Ground

The characteristic impedance of a rod over a flat ground plane of infinite extent is

$$Z_0 = \frac{138.06}{\sqrt{\varepsilon_r}} \log\left[\left(1 + \frac{2h}{d}\right) + 2\sqrt{\frac{h}{d}\left(1 + \frac{h}{d}\right)}\right] \qquad (232)$$

where h is the height of the outside of the rod over the ground plane.

A.6.4 Rod over Flat Ground with Dielectric Layer

For a rod mounted on top of a PWB there is a dielectric layer between the rod and the ground plane. In this case, the characteristic impedance of the rod is

$$Z_0 = \frac{59.96}{\sqrt{\varepsilon_r}} \sqrt{L(L+S)} \qquad (233)$$

where

$$L = \ln\left[1 + \frac{1}{2x}\left(1 + \sqrt{1 + 4x}\right)\right] \qquad (234)$$

$$x = \frac{d}{4h} \qquad (235)$$

$$S = \sum_{n=0}^{\infty} (-E)^{n+1} \ln\left(1 + \frac{2}{n+x}\right) \qquad (236)$$

$$E = \frac{\varepsilon_r' - \varepsilon_r}{\varepsilon_r' + \varepsilon_r} \qquad (237)$$

A.6.5 Rod Between Ground Planes

A round rod centered between two flat ground planes of infinite extent is referred to as slabline. The characteristic impedance of slabline is approximately [5,15]

$$Z_0 \approx \frac{138.06}{\sqrt{\varepsilon_r}} \log\frac{4\left(1 + e^{6.48(d_n - 1.28)}\right)}{\pi d_n} \qquad (238)$$

where d_n is the diameter of the rod normalized to the ground to ground spacing.

$$d_n = \frac{d}{H} \quad (239)$$

The exponential function is a curve fit to the data published by Stracca et. al. The results are accurate to better than 1% for $0.1 < d_n < 0.9$.

A.6.6 Stripline

The impedance of stripline with zero strip thickness is [16]

$$Z_0 = \frac{\eta}{4\sqrt{\varepsilon_r}} \frac{K(k)}{K(k')} = \frac{94.18}{\sqrt{\varepsilon_r}} \frac{K(k)}{K(k')} \quad (240)$$

where

$$k = \operatorname{sech}\left(\frac{\pi w}{2H}\right) \quad (241)$$

$$k' = \tanh\left(\frac{\pi w}{2H}\right) \quad (242)$$

$K(k)$ is the complete integral of the first kind. Hilberg [17] provides closed form expressions for the ratio of the unprimed and primed elliptic integrals that are very precise. For $0 < k < 0.707$

$$\frac{K(k)}{K(k')} = \frac{\pi}{\ln\left(2\frac{1+\sqrt{k'}}{1-\sqrt{k'}}\right)} \quad (243)$$

and for $0.707 < k < 1$

$$\frac{K(k)}{K(k')} = \frac{1}{\pi} \ln\left(2\frac{1+\sqrt{k}}{1-\sqrt{k}}\right) \quad (244)$$

where

$$k' = \sqrt{1-k^2} \quad (245)$$

Wheeler [18] gives the characteristic impedance of stripline with a thick strip.

$$Z_0 = \frac{30}{\sqrt{\varepsilon_r}} \ln\left[1 + \frac{c}{2}\left(c + \sqrt{c^2 + 6.27}\right)\right] \quad (246)$$

where

$$c = \frac{\frac{8}{\pi}}{\frac{w}{H-t} + \frac{\Delta w}{H-t}} \quad (247)$$

$$\frac{\Delta w}{H-t} = \frac{x}{\pi(1-x)}\left\{1 - \frac{1}{2}\ln\left[\left(\frac{x}{2-x}\right)^2 + \left(\frac{0.079x}{w/H+1.1x}\right)^m\right]\right\} \quad (248)$$

$$m = \frac{3}{1.5 + \frac{x}{1-x}} \quad (249)$$

$$x = \frac{t}{H} \quad (250)$$

The accuracy of these expressions is claimed to be better than 0.5% for c>0.25.

A.6.7 Microstrip

With microstrip, a portion of the electric field is in the dielectric and a portion is in the air above the strip. This introduces longitudinal components and the propagation mode is not purely TEM. The resulting dispersion increases as the substrate thickness increases with respect to a wavelength. Modern circuit simulators model these effects using complex analytical formula. In this appendix, static formulas are given that ignore dispersion. The introduced error is less than 2% for a 1-mm thick substrate up to 13 GHz for a relative dielectric constant of 2.2, 7 GHz for a relative dielectric constant of 10, and 5 GHZ for a relative dielectric constant of 30. The frequency for accurate results is inversely proportional to the substrate thickness. $Z_o(o)$ given here [19] refers to the static characteristic impedance for a thin strip.

$$Z_0(0, \varepsilon_r = 1) = 60 \ln\left(\frac{F_1 h}{w} + \sqrt{1 + (2h/w)^2}\right) \quad (251)$$

where

$$F_1 = 6 + (2\pi - 6)e^{-(30.666h/w)^{0.7528}} \quad (252)$$

Then,

$$Z_0(0) = \frac{Z_0(0, \varepsilon_r = 1)}{\sqrt{\varepsilon_{r,eff}}} \quad (253)$$

$$\varepsilon_{r,eff} = \frac{\varepsilon_r + 1}{2} + \frac{\varepsilon_r - 1}{2}\left(1 + \frac{10h}{w}\right)^{-ab} \quad (254)$$

where

$$a = 1 + \frac{1}{49}\ln\left[\frac{(w/h)^4 + (w/52h)^2}{(w/h)^4 + 0.432}\right] + \frac{1}{18.7}\ln\left[1 + (w/18.1h)^3\right] \quad (255)$$

$$b = 0.564\left(\frac{\varepsilon_r - 0.9}{\varepsilon_r + 3.0}\right)^{0.053} \quad (256)$$

the stated accuracy for these formulae is better than 0.04% for $w/h<1000$ and $1<\varepsilon_r<90$. Finite strip thickness decreases the characteristic impedance and the effective dielectric constant.

A.6.8 Twisted-Pair Transmission Line

For high-power applications, coaxial transmission line is often used in broadband transformers. For narrow and moderate bandwidth application, the line impedance need not be optimum and coax with commercially available impedance is used. For low- and medium-power applications, close-spaced or twisted-pair magnet wire twin lead is used.

The static characteristic impedance of twin lead and twisted pair embedded within air is [20]

$$Z_0 = \frac{120}{\sqrt{\varepsilon_{eff}}}\cosh^{-1}\frac{D}{d} = \frac{120}{\sqrt{\varepsilon_{eff}}}\ln\left(\frac{D}{d} + \sqrt{\left(\frac{D}{d}\right)^2 - 1}\right) \quad (257)$$

where

$$\varepsilon_{eff} = 1 + (\varepsilon_r - 1)(0.25 + 0.0004\theta^2) \quad (258)$$

where ε_r is the relative dielectric constant of the wire insulation and θ is the twist angle in degrees

$$\theta = \tan^{-1}(T\pi D) \quad (259)$$

and T is the number of twists per unit length.

Given in Table A.3 are the properties of popular film insulating coatings for copper magnet wire. The last column is the nominal thickness of a single-coating layer. Double, triple, and quad thickness coatings are available. This data is from a variety of sources. Significant variability exists for magnet wire data. Consulting the manufacturer is advised.

Table A.3 Properties of magnet wire insulating films

Brand	Material	ε_r	Single (mm)	Class (temp ^{0}C)
Thermaleze	Polyester-imide	3.58	0.071	200
Soldereze	Modified polyurethane	3.66	0.074	105
Nyleze	Polyurethane	3.75	0.074	130
Polythermaleze	Polyester and polyamide-imide	3.67	0.077	200
Formvar	Polyvinyl formal	3.37	0.074	105

A.7 Helical Resonators

A limitation of transmission line resonators is their extreme length. The advent of modern high-dielectric constant ceramic extends the useful range of distributed resonators down to the UHF frequency range. Helical resonators offer operation down to the HF range with unloaded Q 2 to 3 times that of solenoids. Operation is similar to a shielded solenoid that avoids radiation loss, thus allowing a larger physical size and higher unloaded Q. Formula and nomographs for the design of quarter-wavelength helical resonators such as depicted in Fig. A.17 are given in Zverev [21]. The resonator is coupled electrostatically by capacitive probes near the open end or magnetically by coupling loops near the grounded end.

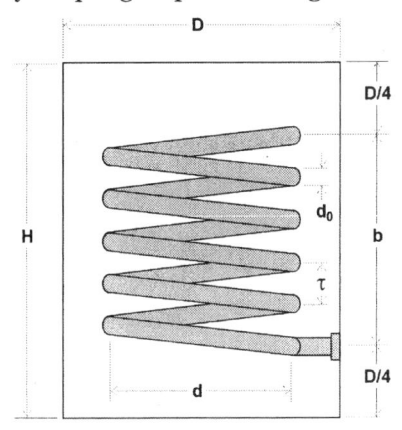

Figure A.17 Helical resonator structure with definition of terms.

The unloaded Q peaks sharply at d/D near 0.53. Zverev's equations are valid for

$$1.0 \le \frac{b}{d} \le 4.0 \qquad (260)$$

$$0.45 \le \frac{d}{D} \le 0.6 \qquad (261)$$

$$0.4 \le \frac{d_0}{\tau} \le 0.6 \ at \ \frac{b}{d} = 1.5 \qquad (262)$$

$$0.5 \le \frac{d_0}{\tau} \le 0.7 \ at \ \frac{b}{d} = 4.0 \qquad (263)$$

$$\tau \le \frac{d}{2} \qquad (264)$$

A circular shield or square shield with inside length S may be used. The frequency is in megahertz and copper metal is assumed. Dimensions here are converted to millimeters from inches in Zverev. Consistent with terminology in this book, the total number of turns is n. Then

$$S \approx \frac{D}{1.2} \qquad (265)$$

$$n = \frac{43{,}688}{f_0(MHz)d} \sqrt{\frac{\log(D/d)}{1-(d/D)^2}} \qquad (266)$$

$$Q_U \approx 1.97 D \sqrt{f_0} \qquad (267)$$

For example, a 20-mm circular shield 300-MHz helical resonator with d equal to 10 mm and b equal to 15 mm requires 9.23 turns of 0.3-mm diameter wire at a pitch of 1.63 turns/mm. The unloaded Q is approximately 682, several times that available using an unshielded solenoid.

These formula conveniently define parameters for the optimum unloaded Q and finding the required number of turns for a desired resonant frequency. However, the helical resonator is a helical transmission line and a direct transmission line model for the characteristic impedance and electrical length for use in circuit simulators is described in previous sections of this appendix.

References

[1] AVX Corporation, www.avx.com.

[2] Modelithics, Inc., www.modelithics.com.

[3] Application Note 1004: Varactor SPICE Models for RF VCO Application, Skyworks Solutions, Inc., December 2005, www.skyworksinc.com.

[4] H.A. Wheeler, "Simple Inductance Formulas for Radio Coils," *Proc. IRE*, Oct 1928, p. 1398 and March 1929, p. 580.

[5] R. Rhea, *HF Filter Design and Computer Simulation*, Scitech Publishing (Noble Publishing), Raleigh, 1994, pp. 73-74.

[6] R.G. Medhurst, "H.F. Resistance and Self-Capacitance of Single-Layer Solenoids," *Wireless Engineer*, February 1947, p. 35, and March 1947, p. 80.

[7] F. Langford-Smith, *Radiotron Designer's Handbook*, Wireless Press, Sydney, 1952, p. 464.

[8] R. Rhea, "A Multimode High-Frequency Inductor Model," *Applied Microwave & Wireless*, November/December 1997, pp. 70-74, 76-78, 80.

[9] F. Langford-Smith, *Radiotron Designer's Handbook*, Wireless Press, Sydney, 1952, p. 445.

[10] C. Snow, "Formulas for Computing Capacitance and Inductance," *National Bureau of Standards Circular #544*, 1954.

[11] ww.tdk.co.jp.

[12] A.G. Bogle, "The Effective Inductance and Resistance of Screened Coils," *Journal of the IEE*, p. 299, 1940.

[13] J. Sevick, *Transmission Line Transformers*, Scitech Publishing (Noble Publishing), Raleigh, NC, 2006.

[14] M. A. R. Gunston, *Microwave Transmission-Line Impedance Data*, SciTech Publishing (Noble Publishing), Raleigh, 1996.

[15] G. Stracca, G. Machiarella and M. Politi, "Numerical Analysis of Various Configurations of Slab Lines," *IEEE Trans. MTT-34*, March 1986, p. 359.

[16] S.B. Cohn, "Characteristic Impedance of the Shielded-Strip Transmission Line," *IEEE Trans. MTT-2*, July 1954, p. 52.

[17] W. Hilberg, "From Approximations to Exact Relations for Characteristic Impedances," *IEEE Trans. MTT-17*, May 1969, p. 259.

[18] H.A. Wheeler, "Transmission Line Properties of a Stripline Between Parallel Planes," *IEEE Trans. MTT-26*, November 1978, p. 866.

[19] E. Hammersted and O. Jensen, "Accurate Models for Microstrip Computer-Aided Design," *IEEE Trans. MTT-28*, May 1980, pp. 407-409.

[20] P. Lefferson, "Twisted Magnet Wire Transmission Line," *IEEE Trans. on Parts, Hybrids and Packaging*, December 1971, pp. 148-154.

[21] A. I. Zverev, *Handbook of Filter Synthesis*, John Wiley & Sons, Hoboken, NJ, 2005, pp. 499-505.

Appendix B: Device Biasing

The initial small-signal bias state of the oscillator sustaining stage is generally less critical than with amplifiers because the steady-state operating voltage and current are largely a function of nonlinear conditions. Nevertheless, oscillator devices are normally initially biased in the linear region. This appendix reviews biasing techniques for bipolar, JFET, MOSFET, and dual-gate MOSFET devices.

B.1 Biasing Bipolar Transistors

With the appropriate use of coupling and bypass reactors, any of the following bias networks may be used for common-emitter (CE), common-base (CB), or common-collector (CC) device topologies. However, proper selection of the bias network minimizes the number of required coupling or bypass reactors.

B.1.1 Bipolar Model for Biasing

A simplified model for biasing a bipolar transistor is shown in Fig. B.1. The collector current is primarily the device forward beta, β, times the base current I_b. The reverse-biased collector-base junction leakage current, I_{cbo}, is generally less than a microamp and for our purpose it is assumed 0. The resulting error is small except at very low collector current or high operating temperature.

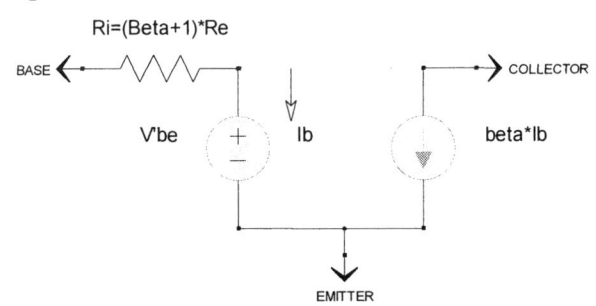

Figure B.1 Simplified bipolar transistor model for bias network design.

A significant factor in bias design is variation in β between devices, which is often 3:1 or more. For example, a typical β specification is 50 minimum and 150 maximum. β is the collector current to base current ratio.

Secondary influences on bias stability are temperature variations in the intrinsic forward base-emitter voltage drop, V'_{be}, and temperature variation in β. The intrinsic forward biased base-emitter junction potential for silicon is approximately 0.79 at 27°C with a negative temperature coefficient.

$$V'_{be}(T) = V'_{be(27°C)} - 0.002(T - 27°C) \qquad (268)$$

β increases with temperature and is approximated by

$$\beta(T) = \beta_{(27°C)}(1 + 0.005(T - 27°C)) \qquad (269)$$

The increase in β and drop in V_{be} with an increase in temperature can lead to thermal runaway in bipolar transistors. If the external resistance in the collector to emitter path is low, an increase in the collector current does not reduce the voltage across the transistor active region adequately and the transistor dissipation increases. This may further increase the current and lead to runaway. This was more of an issue with power devices and earlier germanium transistors.

The base current drive required to sustain the collector current is

$$I_b = \frac{I_c}{\beta} \qquad (270)$$

Since

$$I_e = I_c + I_b \qquad (271)$$

then

$$I_e = \frac{\beta}{\beta + 1} I_c \qquad (272)$$

It is sometimes convenient to use

$$\alpha = \frac{\beta}{\beta + 1} \qquad (273)$$

where α is the collector current to emitter current ratio.

Base current flowing in R_i increases the terminal base-emitter voltage drop. Therefore, the terminal base-emitter voltage, V_{be}, is

$$V_{be} = V'_{be} + R_i I_b \qquad (274)$$

where

Device Biasing

$$R_i = (\beta + 1)r_e \quad (275)$$

and r_e is the intrinsic emitter-base diode resistance. It is approximately

$$r_e = \frac{kT(^\circ K)}{qI_e} = \frac{8.62 \times 10^{-5}(273.15 + T(^\circ C))}{I_e} \quad (276)$$

At 27°C

$$r_e = \frac{26 \times 10^{-3}}{I_e} \quad (277)$$

The base emitter terminal voltage is approximated with a temperature-dependant V'_{be} and a temperature-independent r_e. Then,

$$V_{be}(T) \cong V'_{be}(T) + 0.026 = 0.817 @ 27^\circ C \quad (278)$$

It is interesting to note that with the simplified bias model used in this appendix, V_{be} is independent of the collector current and β. This greatly simplifies the following equations for bias networks.

B.1.2 Common Emitter Bias Networks

Fig. B.2 shows six different bipolar transistor bias networks that use a single-supply voltage. They are an expansion of networks described in Richter [1]. These six topologies are particularly well suited for CE device topologies. The bias network classifications in Fig. B.2 are Bias 1 through Bias 6. In the following formula for resistor values, V_{be} is calculated using Eq. 278, or if temperature can be ignored, V_{be} is fixed at 0.817. In the following sections, formulas are given for finding the bias network resistor values. Nominal β and temperature are used. Then the formula given for the collector current is used to assess the effect of β and temperature variation.

B.1.2.1 Bias 1 Simple

Consider first the Bias 1 network. The designer selects the desired device collector current, I_c, and the collector-emitter voltage, V_{ce}. Then

$$R_{c1} = \frac{(V_{cc} - V_{ce})}{I_c} \quad (279)$$

The designer typically either controls R_c by selecting V_{cc} or accepts R_c based on the required V_{cc}. Then

$$R_{b1} = \frac{\beta(V_{cc} - V_{be})}{I_c} \quad (280)$$

For example, using a nominal β of 80, a desired V_{ce} of 2 volts, I_c of 5 mA and V_{cc} of 2.7 volts, then V_{be} is 0.817 volts, R_{c1} is 140 ohms, and R_{b2} is 30.1K ohms.

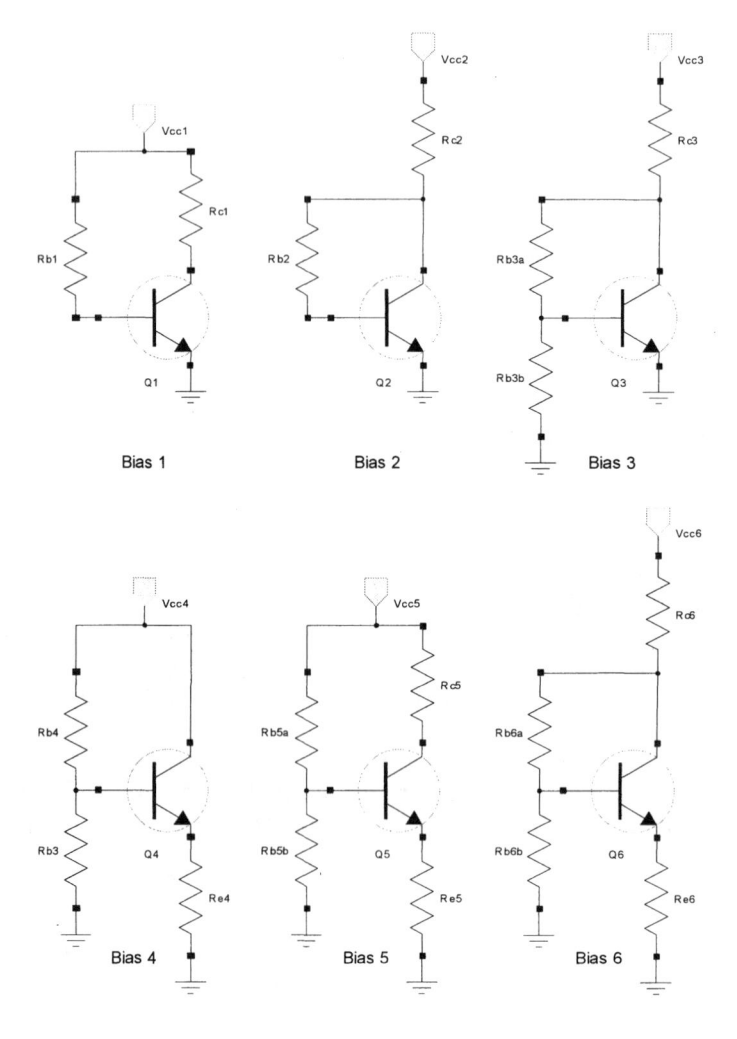

Figure B.2 Bipolar transistor one-battery bias networks for CE topologies.

The stability of the bias network is assessed by considering the collector current as a function of β and temperature. Using the temperature-dependant Eq. 278, then

Device Biasing

$$I_{c1} \cong \frac{\beta(V_{cc} - V_{be})}{R_{b1}} \quad (281)$$

The collector currents with β of 50, 80, and 150 are 3.12, 5, and 9.38 mA, respectively. These -36.4 and +87.6% bias shifts are the result of a nearly direct relationship between β and the collector current with type 1 biasing. The collector currents with a nominal β of 80 and temperatures of 0, 27, and 50°C are 4.20, 5, and 5.71 mA, respectively. Bias shift with temperature is moderate. However, the high sensitivity of the collector current to β suggests the use of a more stable bias network.

B.1.2.2 Bias 2 Simple FB

The Bias 2 network requires only two resistors and feedback insures that the device is biased in the active region at any β and temperature. The base bias is derived from the collector voltage. If the collector current increases, the collector voltage drops, reducing the base drive and thereby reducing the collector current. The bias resistor values are

$$R_{c2} = \frac{\beta(V_{cc} - V_{ce})}{(\beta+1)I_c} \quad (282)$$

$$R_{b2} = \frac{\beta(V_{ce} - V_{be})}{I_c} \quad (283)$$

The collector current is

$$I_{c2} \cong \frac{\beta(V_{cc} - V_{be})}{R_{b2} + (\beta+1)R_{c2}} \quad (284)$$

Using a desired V_{ce} of 2 volts, I_c of 5 mA, and V_{cc} of 2.7 volts, the collector currents with β of 50, 80 and 150 are 3.61, 5, and 7.13 mA, respectively. These -27.8 and +42.6% bias shifts are significantly less than with the Bias 1 network. The collector currents at 0 and 50°C are 4.46 and 5.52 mA, a smaller shift than with Bias 1 networks. The Bias 2 network stability is generally adequate for oscillator design. Stability with β and temperature is improved with a larger voltage drop across R_{c2}.

B.1.2.3 Bias 3 Divided FB

The Bias 3 network adds a base bias resistor to the Bias 2 network. The current flowing in the base-bias resistors is a degree of freedom available to the designer. Increasing this current improves the bias stability with β. In the following equations for the bias resistor values, the current in R_{b3a} is set at 10% of I_c. The current in R_{b3a} must be greater than I_b at the minimum device β.

$$R_{c3} = \frac{V_{cc} - V_{ce}}{1.1 I_c} \qquad (285)$$

$$R_{b3a} = \frac{V_{ce} - V_{be}}{0.1 I_c} \qquad (286)$$

$$R_{b3b} = \frac{V_{be}}{I_c(0.1 - 1/\beta)} \qquad (287)$$

The collector current is then

$$I_{c3} \cong \frac{\beta}{R_{c3}(\beta+1) + R_{b3a}} \left(V_{cc} - V_{be} - \frac{V_{be}}{R_{b3b}}(R_{c3} + R_{b3a}) \right) \qquad (288)$$

Using a desired V_{ce} of 2 volts, I_c of 5 mA, and V_{cc} of 2.7 volts, the collector currents with β of 50, 80, and 150 are 4.47, 5, and 5.51 mA, respectively. The collector currents at 0 and 50°C are 4.08 and 5.80 mA. This is a greater temperature shift than with the Bias 1 and 2 networks, but adequate for oscillator design. As seen later, the Bias 6 network resolves the temperature stability issue.

One of the advantages of the Bias 3 and Bias 6 networks is that the resistors serve the dual purpose of biasing and RF feedback, as illustrated in Section 1.5.7. This works best with relatively low supply voltage. At higher V_{cc}, the current in the resistors becomes excessive.

B.1.2.4 Bias 4 Emitter FB

The Bias 4 network is often used with common-collector Colpitts oscillators. It is also used with common-emitter sustaining stages with an inductive choke between the supply and the collector. The designer selects the current in R_{b4a}. In the following equations it is set at 10% of the collector current. The resistor values are then

$$V_e = V_{cc} - V_{ce} \qquad (289)$$

$$R_{e4} = \frac{\alpha(V_{cc} - V_{ce})}{I_c} \qquad (290)$$

$$R_{b4a} = \frac{V_{ce} - V_{be}}{0.1 I_c} \qquad (291)$$

$$R_{b4b} = \frac{\frac{R_{e4}}{\alpha} + \frac{V_{be}}{I_c}}{(0.1 - 1/\beta)} \qquad (292)$$

The collector current is

Device Biasing

$$I_{c4} = \frac{\dfrac{V_{cc}}{1+R_{b4a}/R_{b4b}} - V_{be}}{\dfrac{R_{e4}}{\alpha} + \dfrac{R_{b4a}}{\beta(1+R_{b4a}/R_{b4b})}} \qquad (293)$$

Using a desired V_{ce} of 2 volts, I_c of 5 mA, and V_{cc} of 2.7 volts, the collector currents with β of 50, 80, and 150 are 4.66, 5, and 5.30 mA, respectively. The collector currents at 0 and 50°C are 4.57 and 5.36 mA. The stability with β and temperature is good. However, Bias 4 through 6 may require capacitor bypass of the emitter resistance. If so, inductance in this path is critical at high frequency, so this emitter resistance is a disadvantage.

B.1.2.5 Bias 5 One-Battery Network

The classic Bias 5 network is often referred to as the one-battery bias network. The designer has two degrees of freedom; the voltage across R_{e4} and the current in R_{b4a}. Setting the voltage across R_{e4} at 10% of V_{cc} and I_{Rb4a} at 10% of the collector current the resistor values are

$$R_{e5} = \frac{0.1\alpha V_{cc}}{I_c} \qquad (294)$$

$$R_{c5} = \frac{V_{cc} - V_{ce} - 0.1V_{cc}}{I_c} \qquad (295)$$

$$R_{b5a} = \frac{V_{cc} - V_{be} - 0.1V_{cc}}{0.1 I_c} \qquad (296)$$

$$R_{b5b} = \frac{0.1V_{cc} + V_{be}}{(0.1 - 1/\beta)I_c} \qquad (297)$$

The collector current is

$$I_{c5} = \alpha \left(\frac{\dfrac{V_{cc} - V_{be}}{R_{b5a}} - \dfrac{V_{be}}{R_{b5b}}}{(1-\alpha) + \dfrac{R_{e5}}{R_{p5}}} \right) \qquad (298)$$

where

$$R_{p5} = \frac{R_{b5a} R_{b5b}}{R_{b5a} + R_{b5b}} \qquad (299)$$

Using a desired V_e of 0.27 volts, V_{ce} of 2 volts, I_c of 5 mA, and V_{cc} of 2.7 volts, the collector currents with β of 50, 80, and 150 are 4.34, 5, and

5.67 mA, respectively. The collector currents at 0 and 50°C are 4.08 and 5.80 mA. The stability with β is fair and the stability with temperature is poor. A higher V_{cc} with a resulting increase in R_{e5} improves the stability of the Bias 5 network significantly.

B.1.2.6 Bias 6 Dual FB

The Bias 6 network uses both emitter and collector-base feedback to stabilize the collector current. The designer has two degrees of freedom; the voltage across R_{e6} and the current in R_{b6a}. Setting R_{e4} at 10% of V_{cc} and I_{Rb4a} at 10% of the collector current the resistor values are given below. Alternatively, the values of R_{b6a}, R_{b6b} and R_{e6} are chosen to set the gain and input and output impedances as described in Section 1.5.7, thus reducing the number of resistors and capacitors. As with the Bias 3 network, this is only practical with low V_{cc}.

$$V_e = 0.1 V_{cc} \tag{300}$$

$$V_c = V_{ce} + V_e \tag{301}$$

$$R_{c6} = \frac{V_{cc} - V_c}{1.1 I_c} \tag{302}$$

$$R_{e6} = \frac{\alpha V_e}{I_c} \tag{303}$$

$$R_{b6a} = \frac{V_c - (V_{be} + V_e)}{0.1 I_c} \tag{304}$$

$$R_{b6b} = \frac{(V_{be} + V_e)}{I_c (0.1 - 1/\beta)} \tag{305}$$

The collector current is then

$$I_{c6} = \frac{V_{cc} - V_{be}(R_{b6a}/R_{b6b} + 1 + R_{c6}/R_{b6b})}{\dfrac{R_{b6a}}{\beta} + \dfrac{R_{e6} + R_{c6}}{\alpha} + \dfrac{R_{e6}(R_{b6a} + R_{c6})}{\alpha R_{b6b}}} \tag{306}$$

Using a desired V_e of 0.27 volts, V_{ce} of 2 volts, I_c of 5 mA, and V_{cc} of 2.7 volts, the collector currents with β of 50, 80, and 150 are 4.59, 5, and 5.37 mA, respectively. The collector currents at 0 and 50°C are 4.40 and 5.51 mA. The stability with β and temperature is good.

B.1.3 Bias 7 Network with Base Diode

The Bias 7 network shown in Fig. B.3 is used for near class-B biasing. Because of the voltage drop in R_i (see Fig. B.1), the base terminal voltage of the transistor is usually higher than the voltage drop of a diode. Therefore, R_{b7} must be rather small to forward bias $Q7$. A more moderate value of R_{b7}, for example 10K ohm, biases $Q7$ just below the active region. A variation of the Bias 7 network uses a stack of multiple diodes at the base and a small resistor in series with the emitter. In this case

$$I_c = \frac{V_{stack} - V_{be}}{\alpha R_e} \tag{307}$$

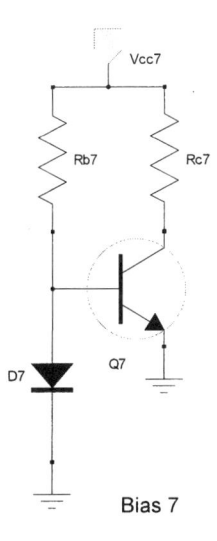

Figure B.3 Bias 7 network with base diode.

B.1.4 Bias 8 Network with Zener

The Bias 8 network utilizes a zener diode in series with a base resistor to provide base drive, as shown in Fig. B.4. The zener is a PN junction specifically designed to be operated in reverse bias. Zener diode operation is influenced by two modes; zener breakdown and avalanche breakdown. Silicon zener diodes up to about 4 volts exhibit a negative temperature coefficient similar to a forward biased diode. At approximately 5.6 volts the temperature coefficient is 2mV/ºC, thus cancelling the temperature coefficient of the transistor. Higher-voltage zeners have an increasingly positive temperature coefficient.

V_{ce} must exceed the zener voltage plus V_{be}. Therefore, this network is not suitable for low supply voltage. The resistor values are

$$R_{c8} = \frac{\alpha(V_{cc}-V_{ce})}{I_c} \quad (308)$$

$$R_{b8} = \frac{\beta(V_{ce}-V_{be}-V_z)}{I_c} \quad (309)$$

Then the collector current is

$$I_c = \frac{V_{cc}-V_{be}-V_z}{\dfrac{R_{b8}}{\beta}+\dfrac{R_{c8}}{\alpha}} \quad (310)$$

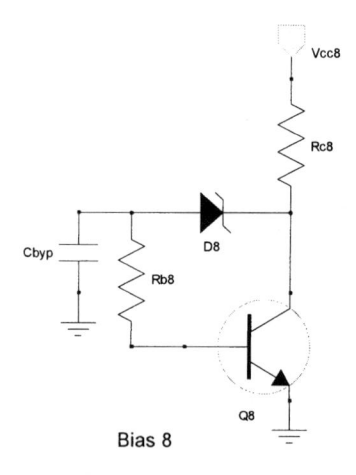

Figure B.4 Bias 8 network with zener diode.

Using a desired V_{ce} of 2 volts, I_c of 5 mA, and V_{cc} of 2.7 volts, the collector currents with β of 50, 80, and 150 are 3.63, 5, and 7.07 mA, respectively. The stability with beta is poor but acceptable for oscillators. The stability with temperature is excellent with proper selection of the zener.

B.1.5 Bias 9 Active Network

The Bias 9 network shown in Fig. B.5 utilizes a PNP sense transistor to control the collector current of the active transistor. The collector current of Q_9 flowing in R_{d9} develops a voltage drop. If the collector current in the active transistor increases, the voltage drop in R_{d9} increases and reduces the sense transistor V_{be} and the base drive to the active transistor. Therefore, the active transistor base drive is determined directly by the collector current. This significantly reduces the sensitivity of collector current to the device β.

Device Biasing

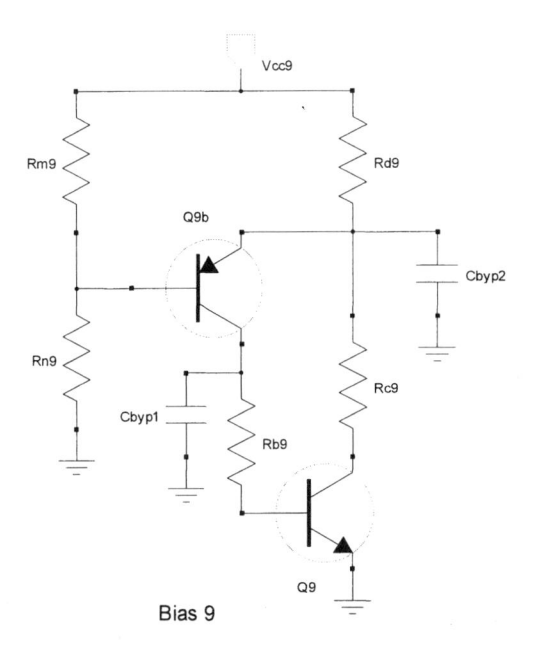

Figure B.5 Bias 9 network with active control of the base drive.

The base current in the sense transistor is approximately $1/\beta^2$ so it is ignored in the following equations. V_e is defined as the voltage at the emitter of the sense transistor. To improve temperature stability, V_e is generally set at least twice V_{be} below V_{cc}. Setting the current in R_{m9} at 10% of I_c, and setting both β and V_{be} of both transistors equal, the resistor values are

$$R_{d9} = \frac{\alpha(V_{cc} - V_e)}{I_c} \tag{311}$$

$$R_{m9} = \frac{V_{cc} - V_e + V_{be}}{0.1 I_c} \tag{312}$$

$$R_{n9} = \frac{V_e - V_{be}}{0.1 I_c} \tag{313}$$

Resistors R_{c9} and R_{b9} do not directly influence the collector current but their values are limited.

$$0 < R_{c9} < \frac{V_e}{I_c} \quad (314)$$

$$0 < R_{b9} < \frac{\beta(V_e - V_{be})}{I_c} \quad (315)$$

Resistor R_{c9} may be set at 0 by using an inductive choke, it may be set at 50 ohms, or it may be set high to avoid shunting signal. However, a resistor reduces V_{ce} of the active transistor.

$$V_{ce} = V_e - I_c R_{c9} \quad (316)$$

Resistor R_{b9} is set high enough to avoid shunting signal at the base of the active transistor. The collector current of the active transistor is

$$I_c = \frac{\alpha\left(V_{cc}\left(1 - \frac{R_{m9}}{R_{m9} + R_{n9}}\right) - V_{be}\right)}{R_{d9}} \quad (317)$$

Using a desired V_{ce} of 2 volts, I_c of 5 mA, and V_{cc} of 2.7 volts, the collector currents with β of 50, 80, and 150 are 4.96, 5, and 5.03 mA, respectively. The collector currents at 0 and 50°C are 4.61 and 5.34 mA. The stability with β is nearly perfect. The stability with temperature is good and improves with a larger voltage drop across R_{d9}. This degree of bias network complexity is justified only if a highly stable oscillator output amplitude is required.

B.1.6 Bias 10 Dual Supply

If both positive and negative supplies are available, the Bias 10 network shown in Fig. B.6 is economic and stable. It is well suited for common-base topologies. Using the absolute values for the voltages of V_{ee} and V_{be}, the value of the emitter resistor is

$$R_{e10} = \frac{\alpha(V_{ee} - V_{be})}{I_c} \quad (318)$$

where both V_{ee} and V_{be} are negative. The value of R_{c9} is confined by the range

$$0 < R_{c10} << \frac{V_{cc}}{I_c} \quad (319)$$

Smaller values of R_{c9} provide higher V_{ce}, which is desirable.

$$V_{ce} = V_{cc} - I_c R_{c10} + V_{be} \quad (320)$$

Device Biasing

However, because the collector impedance of the CB stage is high, an inductive choke may be required at the collector for small V_{cc} to avoid shunting the signal with small values of R_{c10}.

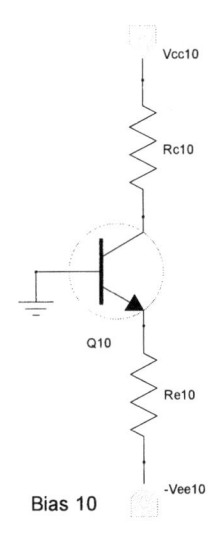

Figure B.6 Bias 10 network with dual supplies.

The collector current is

$$I_c = \frac{\alpha(V_{ee} - V_{be})}{R_{e10}} \qquad (321)$$

Using a desired V_{ee} of -2.7 volts, I_c of 5 mA, and V_{cc} of 2.7 volts, the collector currents with β of 50, 80, and 150 are 4.96, 5, and 5.03 mA, respectively. The collector currents at 0 and 50°C are 4.84 and 5.13 mA. The stability with β is outstanding and with temperature is excellent. The excellent temperature stability is because V_{ee} is several times V_{be}. Bias 10 network is the obvious choice for CB topologies when both positive and negative supplies are available.

B.1.7 Bias 11 Common Collector Network

The Bias 11 network depicted in Fig. B.7 is a variation of the Bias 4 network with the supply and ground nodes exchanged. An NPN device is used with a negative supply or a PNP device with a positive supply. The resistor values and collector current are identical to the Bias 4 network.

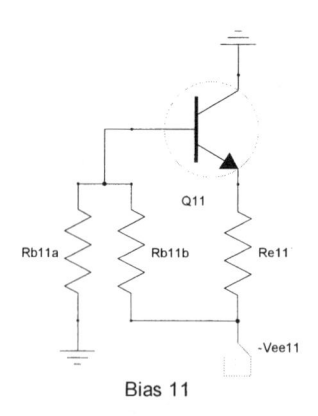

Figure B.7 Bias 11 network for common-collector topologies and a negative supply for NPN transistors.

A variation on the Bias 11 network omits resistor R_{b11b}. V_e is commonly near 50% of V_{ee}. Choosing V_e and I_e, then

$$R_{e11} = \frac{V_{ee} - V_e}{I_e} \quad (322)$$

where the right-side variables are all negative. Then

$$R_{b11a} = \frac{\beta(V_e + 0.817)}{I_e} \quad (323)$$

B.1.8 Bipolar Bias Network Summary

The collector currents with variation of β and temperature for each bias network are repeated in Table B.1. The bias stability performance of the networks is a function of design choices. These examples used a supply voltage of 2.7 volts and emitter resistor drops as low as 0.27 volts. These low values accentuate the effects of β and temperature changes.

Table B.1 Comparison of the stability of I_c for the 11 bipolar bias networks. The nominal current is 5 mA at beta=80 and T=27°C

Type	Name	Beta=50	=150	T=0°C	=50°C
1	Simple	3.12	9.38	4.20	5.71
2	Simple FB	3.62	7.10	4.42	5.48
3	Divided FB	4.47	5.51	4.08	5.80
4	Emitter FB	4.66	5.30	4.57	5.36
5	One Battery	4.34	5.67	4.08	5.80
6	Dual FB	4.59	5.37	4.40	5.51
8	Zener	3.63	7.07	N/A	N/A
9	Active	4.96	5.03	4.61	5.34
10	Dual supply	4.96	5.03	4.84	5.13

Because bias stability is not critical for most oscillator applications, any of the networks of type 3 and higher are adequate. The choice of bias network type is driven by minimization of the number of required resistors, chokes, and bypass capacitors.

B.1.9 Saturated Output Power and Biasing

When designing high-power amplifiers, whether for use as amplifiers or oscillators, the type of biasing is important. Higher-power compression levels are achieved by using a constant base voltage source rather than a current source [2]. In other words, at DC, the source impedance of the bias network should be low. Many bias schemes use high resistor values at the base for biasing to avoid shunting the signal. This is perfectly acceptable when high output power and efficiency are not critical. However, the highest output power and efficiency are achieved by using a low impedance voltage source to bias the base and then using a choke to deliver this voltage so as not to shunt the signal.

B.2 FET Bias Networks

Junction field-effect transistors (JFETs) are voltage-controlled current sources. Material between the source and drain terminals is doped with an abundance of positive charge carriers in the P-type JFET or of negative charge carriers in the N-type JFET. The gate terminal is reverse-biased but the field potential effectively changes the width of the source-drain channel and controls the current. The reverse-biased input gate is high resistance and little current flows into the gate.

Metal oxide silicon FET (MOSFET) devices have an insulating oxide layer between the gate and channel, further increasing the input resistance. MOSFETs typically have a lower transconductance. Today, the "metal" material is generally replaced by a polysilicon layer. VMOS devices are MOSFETs with a "V" shaped gate, allowing greater current in the source to drain channel. GaAs MESFETs, or GASFETs, are also related [3, 4]. In the FET, the drain current, I_d, is controlled by the gate to source voltage, V_{gs}.

$$I_d = I_{dss}\left(1 - \frac{V_{gs}}{V_P}\right)^2 \tag{324}$$

for

$$0 < V_{gs} < V_P \tag{325}$$

where I_{dss} is the drain saturation current and V_P is the pinch-off voltage. The I_{dss} specification for the 2N5484 is 1 to 5 mA and for the J309 is 12 to

30 mA [5]. A wide variation in I_{dss} is typical for FETs. The V_P specification for the 2N5484 is -0.3 to -3 volts and for the J309 is -1 to -4 volts.

B.2.1 Bias 15 Simple FET Network

Fig. B.8 depicts three different FET bias networks. N-type FETs that utilize a positive drain supply voltage are shown. The supply voltage polarities are reversed for P-type FETs.

The Bias 15 network illustrates the simplest form of biasing for the FET. Resistor R_{g15} holds the gate near 0 volts. Since the DC current flowing in the gate is essentially nil, a large value such as 100K ohm is used for R_{g15} to avoid shunting signal. Alternatively, an inductive choke may be used. With a V_{gs}, of 0 volts, I_d, is equal to I_{dss}. The drain resistor is

$$R_d = \frac{V_{dd} - V_d}{I_{dss}} \qquad (326)$$

where V_d is the desired V_{ds} of the FET. I_{dss} has a small negative temperature coefficient so thermal runaway is not an issue.

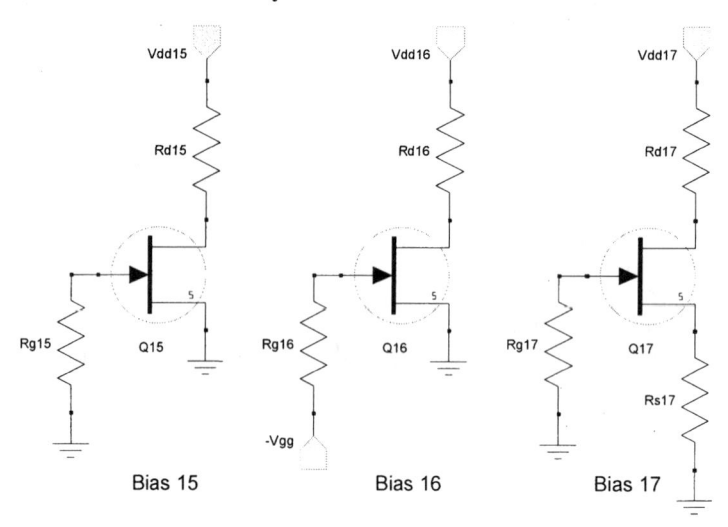

Figure B.8 Bias configurations for N-type FETs.

B.2.2 Bias 16 Gate Voltage

The Bias 16 network utilizes a fixed negative bias voltage at the gate. Again R_{g16} may be either a choke or high value of resistance. Because V_s is zero, V_{gg} equals V_{gs} and from Eq. 139, the gate voltage is

Device Biasing

$$V_{gg} = V_P\left(1 - \sqrt{\frac{I_d}{I_{dss}}}\right) \quad (327)$$

For example, with I_{dss} of 17.2 mA and V_P of -1.72 volts, for a desired I_d of 5 mA, V_{gg}=-0.795 volts. The drain resistor is

$$R_{d16} = \frac{V_{dd} - V_d}{I_d} \quad (328)$$

The drain current as a function of I_{dss} and V_P is then given by Eq. 324. The Bias 16 network provides a method of biasing the FET drain current at a value lower than I_{dss}. However, it does not stabilize the drain current against the high device to device variation of I_{dss}. It also requires a negative supply. A more practical solution is the next FET bias network.

B.2.3 Bias 17 Source FB

The Bias 17 network shown in Fig. B.8 provides a method for stabilizing the drain current and setting it at a specific value below I_{dss}. Drain current flowing in the source resistor biases the source at a positive voltage relative to the gate effectively reverse-biasing the gate. The required value of source resistance is

$$R_{s17} = -\frac{V_P}{I_d}\left(1 - \sqrt{\frac{I_d}{I_{dss}}}\right) \quad (329)$$

For a desired drain to source voltage, V_{ds}, the drain resistor is

$$R_{d17} = \frac{V_{dd} - V_{ds} - R_{s17}I_d}{I_d} \quad (330)$$

The drain current is then

$$I_d = \frac{\left(\frac{2R_{s17}}{V_P} - \frac{1}{I_{dss}}\right) - \sqrt{\left(\frac{2R_{s17}}{V_P} - \frac{1}{I_{dss}}\right)^2 - 4\frac{R_{s17}^2}{V_P^2}}}{2R_{s17}^2/V_P^2} \quad (331)$$

Consider a desired V_{dd} of 5 volts, V_{ds} of 3 volts, and I_d of 8 mA. Using I_{dss} of 17.2 mA and V_P of -1.72, then R_{s17} is 68.37 and R_{d17} is 181.63. Using standard values of 68 and 180 ohms and device I_{dss} values of 12, 17.2, and 30, the drain currents are 6.57, 8.02, and 10.40 mA, respectively. Using these standard values and a nominal I_{dss} of 17.2 mA, the drain current with device pinch-off voltages of -1, -1.72, and -4 volts, the drain currents are 6.01, 8.02, and 11.25 mA, respectively. If capacitor bypass of the source resistance is required, inductance in this path is critical at high frequency, so this source resistance is a disadvantage.

B.2.4 Bias 18 Dual-Gate FET

Dual-gate FETs are popular as amplifiers through the VHF frequency range and as mixers with the RF signal applied to gate 1 and the LO signal applied to gate 2. Dual-gate FETs also offer a convenient method of applying AGC to control the loop gain without limiting, thus linearizing an oscillator [6]. The Bias 18 network for dual-gate FETs is shown in Fig. B.9.

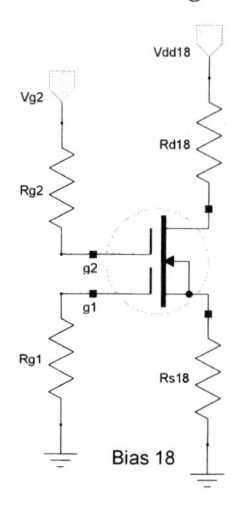

Figure B.9 Bias network for a dual-gate N-channel FET.

Figure B.10 shows drain current curves versus the gate 1 to source voltage for an MRF966 N-channel, dual-gate, depletion-mode GaAsFET. The drain-source voltage is 5 volts and the gate 2-to-source voltage is a running parameter.

Consider the case with a gate 2-to-source voltage of 0 volts. The drain current data from the graph agrees reasonably well with Eq. 324 with an I_{dss} of 48.5 mA and V_P of -5.3 volts. The specification I_{dss} is 30 mA minimum, 50 mA typical and 80 mA maximum. The specification gate 1-to-source V_P is -2.0 volts minimum and -4.5 volts maximum.

The Bias 15 network can be used to operate the dual-gate FET at I_d equal to I_{dss}. Gate 2 is also set at 0 volts to operate at full gain, or AGC is implemented by applying a gate 2 control voltage of 0 to -3 volts.

Device Biasing

MRF966 Dual-Gate Depletion-Mode GaAsFET

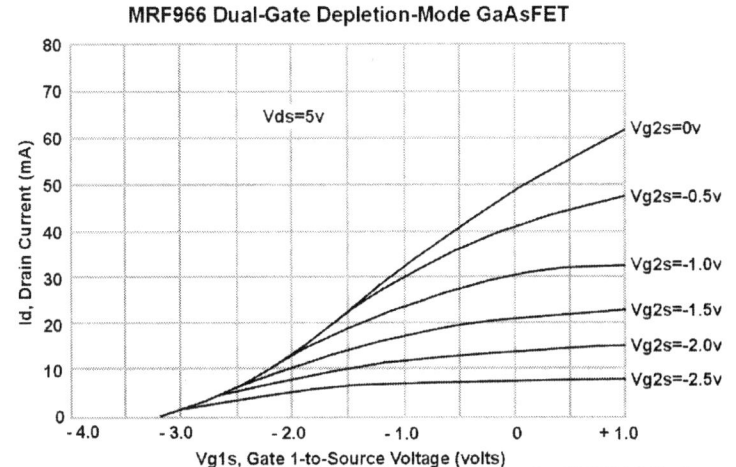

Figure B.10 Drain current versus gate 1-to-source voltage for the MRF966 N-channel, dual-date, depletion-mode GaAsFET with a drain-source voltage of 5 volts and gate 2-to-source running parameter.

Alternatively, the Bias 17 network can be used to operate I_d at less than I_{dss} and stabilize the operating point against variation in I_{dss}. For example, to operate at an I_d of 25 mA, a gate 1-to-source voltage of -1.4 volts is required. This gate bias is achieved using

$$R_{s18} = \frac{-V_{g1s}}{I_d} = \frac{1.4}{0.025} = 56\, ohms \qquad (332)$$

The drain resistor is then given by Eq. 330. If necessary, the source resistance is bypassed and a choke is placed in series with the drain resistor.

B.3 Bias 19 MMIC Gain Block

As indicated in Chapter 1, MMIC gain blocks available from a variety of manufacturers make good oscillator sustaining stages. These device typically include internal bias networks and power only need be applied to the output via a choke or supply dropping resistor. A typical configuration is shown in Fig. B.11.

The MMIC operates at a specific voltage, V_c, and current, I_c. When the supply voltage equals the MMIC operating voltage, R_{c19} is replaced with an inductive choke and the device sinks the specified current. The choke may be a large value to avoid shunting the output signal or it may be a smaller value to act as a lead network to remove some high-frequency phase lag of the MMIC.

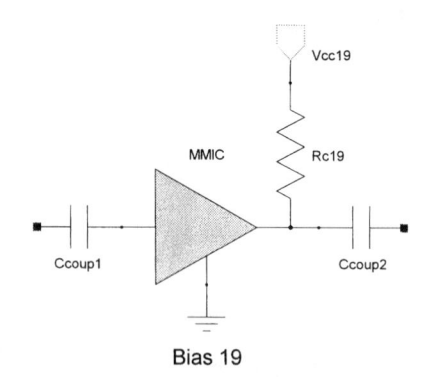

Figure B.11 Bias network for MMIC gain block with internal biasing.

If the supply voltage exceeds the operating voltage of the MMIC then the dropping resistor is

$$R_{c19} = \frac{V_{cc19} - V_c}{I_c} \qquad (333)$$

If R_{c19} is less than approximately 3X Z_o then a choke is placed in series with the resistor.

References

[1] K. Richter, Application Note 1293: A Comparison of Various Bipolar Transistor Biasing Circuits, Avago Technologies, www.avagotech.com.

[2] B. Lee and L. Dunleavy, "Understanding Base Biasing Influence on Large Signal Behavior in HBTs," *High Frequency Electronics*, May 2007, pp. 66, 68-70, 72-73.

[3] W. Hayward, *Introduction to Radio Frequency Design*, American Radio Relay League, Newington, CT, 2004.

[4] R. Pengelly, *Microwave Field-Effect Transistors*, SciTech Publishing (Noble Publishing), Raleigh, NC, 1994.

[5] Fairchild Semiconductor, www.fairchildsemi.com.

[6] J. McLucas, "Stable, 18 MHz Oscillator Features Automatic Level Control," Clean-Sine-Wave Output, *EDN*, June 23, 2005, pp. 82-83.

[7] *RF Device Data*, Motorola, 1983, pp. 6-2.

Constants and Symbols

Note: not all variables are listed. For example, variables commonly used in electronic literature, such as V_{be}, or variables defined locally for use only in nearby equations, are not listed here.

A	area
$\overline{A}$	acceleration vector
A_L	inductance index for toroids
α	attenuation of a transmission line
α	drive-level frequency dependence for piezoelectric material
α	ratio of transistor DC emitter current to the collector current
B	susceptance term of a complex admittance
B	bandwidth term in the noise power $kTBF$
BW_{corr}	SSB phase noise correction due to the SA resolution bandwidth
BW_{3dB}	absolute 3-dB down bandwidth of an amplitude response
$BW_{\Delta frms}$	bandwidth of the spectral density of frequency fluctuations
$BW_{\Delta \phi rms}$	bandwidth of the spectral density of phase fluctuations
$B1$	adjunct to the Rollet stability factor, K
β	ratio of transistor DC collector current to the base current
C	value of capacitance
C	power coupling ratio of a directional coupler
C_i	correction factor of an SSB phase noise test set due to image power
C_V	junction capacitance of a varator diode
D	diameter of an object such as a cylinder
Δ	determinant of a 2-port S-parameter matrix used to calculate K
$\Delta f_{rms}(f_m)$	frequency fluctuation, rms value
Δf	residual FM noise
$\Delta P_{rms}(f_m)$	baseband spectrum analyzer measured noise power at offset f_m
$\Delta \phi_{rms}(f_m)$	phase fluctuation, rms value
$\Delta \phi_{peak}$	peak integrated phase deviation in radians
$\Delta \phi$	residual PM noise
$\Delta \phi(t)$	phase perturbation with time
$\Delta v(f_m)$	voltage transfer function of a delay-line phase discriminator
$\Delta v_{rms}(f_m)$	baseband spectrum analyzer measured noise voltage
$\Delta v(t)$	phase detector output voltage resulting from a phase perturbation
ESR	effective series resistance of a capacitor
ε_0	permittivity of free space 8.8542×10^{-12} *farads/meter*
ε_r	relative permittivity of a material
F	noise factor of the amplifier sustaining stage of an oscillator

F_{buf}	noise factor of a buffer amplifier following the oscillator
f_c	flicker corner frequency of an amplifier
f_m	modulation frequency (also Fourier, offset, or baseband frequency)
F_t	transistor unity current-gain frequency, $h_{21}=1$
G	conductance term of a complex admittance
G	Randall/Hock complex open-loop gain considering mismatch
G_f	target gain of an amplifier with shunt and series feedback
Γ	piezoelectric material acceleration sensitivity vector
$\Gamma(\omega_0\tau)$	impulse sensitivity function (ISF) used in LTV noise theory
Γ_{rms}	rms value of the impulse sensitivity function (ISF)
H_o	magnetic field intensity
h	height of an object such as a substrate or dielectric resonator
I_{dss}	FET drain current with $V_{gs}=0$
$\overline{i_n^2}/\Delta f$	mean squared spectral density of the white noise current
IL	insertion loss of component or circuit, $=-|S_{21}|$
k	Boltzsmann constant $\quad 1.3806\times10^{-23}\ joules/kelvin$
k	Coupling coefficient
kT	$k\times$ room temp (23.8 ºC) $\quad 4.14\times10^{-21}\ joules$
K	kelvin unit of temperature $\quad 0\,K=-273.16\ ^0C$
K	Rollet stability factor, see also $B1$ and Δ
K_p	oscillator frequency pushing caused by a supply voltage change
K_ϕ	phase detector gain constant, volts/radian
K_v	voltage controlled oscillator tuning gain constant
l	length of an object such as a solenoid
L	value of inductance
$L(f_m)$	SSB phase noise as a function of the modulation frequency
L_A	transmission loss, ignoring dissipation, due to reflection
LF	inductance decrement factor of a shield surrounding a solenoid
λ_o	wavelength in a free space
λ_{go}	wavelength in a transmission line medium
m,n	number of turns in inductors and transformers
M	mutual inductance
N	multiplication factor of a frequency multiplier
N	number of inverter sections in a ring oscillator
N_i	difference of the power level of the image and desired signal
N_{corr}	SSB phase noise total correction factor for a spectrum analyzer
NBW_{corr}	noise correction factor due to an SA IF filter bandwidth
ND_{corr}	noise correction factor due to an SA peak detector
η	amplifier or oscillator output power to consumed DC power ratio
η	characteristic impedance of free space $\quad 376.73\ ohms$
P_{1dB}	amplifier output power with 1-dB gain compression
$P_{out,\,max}$	maximum output power of a given class of amplifier

Constants and Symbols

P_s	output power of an oscillator
Ψ	Q factor for a solenoid due to Medhurst
ϕ	transmission phase shift of a network (angle of S_{21})
ϕ_o	a transmission phase shift equal to 0 degrees
q	elementary charge (electron) 1.602×10^{-19} coulomb
q_{max}	maximum charge displacement across the resonator capacitor
Q_{cap}	unloaded Q of a capacitor
Q_{ind}	unloaded Q of an inductor
Q_{ext}	external loaded Q of a resonator
Q_L	loaded Q
Q_U	unloaded Q
Q_R	unloaded Q of a resonator
r	radius of an object such as a solenoid
R	resistance or real term of a complex impedance
R_{enr}	effective noise resistance of a varactor diode
R_L	load resistance
ρ_n	voltage reflection coefficient at port n
S	side length of a square shield for coax and helical resonators
S_{nm}	scattering parameter: example, S_{21} is the forward S-parameter
$S_\phi(f_m)$	spectral density of phase fluctuations
$S_{\Delta f}(f_m)$	spectral density of frequency fluctuations
$S/N_{ultimate}$	demodulated signal to noise ratio with high carrier to noise ratio
t	thickness of an object such as a toroid core
t	time
τ	time that an impulse occurs in LTV noise theory
τ	mean spacing of turns in a helical resonator
τ_d	delay of a delay line
θ	electrical length of a transmission line
θ_d	phase deviation of a carrier
μ_o	permeability of free space $4\pi \times 10^{-7}$ henry/meter
V_{nv}	varactor noise voltage
υ_o	free space velocity of light 2.99792×10^8 meters/second
V_P	FET drain-source voltage that pinches off the channel
$V_{P,pk}$	peak RF voltage across a resonator
VSWR	voltage standing wave ratio
ω	radian frequency, $\omega = 2\pi f$
X	reactance term of a complex impedance
y	gyromagnetic ratio, =2.8 MHz/oerstead for undoped YIG
Z_o	characteristic impedance of a measurement system or line
Z_n	impedance (complex) at port n

About the Author

Randall W. Rhea was born to Noble (Bill) Rhea and Emma Jane Wright, June 10, 1947 in Findlay, Illinois. He was licensed as radio amateur WN9FFO in 1962. He graduated BSEE with honors from the University of Illinois and married Marilynn Sue Thomas in 1969. He attended Seattle University and they both graduated from Arizona State University (ASU) in 1973. His MSEE thesis at ASU was the construction and operation of an earth station that received Unified S-Band communications from the Apollo 16 and 17 Command Modules in lunar orbit.

He worked at the Boeing Company, Goodyear Aerospace, and Scientific Atlanta were he was conferred Principal Engineer. His engineering experiences include amplifiers, antennas, CATV distribution, CATV head-ends, earth stations, filters, modems, oscillators, radar, receivers, synthesizers, and management. In 1985, Marilynn and Randall founded Circuit Busters (later Eagleware Corporation). He wrote the initial releases of the Eagleware circuit simulator and the filter, oscillator, and transmission line synthesis programs. In 1994, they founded Noble Publishing. He is the author of numerous technical papers on antennas, amplifiers, components, filters, matching, modeling, oscillators, simulation, synthesis, and the history of the microwave industry. He is the lecturer in numerous CD-ROM tutorials published by SciTech Publishing. He is the author of the books *Oscillator Design and Computer Simulation*, first edition; *HF Filter Design and Computer Simulation; Oscillator Design and Computer Simulation*, second edition; *Cable Television Signal Distribution;* and Chapter 6 of the *Handbook of Microwave Filter Technology*. Eagleware Corporation was sold to Agilent Technologies in 2005 and Noble Publishing was sold to SciTech Publishing in 2006. He has taught full-day seminars on filter and oscillator design to over a thousand engineers at trade shows, the Georgia Institute of Technology, and companies worldwide.

Randall and Marilynn live in an antebellum plantation home near Thomasville, Georgia, enjoying visits from two adult children, three grandchildren, family, and friends. They spend 2 months of the year at their cabin in New Mexico. In the summer of 2003, Randall toured 48 states by motorcycle. His current amateur radio license is N4HI and he has DXCC and first place Georgia CW operating contest awards. Randall also enjoys technical writing, astronomy, and wine making.

Randall may be reached at *oscillatordesign@gmail.com* for questions, comments, and the reporting of errata. From time to time, an errata sheet may be published and supplemental material may be available.

Index

1-port method, *see* one-port
Acceleration effects, crystal, 339
Acoustic noise, 179, 339
AGC
  dual-gate FET, 436
  example, 211
  HP200A, 129
  intro, 11
  passive, 128
Aging, crystal, 338
Alechno's technique, 63
  example transform, 65
Amplifier, 1, 15
  basic CE, 23
  bipolar, 15
  compression, 86
  differential, 35
  FET, 21
  general purpose, feedback, 29
  MMIC, 34
  nonlinear, 84
  Norton, 33
  resistive feedback, 26
  stability, 11, 19, 21, 29
  statistical analysis, 25
  summary, 39
  sustaining stage, 15
  transformer feedback, 33
  type A and B, 29
Analysis methods, 1
  1-port, 57
  open-loop 2-port, 1
  negative-conductance, 69
  negative-resistance, 58
  analyzing existing oscillators, 74
  statistical, 78
Biasing, Device (appendix), 419
  biasing for high power, 433
  bipolar bias networks,
    active (type 9), 428
    base diode (type 7), 426
    common collector (type 11), 431
    divided FB (type 3), 423
    dual FB (type 6), 426
    dual supply (type 10), 430
    emitter FB (type 4), 424
    model, simplified, 419
    one-battery, common (type 5), 425
    simple (type 1), 421
    simple with FB (type 2), 423
    summary, bipolar bias, 432
    zener, (type 8), 427
  FET bias networks, 433
    dual-gate FET (type 18), 436
    gate voltage (type 16), 434
    simple (type 15), 434
    source FB (type 17), 435
  MMIC bias network (type 19), 437
Bifilar transformers, 404
Bode response, 1
Buffer noise, 202
Butler oscillator
  ceramic piezoelectric, 383
  fundamental mode, 355
  overtone mode, 364
  overtone mode, multiplied, 369
  SAW oscillator, 377
Capacitance, effective input, 62
Capacitor models, 389
Cascade, *see* open-loop
Cayenne (simulator), 152
  frequency-dependant models, 153
  numerical precision, 153
Ceramic resonators (TEM coaxial), 284
Ceramic resonators (piezoelectric), 380
  see also Piezoelectric Oscillators
Characteristic impedance,
  see transmission lines
Clapp oscillator, 242
Class A, B, C...*see* nonlinear amplifiers
Closed-loop
  gain, 145
  nonlinear introduction, 112
CMOS inverter crystal oscillator, 351
Coaxial line model, 409
Coaxial resonator, 268, 284
Colpitts oscillator, 14
  harmonic balance example, 109
  L-C examples, 236
  resonator, 49
  SAW example, 374
  vacuum tube, 14
Compression, amplifier, 86
Constants (list), 439
Conversion efficiency
  example oscillator, 305
  versus class, *see* nonlinear amplifiers
  versus path resistance, 105
Coupled parallel resonator oscillator, 254
Coupled series resonator oscillator, 251
Couplers (directional), 407
Coupling, *see* output, resonator or
    varactor coupling
Coupling test by modulation, 293
Crystal, *see* **Piezoelectric Oscillators**
Cuts, *see* **Piezoelectric Oscillators**

Delay line oscillator (ring), 230
Delay line resonator, 41
Device selection
   negative-resistance oscillator, 61
   open-loop cascade, 11
Directional coupler
   model, 407
   output coupling example, 127
Distributed (chapter) examples
   coaxial, probe coupled, bipolar, 279
   coaxial, hybrid, MMIC, 277
   dielectric resonators, 284
      see also DRO
   DRO, cascade, MMIC, 292
   DRO, negative-R, bipolar, 289
   half-wave, end-coupled, bipolar, 281
   helical transmission line, bipolar, 282
   negative-R high power, bipolar, 274
   negative-R hybrid, bipolar, 273
   quarter-wave hybrid, bipolar, 276
   resonator technologies, 263
Distributed/lumped equivalents, 264
Driscoll crystal oscillator, 357
Drive level, crystal, 334
DRO, 284
   coupling, 287
   coupling test by modulation, 293
   examples, 289
   resonant frequency, 286
   resonator properties, 284
   simulation, 290
   unloaded Q, 286
Efficiency, see conversion efficiency
Examples with measured data
   Butler ceramic resonator oscillator, 384
   ceramic (piezoelectric), Butler, 383
   ceramic (piezoelectric) resonator
      model, 380
   coaxial (probe-coupled) oscillator, 280
   coaxial resonator VCO, MMIC, 310
   coaxial resonator VCO on FR4, 314
   Driscoll crystal oscillator, 359
   helical coaxial line model, 404
   Gumm oscillator, 256
   high-power 1-GHz oscillator, 275
   HP8640B signal generator noise, 196
   inductor (solenoid) model, 397
   loaded microstrip on FR4, MMIC, 312
   Modelithics capacitor model, 393
   negative-conductance oscillator, 307
   negative-resistance oscillator, 210
   oscillator with AGC, 129
   overtone crystal model, 361
   permeability-tuned FET Colpitts, 299
   permeability-tuned inductors, 298
   Pierce crystal oscillator, 349
   power amplifiers, 95, 102, 107

quartz-resonator response, 325
Rhea oscillator, 155
solenoid inductor model, 397
Existing designs, analyzing, 74
FBAR resonators, 386
Ferrite bead model, 401
Flicker noise
   crystal resonators, 338
   devices, 199
   introduction, 178
   Leeson's equation, 180
   PLL, 214
Frequency accuracy/stability, 380
Frequency discriminator, noise
   measurement, 192
Frequency multiplication noise, 183
Gain, 1
   margin, 1
   peak, 11
   true, self-terminated, 10
General Purpose (chapter) examples
   Clapp, CC bipolar, 242
   Colpitts, CB bipolar, 238
   Colpitts, CC bipolar, 236
   Colpitts, CC bipolar, neg. supply, 238
   Colpitts, CD FET, 240
   Colpitts, CG FET, 241
   coupled series resonator, CE bip, 251
   coupled parallel resonator, CE bip, 254
   Gumm, differential bipolar, 256
   Gumm, simplified, FET, 257
   Hartley, CB bipolar, 243
   Hartley, CD FET, 245
   Hartley, CD FET, simple, 247
   Hartley, CG FET, 248
   multivibrator, 227
   Pierce, CE bipolar, 249
   Pierce, CS FET, 251
   R-C phase shift, bipolar highpass, 224
   R-C phase shift, bipolar lowpass, 220
   R-C phase shift, opamp, 225
   Rhea, CE bipolar, 253
   ring, TTL, 230
   Seiler, CC bipolar, 243
   twin-T, opamp, 232
   twin-T, bipolar, 232
   Wien bridge, opamp, 226
Gumm oscillator
   differential bipolar, 256
   simplified FET, 257
Harbec, 83
   see also harmonic balance
Harmonic balance analysis, 83
   Colpitts example, 109
   excitation current vs. parameters, 115
   excitation source, 113
   negative-resistance example, 116

Index

nonlinear noise example, 206
Harmonics
  versus coupling reactance, 125
  versus drive level, 90
  versus number balanced, 90
Hartley oscillators, 243
Helical
  resonator model, 415
  transmission line model, 402
Heterodyne/counter, noise measurement, 190
High loaded Q simulation, 162
Holder, crystal, 341
HP 200A audio generator, 129
Impedance
  introduction, 4
  nonlinear shift, 87
Inductor
  models, 394
  mutually coupled, 402
  unloaded Q, 235
Insertion loss, resonator, 47
Inverted-mesa
  crystal resonator, 328
  Pierce oscillator, 359
Jitter, 174
Leeson noise
  definition, 79
  reducing, 197
Line, *see* transmission line
Linear Techniques (chapter), 1
  summary, 80
Linear time invariant theory (LTI), 180
Linear time variant theory (LTV), 184
Loaded Q, 45
  definitions, 45
  from group delay, 46
  see also unloaded Q
Load pulling, 121
  versus loaded Q, 122
  versus coupling reactance, 125
Lumped/distributed equivalents, 264
Master noise equation, 183
Match, 3
  example, 7
  impedance relationship, 4
  mismatch error, 5
  mismatch error table, 6
  Randall/Hock correction, 9
  requirements, 3
  summary, 10
Measured oscillator data,
  see examples with measured data
Measuring phase noise, *see* noise
MEMS resonators, 386
Microstrip model, 413
Miller crystal oscillator, 343

Mismatch error, 5
Modeling (appendix), 389
  capacitor models, 389
  ferrite beads, 401
  helical transmission lines, 402
  inductors, 394
    core, cylindrical, 396
    inductance (Wheeler), 207
    mutually coupled, 402
    toroid, 400
    transmission-line model, 397
    unloaded Q, 235, 396
  varactors, 393
Multivibrator oscillator, 227
Mutually coupled inductors, 402
Naming conventions, 13
Negative-conductance, 69
  examples
    simple, 70
  improving the device, 71
  starting criteria, 70
  resistance or conductance?, 72
  see also one-port
Negative-resistance oscillator, 58
  Alechno's technique, 63
  coupling capacitor, 67
  device selection, 61
  effective input capacitance, 62
  examples
    900 MHz, 116
    simple, 58
  harmonic balance example, 116
  improving the device, 66
  loaded Q, 64
  measurement alternatives, 60
  nonlinear noise example, 206
  resistance or conductance?, 72
  starting criteria, 59
  see also one-port method
Noise (chapter), 173
  acoustic, 179
  buffer, 183
  definitions, 173
  designing for low phase noise
    estimating the dominant source, 197
    reducing buffer noise, 202
    reducing Leeson noise, 197
    reducing pushing noise, 201
    reducing varactor modulation, 203
    summary, reducing noise, 204
  flicker, 178
  frequency domain, 174
  frequency multiplication, 183
  jitter, 174
  Leeson, 179
  linear oscillator noise example, 211
  linear time invariant, 180

linear time variant, 184
master noise equation, 183
measuring, 186
 example system, 194
 frequency discriminator, 192
 heterodyne/counter, 189
 reference oscillator, 190
 selective receiver, 188
 spectrum analyzer, direct, 186
negative-resistance example, 206
nonlinear noise simulation, 205
PLL, 213
predicting, 179
pushing induced, 182
reducing, *see* designing for low noise
residual FM and PM, 175, 177
SSB phase noise, 176
two-port, 178
varactor modulation, 182
versus out power, example, 308
Nonlinear amplifiers, 84
 compression, 86
 conversion efficiency, 92
 harmonics, 90
 impedance shift, 87
 operating class, 93
  class A, 93, 101, 107
  class AB example, 103
  class B, 95
  class C, 97, 103, 108
  class E example, 104
  class, other, 98
 output spectrum, 89
 output waveform, 91
 phase shift, 88
 power amplifier case study, 99
 test circuit, CE, 85
Nonlinear noise simulation, 205
Nonlinear open-loop cascade, 109
 example 1, 109
 example 2, 111
Nonlinear Techniques (chapter), 83
 harmonic balance overview, 83
Norton amplifier, 33
Open-loop, 1
 cascade, 2
 cascade, nonlinear, 109
 response, 2
 starting conditions, 2
Operating class, 93
  see also nonlinear amplifiers
One-port method, 57
 resistance or conductance?, 72
 see also negative-conductance
 see also negative-resistance
Optimizing the design, 76
Output coupling, 120

degree of coupling, 123
 example 1, 123
 example 2, 125
 loaded Q versus coupling, 124
 load pulling, 121
 summary, 128
Parametric spurious modes, 136
Passive level control, 128
 example, 130
 HP 200A, 129
Phase zero-crossing (ϕ_0), 1
Phase-lead compensation, 37
Pierce oscillators
 crystal, 347
 L-C oscillator, 249
 SAW, 378
Piezoelectric Oscillators (chapter), 321
 acceleration effects, 339
 aging, 338
 bulk quartz resonator cuts, 322
 calculating model parameters, 325
 ceramic (piezoelectric) oscillators, 381
  Butler, bipolar, 383
  Colpitts, FET, 382
  Pierce, CMOS, 381
 ceramic (piezoelectric) resonators, 380
  accuracy/stability, 380
  models, 380
 current probes, 336
 drive level, 334
 examples
  Butler, fundamental, bipolar, 355
  Colpitts, bipolar, 344
  Colpitts, JFET, 346
  Driscoll, bipolar, 357
  frequency multiplied, Butler, 369
  inverted-mesa, Pierce, bipolar, 359
  Miller, JFET, 343
  overtone, Butler, CB bipolar, 364
  overtone, Butler, CC bipolar, 366
  overtone, Colpitts, bipolar, 361
  Pierce, bipolar, 347
  Pierce, CMOS, 351
  Pierce, MMIC, 250
  Pierce, TTL, 350
  example summary, 366
 flicker noise, 338
 frequency accuracy/precision, 332
 holders, 341
 inverted-mesa resonators, 328
 model, 324
 oscillator operating modes, 329
 pulling, 326
 response, 325
 resonator spurious modes, 337
 SAW examples, 374
  Butler, 1-port, bipolar, 377

Index

Colpitts, 1-port, bipolar, 375
Pierce, 2-port, MMIC, 378
SAW models, 372
SAW resonators, 371
temperature effects
  bulk quartz cuts, 322
  SAW, 334
vibrational modes, 323
PLL noise, 213
Phase slope, 10
  aligning maximum, 10
  loaded Q, 45
Pierce oscillator, 15
Power Amplifier, 99
  summary, 108
  see also nonlinear amplifier
Power supply pushing, 132
Pulling, *see* load pulling
Pushing
  definition, power supply, 132
  induced noise, 182
Q
  loaded, 45
  measuring, 55
  unloaded, 46
Radially scaled parameters, 4
  table of, 4
Randall/Hoch correction, 9
Random resonator and amplifier
  combination, 12
R-C phase shift oscillators, 220
Reflection coefficient, 4
Relaxation spurious modes, 135
Residual FM and PM, 175, 177
Resistive feedback, 26
Resonator, 40
  capacitor loaded, distributed, 271
  ceramic loaded coaxial, 268
  Colpitts, 49
  coupled resonator example, 56
  coupling, 51
  delay line, 41
  helical, 415
  loaded Q, 45
  loss, 47
  L-C, simple, 43
  matching with, 52
  quarter-wavelength, 268
  R-C phase shift, 40
  summary, 56
  technologies (types), 263
  voltage, 166, 297
Rhea type oscillator, 155, 253
Ring oscillator, 230
Rod-over-ground model, 411
Round-rod line model, 411
Ruthroff transformers, 405

S-parameters, 1
SAW
  examples, 374
    Butler, 1-port, bipolar, 377
    Colpitts, 1-port, bipolar, 375
    Pierce, 2-port, MMIC, 378
  frequency stability, 374
  models, 372
  oscillators, 374
  resonators, 371
Seiler oscillator
  coaxial resonator, 314
  L-C bipolar, 243
Selective receiver, noise
  measurement, 188
Signal control devices, 404
  bifilar transformers, 404
  Ruthroff transformers, 405
  couplers (directional), 407
Simulation
  high loaded Q, 162
  transient (Cayenne and SPICE), 151
  transient set up, 155
  also see harmonic balance
Spectrum analyzer, direct noise
  measurement, 186
Spurious modes, 134
  quartz crystal resonator, 337
  multiple phase zero crossings, 135
  multiple resonances, 137
  parametric modes, 136
  relaxation modes, 135
  unstable amplifiers, 134
  summary, 138
Smith chart, 6
SPICE, 152
Stability, *see* amplifier
Starting conditions, 2
  amplifier stability, 11
  gain, 11
  gain peak, 11
  match requirements, 3
  open-loop 2-port, 3
  negative-resistance, 59
  negative-conductance, 70
Starting
  case study, 158
  criteria
    negative conductance, 70
    negative resistance, 59
    open-loop cascade, 2
  examples
    simple, 149
    example 2, 155
  high loaded Q techniques, 162
  modes, 144
  noise mode, 146

simulator set up, 155
simulation techniques, 151
starting time, 146
supply step time constant, 148
transient mode, 147
triggering, 160
Statistical analysis, 25, 78
Stripline model, 412
Summaries
  amplifiers, 39
  bipolar biasing, 432
  examples, crystal, 366
  L-C and *R-C* topology selection, 258
  linear techniques, 80
  match, 10
  noise, reducing, 204
  output coupling, 128
  power amplifiers, 108
  resonators, 56
  spurious modes, 138
Supply pushing, 132
Symbols (list), 439
Temperature effects
  biasing, 419
  bulk quartz resonator cuts, 322
  SAW resonators, 334
Toroid model, 400
Transformer, *see* bifilar transformer
Transformer feedback amplifier, 33
Transient Techniques (chapter), 143
  introduction, 143
Transmission lines
  characteristic impedance, 409
    coax, 409
    coax, square ground, 410
    microstrip, 413
    rod between ground planes, 411
    rod over ground, 411
    twisted parallel pair, 414
    stripline, 412
  helical model, 402
    also see distributed
Transmission loss from reflection, 4
Triggering (reduced starting time), 160
TTL gate model and oscillator, 230
TTL inverter crystal oscillator, 350
Tuned Osc (chapter) examples, 295
  coaxial resonator, MMIC, 308
  Colpitts, permeability tuned, JFET, 299
  loaded quarter-wave, MMIC, 312
  negative-*G*, bipolar, 305
  negative-*R*, capacitor transformed, 304
  negative-*R*, hybrid, bipolar, 301
  negative-*R*, transformer, bipolar, 304
  permeability tuning, 298
  resonator voltage, 166, 297
  Seiler coaxial resonator, bipolar, 314

tuning bandwidth, 295
Vacker, JFET, 300
YIG resonator oscillators, 316
Tuning bandwidth, 295
Twin-T oscillators, 232
Twisted-pair parallel-line model, 414
Ultimate test, 139
Unloaded Q, 46
  definitions, 46
  measuring, 55
  resonator, 47
  see also loaded Q
Varactor
  coupling, 168
  models, 393
  modulation noise, 182
  RF voltage, 297
Vibrational modes
  bulk quartz, 321, 337
  SAW, 371
Voltage, RF across resonator, 166, 297
VSWR, 4
Waveforms, steady-state, 164
  Clapp oscillator, 164
  derived from spectrum, 169
  resonator voltage, 166
Wien bridge oscillator, 226
Wire-wound transformers,
      see signal control devices
Vacker oscillator, 300

Recent Titles in the Artech House Microwave Library

Active Filters for Integrated-Circuit Applications, Fred H. Irons

Advanced Techniques in RF Power Amplifier Design, Steve C. Cripps

Automated Smith Chart, Version 4.0: Software and User's Manual, Leonard M. Schwab

Behavioral Modeling of Nonlinear RF and Microwave Devices, Thomas R. Turlington

Broadband Microwave Amplifiers, Bal S. Virdee, Avtar S. Virdee, and Ben Y. Banyamin

Computer-Aided Analysis of Nonlinear Microwave Circuits, Paulo J. C. Rodrigues

Designing Bipolar Transistor Radio Frequency Integrated Circuits, Allen A. Sweet

Design of FET Frequency Multipliers and Harmonic Oscillators, Edmar Camargo

Design of Linear RF Outphasing Power Amplifiers, Xuejun Zhang, Lawrence E. Larson, and Peter M. Asbeck

Design Methodology for RF CMOS Phase Locked Loops, Carlos Quemada, Guillermo Bistué, and Iñigo Adin

Design of RF and Microwave Amplifiers and Oscillators, Second Edition, Pieter L. D. Abrie

Digital Filter Design Solutions, Jolyon M. De Freitas

Discrete Oscillator Design Linear, Nonlinear, Transient, and Noise Domains, Randall W. Rhea

Distortion in RF Power Amplifiers, Joel Vuolevi and Timo Rahkonen

EMPLAN: Electromagnetic Analysis of Printed Structures in Planarly Layered Media, Software and User's Manual, Noyan Kinayman and M. I. Aksun

Essentials of RF and Microwave Grounding, Eric Holzman

FAST: Fast Amplifier Synthesis Tool—Software and User's Guide,
Dale D. Henkes

Feedforward Linear Power Amplifiers, Nick Pothecary

Foundations of Oscillator Circuit Design, Guillermo Gonzalez

Fundamentals of Nonlinear Behavioral Modeling for RF and Microwave Design, John Wood and David E. Root, editors

Generalized Filter Design by Computer Optimization, Djuradj Budimir

High-Linearity RF Amplifier Design, Peter B. Kenington

High-Speed Circuit Board Signal Integrity, Stephen C. Thierauf

Intermodulation Distortion in Microwave and Wireless Circuits, José Carlos Pedro and Nuno Borges Carvalho

Introduction to Modeling HBTs, Matthias Rudolph

Lumped Elements for RF and Microwave Circuits, Inder Bahl

Lumped Element Quadrature Hybrids, David Andrews

Microwave Circuit Modeling Using Electromagnetic Field Simulation, Daniel G. Swanson, Jr. and Wolfgang J. R. Hoefer

Microwave Component Mechanics, Harri Eskelinen and Pekka Eskelinen

Microwave Differential Circuit Design Using Mixed-Mode S-Parameters, William R. Eisenstadt, Robert Stengel, and Bruce M. Thompson

Microwave Engineers' Handbook, Two Volumes, Theodore Saad, editor

Microwave Filters, Impedance-Matching Networks, and Coupling Structures, George L. Matthaei, Leo Young, and E.M.T. Jones

Microwave Materials and Fabrication Techniques, Second Edition, Thomas S. Laverghetta

Microwave Mixers, Second Edition, Stephen A. Maas

Microwave Radio Transmission Design Guide, Second Edition,
Trevor Manning

Microwaves and Wireless Simplified, Third Edition,
Thomas S. Laverghetta

Modern Microwave Circuits, Noyan Kinayman and M. I. Aksun

Modern Microwave Measurements and Techniques, Second Edition,
Thomas S. Laverghetta

Neural Networks for RF and Microwave Design, Q. J. Zhang and
K. C. Gupta

Noise in Linear and Nonlinear Circuits, Stephen A. Maas

Nonlinear Microwave and RF Circuits, Second Edition,
Stephen A. Maas

*QMATCH: Lumped-Element Impedance Matching, Software and
User's Guide,* Pieter L. D. Abrie

Practical Analog and Digital Filter Design, Les Thede

Practical Microstrip Design and Applications, Günter Kompa

*Practical RF Circuit Design for Modern Wireless Systems, Volume I:
Passive Circuits and Systems,* Les Besser and Rowan Gilmore

*Practical RF Circuit Design for Modern Wireless Systems, Volume II:
Active Circuits and Systems,* Rowan Gilmore and Les Besser

*Production Testing of RF and System-on-a-Chip Devices for Wireless
Communications,* Keith B. Schaub and Joe Kelly

Radio Frequency Integrated Circuit Design, Second Edition,
John W. M. Rogers and Calvin Plett

RF Bulk Acoustic Wave Filters for Communications, Ken-ya
Hashimoto

RF Design Guide: Systems, Circuits, and Equations, Peter Vizmuller

RF Measurements of Die and Packages, Scott A. Wartenberg

The RF and Microwave Circuit Design Handbook, Stephen A. Maas

RF and Microwave Coupled-Line Circuits, Rajesh Mongia, Inder Bahl, and Prakash Bhartia

RF and Microwave Oscillator Design, Michal Odyniec, editor

RF Power Amplifiers for Wireless Communications, Second Edition, Steve C. Cripps

RF Systems, Components, and Circuits Handbook, Ferril A. Losee

The Six-Port Technique with Microwave and Wireless Applications, Fadhel M. Ghannouchi and Abbas Mohammadi

Solid-State Microwave High-Power Amplifiers, Franco Sechi and Marina Bujatti

Stability Analysis of Nonlinear Microwave Circuits, Almudena Suárez and Raymond Quéré

System-in-Package RF Design and Applications, Michael P. Gaynor

TRAVIS 2.0: Transmission Line Visualization Software and User's Guide, Version 2.0, Robert G. Kaires and Barton T. Hickman

Understanding Microwave Heating Cavities, Tse V. Chow Ting Chan and Howard C. Reader

For further information on these and other Artech House titles, including previously considered out-of-print books now available through our In-Print- Forever® (IPF®) program, contact:

Artech House Publishers
685 Canton Street
Norwood, MA 02062
Phone: 781-769-9750
Fax: 781-769-6334
e-mail: artech@artechhouse.com

Artech House Books
16 Sussex Street
London SW1V 4RW UK
Phone: +44 (0)20 7596 8750
Fax: +44 (0)20 7630 0166
e-mail: artech-uk@artechhouse.com

Find us on the World Wide Web at: www.artechhouse.com